IRON METABOLISM

IRON METABOLISM

IVÁN BERNÁT, M.D., D.Sc.
PROFESSOR OF HEMATOLOGY

SPRINGER SCIENCE+BUSINESS MEDIA, LLC

TRANSLATED BY

ÉVA GOSZTONYI

A DIVISION OF

PLENUM PUBLISHING CORPORATION

233 SPRING STREET, NEW YORK, N.Y. 10013

DOI 10.1007/978-1-4615-7308-1

JOINT EDITION PUBLISHED WITH

AKADÉMIAI KIADÓ, BUDAPEST, HUNGARY

CONTENTS

Chapter 1. *The Distribution of Iron in Nature* 9

The basic physical, chemical, and biochemical properties of iron 9

Bibliography 12

Chapter 2. *Biochemical Evolution of the Heme-Type Enzymes* 15

Bibliography 17

Chapter 3. *The Biological Significance of Iron-Containing Compounds* 19

Bibliography 22

Chapter 4. *Distribution and Function of the Iron-Containing Complexes of the Human Organism* 23

Heme iron compounds 23

Storage iron 24

Transport iron 24

Other non-heme iron compounds 24

Bibliography 26

Chapter 5. *Dietary Iron* 27

Iron intake 27

Bibliography 34

Chapter 6. *Iron Absorption* 37

Factors influencing iron absorption 38

The control of iron absorption 47

Measurement of iron absorption 60

Bibliography 64

Chapter 7. *Iron Transport* 71

Plasma iron 71

Transferrin 78

Bibliography 85

Chapter 8. *Storage Iron* 91

Ferritin and hemosiderin 91
Deposition and mobilization of iron 95
Quantitative aspects of iron stores 97
Clinical methods used for the estimation of available iron stores 103
Bibliography 108

Chapter 9. *Iron Loss and Iron Requirement* 113

Iron loss in women 119
Iron requirements for newborns, infants, and children 124
Bibliography 134

Chapter 10. *Erythropoiesis* 141

Bibliography 145

Chapter 11. *Hemoglobin Synthesis* 147

Biosynthesis of heme 149
Biosynthesis of globin 155
Bibliography 155

Chapter 12. *Red Cell Destruction* 159

Bibliography 160

Chapter 13. *Hemoglobin Catabolism* 161

Bibliography 164

Chapter 14. *Ferrokinetics* 167

Survey 167
Plasma iron clearance and plasma iron transport rate 170
Incorporation of radioiron into the erythroblasts and reticulocytes. Iron utilization in the course of red cell production. Effective erythropoiesis 173
Distribution of iron among the bone marrow, liver, and spleen 176
Bibliography 179

Chapter 15. *Erythrokinetics* 183

Red cell production 184
Red cell destruction 185
Bibliography 197

Chapter 16. Cytochemical Stains and Microscopy 201
Siderocytes 201
Sideroblasts 202
Sideromacrophages 203
Bibliography 203

Chapter 17. Electron Microscopic Investigations 205
Ferritin 205
Hemosiderin 205
Erythrophagocytosis 206
Iron transport 207
Rhopheocytosis 209
Sideroblasts and siderocytes 211
Bibliography 213

Chapter 18. Iron Deficiency 215
Incidence 215
The clinical picture of iron deficiency 218
Laboratory findings 233
Ferrokinetics 238
The diagnosis of iron deficiency 239
Differential diagnosis of iron deficiency 242
Etiology and pathogenesis of iron deficiency 242
Therapy of iron deficiency 245
Acute iron intoxication 257
Bibliography 258

Chapter 19. Anemia of Infection 275
Differential diagnosis 279
Treatment 280
Bibliography 280

Chapter 20. Anemia of Thermal Injury 285
Development, type, and course of the anemia of thermal injury 285
Therapy 295
Bibliography 295

Chapter 21. Protein-Deficiency Anemia 299
Vitamin E deficiency 300
Bibliography 300

Chapter 22. Pernicious Anemia 301
Bibliography 303

Chapter 23. Hemolytic Anemias 305
Bibliography 306

Chapter 24. Refractory Hypochromic Anemias 309
The sideroblastic anemias 309
Pyridoxine-responsive anemias 313
The disturbance of heme synthesis in thalassemia 315
Pathological heme synthesis associated with lead and other toxic substances 317
Sideroblastic anemias arising in connection with anti-tuberculous drugs 319
Shahidi–Nathan–Diamond anemia 319
Fanconi's anemia 320
Genetically determined microcytic hypochromic anemias 320
Bibliography 321

Chapter 25. Disturbed Iron Metabolism in Acute Radiation Injury 327
Bibliography 333

Chapter 26. Iron Metabolism in Polycythemia Vera and Secondary Polycythemias 335
Bibliography 337

Chapter 27. Iron Overload 339
Idiopathic hemochromatosis (Iron storage disease) 340
Secondary hemochromatosis associated with cirrhosis of the liver 356
Congenital atransferrinemia 358
Congenital (familial) hypersiderosis 360
Nutritional siderosis – Bantu siderosis 360
Siderosis developing in refractory anemias associated with ineffective erythropoiesis 364
Transfusional siderosis 364
Renal hemosiderosis 367
Idiopathic pulmonary hemosiderosis 368
Goodpasture syndrome 370
Bibliography 371

Author Index 383

Subject Index 399

CHAPTER 1

THE DISTRIBUTION OF IRON IN NATURE

THE BASIC PHYSICAL, CHEMICAL, AND BIOCHEMICAL PROPERTIES OF IRON

Iron, in the form of various combined ores, is one of the most common elements, constituting about 5% of the earth's crust. The most important iron-containing minerals are the oxides and sulfides. Hematite (red iron ore, Fe_2O_3), magnetite (loadstone, Fe_3O_4), and goethite (hydrous iron oxide, FeO_2H) belong to the former group, whereas pyrite (FeS_2) and marcasite (formerly crystallized iron pyrite, FeS_2) belong to the latter. Iron is also present in meteorites, in other planets, and in the sun. Iron is found in both sea and fresh water but only those springs whose water contains at least 10 mg/kg of iron are regarded as medicinal iron springs. Euthermic or hyperthermic springs of high iron content in which blue algae and iron bacteria are present are classified as iron thermae (siderophytathermae or F-thermae), e.g., Yamagataken, Yiraka, Yamazaki in Japan.

Pure metallic iron is rare in nature; it is bluish white and strongly magnetic. It is unstable, being di-, tri-, or occasionally sexvalent. Its atomic number is 26, its atomic weight 52–61. Thus its nucleus contains 26 protons and 26–35 neutrons. The four stable iron isotopes have an atomic weight of 54, 56, 57, and 58, giving an atomic weight of the naturally occurring iron of 55.847. Around the atomic nucleus of the iron $2+8+14+2=26$ electrons circulate in four "shells."

Six of the ten isotopes of iron are radioactive; ^{52}Fe has a half-life of 8.4 hours, ^{53}Fe – 9 minutes, ^{55}Fe – 2.6 years, ^{59}Fe – 45.1 days, ^{60}Fe – 3.10^5 years, ^{61}Fe – 6.1 minutes. ^{59}Fe, ^{55}Fe and ^{52}Fe are all useful in medical and biological studies, the first two being the most widely used.

Iron derivatives may be divalent ferrous compounds, e.g., $FeSO_4$, trivalent ferric compounds, e.g., $Fe_2(SO_4)_3$, or complex iron compounds, e.g., $K_4[Fe(CN_6)]$, in which the iron is a part of a complex anion. The ferrous salts are white in the dehydrated form while their hydrates and solutions are light green. The ferric salts in the dehydrated form are white or light violet, their hydrates and solutions being yellow or brown.

Iron, owing to its oxidoreduction and to its complex forming properties, is a central constituent of the enzymes that regulate the oxidoreduction processes of tissues (iron porphyrin proteids, "tissue hemins"). These enzymes were probably among the first intracellular compounds developed in primitive organisms, and hence they have a general biological significance. The iron porphyrin proteids of hemoglobin and myoglobin were developed only at a later stage of evolution.

In addition to its role in tissue respiration, iron is also involved in oxidative phosphorylation, porphyrin metabolism, collagen synthesis, lymphocyte and granulocyte function, tissue growth, and neurotransmitter synthesis and catabolism (Pollitt and Leibel, 1976; Leibel et al., 1978).

Iron may also be involved in the nonspecific defense reactions of the organism. In the past three decades several iron-containing metabolites (siderochromes) have been isolated from cultures of microorganisms (Bickel et al., 1960; Prelog, 1964). Most of these, even in high dilution, possess considerable biological activity; some promote growth of bacteria (sideramines), others have an antibiotic effect (sideromycins).

The first sideromycin (Grisein) was discovered by Reynolds, Schatz, and Waksman in 1947, and albomycin was isolated by Gause and Brazhnikova in 1951.

In 1952 the isolation of several sideramines was reported. Neilands (1952) described ferrichrome, Hesseltine et al. (1952) coprogen, Lochhead and his team (1952) the terregens factor. The isolation of the ferrioxamines from Actinomyces cultures was achieved by Bickel and his collaborators (1960a, b, c, d). Zähner et al. (1960) recognized the antagonism between the sideramines and sideromycins. This recognition had a significant influence on the further investigation of these compounds. Between 1961 and 1963 the structure of the ferrioxamines was successfully established, and this enabled the complete or partial synthesis of these compounds (Bickel et al., 1960; Keller-Schierlein and Prelog, 1962; Prelog and Walser, 1962). They proved to be ferric complexes with three hydroxamic acid [CO—N(OH)] groups (Fig. 1/1 and Table 1/1).

Table 1/1
Siderochromes

Sideramines		Sideromycins
Ferrichrome	Ferrioxamine A	Grisein
Coprogen	Ferrioxamine B	Albomycin
Terregens factor	Ferrioxamine C	
	Ferrioxamine D_1	Ferrimycin A_1
	Ferrioxamine D_2	Ferrimycin A_2
	Ferrioxamine E	Ferrimycin B
	Ferrioxamine F	ETH 22765
	Ferrioxamine G	
	Ferrichrysin	LA 5352
	Ferricrocin	LA 5937
	Ferrirhodin	
	Ferrirubine	

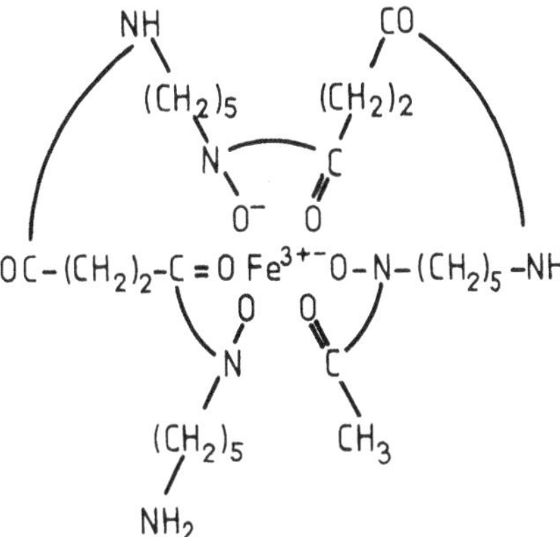

Fig. 1/1. Structural formula of ferrioxamine-B (after Prelog, V., in: Gross, F.: Iron Metabolism. Springer, Berlin–Göttingen–Heidelberg 1964)

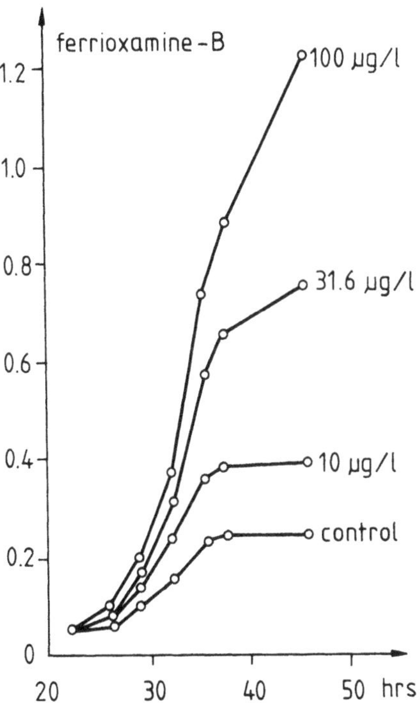

Fig. 1/2. Effect of different concentrations of ferrioxamine-B upon the growth of Microbacterium lacticum ATCC 8181 (after Prelog, V., in: Gross, F.: Iron Metabolism. Springer, Berlin–Göttingen–Heidelberg 1964)

Siderochromes can be found in many microorganic cultures; some strains produce iron-free acids (desferrisiderochromes). Siderochromes have so far not been found in higher class plants or animals.

The growth-promoting activity of sideramines can be demonstrated directly in the sideramine-heterotrophic strains of certain microorganisms. Figure 1/2 shows the results of such an experiment using Microbacterium lacticum ATCC 8181, and ferrioxamine-B. Technically, an indirect method of measuring the activity based on the antagonism between sideramines and sideromycins is simpler (for details see Prelog's study, 1964).

The desferrisideramines form very stable and complex compounds with ferric ions. The stability constant of desferrioxamine $+Fe^{3+}$ is $10^{30.6}$ compared with the complex of iron with EDTA, whose stability constant is $10^{25.1}$. The complexes of desferrioxamine-B with other biologically important metals such as calcium, cobalt, zinc, and copper are less stable than those formed by EDTA with the same elements. This selectivity of DFO combined with its low toxicity and rapid excretion makes it a valuable diagnostic and therapeutic tool.

BIBLIOGRAPHY

BICKEL, H., BOSSHARDT, R., GÄUMANN, E., RENSSER, P., VISCHER, E., VOSER, W., WETTSTEIN, A., ZÄHNER, H.: Stoffwechselprodukte von Actinomyceten. Über die Isolierung und Charakterisierung der Ferrioxamine A–F, neuere Wuchsstoffe der Sideramin-Gruppe. Helv. chim. Acta *43,* 2118 (1960a).

BICKEL, H. et al.: Stoffwechselprodukte von Actinomyceten. Über die Isolierung und Synthese des 1-amino-5-hydroxylamino-pentans, eines wesentlichen Hydrolyseproduktes der Ferrioxamine und der Ferrimycine. Helv. chim. Acta *43,* 901 (1960b).

BICKEL, H., GÄUMANN, E., KELLER-SCHIERLEIN, W., PRELOG, V., VISCHER, E., WETTSTEIN, A., ZÄHNER, H.: Über eisenhaltige Wachstumsfaktoren, die Sideramine, und ihre Antagonisten, die eisenhaltigen Antibiotika, Sideromycine. Experientia *16,* 129 (1960c).

BICKEL, H., GÄUMANN, E., NUSSBERGER, G., RENSSER, P., VISCHER, E., VOSER, W., WETTSTEIN, A., ZÄHNER, H.: Stoffwechselprodukte von Actinomyceten. Über die Isolierung und Charakterisierung der Ferrimycine A_1 und A_2, neuere Antibiotika der Sideromycin-Gruppe. Helv. chim. Acta *43,* 2105 (1960d).

FRIEDEN, E.: The evolution of metals as essential elements (with special reference to iron and copper). Advanc. Exp. Med. Biol. *48,* 1 (1974).

GAUSE, G. F., BRAZHNIKOVA, M. G.: Nov. Med. Akad. Med. Nauk SSSR *23,* 3 (1951).

HESSELTINE, C. W. et al.: J. Amer. Chem. Soc. *74,* 1362 (1952).

HUNT, J., RICHARDS, R. J., HARWOOD, R., JACOBS, A.: The effect of desferrioxamine on fibroblasts and collagen formation in cell cultures. Brit. J. Haemat. *41,* 69 (1979).

KELLER-SCHIERLEIN, W., PRELOG, V.: Stoffwechselprodukte von Actinomyceten. Über das Ferrioxamin E; ein Beitrag zur Konstitution des Nocardamins. Die Konstitution des Ferrioxamins D_1. Ferrioxamin G. Helv. chim. Acta *44,* 709 (1961); *45,* 590 (1962).

LEIBEL, R. L., GREENFIELD, D., POLLITT, E.: In: WINICK, M. (ed.): Nutrition: Pre- and Postnatal Development. Plenum Press, New York 1978.

Lochhead, A. G., Burton, M. O., Thexton, R. H.: A bacterial growth-factor synthetized by a soil bacterium. Nature *170*, 282 (1952).
Neilands, J. B.: J. Amer. Chem. Soc. *74*, 4846 (1952).
Pollitt, E., Leibel, R. L.: Iron deficiency and behavior. J. Pediat. *88*, 372 (1976).
Prelog, V.: Iron-containing compounds in micro-organisms. In: Gross, F. (ed.): Iron Metabolism, p. 79. Springer, Berlin–Göttingen–Heidelberg 1964.
Prelog, V., Walser, A.: Helv. chim. Acta *45*, 631 (1962).
Reynolds, D. M., Schatz, A., Waksman, S. A.: Grisein, a new antibiotic produced by strain of Streptomyces griseus. Proc. Soc. exp. Biol. Med. *64*, 50 (1947).
Zähner, H., Hütter, R., Bachmann, E.: Metabolites of Actinomycetes, Part 23. On a study of the effect of sideromycin. Arch. Mikrobiol. *36*, 325 (1960).

CHAPTER 2

BIOCHEMICAL EVOLUTION OF THE HEME-TYPE ENZYMES

The biological significance of iron compounds lies in the fact that they are capable of reversible oxidoreduction, which is such a basic function of living organisms that the first protoplasms to evolve must have possessed this ability (Granick, 1953). It is unlikely that the ability for oxidation would have developed only with the development of the heme molecule.

There are two fundamental porphyrin complexes: chlorophyll in the vegetable kingdom is a magnesium-containing green pigment, and hemin found in the animal kingdom is an iron-containing red pigment. The simpler forms of iron-containing enzymes are perhaps as old as life itself (Schapira, 1964). Even the simple inorganic iron compounds (hydrated iron ions) possess certain catalytic properties, e.g., they

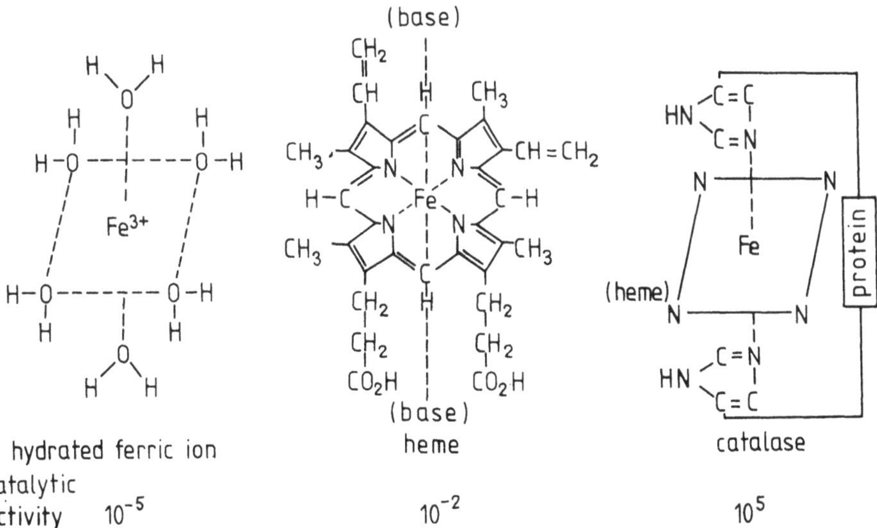

Fig. 2/1. Increase in the catalytic activity of iron (after Calvin, M.: Advanc. biol. med. Phys. *8*, 322, 1962)

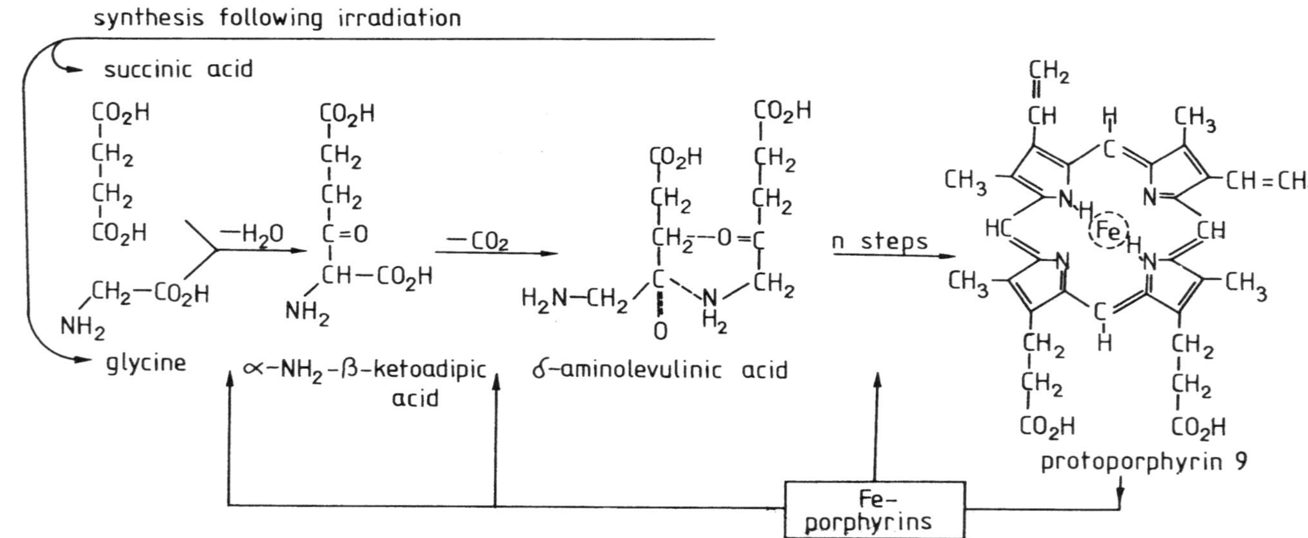

Fig. 2/2. Possible points for catalytic activities of iron (after Calvin, M.: Advanc. biol. med. Phys. *8*, 322, 1962)

are able like peroxidase to break down hydrogen peroxide. The difference between the simple inorganic iron compounds and the complex iron-porphyrin proteids lies in the greater catalytic activity of the latter. The catalytic activity of the hydrated ferric ion is slight (10^{-5}), but that of iron porphyrin compounds (heme) is 10^{-2}. If the heme is incorporated into a protein structure, the activity may be increased to 10^5 (Calvin, 1962) (Fig. 2/1).

How did these very complex iron-containing compounds evolve? According to Calvin (1962), the development is essentially the result of autocatalysis and self-selection. He postulates that simple iron organic compounds were developed as a result of radiation energy, and others, such as succinic acid and glycine, evolved through condensation, decarboxylation, further condensation, and a series of oxidations finally into tetrapyrols. One or other of these reactions was probably catalyzed by iron (Fig. 2/2). Iron porphyrin structures formed that proved to be more active catalysts than the iron itself, so that the transformation of succinic acid and glycine into delta-aminolevulinic acid, etc., and finally the development of porphyrins was enhanced by autocatalysis and self-selection.

BIBLIOGRAPHY

CALVIN, M.: Advanc. biol. med. Phys. *8*, 322 (1962).

GRANICK, S.: Amer. Naturalist *87*, 65 (1953).

SCHAPIRA, G.: Iron metabolism, past, present, and future. In: GROSS, F. (ed.): Iron Metabolism. Springer, Berlin–Göttingen–Heidelberg 1964.

CHAPTER 3

THE BIOLOGICAL SIGNIFICANCE OF IRON-CONTAINING COMPOUNDS

Iron is essential for normal metabolism in both plants and animals. In the former, iron deficiency results in a lack of production of the green pigment chlorophyll, and the condition is described as "chlorosis" (Beutler, 1964); animals develop tissue changes and anemia.

The living cell performs work, the energy for which is obtained from metabolism of nutrients; tissue respiration requires oxygen, but fermentation and glycolysis do not.

Hemoglobin in the red blood corpuscles carries oxygen from the lungs to the tissues and conveys about 700–1000 liters of oxygen/24 hours to the cells. Both oxygen uptake in the lungs and oxygen release in the tissues are regulated by physical and physicochemical mechanisms. The oxyhemoglobin binding is loose and reversible; in arterial blood, at an oxygen pressure of 100 mm Hg, the saturation of oxygen is almost 100%, whereas in venous blood it is only 74% at an oxygen pressure of 40 mm Hg (Fig. 3/1).

Functionally, myoglobin takes an intermediary position between hemoglobin and the tissue hemins. The affinity for oxygen of myoglobin exceeds that of hemoglobin, and the formation of oxymyoglobin occurs 2.5–5 times as fast as the formation of oxyhemoglobin from hemoglobin. At an oxygen pressure of 40 mm Hg the oxygen saturation of myoglobin is still over 90%, compared with the 74% of hemoglobin (see Fig. 3/1). Myoglobin obtains the oxygen from hemoglobin, but activation of the oxygen takes place within the cell by oxidoreduction systems.

Tissue respiration is responsible for the inflow of oxygen and nutrients into cells and elimination of end products of metabolism; for both of these the physical processes of flow and diffusion are important. The respiration is also partly responsible for the liberation, storage, and utilization of the energy of the cell nutrients; these functions involve predominantly chemical processes.

Catalyzers are necessary for the metabolic processes of the cell involving molecular oxygen and nutrients, and complex bindings may develop between the substrate and the enzyme, or highly reactive intermediary products may be formed. The enzymes ensure that the decomposition of cell nutrients proceeds at a definite speed and in an orderly succession of several stages. An anaerobic and an aerobic phase can be distinguished, the latter being the more important from the point of view of liberation of energy.

Electron transfer is the essence of the oxidoreductive processes (Haurowitz, 1958). The electron donor exerts a reducing effect by being itself oxidized; the electron acceptor, on the other hand, is reduced and hence exerts an oxidizing effect. The electron of the substrate is transferred by the catalyzer to the acceptor:

$$\text{I} \quad D^{-e} + C \rightleftharpoons D \quad + C^{-e}$$

$$\text{II} \quad C^{-e} + A \rightleftharpoons A^{-e} + C$$

(D = donor, C = catalyzer, A = acceptor, $^{-e}$ = electron).

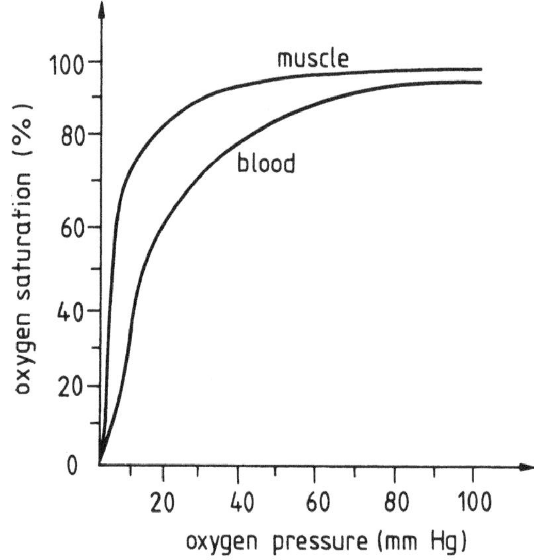

Fig. 3/1. Dissociation curve of oxyhemoglobin and oxymyoglobin

The hydrogen of the substrate is taken up by the dehydrogenases and transferred via coenzyme I (diphosphopyridine nucleotide — DPN) and coenzyme II (triphosphopyridine nucleotide — TPN) to the flavoenzymes. The hydrogen of the substrate reaches the tissue hemins (the iron porphyrin proteids) via these flavoenzymes. A small amount of the activated hydrogen reaches the iron-containing tissue enzymes directly through the dehydrogenases.

The prosthetic groups of the above-mentioned enzymes are bound to specific proteins, which fact results in the multiplication of the redox potential of the enzymes.

The electrons, which reach the iron-containing respiratory enzymes through the flavoenzymes, first of all reduce the ferric cytochrome C to ferrous cytochrome C:

$$2e^- + 2\,\text{cyt.-C}(Fe^{3+}) \rightleftharpoons 2\,\text{cyt.-C}(Fe^{2+})\,.$$

The ferrous cytochrome C reduces ferric cytochrome A to ferrous cytochrome A, which reduces the oxidized cytochrome oxidase to reduced cytochrome oxidase:

$$2\,\text{cyt.-C}(Fe^{2+}) + 2\,\text{cyt.-A}(Fe^{3+}) \rightleftharpoons 2\,\text{cyt.-C}(Fe^{3+}) + 2\,\text{cyt.-A}(Fe^{2+})\,,$$

$$2\,\text{cyt.-A}(Fe^{2+}) + 2\,\text{cyt.-oxidase}(Fe^{3+}) \rightleftharpoons 2\,\text{cyt.-A}(Fe^{3+}) + 2\,\text{cyt.-oxidase}(Fe^{2+})\,.$$

Cytochrome oxidase is capable of autooxidation and can transfer its electrons directly to molecular O_2:

$$2\,\text{cyt.-oxidase}(Fe^{2+}) + \tfrac{1}{2}\,O_2 \rightleftharpoons 2\,\text{cyt.-oxidase}(Fe^{3+}) + O^{2-}\,.$$

Then the oxygen ion (O^{2-}) combines with 2 H^+ ions to water.

Cytochrome B, according to Slater (1957), is situated between succinic dehydrogenase and cytochrome C. This part of the oxidoreduction processes avoids the DPN-flavoenzyme line.

The catalases and peroxidases decompose the peroxides that develop continuously as side products of metabolic processes.

According to Theorell et al. (cited by Opitz and Lübbers, 1957), the reaction is presumed to take place in the following way:

$$\begin{matrix} \text{HO—Fe}\cdot\text{Fe—OH} \\ \cdot\quad\cdot \\ \text{HO—Fe}\cdot\text{Fe—OH} \end{matrix} \quad + H_2O_2 \quad \begin{matrix} \text{HO—Fe}\cdot\text{Fe—OOH} \\ \cdot\quad\cdot \\ \text{HO—Fe}\cdot\text{Fe—}\dot{\text{O}}\text{H} \end{matrix} \quad + H_2O$$

If more than one H_2O_2 falls to the 4 Fe atoms of the catalase molecule, the course of the reaction is as follows:

$$\begin{matrix} \text{HO—Fe}\cdot\text{Fe—OOH} \\ \cdot\quad\cdot \\ \text{HO—Fe}\cdot\text{Fe—OH} \end{matrix} \quad + H_2O_2 \quad \begin{matrix} \text{HO—Fe}\cdot\text{Fe—OH} \\ \cdot\quad\cdot \\ \text{HO—Fe}\cdot\text{Fe—OH} \end{matrix} \quad + O_2 + H_2O$$

The iron-containing respiratory enzymes play a key role in the oxidoreduction processes and are indispensable for the normal course of biological processes.

Reduction in the activity of respiratory enzymes leads to disturbed cell function and gives rise to a wide range of clinical syndromes. Indeed, some of the most interesting features of hyposiderosis are those due to decrease in tissue hemins rather than anemia (Waldenström, 1938, 1946; Beutler, 1964; Jasinski and Roth, 1954; Dallman, 1975; Dallman et al., 1978; Ohira et al., 1979) (see Chapter 18). The tissue changes have perhaps received less attention than the hematological changes, although they were recognized even in the last century (Laache, 1883; Sahli, 1908; Naegeli, 1908; Morawitz, 1910).

BIBLIOGRAPHY

Beutler, E.: Tissue effects of iron deficiency. In: Gross, F. (ed.): Iron Metabolism. Springer, Berlin–Göttingen–Heidelberg 1964.

Dallman, P. R.: Effects of iron deficiency exclusive of anaemia. Brit. J. Haemat. *30*, 179 (1975).

Dallman, P. R., Beutler, E., Finch, C. A.: Effects of iron deficiency exclusive of anaemia. Brit. J. Haemat. *40*, 179 (1978).

Haurowitz, F.: Fortschritte der Biochemie, Karger, Basel 1958.

Jasinski, B., Roth, O.: Die larvierte Eisenmangelkrankheit. Schwabe, Basel 1954.

Laache, S.: Die Anämie. Walling, Christiania 1883.

Morawitz, P.: Untersuchungen über Chlorose. Münch. med. Wschr. *57*, 1425 (1910).

Naegeli, O.: Blutkrankheiten und Blutdiagnostik. Von Veit, Leipzig 1908.

Ohira, Y., Edgerton, V. R., Gardner, G. W., Senewiratre, B., Bernard, R. J., Simpson, D. R.: Work capacity, heart rate and blood lactate responses to iron treatment. Brit. J. Haemat. *41*, 365 (1979).

Sahli, H.: Lehrbuch der klinischen Untersuchungs-Methoden, p. 993. Deuticke, Leipzig–Wien 1908.

Slater, L., cit. Opitz, E., Lübbers, D.: Allgemeine Physiologie der Zell- und Gewebsatmung. In: Büchner, F. et al. (eds.): Handbuch der allgemeinen Pathologie, Vol. 4/2. Springer, Berlin–Göttingen–Heidelberg 1957.

Theorell, H.: cit. Opitz, E., Lübbers, D.: Allgemeine Physiologie der Zell- und Gewebsatmung. In: Büchner, F. et al. (eds.): Handbuch der allgemeinen Pathologie, Vol. 4/4. Springer, Berlin–Göttingen–Heidelberg, 1957.

Waldenström, J.: Iron and epithelium. Some clinical observations. Acta med. scand. Suppl. *90*, 380 (1938).

Waldenström, J.: The incidence of "iron deficiency" (sideropenia) in some rural and urban populations. Acta med. scand. *170*, 252 (1946).

CHAPTER 4

DISTRIBUTION AND FUNCTION OF THE IRON-CONTAINING COMPLEXES OF THE HUMAN ORGANISM

HEME IRON COMPOUNDS

The healthy human adult contains on an average 4–5 g of iron (Fig. 4/1), but individual values depend on body weight, on the total amount of the circulating hemoglobin, and on iron stores, so that the normal variation may be between 3 and 6 g.

The bulk of the iron (2.5–3.0 g) is in the hemoglobin (Hb), which contains 0.34% of iron by weight. Thus 1 ml of packed erythrocytes contains approximately 1 mg of iron (see Chapter 11).

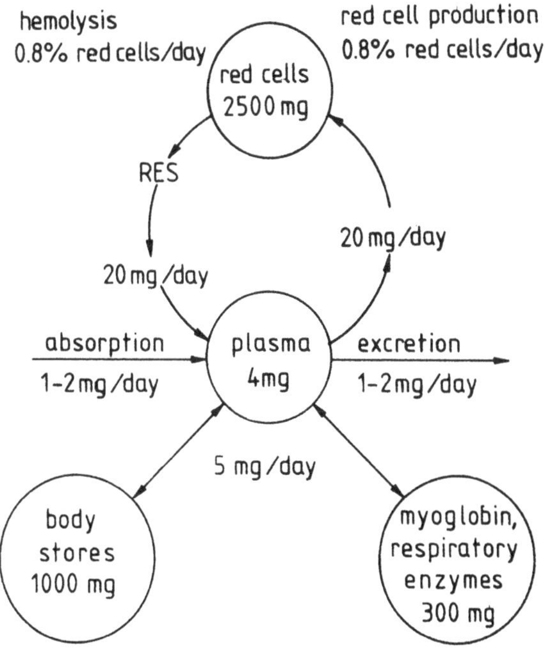

Fig. 4/1. Schematic representation of iron turnover (after Conrad, M. E., in: Hallberg, L. et al.: Iron Deficiency. Academic Press, London–New York 1970)

Myoglobin is structurally similar to hemoglobin except that it is monomeric, each myoglobin molecule consisting of a heme group nearly surrounded by loops of a long strand of protein containing approximately 150 amino acid residues. Its molecular weight is 17,000 and it contains 0.34% of iron by weight. The iron in myoglobin accounts for 0.1–0.3 g of iron.

Iron-containing tissue enzymes constitute a small but important compartment of iron amounting to 4–8 mg, and these enzymes are sensitive to changes in the total body iron content (Fairbanks and Beutler, 1977; Dallman et al., 1967).

All these compounds are hemins, and they are responsible for oxygen uptake, transport, and utilization.

STORAGE IRON

Storage iron in the form of ferritin and hemosiderin varies in the adult male between 800 mg and 1.5 g, but the stores of women are in general much less. The bulk of the storage iron is found in the liver, bone marrow, and spleen.

TRANSPORT IRON

Iron is carried in the blood by a $beta_1$ globulin, transferrin. The total amount of transport iron is only 3–4 mg, i.e., only about 1‰ of the total body iron.

OTHER NON-HEME IRON COMPOUNDS

Non-heme iron compounds such as flavoproteins are also found in the tissues in minimal amounts, and they participate in cell respiration. One such compound is succinodehydrogenase, which is found in the mitochondria. Its molecular weight is between 200,000 and 240,000, and it contains four iron atoms per molecule. DPNH-cytochrome C reductase is another such enzyme with a molecular weight of about 80,000 and again four iron atoms/molecule.

Only some of the flavoproteins have iron-containing prosthetic groups; others contain copper or molybdenum (Green et al., 1944; Mahler, 1955; Richert and Westerfeld, 1954; Takeda and Hara, 1955).

About 500 mg of other non-heme-type iron compounds have been found in the muscles (Schapira and Dreyfus, 1947, 1948; Schapira et al., 1947; Dreyfus, 1950). These compounds can be only partially extracted from the muscles, and they are bound to the contractile protein actomyosin, and are thought to have a role in muscle function (Buchthal et al., 1961).

Estimates from the literature of the amounts of the various iron-containing compounds are shown in Table 4/1.

Table 4/1

The iron-containing compounds of the human organism

Name of compound	Total amount (g)	Iron content (g)	In per cent of total iron content of organism	References*
Hemoglobin	800	2.67	66.7	(1), (2)
	900	3.06	approx. 70	(3)
	650–750	2.1–2.5	65	(4)
		2.5	67	(5)
Myoglobin	40	0.14	3.3	(1), (2), (3)
	40	0.13	3.5	(4), (5)
Cytochromes	0.8	0.0034	0.08	(3)
Cytochrome C	0.8	0.004	0.1	(4)
Cytochrome C, A, A_3, B, C_1	0.8	0.0034	0.08	(1), (2)
Catalase	5.0	0.004	0.1	(4)
	5.0	0.0045	0.11	(1), (2), (3)
Peroxidase	—	—	—	(1), (2)
Tissue hemins (total)		0.008	0.2	(5)
Ferroflavoproteins (succinodehydrogenase, xanthinoxidase, DPNH-cytochrome C reductase)	—	—	—	(4)
Ferritin	3.0	0.7–1.5	30.0	(1), (2)
Ferritin + hemosiderin		1.2–1.5		
Ferritin + hemosiderin	3.0	0.69	16.4	(3)
Ferritin + hemosiderin	—	0.8–1.5	30.0	(4)
Ferritin + hemosiderin		1.0	27.0	(5)
Transferrin	7.5	0.003	0.07–0.08	(1), (2), (3), (5)
	10.0	0.004	0.1	(4)
		0.003	0.08	(5)
Labile pool		0.08	2.2	(5)
Non-heme iron in muscles	—	0.5	10–12	(3)
Total		4.0	100	(1), (2)
		4.4	100–102	(3)
		3.0–4.0	100	(4)

* (1) Drabkin, 1951; (2) Granick, 1958; (3) Dreyfus and Schapira, 1958; (4) Bothwell and Finch, 1962; (5) Fairbanks and Beutler, 1977.

BIBLIOGRAPHY

BOTHWELL, T. H., FINCH, C. A.: Iron Metabolism. Little, Brown and Co., Boston 1962.

BUCHTHAL, F., DEUTSCH, A., KNAPPEIS, G. G., MUNCH-PETERSON, A.: Further investigations on the phosphorus and nucleotide uptake of actomyosin. Acta physiol. scand. *24*, 368 (1961).

CONRAD, M. E.: Factors affecting iron absorption. In: HALLBERG, L., HARWERTH, H. G., VANNOTTI, A. (eds.): Iron Deficiency. Academic Press, London–New York 1970.

DALLMAN, P. R., SINSHINE, P., LEONARD, Y.: Intestinal cytochrome response with repair of iron deficiency. Pediatrics *39*, 863 (1967).

DRABKIN, D. L.: Metabolism of the hemin chromoproteins. Physiol. Rev. *31*, 245 (1951).

DREYFUS, J. C.: Recherches sur le fer musculaire. Thèse Sciences, Paris 1950.

DREYFUS, J. C., SCHAPIRA, G.: Le fer. L'Expansion Scientifique Française, Paris 1958.

FAIRBANKS, V. F., BEUTLER, E.: Iron metabolism. In: WILLIAMS, W. J., BEUTLER, E., ERSLEV, A. J., RUNDLES, R. W. (eds.): Hematology. McGraw-Hill Inc., New York 1977.

GRANICK, S.: Iron metabolism. In: Trace Elements. Academic Press, New York 1958.

GREEN, D. E., MOORE, D. H., NOCITO, V., RATNER, S.: A flavoprotein enzyme. J. biol. Chem. *156*, 383 (1944).

MAHLER, H. R.: Metalloproteins and electron transport. 3rd Internat. Congr. Biochem., p. 103. Bruxelles 1955.

MALMSTRÖM, B. G.: Biochemical functions of iron. In: HALLBERG, L., HARWERTH, H. G., VANNOTTI, A. (eds.): Iron Deficiency. Academic Press, London–New York 1970.

RICHERT, D. A., WESTERFELD, W. W.: The relationship of iron to xanthine oxidase. J. biol. Chem. *209*, 179 (1954).

SCHAPIRA, G., DREYFUS, J. C.: Sur une fraction nouvelle du fer musculaire. C. R. Soc. Biol. (Paris) *141*, 155 (1947).

SCHAPIRA, G., DREYFUS, J. C.: Recherches sur le fer musculaire. I. Sur une nouvelle fraction de fer non héminique musculaire. Bull. Soc. Chim. Biol. *30*, 82 (1948).

SCHAPIRA, G., DREYFUS, J. C., LEAU, O.: Repartition de deux fractions de fer non héminique musculaire. C. R. Soc. Biol. (Paris) *141*, 704 (1947).

SCHMIDT, K. P., REIN, H.: Eisenbestimmungen in biologischen Materialen mit der flammenlosen Atomabsorptionsspektroskopie. Blut *37*, 119 (1978).

TAKEDA, Y., HARA, M.: Significance of ferrous iron and ascorbic acid in the operation of the tricarboxylic acid cycle. J. biol. Chem. *214*, 657 (1955).

CHAPTER 5

DIETARY IRON

IRON INTAKE

In many parts of the world iron deficiency is the commonest form of nutritional deficiency (Wretlind, 1970). There is a tendency, particularly in the developing countries, both for traditional and economic reasons, for the diet to be based on foods with a low iron content. Iron deficiency has a predilection for those individuals who have a low energy requirement but a high iron requirement, e.g., women in the reproductive period.

Physiological iron requirements vary considerably between the sexes and at different ages (see Table 5/2). The optimal iron intake should cover not only an average requirement but the physiological requirements of every member of a particular community. The recommended amount should also provide for storage iron. Pathological conditions leading to increased requirements, e.g., hemorrhage from the gastrointestinal tract, usually cannot be compensated for by diet alone.

The recommendations for iron intake in various countries are given in Table 5/1, and that recommended for different ages in Table 5/2.

The iron content of different foods varies widely, and even that in the same food may differ depending on the site and conditions of cultivation (Asenjo, 1962). The higher the temperature, humidity, and organic content of the soil, the higher the iron concentration of agricultural products. The older the plants, the lower their iron content. The actual iron content of the soil is less important (Elwood, 1965), since iron is present in excess in most soils (Wretlind, 1970). Iron-rich soil, however, may increase the iron content of the diet by contamination (Blix, 1965). Food processing may also add to the iron by contamination, but since the introduction of stainless steel equipment into the agricultural industry, contamination of iron from these sources has dropped considerably (Blix, 1965).

Methods of preparation may also influence the iron in some articles of diet, the most well-known example being the Kaffir beer of the Bantu, which is traditionally brewed in iron pots (Walker and Arvidsson, 1953; Charlton et al., 1973). MacDonald (1963) investigated the iron content of various beers and wines and found it to vary from 2 to 16 mg/liter. Even water may show considerable variation in iron content depending on whether it is pumped directly from a deep, bored well or originates from water mains. Water from the former may contain as much as

Table 5/1
Daily iron intake recommended for adults
(from Food and Nutrition Board at the National Academy of Sciences–National Research Council of the USA, 1968; Swedish National Institute of Public Health, 1969)

Country	Sex	Age (years)	Body weight (kg)	Iron (mg/day)
Australia	male	25	70	10
	female	25	58	10
Canada	male	25	72	6
	female	25	57	10
Central America and Panama	male	25	55	10
	female	25	50	10
Japan	male	26–29	56	10
	female	26–29	49	10
The Netherlands	male	20–29	70	10
	female	20–29	60	12
Norway	male	25	70	12
	female	25	60	12
South Africa	male	—	73	9
	female	—	60	12
Sweden	male	22	70	10
	female	22	58	18
United Kingdom	male	20	65	12
	female	20	56	12
USA	male	22	70	10
	female	22	58	18

5 mg/liter of iron (Taylor, 1958), whereas the water from the latter may be almost iron-free (Wretlind, 1970).

The iron content of some common foodstuffs is given in Table 5/3. Liver and kidney are particularly rich in iron. Pork and chicken liver are richer in iron than beef liver. Other forms of meat and fish are all rich in iron.

Eggs have a relatively high iron content (16–17 mg/1000 kcal*), but it is in a poorly available form (see below). Among other animal products, milk and dairy products are relatively low in iron.

* 1 cal = 4.19 J (joule).

Table 5/2

Recommended dietary allowances of energy and iron according to Swedish National Institute of Public Health (1969), as well as corresponding calculated content of iron per 1000 kcal

Category	Age (years)	Recommended energy (kcal/day)	Recommended amount of iron (mg/day)	Iron content to 1000 kcal (mg)*
Infants**	0– 1		10	
Children	3– 4	1400	10	7
	4– 6	1600	10	6
	6– 8	2000	10	5
	8–10	2200	10	5
Adolescent boys and adult males	10–12	2500	10	4
	12–14	2700	18	7
	14–18	3000	18	6
	18–22	2800	10	4
	22–35	2800	10	4
	35–55	2600	10	4
	55–75+	2400	10	4
Adolescent girls and adult females	10–12	2250	18	8
	12–14	2300	18	8
	14–16	2400	18	8
	16–18	2300	18	8
	18–22	2000	18	9
	22–35	2000	18	9
	35–55	1850	18	10
	55–75+	1700	10	6
Pregnant women**			30	

* The values have been rounded off to the nearest whole numbers.
** Fairbanks and Beutler, 1977.

Most fruits other than black currants, strawberries, and dried fruits contain less than 10 mg/1000 kcal of iron. Cereals are rich in iron, but through purification and milling their iron content is reduced to a quarter or a half of the original, and in some countries such as the United States, Great Britain, and Sweden flour is enriched with iron in order to restore its original iron content.

Some vegetables, e.g., spinach and parsley, are very rich in iron in relation to their caloric value, but because of their bulk they do not make a large contribution to the total dietary iron.

Although the tables indicating the iron content of foods can give an overall indication of the total dietary iron, they have a limited value (Elwood, 1966, and others), since direct determinations of dietary iron are often at considerable variance

Table 5/3

Iron content (mg/100 g and mg/1000 kcal) of the various foods, and the weight of food corresponding to 1000 kcal* (from Swedish National Institute of Public Health, 1967)

Foods	Iron (mg/100 g)	Iron (mg/1000 kcal)	Weight (g/1000 kcal)
Vegetables			
String beans	1.1	32.4	2941
Dry beans	7.0	20.2	288
White cabbage	0.5	12.5	3125
Cauliflower	0.6	24.0	4000
Kale	2.2	59.4	2703
Garlic	2.0	76.9	3846
Onion	0.4	11.1	2778
Parsley	8.0	186.0	2326
Green pepper	0.7	31.8	4545
Red pepper	0.6	19.0	3226
Green peas	1.9	22.1	1163
Split peas	5.0	16.1	323
Spinach	3.0	142.9	4762
Tomato	0.5	25.0	5000
Fruits			
Apple	0.3	4.8	1613
Apricot (dried)	4.5	18.2	405
Banana	0.5	5.4	1075
Black currant	1.3	34.2	2632
Prunes	3.4	12.8	376
Orange	0.3	6.1	2041
Pear	0.2	3.5	1754
Raisin	2.1	7.3	346
Strawberry	0.7	17.9	2564
Walnut, hazelnut, almond	4.2	6.7	159
Other plants			
Carrot	0.6	15.8	2632
Potato	0.8	9.6	1205
Radish	1.4	70.0	5000
Milk and dairy products			
Milk	0.1	1.6	1613
Whey-cheese	5.0	16.6	332
Cheese	0.3	0.9	301
Butter	0.2	0.3	130

Table 5/3 (cont'd)

Foods	Iron (mg/100 g)	Iron (mg/1000 kcal)	Weight (g/1000 kcal)
Cream	0.1	0.4	374
Ice cream	0.1	0.6	571
Meat, fish, egg, etc.			
Pork	2.3	5.9	256
Veal	2.9	16.2	559
Beef	2.9	13.3	459
Chicken	2.0	17.1	855
Meat of other poultry	3.2	24.6	769
Beef heart	4.0	37.0	926
Calf liver	10.6	75.2	709
Pork liver	18.0	134.3	746
Liver paste	6.3	16.8	267
Veal kidney	15.0	121.0	806
Pork kidney	8.0	70.2	877
Blood	35.0	466.7	1333
Fish	0.9–1.5	7.5–12.7	498–1389
Crab	1.8	18.4	1020
Oysters	6.0	101.7	1695
Eggs	2.5	16.0	641
Cereals			
Wheat flour	5.0	14.2	284
Bread	3.3–3.8	10.3–12.9	272–392
Pastry	1.0	2.1	209
Other			
Sugar	0	0	244
Chocolate	3.1	5.4	174
Margarine	0	0	130
Fruit juice	1.5	4.6	308

* Values are means of serial determinations.

with those calculated from tables. Wretlind (1970) showed that direct measurements usually gave a value higher than that calculated, the average being 135%. In only 35% were the differences less than $\pm 10\%$. This is due partly to the variation in iron content of individual foodstuffs and partly to inaccurate estimation of the weight of foods.

In many parts of the world, but particularly in the developing countries, the dietary habits are such that the staple foods tend to be low in iron content, although in some areas, e.g., Ethiopia, where actual measurements have been made, the daily iron intake has been surprisingly high, partly at least from contamination. Even in economically developed countries the ratio of foodstuffs that are poor in iron is too

Table 5/4

The average daily caloric and iron intake, and the iron content (mg) to 1000 kcal of the food

Category	Average intake: energy (kcal)	Average intake: iron (mg)	Iron content of food (mg/1000 kcal)	References
Italy Rural population around Modena	3760	11.2	3.0	Vecchi et al. (1959)
Finland Rural population	2333	13.0	4.8	Pekkarinen and Roine (1964)
USA Teenage girls	1960	9.7	4.9	Hampton et al. (1967)
USA Teenage boys	2796	14.1	5.0	Hampton et al. (1967)
The Netherlands Females	2770	14.5	5.2	Mantz and den Hartog (1962)
USA 16- to 18-year-old schoolgirls	2045	11.0	5.4	Wharton (1963)
USA 16- to 17-year-old schoolboys	2784	14.8	5.3	Wharton (1963)
The Netherlands Schoolgirls	2230	12.8	5.7	Mantz and den Hartog (1962)
The Netherlands Schoolboys	2320	13.4	5.8	Mantz and den Hartog (1962)
Nigeria Schoolchildren	2350	14.6	6.2	Hauck (1961)
Uruguay Civilian population	2614	17.0	6.5	Valassi and Reynolds (1966)
Angola Rural population	1721	12.7	7.4	Strangeway (1961)
Guatemala 3- to 4-year-old children	963	9.7	10.1	Flores et al. (1966)

high in relation to iron-rich foods. Wretlind (1970) reported that foods that make up 48% of the total caloric intake contain only 6% of the daily iron needed. The situation in other countries is similar (Table 5/4).

Within a population the dietary iron varies considerably. Only 4% of 872 Swedish women ate a diet that contained more than 8 mg/1000 kcal iron, and in 50% the iron

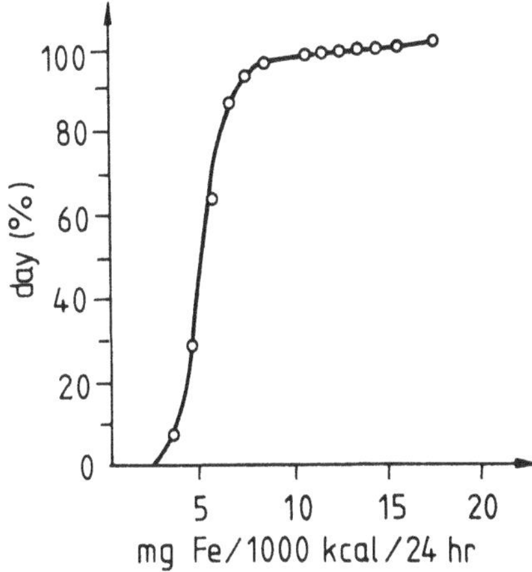

Fig. 5/1. Iron content of the daily diet/1000 kcal in Swedish women (after Wretlind, A., in: Hallberg, L. et al.: Iron Deficiency. Academic Press, London–New York 1970)

Table 5/5
The amount of iron split from various nutrients by gastric juice (from Heilmeyer and Mutius, 1943)

Nutrient	Iron (μg%)
Liver	1926
Boiled and fried meat	121–184
Bread	192–207
Boiled chicken	169
Brussels sprouts	139
Egg yolk	96
Fried potatoes, mashed potatoes	56–84
Lettuce	56
Butter	48
Boiled potatoes	32
Boiled carrots	31
Semolina	28
White cabbage	16
Milk	11–15
Noodles	14
Rice	11

content of the diet hardly exceeded 5 mg/1000 kcal (Wretlind, 1970) (Fig. 5/1). The energy requirement of the majority of the women was not more than 1400–2200 kcal, so that the daily iron intake was considerably below the desired level. Elderly subjects have a lower caloric requirement, which restricts still further the iron intake (Hallberg, 1970).

Although the total iron content of the diet is important in relation to iron nutrition, there is a further factor to be considered, namely, the variability in iron absorption from different foods.

The proportion of iron liberated from its organic bond during digestion from different foods is variable (Heilmeyer and Mutius, 1943) (Table 5/5). Of that which is solubilized and made available, again a variable proportion is absorbed because of the interaction between foods. Some foods promote the absorption of iron from other foods; others may inhibit absorption (see Chapter 6).

BIBLIOGRAPHY

Asenjo, C. F.: Variations in the nutritive values of foods. Amer. J. clin. Nutr. *11,* 368 (1962).

Blix, G.: A study on the relation between total calories and single nutrients in Swedish food. Acta Soc. Med. Uppsala *70,* 117 (1965).

Callender, S. T.: Food iron utilization. In: Hallberg, L., Harwerth, H. G., Vannotti, A. (eds.): Iron Deficiency, p. 75. Academic Press, London–New York 1970.

Callender, S. T., Marney, S. R., Warner, G. T.: Eggs and iron absorption. Brit. J. Haemat. *19,* 657 (1970).

Charlton, R. W., Bothwell, T. H., Seftel, H. C.: Dietary iron overload. In: Callender, S. T. (ed.): Iron Deficiency and Iron Overload (Clinics in Haematology), Vol. 2/2. Saunders, London–Philadelphia–Toronto 1973.

Elwood, P. C.: Bread and other foods of plant origin as a source of iron. Proc. Nutr. Soc. *24,* 112 (1965).

Elwood, P. C.: Utilization of food iron — an epidemiologist's view. Nutr. et Dieta *8,* 210 (1966).

Fairbanks, V. F., Beutler, E.: Iron metabolism. In: Williams, W. J., Beutler, E., Erslev, A. J., Rundles, R. W. (eds.): Hematology. McGraw-Hill Inc., New York 1977.

Flores, M., Flores, Z., Lara, M. Y.: Food intake of Guatemalan Indian children, ages 1 to 5. J. Amer. diet. Ass. *48,* 480 (1966).

Food and Nutrition Board at the National Academy of Sciences. National Research Council of the USA. Recommended Dietary Allowances, 1968.

Hallberg, L.: Prevalence of iron deficiency in Sweden. In: Hallberg, L., Harwerth, H. G., Vannotti, A. (eds.): Iron Deficiency. Academic Press, London–New York 1970.

Hampton, M. C., Huenemann, R. L., Shapiro, L. R., Mitchell, B. W.: J. Amer. diet. Ass. *50,* 385 (1967).

Hauck, H. M.: Dietary study in a Nigerian secondary school. J. Amer. diet. Ass. *39,* 467 (1961).

Heilmeyer, L., Mutius, I.: Untersuchungen über die Herauslösung von Eisen aus Nahrungsmitteln durch Magensaft und Galle. Z. ges. exp. Med. *112,* 192 (1943).

Heinrich, H. C. et al.: Die intestinale Resorption des Nahrungs-Eisens aus dem Hämoglobin, der Leber und Muskulatur bei Menschen mit normalen Eisenreserven und Personen mit prälatentem/latentem Eisenmangel. Klin. Wschr. *47,* 309 (1969).

Heinrich, H. C. et al.: Nahrungs-Eisenresorption aus Schweine-Fleisch, Leber und Hämoglobin bei Menschen mit normalen und erschöpften Eisenreserven. Klin. Wschr. *49,* 819 (1971).

Jacobs, A.: In: Hallberg, L., Harwerth, H. G., Vannotti, A. (eds.): Iron Deficiency, p. 81. Academic Press, London–New York 1970.

MacDonald, R. A.: Idiopathic hemochromatosis. Genetic or acquired? Arch. Intern. Med. *112,* 184 (1963).

Mantz, J. J. C., den Hartog, C.: Some data on diet in the Netherlands. Nutr. et Dieta *4*, 81 (1962).

Moore, C. V., Dubach, R.: Metabolism and requirements of iron in the human. J. Amer. med. Ass. *162*, 197 (1956).

Pekkarinen, M., Roine, P.: Studies on the consumption of the rural populations in East and West Finland. Ann. Med. exp. Fenn. *42,* 93 (1964).

Strangeway, A. K.: Malnutrition in Angola. J. Amer. diet. Ass. *39*, 585 (1961).

Swedish National Institute of Public Health. Tables of contents of nutrients in foods for automatic electronic data processing, 1967.

Swedish National Institute of Public Health. Recommended dietary allowances, 1969.

Taylor, A.: In: Thresh, Beale, Suckling (eds.): The Examination of Water and Water Supplies, p. 699. 7th ed., Little, Brown and Co., Boston 1958.

Valassi, K. V., Reynolds, J. W.: Dietary studies of the civilian population of Uruguay. Amer. J. clin. Nutr. *18*, 203 (1966).

Vecchi, G. P., Rubbiani, V., Saetti, G. C., Guarient, F.: Dietary consumption in patients with myocardial infarction. Notes on a nutritional investigation. Nutr. et Dieta *1*, 27 (1959).

Walker, A. R. P., Arvidsson, U. B.: Iron "overload" in the South African Bantus. Trans. roy. Soc. trop. Med. Hyg. *47*, 536 (1953).

Wharton, M. A.: Nutritive intake of adolescents. A study in Southern Illinois. J. Amer. diet. Ass. *42,* 306 (1963).

Wretlind, A.: Food iron supply. In: Hallberg, L., Harwerth, H. G., Vannotti, A. (eds.): Iron Deficiency. Academic Press, London–New York 1970.

CHAPTER 6

IRON ABSORPTION

A proportion of the food iron is liberated from its organic bond and reduced to ferrous iron in the stomach. The ferrous iron enters the cells of the duodenal and jejunal mucosa and either passes directly into the bloodstream or forms a complex of ferric compounds with apoferritin in the mucosal cells, in which case it remains in the cell and is later desquamated into the gut lumen. The iron that passes across the serosal surface of the mucosal cell into the blood is bound to transferrin and transported to the tissues (Fig. 6/1).

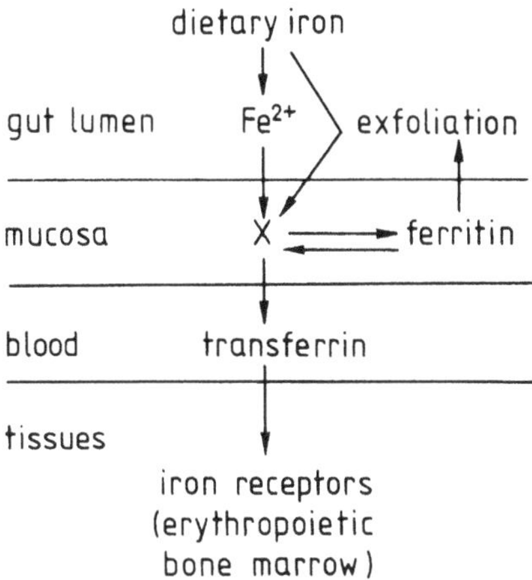

Fig. 6/1. Schematic representation of iron absorption

FACTORS INFLUENCING IRON ABSORPTION

Ferrous iron can be absorbed from any part of the gastrointestinal tract if it comes into direct contact with the mucous membrane in a soluble form. Absorption from the stomach is, however, minimal: less than 1–2% (Dagg et al., 1967). Following gastric surgery there may be some reduction of iron absorption (Baird and Wilson, 1959; Stevens et al., 1959; Turnbull, 1965), but this is probably related less to the removal of part of the stomach than to disturbances in gastric secretion. The normal gastric juice reduces the food iron and prepares it for absorption. Although iron can be absorbed in the absence of acid secretion, absorption can be improved by the administration of hydrochloric acid (Cook et al., 1964; Jacobs et al., 1964). Jacobs et al. (1966) have shown that an acid medium is necessary for iron to be split from the complex iron protein compounds during digestion in the stomach. The absorption of iron from heme compounds is not reduced by a lack of gastric secretion (Callender, 1964). This is probably because the heme molecule is absorbed directly into the intestinal mucosal cells, and the iron is split off within the cells (Conrad, 1970).

Davis and his collaborators (1966) claimed to have found an inhibitor to iron absorption in gastric juice, which they termed gastroferrin. They found less in iron-deficiency anemia and in idiopathic hemochromatosis, and in the gastric juice of relatives of patients with hemochromatosis (Luke et al., 1967; Davis et al., 1966; Deller et al., 1969). Other workers have been unable to confirm these observations. Wynter and Williams (1968a, b) found no difference between the iron-binding capacity of the gastric juice of healthy individuals and patients suffering from hemochromatosis, and Smith et al. (1969) failed to find a factor in the gastric juice that would potentiate iron absorption in hemochromatotics.

There are, however, iron-binding mucopolysaccharides in the gastric juice that are capable of inhibiting iron precipitation in the more alkaline medium of the small intestine (Jacobs and Miles, 1969). Rudzki and Deller (1973) have isolated from human gastric juice a glycoprotein, containing approximately 90% sugar residues and 10% amino acid residues, which appears to maintain colloidal iron particles in suspension.

Iron absorption is maximal in the duodenum and the first part of the small gut. This has been repeatedly demonstrated in animal experiments with the use of radioiron (Fig. 6/2) (Brown and Justus, 1958; Wack and Wyatt, 1959), and this is also the site at which ferritin accumulates (Gabrio and Salomon, 1950; Wöhler et al., 1957).

The part played by pancreatic secretion in iron absorption is not entirely clear. Taylor et al. (1931, 1935) observed the development of hypersiderosis after the ligation of the pancreatic duct or removal of the pancreas in experimental animals. Kinney et al. (1955), in ethionine-induced pancreatic necrosis, and Kaufman et al. (1958), in animals fed a fat-rich protein-poor diet, found iron absorption to be increased. Andersen noted siderosis of the tissues in patients dying of mucoviscidosis, and Banwell et al. (1964) found increased iron in the tissues in pancreatic disease. Davis and Badenoch (1962) made some studies of radioactive iron absorption in patients with chronic pancreatitis and found significantly increased iron absorption and increased iron in the liver in this condition (Fig. 6/3).

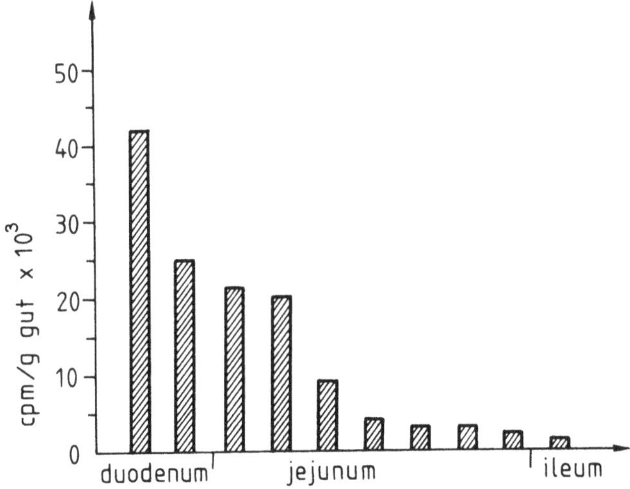

Fig. 6/2. Iron absorption from various gastrointestinal segments (after Brown, E. B. Jr. and Justus, B. W.: Amer. J. Physiol. *194*, 319, 1958)

In cirrhosis of the liver Callender and Malpas (1963) demonstrated increased iron absorption, which they attributed to accompanying pancreatic disease since the absorption could be reduced by the addition of pancreatic extract (Fig. 6/4). Davis and Biggs made similar observations, but Balcerzak and his collaborators (1967) did not find any increased iron absorption in chronic pancreatitis, and Kavin et al. (1967) found variable results, some patients showing an increased absorption, others normal absorption. The role of the pancreas in the regulation of iron absorption therefore remains questionable although the high phosphate and bicarbonate concentration of normal pancreatic juice probably limits iron absorption by precipitating a portion of the inorganic iron.

The role of the liver in the regulation of iron absorption is also questionable. Although increased iron absorption has been described in both cirrhosis of the liver (Callender and Malpas, 1963; Greenberg et al., 1964), and acute hepatitis (Turnberg, 1966; Bolin and Davis, 1968), and the relationship between hypersiderosis and cirrhosis of the liver is also well known (Herbut and Tamaki, 1964), it does not follow that the liver itself regulates iron absorption, as suggested by Murray and Stein (1966). In addition to the increased absorption being ascribed to associated

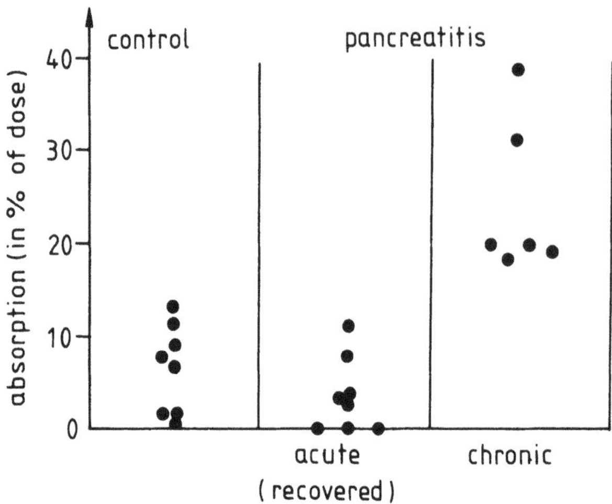

Fig. 6/3. Iron absorption in patients recovered from acute pancreatitis and in patients suffering from chronic pancreatitis (after Davis, A. E. and Badenoch, J.: Lancet, *2*, 6, 1962)

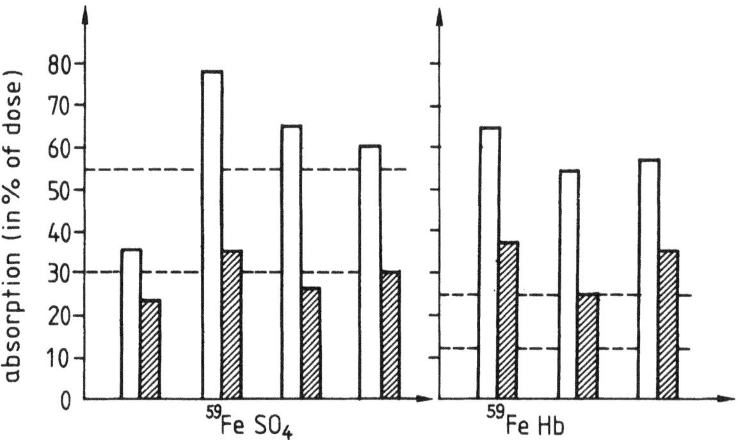

Fig. 6/4. Iron absorption from radioiron-labeled ferrous sulfate and labeled hemoglobin in cirrhosis of the liver with (▨) and without (□) the addition of pancreatic extract

pancreatic damage (Callender and Malpas, 1963; Sobel and Waye, 1963), it has also been attributed to enhanced function of the bone marrow (Deller, 1965) and to the effect of alcohol (Charlton et al., 1964).

The transfer of iron from the intestinal lumen to the plasma is mediated by the mucosal cells. If large doses of iron are repeatedly introduced into the gut, a small fraction of the iron will enter the phagocytes of the submucosa and peripheral lymph nodes (Wöhler et al., 1957; Cattau et al., 1965), and then the blood through the thoracic duct (Everett et al., 1954). Cattau and co-workers (1965) considered that the macrophages of the small intestinal villi play some part in the regulation of iron absorption. However, radioactive iron studies (Table 6/1) suggest that this can be of

Table 6/1
Iron content of the rat intestinal mucosa following the administration of radioactive iron (in percent of the absorbed iron) (from Finch, 1961)

Dose (μg)	Iron content of the mucosa			
	2	4	8	24
	days after the administration of radioactive iron			
30	7.6	4.3	2.1	2.0
300	7.4	3.0	3.1	1.8
3000	26.0	9.3	4.2	2.3

only minor significance, since following a dose of radioactive iron the residual activity in the submucosa and peripheral lymph nodes is slight (Finch, 1961). At autopsy an average of only 0.3% of 10 mg of iron given to patients 12 or more hours before death was recovered from the wall of the small intestine (Bothwell, 1961, cited by Bothwell and Finch, 1962).

Ferrous iron salts are better absorbed than ferric compounds (Moore et al., 1944; Hahn et al., 1945; Brise and Hallberg, 1962) (see Fig. 18/20), but in the case of small amounts, of the order of 1 μg/kg, the difference between the absorption from the ferrous and ferric compounds becomes indistinct (Bonnet et al., 1960). With larger doses the absolute amount of iron absorbed increases, but the percent absorption decreases (Bothwell et al., 1958; Smith and Pannacciulli, 1958) (Fig. 6/5). With increasing doses of iron in food, however, both the actual amount and percent absorption appear to increase (Josephs, 1958).

Ascorbic acid enhances iron absorption (Bernát and Kovács, 1956b; Bothwell et al., 1958; Brise and Hallberg, 1962) (Fig. 6/6); this is due partly to its reducing action but also to the fact that it forms a soluble ligand with iron; thus it will enhance the absorption of both ferric and ferrous compounds (Bernát and Kovács, 1956b; Bothwell and Finch, 1962; Bothwell et al., 1958; Brise and Hallberg, 1962). Other substances that increase iron absorption are succinic acid (Brise and Hallberg, 1962; Hallberg, 1963) (Fig. 6/7), cysteine, histidine, and fructose (Jacobi et al., 1956; Charley et al., 1963), and sodium dioctylsulfosuccinate (Bernát, 1968). Other substances, e.g., phosphates and phytates, decrease iron absorption by forming insoluble compounds (Turnbull and Finch, 1961). A phosphate-poor diet results in

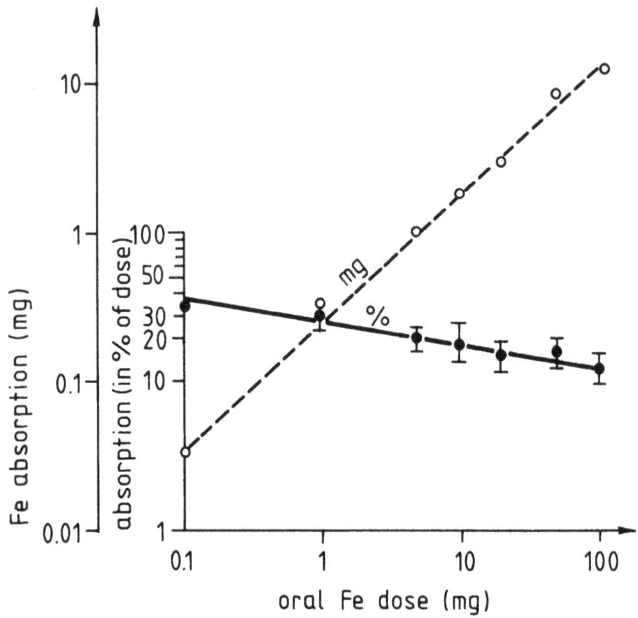

Fig. 6/5. The total and percent absorption from increasing doses of oral iron (after Smith, M. D. and Pannacciulli, I. M.: Brit. J. Haemat. *4*, 428, 1958)

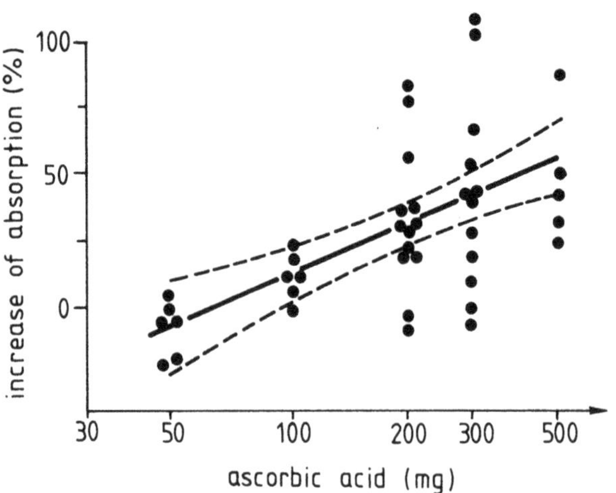

Fig. 6/6. The effect of varying doses of ascorbic acid on absorption of iron (after Brise, H. and Hallberg, L.: Acta med. scand. *171* [Suppl. 376], 59, 1962)

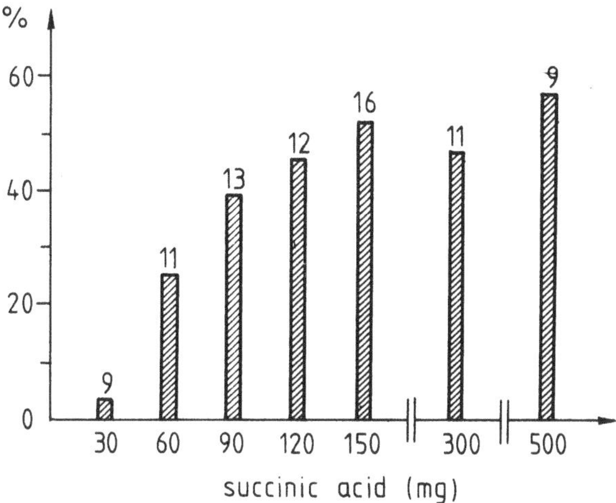

Fig. 6/7. The effect of varying doses of succinic acid on iron absorption (after Hallberg, L., in: Keiderling, W., Hoffmann, G.: Radioisotope in der Hämatologie. Schattauer, Stuttgart 1963)

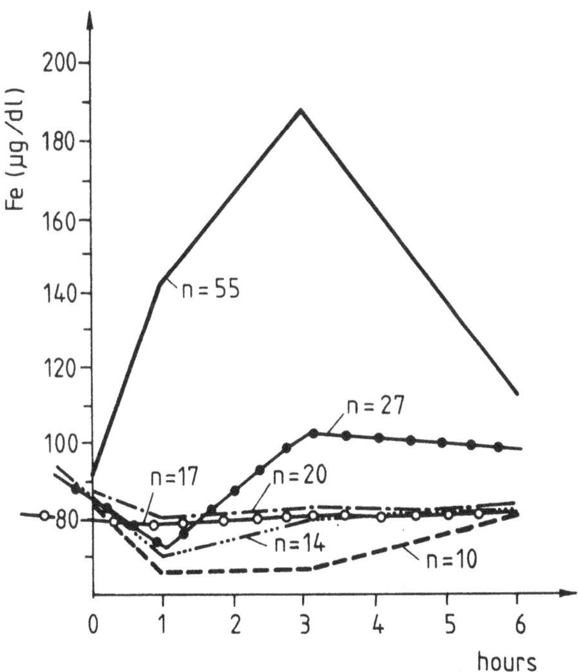

Fig. 6/8. The effect of oral iron loading on the plasma iron level fasting, and during, and after a meal (after Bernát, I. and Kovács, E., 1956b). ——— fasting; – ● – during a meal; – · – immediately after a meal; – ··· – 30 min; – ○ – 90 min after a meal; – – – spontaneous diurnal variations

increased iron absorption (Hegsted et al., 1949, 1952). Chelates such as desferrioxamine decrease the absorption of inorganic iron in food but not heme iron (Bannerman and Malpas, 1965; Hwang and Brown, 1965; Kuhn et al., 1968).

Therapeutic doses of iron are influenced by food intake. If inorganic iron salts, particularly ferrous compounds, are taken on an empty stomach, they are well absorbed. If they are given during or after a meal, absorption is reduced (Bernát and Kovács, 1956b; Chodos et al., 1957; Schulz and Smith, 1958; Pirzio-Biroli et al., 1958). Thus the plasma iron level increases to about double the original level following a dose of 170 mg of ferrous iron to a healthy, fasting subject; the same dose consumed during or after a meal produces hardly any elevation of plasma iron (Fig. 6/8).

In relation to food iron, certain foods, e.g., egg yolk, will impede iron absorption from other foods, e.g., bread, whereas citrus fruits will enhance the iron absorption from other foods (Callender et al., 1970). From a meal containing 5 mg of iron labeled with a tracer dose of radioactive iron, only about half the amount of iron is absorbed compared with that from a similar dose of iron given alone (Pirzio-Biroli et al., 1958). In children, 200 ml of milk was found by Schulz and Smith (1958) to reduce the absorption of iron from 27% to 17%. This effect has been attributed by some to the composition of the food, particularly in relation to the ratios of iron and phosphate and phytate content, or alternatively to the bulk of food consumed. Whatever the reason for the effect of food on iron absorption, however, it is recommended that therapeutic iron should be administered, preferably on an empty stomach (Bernát and Kovács, 1956b).

The question has been raised as to whether previous dietary habits have any influence on the absorption of iron in man. Norrby and Sölvell (1972) looked at the absorption of a 0.56-mg $^{59}Fe^{2+}$ test dose given after a week on an iron-deficient diet and compared it with that after a week on an iron-rich diet. The average absorption after the iron-rich diet was about 20% and that after the iron-deficient diet was 30%. These findings may be of importance when evaluating absorption data from small test doses in man.

The factors influencing the absorption of iron may be

(1) intraluminal

(2) mucosal

(3) corporeal

(Table 6/2).

As already indicated, iron absorption from different foods varies and is not proportional to the total iron content of the food. In the first place, it is influenced by the chemical bond of the iron, and in particular whether the iron is in the form of a heme compound or an inorganic iron complex.

Heme iron from hemoglobin and myoglobin in the diet represents an important source of iron in the economically developed countries. The heme is split off in the duodenum and enters the mucosal cell, where the iron is liberated (Turnbull et al., 1962; Bannerman and Malpas, 1965; Conrad et al., 1966, 1967) (Fig. 6/9). The

Table 6/2

Factors affecting iron absorption (from Conrad, 1970 – slightly modified)

		Absorption	
		increased	decreased
Intraluminal factors	Dietary iron content	Iron-replete diet	Iron-deficient diet
	Chemical form of dietary iron	Fe^{2+} Heme-iron	Fe^{3+}
	Food intake		+
	Dietary constituents	Certain carbohydrates; amino acids; ascorbic acid; succinic acid	Phytates; phosphates; carbonates; oxalates; hypoproteinemia; the majority of organic acids (lactic acid, citric acid, acetic acid)
	Intestinal secretions	HCl; ascorbic acid; bile; intrinsic factor; certain proteolytic and carbohydrate-splitting enzymes	Achlorhydria; pancreatic juice (bicarbonates, phosphates)
	Intestinal motility	Reduced (e.g. atropine, reserpine, dibenamine)	Increased (e.g. diarrhea)
	Stable chelators		EDTA, DTPA, DFO, gastroferrin
	Metallic cations		Co, Mn, Ni
Mucosal factors	Anatomic and histologic		Gastrectomy, gastric atrophy, resection of the small intestine, bypass, sprue
	Mucosal iron content	Decreased	Increased
Corporeal factors	Iron content of the body (extent of iron stores)	Iron deficiency	Iron overload
	Erythropoiesis	Acute blood loss, hemolysis, polycythemia vera, hypoxia, erythropoietin, parenteral cobalt	Aplastic anemia, transfusional erythrocytosis
	Increase in iron turnover due to other causes	Thalassemia, sideroblastic anemia, certain cases of liver cirrhosis	Endotoxin
	Other	Idiopathic hemochromatosis, atransferrinemia, pituitary extracts, testosterone	

absorption of heme iron is not influenced by gastric acid (Biggs et al., 1961), nor is it affected by associated ascorbic acid or phytates (Turnbull et al., 1962). Weintraub et al. (1968) demonstrated a substance in the homogenate of intestinal mucosal cells of the dog that was able to split iron from the heme molecule, and presumably the cells of the human intestinal mucosa contain a similar substance (Dagg et al., 1971). The atomic weight of the heme-splitting substance obtained from the dog exceeded

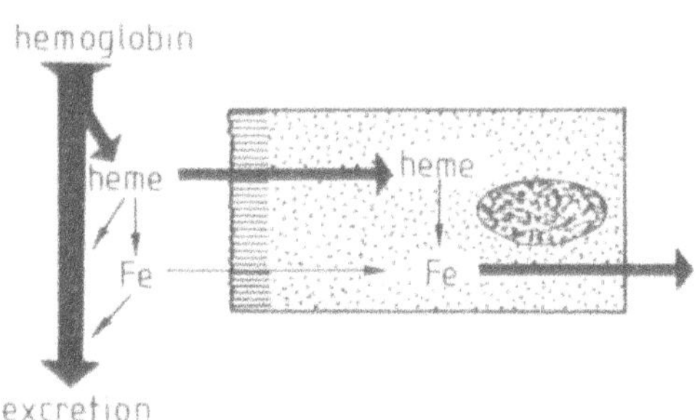

Fig. 6/9. Schematic representation of the absorption of hemoglobin iron. Hemoglobin is split to heme and globin in the duodenum, and the heme is absorbed as such into the mucosal cell where release of iron takes place. Subsequently iron enters the circulation depending upon requirement (after Conrad, M. E. et al.: Gastroenterology *53*, 5, 1967)

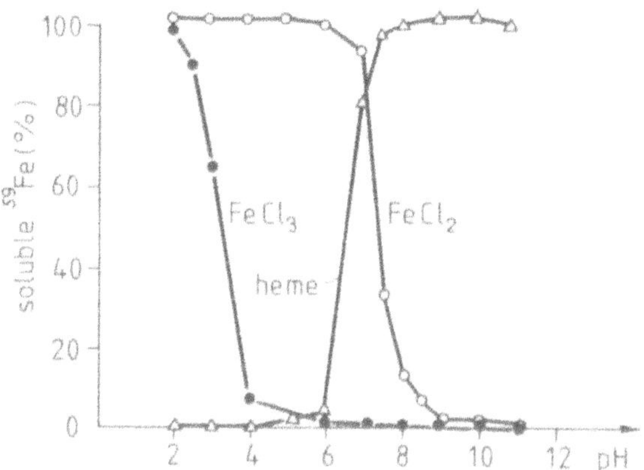

Fig. 6/10. Solubility of $FeCl_3$, $FeCl_2$ and heme at different pH values (after Conrad, M. E., in: Hallberg, L. et al.: Iron Deficiency. Academic Press, London–New York 1970)

64,000, and it behaved like an enzyme (Weintraub et al., 1968). Dawson et al. (1970) believed this substance to be xanthinoxidase.

In contrast to inorganic iron, heme remains in solution better in an alkaline medium than at acid pH, and the conditions in the duodenum therefore favor heme absorption (Fig. 6/10). Polymerization of the heme molecules reduces absorption (Conrad et al., 1966), but globin degradation products hinder the polymerization and thus promote heme absorption.

A two-pool model has been suggested for food iron absorption: a non-heme and a heme iron pool. These pools can be independently labeled with an inorganic radioiron tracer and with hemoglobin labeled with radioiron, using the two isotopes ^{55}Fe and ^{59}Fe. The total absorption of iron from these pools, i.e., from the whole diet, can be calculated from measurement of the whole-body retention of radioiron using a Whole-Body Counter, and the determination of the ratio of the two iron isotopes in a blood sample (Hallberg and Bjorn Rasmussen, 1972). In general, iron from animal foods is better absorbed than that from vegetable foods (Layrisse et al., 1969).

THE CONTROL OF IRON ABSORPTION

The equilibrium of iron metabolism is maintained primarily by changes in the amount of iron absorbed from the gut. If the requirement is great and the iron stores are empty, absorption is increased, but if the requirement is small and the iron stores of the body are considerable, the absorption decreases (see Fig. 6/11). The two principal factors that influence absorption are the extent of the iron stores (Bothwell et al., 1958), and the rate of erythropoiesis (Heinrich, 1966).

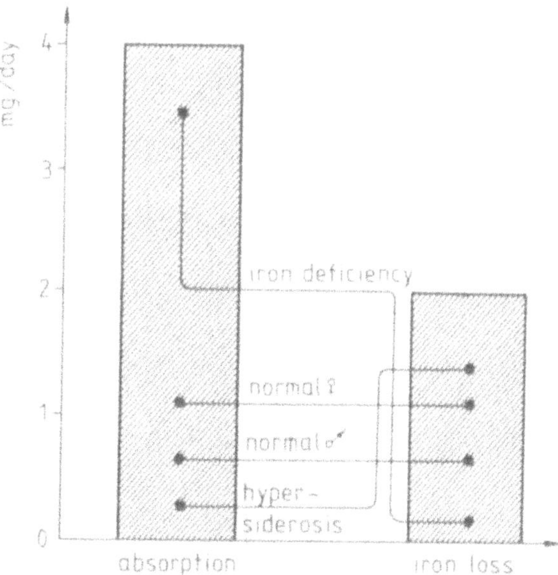

Fig. 6/11. Physiological limits of iron absorption and iron excretion showing a wider range in absorption than excretion (after Bothwell, T. H. and Finch, C. A.: Amer. J. dig. Dis. *2*, 145, 1957)

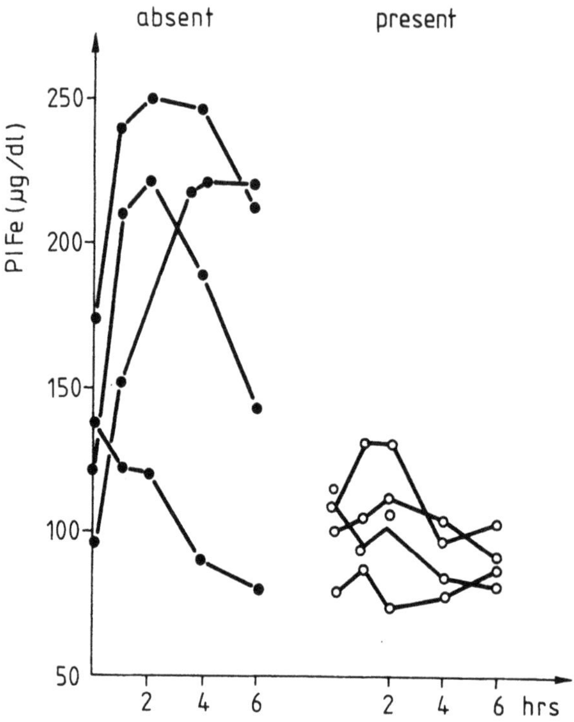

Fig. 6/12. Iron loading tests in the presence and absence of storage iron (after Pirzio-Biroli, G. and Finch, C. A.: J. Lab. clin. Med. *55*, 216, 1960)

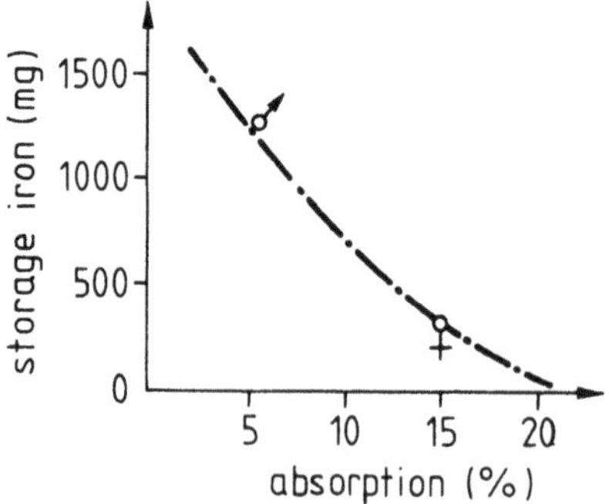

Fig. 6/13. Correlation between the amount of storage iron and the rate of absorption. Due to the lower iron reserves of the female, iron absorption is greater in women than in men (after Finch, C. A., in: Hallberg, L. et al.: Iron Deficiency. Academic Press, London–New York 1970)

Iron excretion is minimal and can vary only within very narrow limits. Thus the regulation is unidirectional, unlike other metabolic processes in which excretion plays a role in regulation.

In the normal individual, iron absorption is a sensitive index of the state of iron stores, the absorption of both inorganic iron salts and food iron being increased where the stores are diminished (Hahn et al., 1943; Dubach et al., 1948; Bothwell et al., 1958; Pirzio-Biroli et al., 1958; Moore, 1955; Pirzio-Biroli and Finch, 1960) (Figs. 6/11 and 6/12). The variation in absorption observed in healthy individuals

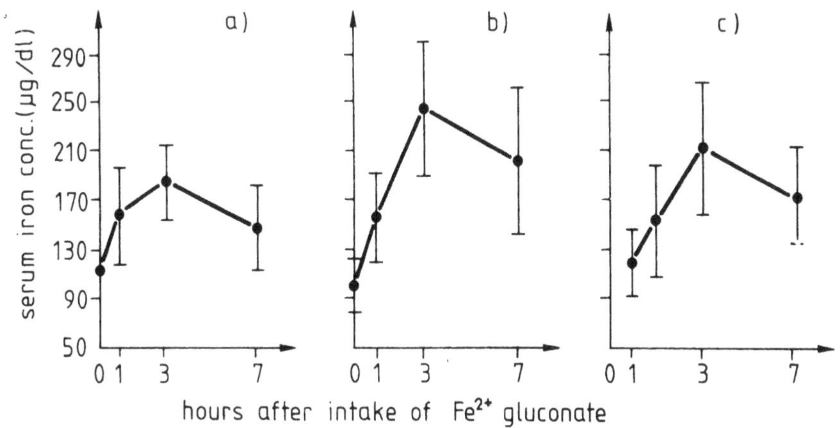

Fig. 6/14. Oral iron loading tests a) in 11 healthy humans; b) in 16 untreated blood donors; and c) in 15 blood donors after 2-month iron therapy (after Berde, B. et al.: Schweiz. med. Wschr. *85*, 936, 1955)

probably is largely attributable to the state of iron stores (Jasinski and Roth, 1954; Bothwell and Finch, 1962; Weinfeld, 1970; Hallberg, 1963) (Figs. 6/13 and 6/14).

If hemopoiesis is stimulated either by hemolysis (Stewart et al., 1953; Chappelle et al., 1955; Bothwell et al., 1958; Krantz et al., 1959), hemorrhage (Bothwell et al., 1958), or anoxia or by administration of cobalt, iron absorption is enhanced (Reynafarje et al., 1959; Bothwell et al., 1958) (Fig. 6/15). On the contrary, reduced erythropoiesis results in reduced iron absorption (Bothwell et al., 1958; Reynafarje et al., 1959; Krantz et al., 1959). The effect of the state of erythropoiesis on absorption can be modified by the state of iron stores (Bothwell et al., 1958). The precise mechanism by which the intestinal mucosal cells are stimulated to adjust the amount of iron absorbed has not been elucidated. Searches for a humoral factor in liver or kidney or in plasma have not been rewarding (Beutler and Buttenwieser, 1960). The stimulus to the mucosa does not appear to be erythropoietin (Krantz et al., 1959; van Dyke et al., 1961); neither is the transferrin concentration of the plasma a direct regulating factor (Bothwell et al., 1958; Greenberg et al., 1960).

Other workers have found some evidence of a humoral factor in the serum of iron-deficient animals or animals made hypoxic. Thus Fischer and Price (1963) found increased iron absorption following injection of serum from iron-deficient rats, and Brittin et al. (1968) showed that hypoxia induced in one of a parabiotic pair of animals enhanced iron absorption in the partner. Apte and Brown (1969) found a thermostable factor in the plasma of pregnant women that would enhance the intestinal radioiron uptake in rats.

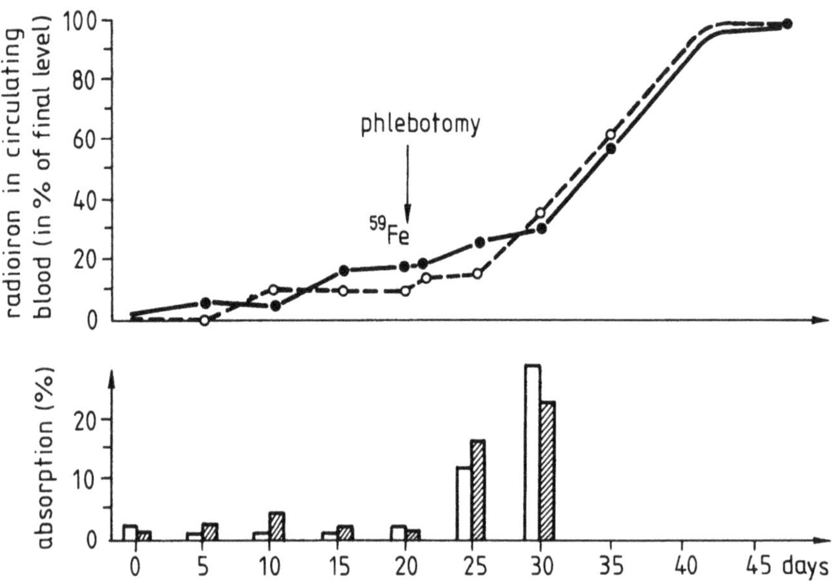

Fig. 6/15. Effect of increased erythropoiesis upon iron absorption. After the withdrawal of 600 ml of blood (↓) there is a rise in iron absorption due to the acceleration of red cell production (after Bothwell, T. H. et al.: J. Lab. clin. Med. *51*, 24, 1958)

Although absorption of physiological amounts of iron is not influenced directly by the iron-binding capacity of the plasma, absorption of therapeutic doses is restricted by the capacity of the plasma to bind iron and by the capacity of the tissue receptors. Following therapeutic doses the iron-binding capacity may be saturated, and this results in decrease in iron absorption (Hallberg and Sölvell, 1960b) (Fig. 6/16); direct iron transfer from the mucosa to the plasma is minimal when the transferrin is saturated (Sölvell, 1960), but the ultimate amount of iron absorbed is not reduced (Bothwell et al., 1953, 1958) because of the capacity of the mucosal cells to store the iron temporarily and transfer it to the plasma later (Hallberg and Sölvell, 1960a).

Iron transfer through the mucosal cells is the result of an active metabolic process (Dowdle et al., 1960), but the precise mechanism is still not clear (Bothwell, 1968). The copper-containing enzymes (Chase et al., 1952) and xanthinoxidase (Mazur et al., 1958; Cheney and Finch, 1960) appear to play a role in the process.

Earlier workers considered that the regulatory role of the intestinal mucosa was exercised by the degree of saturation of the iron acceptor protein apoferritin in the

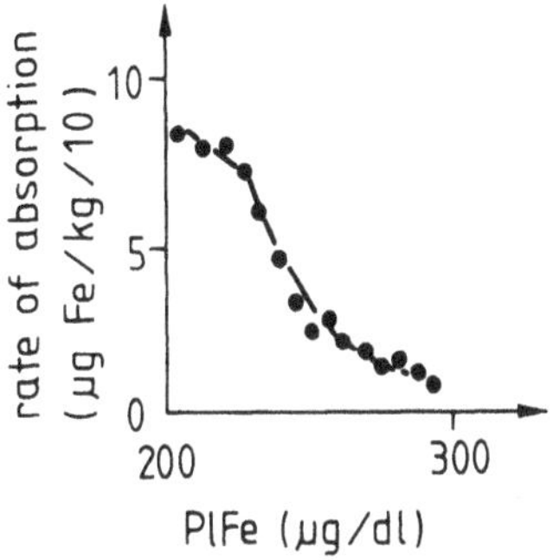

Fig. 6/16. Decline in iron absorption associated with a rise in plasma iron level (after Hallberg, L. and Sölvell, L.: Acta med. scand. *168* [Suppl. 358], 43, 1960b)

mucosal cells. Hahn et al. (1943) found the absorption of radioactive iron to be reduced if the experimental animals received stable iron some hours earlier. This observation led to Granick's forming the "mucosal block" hypothesis (Granick, 1956), according to which, when the ferritin-apoferritin apparatus of the cell was saturated, no further iron could be absorbed; when the iron was released into the plasma, the apoferritin could take up further iron. Granick's suggestion seemed to find support from various experimental studies; for example, Stewart et al. (1950) gave radioactive iron to dogs at various intervals following the oral administration of 100 mg of iron, and found that the shorter the time that had elapsed since the dose of stable iron, the poorer the absorption of radioactive iron. The diminished absorption of radioactive iron with the earlier doses could, however, be attributed in part to the fact that stable iron was still present in the gastrointestinal tract and could by dilution reduce the absorption of the radioisotope.

Granick's "mucosal block" theory was, however, based on unphysiological experiments, and in its original form it is no longer tenable. Furthermore, even at the unphysiological doses the block is not complete. Bernát and Kovács (1956a) showed that the plasma iron concentration rises after a large dose of ferrous sulfate, when a second dose is given 1 hour or 3 hours after the first iron load, although the increase in plasma iron is greater with the longer interval (Figs. 6/17 and 6/18). Brüschke (1962) made similar observations in healthy individuals and iron-deficient patients with about half the dose of iron (1.5 mg/kg) and a 4-hour interval (cited by Brüschke, 1964).

Brown et al. (1958) gave 5 mg of radioactive iron at various intervals to healthy and iron-deficient individuals following a dose of 4–5 mg/kg of iron, and found that although the absorption of radioactive iron was reduced for 6 hours in healthy subjects and for 3 hours in iron-deficient patients, there was no evidence of complete mucosal block.

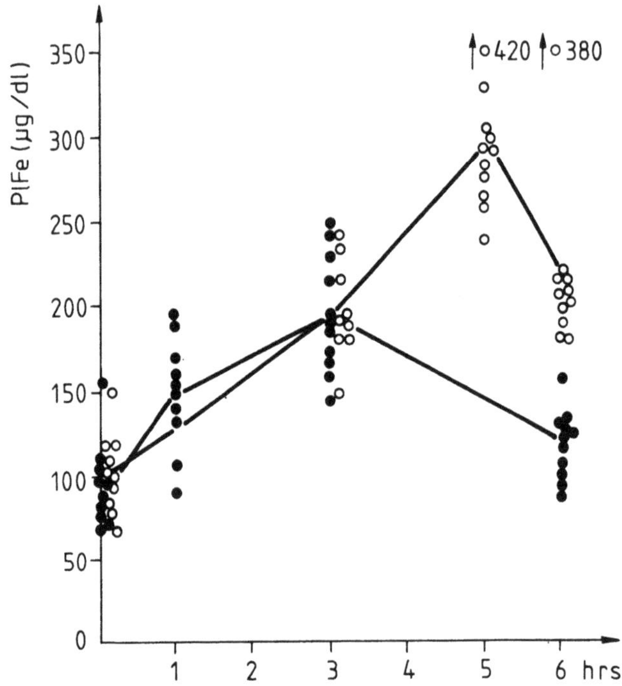

Fig. 6/17. Comparison of the results of double oral iron loading experiments (○) and single iron loads (●) (after Bernát, I. and Kovács, E., 1956a)

Estimations of mucosal ferritin suggested that a high ferritin content was, in fact, associated with a high absorption, and a low ferritin content with a low absorption of iron (Heilmeyer et al., 1957). Wöhler et al. (1957) concluded from this type of experiment that ferritin was not an inhibiting factor of absorption but rather a mediator of iron absorption.

Hallberg and Sölvell (1960b) showed that when large doses of iron were continuously infused into the stomach, the plasma iron concentration rose gradually and remained elevated during the observations (Fig. 6/19).

Crosby et al. (1963) have reinterpreted the part played by ferritin in the mucosal cells in the regulation of iron absorption:

"The duodenal absorptive cells act as an iron pool located between the intestinal lumen and the plasma. The quantity of iron in these mucosal cells remains in

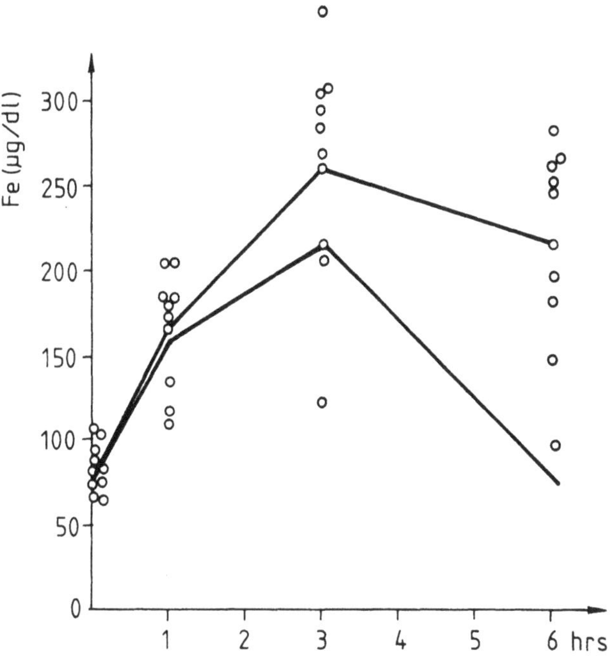

Fig. 6/18. Results of double oral iron loading experiments (the upper curve represents the mean values, the circles are individual values). The second iron dose was administered 1 hr after the intake of the first dose. The lower curve represents the mean values of single loadings (after Bernát, I. and Kovács, E., 1956a)

equilibrium with labile body iron stores. If iron demands in the body are great, the intestinal cells are depleted of iron. On the other hand, if a surplus of body iron exists to meet iron requirements, more iron is incorporated into the mucosal cells. Since these cells are shed into the lumen of the gut at the end of 2–3 day life-span, this can serve as an effective and selective excretory mechanism" (Fig. 6/20).

That the ferritin content of the epithelial cells of the mucosa depends on the iron reserves of the organism is suggested by the studies of Fineberg and Greenberg (1955) and Smith et al. (1968). The intestinal mucosa of experimental animals with high iron stores contains significantly more ferritin than the mucosa of normal or iron-deficient animals (Cumming et al., 1970), but it cannot necessarily be inferred that the differences in the ferritin content of the mucosal cells actually play a role in the regulation of iron absorption. On the other hand, Crosby's proposals are supported by the investigations of Boender and Verloop (1969), who showed that a

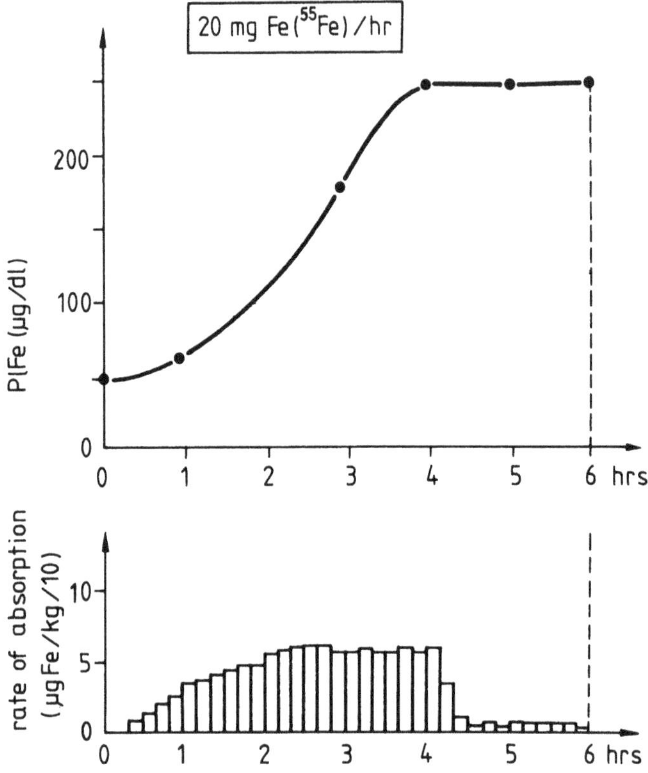

Fig. 6/19. Iron absorption during the infusion of 20 mg/hr of iron (^{55}Fe) into the stomach for 4 hours. By this time the plasma iron level is markedly elevated and remains high for a further 2 hours (after Hallberg, L. and Sölvell, L.: Acta med. scand. *168* [Suppl. 358], 43, 1960b)

proportion of the iron retained by the mucosa in healthy subjects and patients with secondary siderosis is later excreted in the feces; in contrast, in iron-deficiency states there is no late excretion of iron.

Manis and Schachter (1964) and Jacobs et al. (1966) showed that iron absorption takes place in two phases, viz., initial uptake and subsequent transfer through the cells. In the experimental model the uptake of iron can be shown to be uniform throughout the entire length of the small intestine, but iron transfer is restricted essentially to the duodenum. Howard and Jacobs (1972), using everted sacs of rat small intestine, showed that both mucosal uptake and serosal transfer of radioiron were maximal in the duodenum and decreased progressively toward the ileum. Transport of iron did not occur against a concentration gradient in any part of the

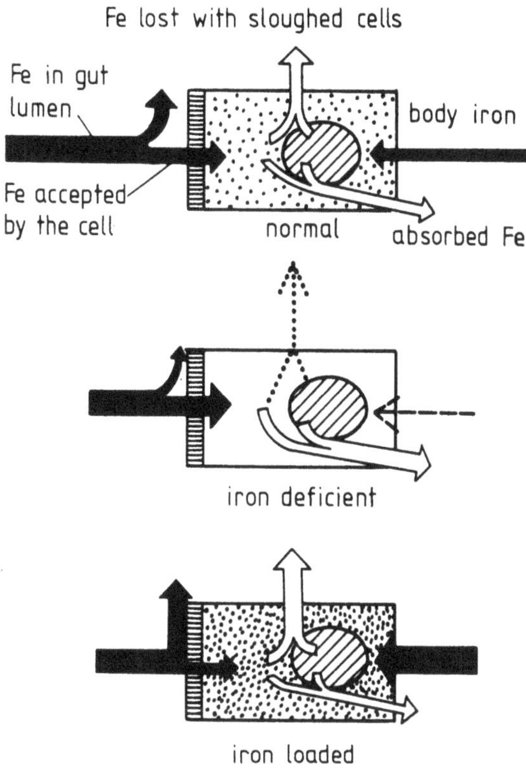

Fig. 6/20. Crosby's concept of the regulation of iron absorption (after Crosby, W. H.: Blood, *22*, 441, 1963)

intestine. In iron-deficient rats mucosal uptake of iron was increased, but in iron-loaded rats it remained normal. The results suggested that in the rat intestine there is a carrier mechanism that responds adaptively to iron deficiency but not to iron overload.

The first of the two phases of iron absorption occurs rapidly (Hallberg and Sölvell, 1960a), beginning 10 minutes after the intake of iron and lasting for 3–4 hours. The second phase is slower and may last longer. If the iron reserves are low, both phases are rapid. Unsaturated transferrin is bound to the serosal surface of the intestinal epithelial cell and may play a role in the regulation of the passage of recently ingested iron from the cell to the plasma (Levine et al., 1972). According to Pinkerton (1969), one carrier takes the iron from the intestinal lumen to the mucosal cell, while another, possibly iron transferase, transports the iron from the intestinal cell to the plasma.

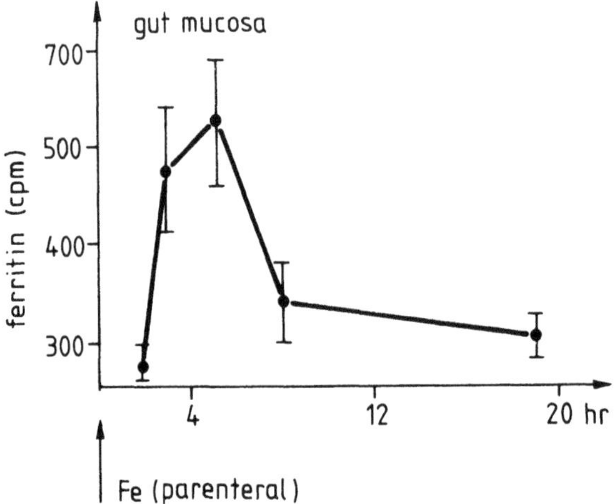

Fig. 6/21. Synthesis of apoferritin in the cells of the rat intestinal mucosa following the injection of iron. For labeling the protein the animal received ^{14}C-leucine (5 μCi/100 g body weight) 2 hours before sacrifice. Mean values ± 2 S. D. in 6 rats (after Millar, J. A. et al., in: Hallberg, L. et al.: Iron Deficiency. Academic Press, London–New York 1970)

Iron given by injection or by mouth initiates apoferritin synthesis in the cells of the intestinal mucosa (Fig. 6/21). The protein synthesis reaches its peak after 3–5 hours and thereafter declines gradually; parenteral iron injection is more effective than oral iron. Millar et al. (1970) found considerable differences in the apoferritin synthesis in the mucosal cells of iron-deficient rats as compared with the iron-overloaded animals (Fig. 6/22), but none between the iron-deficient and control animals. From these experiments Millar et al. concluded that the extent of apoferritin/ferritin synthesis is determined by

(1) the amount of iron available and/or

(2) the iron content of the animal.

These experiments not only corroborate Crosby's hypothesis, but also give an explanation as to how the differences in the body iron reserves regulate iron absorption and iron excretion.

According to Jacobs (1973), the control of iron absorption can be summarized in the following manner: The body's need for iron is reflected by the epithelial cells' need for iron, and this is the major factor controlling passage of iron from the intestinal lumen into the cells. The delivery of iron from the epithelial cell to the plasma appears to be controlled by the size of the soluble nonferritin pool of the cell, saturation of transferrin with iron, and the rate of iron clearance from the plasma.

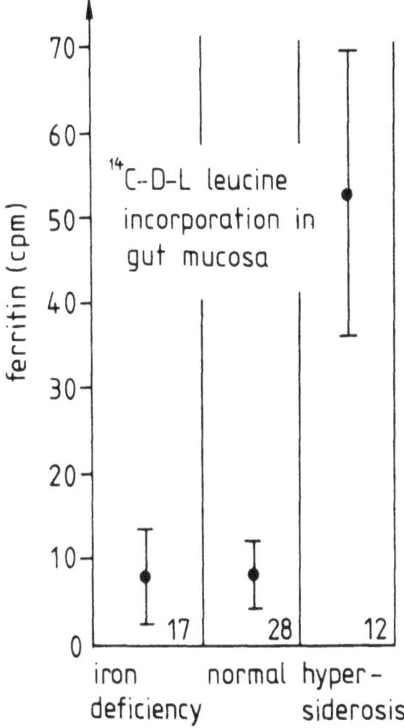

Fig. 6/22. Apoferritin synthesis in iron-deficient, normal and hypersiderotic rat intestinal mucosa (measured by the incorporation of ^{14}C-leucine) (after Millar, J.A. et al., in Hallberg, L. et al.: Iron Deficiency. Academic Press, London–New York 1970)

However, in human subjects with low iron stores, there is an increase in absorption even when transferrin saturation is normal. Ferritin itself plays no fundamental role in regulating iron absorption. It is possible that some of the ferritin iron formed in the epithelial cells may be mobilized into the carrier pool for transfer across the membrane, but it is likely that most of it will be lost along with other cellular iron at the end of the cell's life-span.

Croft (1970) studied the iron loss associated with desquamation of epithelium and calculated that the loss from the small intestine amounted to $60{,}000 \times 10^6$ cells daily, corresponding to 287 grams. Croft suggested that the iron from these cells was either partially or completely reabsorbed and not lost as postulated by Crosby. The long-term studies on iron loss by Green et al. (1968) using ^{55}Fe indicated a total daily loss of considerably less than 1 mg, whereas Crosby et al. (1963) had estimated the loss to be of the order of 4 mg daily.

Croft studied the dynamics of iron loss following an intravenous injection of ^{59}Fe transferrin. He lavaged the intestinal tract of rats at various intervals and measured the radioactivity of the washings. In the first 12 hours he concluded that the iron was lost exclusively through exudation, while later, at about 36 hours, the loss was through exfoliation, the turnover time of the intestinal mucosa being about 36 hours in rats (Fig. 6/23). A close correlation was found between the ^{59}Fe activity and the DNA content, suggesting that the loss was indeed due to exfoliation (Fig. 6/24).

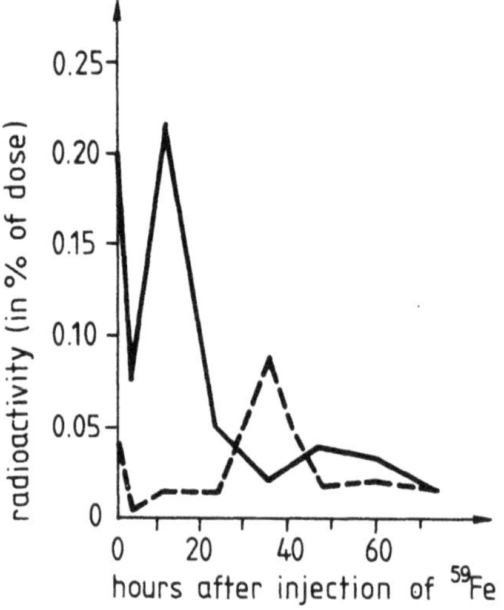

Fig. 6/23. ^{59}Fe activity in the small intestinal content of the rat (after Croft, D. N.: Proc. roy. Soc. Med. *63,* 1221, 1970). – – – normal rats; ——— rats with rapid turnover of mucosal cells

The mechanism of iron loss is shown in Fig. 6/25. Iron is incorporated into the newly formed epithelial cells in the crypts; the cells migrate toward the apex of the villi and are then exfoliated. Some of the cellular iron may be reabsorbed. A small fraction of the iron may reach the intestinal lumen by exudation, but this too may be reabsorbed. According to Croft, the processes of exudation and exfoliation may have a clinical significance in pathological conditions such as atrophic gastritis and

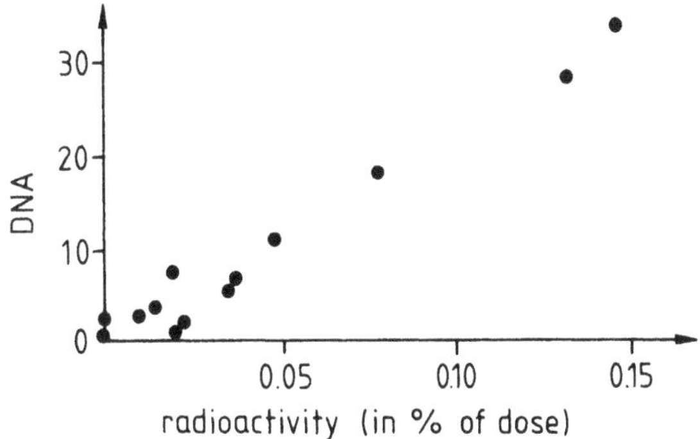

Fig. 6/24. Correlation between DNA (cell) loss and iron loss in the washings from the small intestine measured 36–42 hours after the intravenous injection of ^{59}Fe (after Croft, D. N.: Proc. roy. Soc. Med. *63*, 1221, 1970)

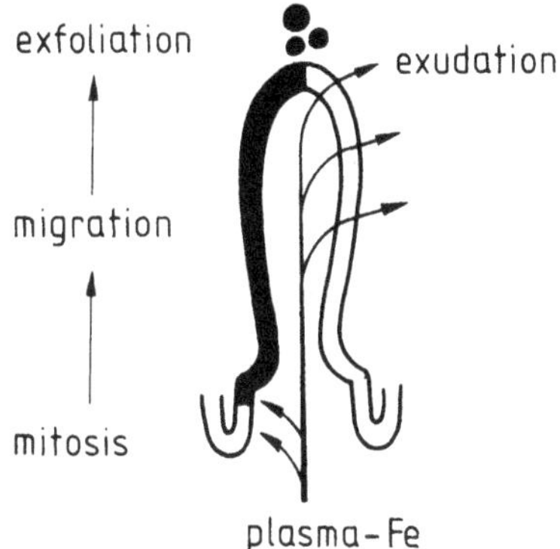

Fig. 6/25. Iron loss is attributable in part to exudation and in part to exfoliation (after Croft, D. N.: Proc. roy. Soc. Med. *63*, 1221, 1970)

coeliac disease, where the cellular turnover is accelerated and exudation enhanced but are unlikely to play a significant role in the regulation of normal iron metabolism. In pathological conditions, however, iron loss may be sufficient to upset the equilibrium and may be a factor in leading to iron deficiency.

MEASUREMENT OF IRON ABSORPTION

ORAL IRON LOADING CURVES

The earliest method for measuring iron absorption depended upon the measurement of plasma iron following an oral iron load. Initially it was presumed that the more rapid or more pronounced the elevation of plasma iron, the greater the iron absorption, while a slow or small rise suggested diminished absorption. However, the plasma iron concentration does not depend only on absorption but is determined also by the free transferrin concentration in the plasma and the rate of outflow of iron, the plasma iron curve being the result of the combined effect of these factors (Finch and Ross, 1952). Thus the curve may be flat if absorption is

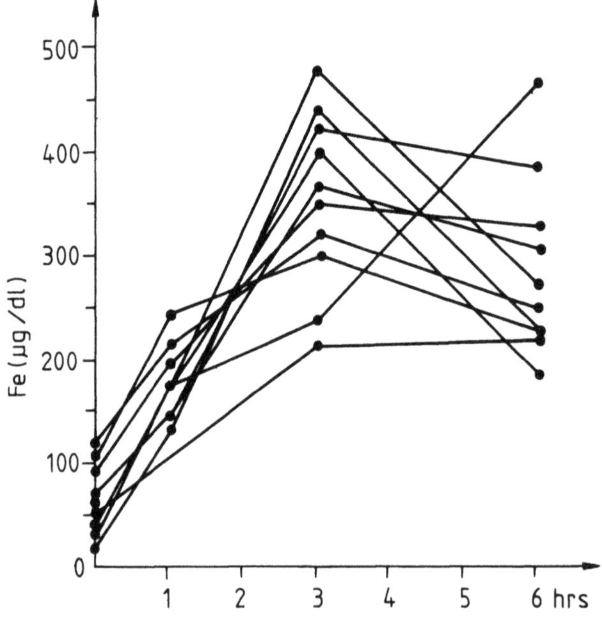

Fig. 6/26. Oral iron loading curves in latent iron deficiency

diminished, but also if the free iron-binding capacity is low or saturated, as in aplastic anemias, hemochromatosis (see Fig. 27/3), and some hemolytic anemias. The curve may also be flat if the outflow of iron from the plasma is accelerated, as for example in infection or burn injury. There is a pronounced rise in plasma iron concentration following an oral load, when the free transferrin concentration is high and the rate of iron outflow is relatively little enhanced, as in latent iron deficiency (Fig. 6/26). In more severe iron deficiency the outflow may be more rapid, resulting in a less marked rise in plasma iron.

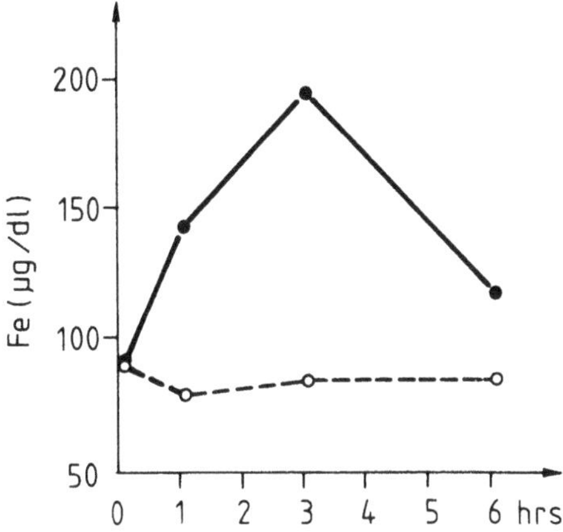

Fig. 6/27. Plasma iron level after oral loading in fasting (———) and after a meal (– – –) (after Bernát, I. and Kovács, E., 1956b)

Relatively large doses of iron are used for the test (100–170 mg Fe), and the test is therefore unphysiological from the point of view of both the dose and the form in which the iron is given. The physiological amounts of iron in food do not increase the plasma iron concentration appreciably (Fig. 6/27). An intake of at least 50 mg of inorganic iron is normally required to raise the plasma iron level appreciably (Josephs, 1958). The oral iron loading test, however, has a limited application, particularly, for example, to compare the absorption of various iron preparations in healthy subjects, but because of the great individual variation in absorption, repeated tests have to be carried out either in the same subjects or in a large number of individuals.

MEASUREMENT OF ABSORPTION WITH RADIOACTIVE IRON

Since the introduction of the radioisotopes of iron, various methods have been devised for measuring iron absorption.

The proportion of a dose of radioactive iron that is absorbed is assessed from the amount of unabsorbed radioactivity recovered in the feces. A measure of 3–5 mg of ferrous sulfate, ferrous chloride, or ferrous citrate containing about 5 μCi* of ^{59}Fe is diluted in 100 ml of water; 1–2 ml of the solution is used as a standard, and the remainder is given to the fasting subject. All feces are collected after the dose, and the collection is continued until the radioactivity of two successive stools is less than 1% of the orally administered dose. This usually takes 6–8 days. Absorption is calculated as follows:

$$\text{percent of daily dose excreted} =$$

$$= \frac{\text{counts/minute in feces}}{\text{standard counts/minute/ml} \times \text{volume of dose (ml)}} \times 100$$

The percent absorption = 100 – total percent activity in the feces.

Dubach et al. (1948) found that 20–30% of the iron was absorbed in the healthy subject, and 50–80% in patients with iron-deficiency anemia. However, the balance procedure is not always reliable since it depends on the meticulous collaboration of the subject, and even when the feces are apparently carefully collected, 20–40% of the oral radioiron dose cannot always be accounted for (Strohmeyer, 1963).

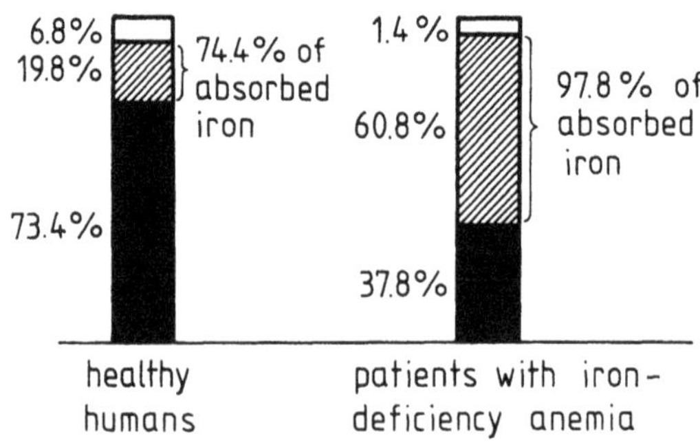

Fig. 6/28. Iron absorption in healthy humans and in patients with iron deficiency anemia.
■ radioactivity in feces (%)
▨ radioactivity in blood (%)
□ unaccountable loss of radioactivity
} rate of absorption

* 1 mCi = 37 MBq (becquerel).

A proportion of the iron that is absorbed from an oral dose of ^{59}Fe appears in the blood 7–10 days after the dose. The total activity in the blood can be calculated on the basis of the blood volume. Normally about 70–85% of the absorbed iron is incorporated into the red cells; in iron deficiency the amount is increased to about 90–100% (Fig. 6/28). The iron appearing in the red cells therefore gives a reasonable measurement of absorption in iron-deficient subjects. In pathological conditions it is only an approximation.

For this reason Saylor and Finch (1953) introduced a double-isotope method, which is based on the assumption that the radioiron absorbed from the gastrointestinal tract and that from an intravenous dose of iron are utilized in the course of erythropoiesis in a similar ratio (Fig. 6/29). If, for example, 80% of the parenterally introduced iron is found subsequently in the red cells, it is assumed that 80% of the iron absorbed from the oral dose is also utilized for erythropoiesis.

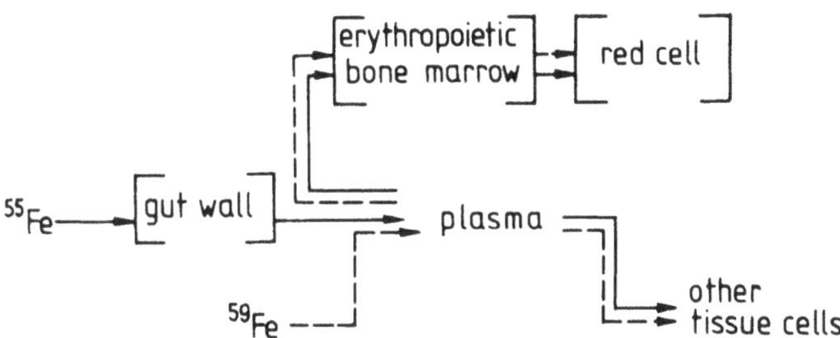

Fig. 6/29. Schematic representation of double-isotope labeling. The absorbed iron (^{55}Fe) and the intravenously injected iron (^{59}Fe) enter a common plasma pool. Thereafter, the distribution of iron isotopes is similar in the organism (after Saylor, L. and Finch, C. A.: Amer. J. Physiol. *172*, 372, 1953)

One of the isotopes of iron (^{59}Fe or ^{55}Fe) is injected intravenously and the other is given orally, and 10–12 days later the activity of the two isotopes, which have different radiation characteristics, is measured in the blood. The fraction of the intravenous dose that appears in the red cells is measured, and the activity from the oral dose is taken to be a similar fraction of the absorbed dose.

WHOLE-BODY COUNTING METHODS

A number of whole-body counters have been devised and used for measurement of iron absorption, the great advantage of the method being that it does not depend on the cooperation of subjects to collect feces, and it gives an absolute measurement of retention of iron. After the background has been counted, the subject takes the dose of iron labeled with ^{59}Fe, and the 100% activity is measured shortly after. The retained activity can be measured as soon as the nonabsorbed iron has been excreted in the feces. This is usually at the end of a week, but if the subject is constipated, it is wiser to wait for 10–14 days before the second measurement is made (van Hoek and Conrad, 1961; Strohmeyer, 1963; Heinrich, 1966).

The absorption in different individuals depends upon many factors, particularly the iron status and the dose of iron used in the test.

MEASUREMENT OF IRON ABSORPTION FROM THE DIET

Several studies have been made in recent years comparing the absorption of iron from food labeled biologically, i.e., with an intrinsic tag, with the absorption of iron from the same food labeled with an extrinsic tag of radioactive iron. Using two isotopes of iron (^{59}Fe and ^{55}Fe) it has been shown that the ratio of absorption from the extrinsically and intrinsically tagged foods is close to unity over a wide range, suggesting that the extrinsic tag becomes intimately mixed with the iron in the food and serves as a true label of the non-heme iron pool. A similar method can be used to label the heme iron pool in the diet (Layrisse and Martinez-Torres, 1972).

BIBLIOGRAPHY

Apte, S. V., Brown, E. B.: Effects of plasma from pregnant women on iron absorption in the rat. Gastroenterology *57*, 126 (1969).

Badenoch, J., Callender, S. T.: Iron metabolism in steatorrhea. The use of radioactive iron in studies of absorption and utilisation. Blood *9*, 123 (1954).

Baird, I. M., Wilson, G. H.: The pathogenesis of anaemia after partial gastrectomy. II. Iron absorption after partial gastrectomy. Quart. J. Med. N. S. *28*, 35 (1959).

Baird, I. M., Podmore, D. A., Wilson, G. H.: Changes in iron metabolism following gastrectomy and other surgical operations. Clin. Sci. *16*, 463 (1957).

Balcerzak, S. P., Peternel, W. W., Heinle, E. W.: Iron absorption in chronic pancreatitis. Gastroenterology *53*, 257 (1967).

Bannerman, R. M., Malpas, J. S.: Studies on desferrioxamine in relation to the absorption of iron. Brit. J. Haemat. *11*, 15 (1965).

Banwell, J. G., Hutt, M. S. R., Marsden, P. D., Blackman, V.: Malabsorption syndromes amongst African people in Uganda. E. Afr. med. J. *41*, 188 (1964).

Berde, B. et al.: Schweiz. med. Wschr. *85*, 936 (1955).

Bernát, I.: Unpublished results (1968).

Bernát, I., Kovács, E.: Vizsgálatok a normális és kóros vasanyagcsere köréből. II. Adatok a "mucosa block" kérdéséhez (Investigations on the normal and pathological iron metabolism. II. Contributions to the question of the "mucosa block"). Katonaorvosi Szle *8*, 725 (1956a).

Bernát, I., Kovács, E.: Vizsgálatok a normális és kóros vasanyagcsere köréből. III. A vasfelszívódást és a vaskezelés hatásosságát befolyásoló egyes tényezőkről (Investigations on the normal and pathological iron metabolism. III. Factors affecting the iron absorption and the effectivity of iron treatment). Katonaorvosi Szle *8*, 882 (1956b).

Bernát, I. Jr., Magyar, J., Bernát, I., Mihályi, L.: A plasma vaskoncentrációjának alakulása különféle vaskészítményekkel végzett orális terhelés után (Plasma iron concentrations after oral administration of various iron preparations). Transfusio *11*, 5 (1978).

Beutler, E., Buttenwieser, E.: The regulation of iron absorption. I. A search for humoral factors. J. Lab. clin. Med. *55*, 274 (1960).

Bezwoda, W. R., Bothwell, T. H. et al.: The relationship between marrow iron stores, plasma ferritin concentrations and iron absorption. Scand. J. Haemat. *22*, 113 (1979).

Bickel, H., Hall, G. E., Keller-Schierlein, W., Prelog, V., Vischer, E., Wettstein, A.: Stoffwechselprodukte von Actinomyceten, 27. Über die Konstitution von Ferrioxamin-B. Helv. chim. Acta *43*, 2129 (1960).

Biggs, J. C., Bannerman, R. M., Callender, S. T.: Proc. 8th Congr. Europ. Soc. Haemat., Vienna 1961.

Boender, C. A., Verloop, M. C.: Iron absorption, iron loss and iron retention in man; studies after oral administration of a tracer dose of $^{59}FeSO_4$ and $^{131}BaSO_4$. Brit. J. Haemat. *17*, 45 (1969).

Bolin, T., Davis, A. E.: Iron absorption in infectious hepatitis. Amer. J. dig. Dis. *13*, 16 (1968).

Bonnet, J. D., Hagedorn, A. B., Owen, C. A.: A quantitative method for measuring the gastrointestinal absorption of iron. Blood *15*, 36 (1960).

Bothwell, T. H. (1961), cit. Bothwell, Finch, 1962.

Bothwell, T. H.: The control of iron absorption. Brit. J. Haemat. *14*, 453 (1968).

Bothwell, T. H., Finch, C. A.: The intestine in iron metabolism. Its role in normal and abnormal states. Amer. J. dig. Dis. *2*, 145 (1957).

Bothwell, T. H., Finch, C. A.: Iron Metabolism. Little, Brown and Co., Boston 1962.

Bothwell, T. H., Pirzio-Biroli, G., Finch, C. A.: Iron absorption. I. Factors influencing absorption. J. Lab. clin. Med. *51*, 24 (1958).

Bothwell, T. H., van Doorn-Wittkampf, H. W., du Preez, M. L., Alper, T.: The absorption of iron. Radioiron studies in idiopathic haemochromatosis, malnutritional cytosiderosis, and transfusional hemosiderosis. J. Lab. clin. Med. *41*, 836 (1953).

Brise, H., Hallberg, L.: Iron absorption studies. II. Effect of succinic acid on iron absorption. Acta med. scand. *171* (Suppl. 376), 59 (1962).

Brittin, G. M., Haley, J., Brecher, G.: Enhancement of intestinal iron absorption by a humoral effect of hypoxia in parabiotic rats. Proc. Soc. exp. Biol. (N. Y.) *128*, 178 (1968).

Brown, E. B., Jr., Justus, B. W.: In vitro absorption of radioiron by everted pouches of rat intestine. Amer. J. Physiol. *194*, 319 (1958).

Brown, E. B., Jr., Dubach, R., Moore, C. V.: Studies on iron transportation and metabolism. XI. Critical analysis of mucosal block by large doses of inorganic iron in human subjects. J. Lab. clin. Med. *52*, 335 (1958).

Brüschke, G. 1962, cit. Brüschke, G.: Der Eisenstoffwechsel, pp. 22–23. Steinkopff, Dresden–Leipzig 1964.

Callender, S. T.: Iron absorption. Brit. med. Bull. *15*, 5 (1959).

Callender, S. T.: Digestive absorption of iron. In: Gross, F. (ed.): Iron Metabolism. Springer, Berlin–Göttingen–Heidelberg 1964.

Callender, S. T., Andrews, St., Warner, G. T.: Iron absorption from bread. Haematologia *5*, 369 (1971).

Callender, S. T., Malpas, J. S.: Absorption of iron in cirrhosis of the liver. Brit. Med. J. *2*, 1516 (1963).

Callender, S. T., Marney, S. R., Warner, G. T.: Eggs and iron absorption. Brit. J. Haemat. *19*, 657 (1970).

Cattau, D., Debray, Ch., Jori, J. P., Marche, C.: Considérations sur le mécanisme régulateur de l'absorption intestinale du fer. Importance des macrophages du chorion villositaire. Une nouvelle théorie. Soc. Méd. Hôp. (Paris) *116*, 1653 (1965).

Chappelle, E., Gabrio, B. W., Stevens, A. R. Jr., Finch, C. A.: Regulation of body iron content through excretion in the mouse. Am. J. Physiol. *182*, 390 (1955).

Charley, P. J., Stitt, C. F., Shore, E., Saltman, P.: Studies on the regulation of iron absorption. J. Lab. clin. Med. *61*, 397 (1963).

CHARLTON, R. W., JACOBS, P., SEFTEL, H., BOTHWELL, T. H.: Effect of alcohol on iron absorption. Brit. Med. J. *2,* 1427 (1964).

CHASE, M. S., GUBLER, C. J., CARTWRIGHT, G. E., WINTROBE, M. M.: Studies on copper metabolism. IV. The influence of copper on the absorption of iron. J. biol. Chem. *199,* 757 (1952).

CHENEY, B., FINCH, C. A.: The effect of inosine on iron absorption in rats. Proc. Soc. exp. Biol. Med. *103,* 37 (1960).

CHODOS, R. B., ROSS, J. F., APT, L., POLLYCOVE, M., HALKETT, J. A. E.: The absorption of radioiron labeled foods and iron salts in normal and iron deficient subjects and in idiopathic hemochromatosis. J. clin. Invest. *36,* 314 (1957).

CONRAD, M. E.: Factors affecting iron absorption. In: HALLBERG, L., HARWERTH, H. G., VANNOTTI, A. (eds.): Iron Deficiency, pp. 87–114. Academic Press, London–New York 1970.

CONRAD, M. E., CROSBY, W. H.: Intestinal mucosal mechanisms controlling iron absorption. Blood *22,* 406 (1963).

CONRAD, M. E., BENJAMIN, B. I., WILLIAMS, H. L., FOY, A. L.: Human absorption of haemoglobin iron. Gastroenterology *53,* 5 (1967).

CONRAD, M. E., CORTELL, E., WILLIAMS, H. L., FOY, A. L.: Polymerisation and intraluminal factors in the absorption of hemoglobin iron. J. Lab. clin. Med. *68,* 659 (1966).

CONRAD, M. E., WEINTRAUB, L. R., SEARS, D. A., CROSBY, W. H.: Absorption of hemoglobin iron. Amer. J. Physiol. *211,* 1123 (1966).

COOK, J. D., BROWN, G. M., VALBERG, L. S.: The effect of achylia gastrica on iron absorption. J. clin. Invest. *43,* 1185 (1964).

COOK, J. D., LIPSCHITZ, D. A., MILES, L. E. M., FINCH, C. A.: Serum ferritin as a measure of iron stores in normal subjects. Amer. J. clin. Nutr. *27,* 681 (1974).

CROFT, D. N.: Body iron loss and cell loss from epithelia. Proc. roy. Soc. Med. *63,* 1221 (1970).

CROSBY, W. H.: The control of iron balance by the intestinal mucosa. Blood *22,* 441 (1963).

CROSBY, W. H., CONRAD, M. E., WHEBY, M. S.: The rate of iron accumulation in iron storage disease. Blood *22,* 429 (1963).

CUMMING, R. L. C., SMITH, J. A., MILLAR, J. A., GOLDBERG, A.: The relationship between body iron stores and the rate of ferritin synthesis in rat liver and intestinal mucosa. Brit. J. Haemat. *18,* 653 (1970).

DAGG, J. H., CUMMING, R. L. C., GOLDBERG, A.: Disorders of iron metabolism. In: GOLDBERG, A., BRAIN, M. C. (eds.): Recent Advances in Haematology. Churchill–Livingstone, Edinburgh–London 1971.

DAGG, J. H., KUHN, I. N., TEMPLETON, F. E., FINCH, C. A.: Gastric absorption of iron. Gastroenterology *53,* 918 (1967).

DAVIS, A. E., BADENOCH, J.: Iron absorption in pancreatic disease. Lancet *2,* 6 (1962).

DAVIS, P. S., LUKE, C. G., DELLER, D. J.: Reduction of gastric iron-binding protein in haemochromatosis. Lancet *2,* 1431 (1966).

DAWSON, R. B., RAFAL, S., WEINTRAUB, L. R.: Absorption of hemoglobin iron: the role of xanthine oxidase in the intestinal heme-splitting reaction. Blood *37,* 94 (1970).

DELLER, D. J.: $Iron^{59}$ absorption measurements by whole body counting. Studies in alcoholic cirrhosis, haemochromatosis with pancreatitis. Amer. J. dig. Dis. *10,* 249 (1965).

DELLER, D. J., WITTS, L. J.: Changes in the blood after partial gastrectomy with special reference to vitamin B_{12}. I. Serum vitamin B_{12}, haemoglobin, serum iron and marrow. Quart. J. Med. N. S. *31,* 71 (1962).

DELLER, D. J., EDWARDS, R. G., DART, G., LUKE, C. G., DAVIS, P. S.: Gastric iron binding substance (gastroferrin) in a family with haemochromatosis. Aust. Ann. Med. *18,* 36 (1969).

DOWDLE, E. B., SCHACHTER, D., SCHENKER, H.: Active transport of Fe^{59} by everted segments of rat duodenum. Amer. J. Physiol. *198,* 609 (1960).

DUBACH, R., CALLENDER, S. T., MOORE, C. V.: Studies in iron transportation and metabolism. VI. Absorption of radioactive iron in patients with fever and with anaemias of varied etiology. Blood *3,* 526 (1948).

EVERETT, N. B., GARRETT, W. E., SIMMONS, B. S.: Lymphatics in iron absorption and transport. Amer. J. Physiol. *178*, 45 (1954).

FINCH, C. A. (1961), cit. BOTHWELL, FINCH, 1962.

FINCH, C. A.: Diagnostic value of different methods to detect iron deficiency. In: HALLBERG, L. et al. (eds.): Iron Deficiency. Academic Press, London–New York 1970.

FINCH, S. C., ROSS, J. F.: The significance of plasma iron turnover. J. clin. Invest. *31*, 627 (1952).

FINEBERG, R. A., GREENBERG, D. M.: Ferritin biosynthesis. III. Apoferritin, the initial product. J. biol. Chem. *214*, 107 (1955).

FISCHER, D. S., PRICE, D. S.: A possible humoral regulator of iron absorption. Proc. Soc. exp. Biol. (N. Y.) *112*, 228 (1963).

GABRIO, B. W., SALOMON, K.: Distribution of total ferritin in intestine and mesenteric lymph nodes of horses after iron feeding. Proc. Soc. Exp. Biol. Med. *75*, 124 (1950).

GREEN, R., CHARLTON, R., SEFTEL, H., BOTHWELL, T. H., MAYET, F., ADAMS, B., FINCH, C. A., LAYRISSE, M.: Body iron excretion in man. Amer. J. med. *45*, 336 (1968).

GREENBERG, M. S., STROHMEYER, G., HINE, G. J., KEENE, W. R., CURTIS, G., CHALMERS, T. C.: Studies in iron absorption. III. Body radioactivity measurements of patients with liver disease. Gastroenterology *46*, 651 (1964).

GREENBERG, M. S., WONG, H., MILLER, S. A., SCARLATA, R. W., CHALMERS, T. C.: Iron absorption and turnover in hypoxia. J. clin. Invest. *39*, 992 (1960).

GROSS, F.: Iron Metabolism. Springer, Berlin–Göttingen–Heidelberg 1964.

HAHN, P. F., BALE, W. F., ROSS, J. F., BALFOUR, W. M., WHIPPLE, G. H.: Radioactive iron absorption by the gastrointestinal tract. Influence of anaemia, anoxia and antecedent feeding. Distribution in growing dogs. J. exp. Med. *78*, 169 (1943).

HAHN, P. F., JONES, E., LOWE, R. C., MENEELY, G. R., PEACOCK, W.: The relative absorption and utilization of ferrous and ferric iron in anemia as determined with the radioactive isotope. Amer. J. Physiol. *143*, 191 (1945).

HAHN, P. F. et al.: Iron metabolism in human pregnancy as studied with the radioactive isotope Fe^{59}. Amer. J. Obstet. Gynec. *61*, 477 (1951).

HALKETT, J. A. E., CHODOS, R. B., ROSS, J. F.: The labeling of human foods with radioactive iron (Fe^{59}). J. Lab. clin. Med. *53*, 816 (1959).

HALKETT, J. A. E., PETERS, T. JR., ROSS, J. F.: Studies on the deposition and nature of egg yolk iron. J. biol. Chem. *231*, 187 (1958).

HALLBERG, L.: Die Eisenresorption — einige neuere physiologische und therapeutische Erkenntnisse. In: KEIDERLING, W., HOFFMANN, G. (eds.): Radioisotope in der Hämatologie. Schattauer, Stuttgart 1963.

HALLBERG, L., BJORN RASMUSSEN, E.: Determination of iron absorption from whole diet. A new two-pool model using two radioiron isotopes given as haem and non-haem iron. Scand. J. Haemat. *9*, 193 (1972).

HALLBERG, L. et al.: Absorption from iron tablets given with different types of meals. Scand. J. Haemat. *21*, 215 (1978).

HALLBERG, L., SÖLVELL, L.: Absorption of a single dose of iron in man. Acta med. scand. *168* (Suppl. 358), 19 (1960a).

HALLBERG, L., SÖLVELL, L.: Iron absorption during constant intragastric infusion of iron in man. Acta med. scand. *168* (Suppl. 358), 43 (1960b).

HEGSTED, D. M., FINCH, C. A., KINNEY, T. D.: The influence of diet on iron absorption. J. exp. Med. *90*, 147 (1949), *96*, 115 (1952).

HEILMEYER, L.: Blut und Blutkrankheiten. 5th ed., Part 1. Springer, Berlin–Heidelberg–New York 1968.

HEILMEYER, L., BEGEMANN, H.: Blut und Blutkrankheiten. Springer, Berlin–Göttingen–Heidelberg 1951.

HEILMEYER, L., KEIDERLING, W., WÖHLER, F.: Existiert bei der Eisenresorption im Dünndarm ein Mucosablock? Klin. Wschr. *35*, 690 (1957).

HEINRICH, H. C.: Bestimmung und Normalbereich der bei hämatologischen Erkrankungen gemessenen Eisenresorption bei Verwendung des Fe^{59}-Resorptions-Gesamtkörper-Retentionstestes. 12. Tagung der Deutschen Gesellschaft für Hämatologie, Berlin 1966.
HERBUT, P. A., TAMAKI, H. T.: Cirrhosis of liver and diabetes as related to hemochromatosis. Amer. J. clin. Path. *16*, 640 (1964).
HOBBS, J. R.: Iron deficiency after partial gastrectomy. Gut *2*, 141 (1961).
HOWARD, J., JACOBS, A.: Iron transport by rat small intestine in vitro: effect of body iron status. Brit. J. Haemat. *23*, 595 (1972).
HWANG, Y. F., BROWN, G. B.: Effect of desferrioxamine on iron absorption. Lancet *1*, 135 (1965).
IWAO, T., KURIHARA, M.: Microchemical investigation of iron absorbed from intestinal canal. Tr. Soc. path. Jap. *23*, 196 (1933).
JACOBI, H., PFLEGER, R., RUMMEL, W.: Komplexbildner und aktiver Eisentransport durch die Darmwand. Arch. exp. Path. Pharmak. *229*, 198 (1956).
JACOBS, A.: The mechanism of iron absorption. In: CALLENDER, S. T. (ed.): Clinics in Haematology, Vol. 2/2, pp. 323–337. Saunders, London–Philadelphia–Toronto 1973.
JACOBS, A., MILES, P. M.: Intraluminal transport of iron from stomach to small intestinal mucosa. Brit. med. J. *iv*, 778 (1969a).
JACOBS, A., MILES, P. M.: The iron binding properties of gastric juice. Clin. chim. Acta *24*, 87 (1969b).
JACOBS, A., WORWOOD, M.: Ferritin in serum. New Engl. J. Med. *292*, 951 (1975a).
JACOBS, A., WORWOOD, M.: Iron absorption: present state of the art. Brit. J. Haemat. Suppl. *31*, 89 (1975b).
JACOBS, P., BOTHWELL, T. H., CHARLTON, R. W.: Role of hydrochloric acid in iron absorption. J. appl. Physiol. *19*, 187 (1964).
JACOBS, P., BOTHWELL, T. H., CHARLTON, R. W.: Intestinal iron transport; studies using a loop of gut with an artificial circulation. Amer. J. Physiol. *210*, 694 (1966).
JASINSKI, B., ROTH, O.: Die larvierte Eisenmangelkrankheit. Schwabe, Basel 1954.
JOHNSTON, F. A., CLARK, S. J.: Reduction of ferric iron to the ferrous form during digestion in vitro with saliva. Amer. J. clin. Nutr. *7*, 203 (1959).
JOSEPHS, H. W.: Absorption of iron as a problem in human physiology. Blood *13*, 1 (1958).
KAUFMAN, N., KLAVINS, J. V., KINNEY, T. D.: Excessive iron absorption in rats fed low-protein, high-fat diets. Lab. Invest. *7*, 369 (1958).
KAVIN, H., CHARLTON, R. W., JACOBS, P.: Effect of the exocrine pancreatic secretions on iron absorption. Gut *8*, 556 (1967).
KINNEY, T. D., KAUFMAN, N., KLAVINS, J. V.: Effect of ethionine-induced pancreatic damage on iron absorption. J. exp. Med. *102*, 151 (1955).
KRANTZ, S., GOLDWASSER, E., JACOBSON, L. O.: Studies on erythropoiesis. XIV. The relationship of humoral stimulation to iron absorption. Blood *14*, 654 (1959).
KUHN, I. N., MONSEN, E. R., COOK, J. D., FINCH, C. A.: Iron absorption in man. J. Lab. clin. Med. *71*, 715 (1968).
LAYRISSE, M., COOK, J. D., MARTINEZ-TORRES, C., ROCHE, M., KUHN, I. N., WALKER, R. B., FINCH, C. A.: Food iron absorption: a comparison of vegetable and animal foods. Blood *33*, 430 (1969).
LAYRISSE, M., MARTINEZ-TORRES, C.: Model for measuring the dietary absorption from heme iron. Test with a complete meal. Amer. J. clin. Nutr. *25*, 401 (1972).
LEVINE, P. H., LEVINE, A. J., WEINTRAUB, L. R.: The role of transferrin in the control of iron absorption: studies on a cellular level. J. Lab. clin. Med. *80*, 333 (1972).
LUKE, C. G., DAVIS, P. A., DELLER, D. J.: Change in gastric iron binding protein (gastroferrin) during iron deficiency anaemia. Lancet *1*, 926 (1967).
MANIS, J., SCHACHTER, D.: Active transport of iron by the intestine: mucosal iron pools. Amer. J. Physiol. *207*, 893 (1964).
MARTINEZ-TORRES, C., LAYRISSE, M.: Effect of amino-acids on iron absorption from a staple vegetable food. Blood *35*, 669 (1970).

MAZUR, A., GREEN, S., SAHA, A., CARLETON, A.: Mechanism of release of ferritin iron in vivo by xanthine oxidase. J. clin. Invest. *37,* 1809 (1958).
MILDER, M. S., COOK, J. D., FINCH, C. A.: Influence of food iron absorption on plasma iron level in idiopathic hemochromatosis. Acta haemat. *60,* 65 (1978).
MILLAR, J. A., GOLDBERG, A., CUMMING, R. L. C.: Studies on ferritin synthesis in relation to iron absorption. In: HALLBERG, L., HARWERTH, H. G., VANNOTTI, A. (eds.): Iron Deficiency, p. 121. Academic Press, London–New York 1970.
MOORE, C. V.: The importance of nutritional factors in the pathogenesis of iron deficiency anemia. Amer. J. clin. Nutr. *3,* 3 (1955).
MOORE, C. V., DUBACH, R.: Observations on the absorption of iron from foods tagged with radioiron. Trans. Ass. Amer. Physicians *64,* 245 (1951).
MOORE, C. V., DUBACH, R., MINNICH, V., ROBERTS, H. K.: Absorption of ferrous and ferric radioactive iron by human subjects and by dogs. J. clin. Invest. *23,* 755 (1944).
MURRAY, M. J., STEIN, N.: Does the pancreas influence iron absorption? A critical review of information to date. Gastroenterology *51,* 694 (1966).
NORRBY, A., SÖLVELL, L.: Effect of dietary iron on iron absorption in man. Scand. J. Haemat. *9,* 396 (1972).
PINKERTON, P. H.: The control of iron absorption by the intestinal epithelial cell. Ann. intern. Med. *70,* 401 (1969).
PIRZIO-BIROLI, G., FINCH, C. A.: Iron absorption. III. The influence of iron stores on iron absorption in the normal subject. J. Lab. clin. Med. *55,* 216 (1960).
PIRZIO-BIROLI, G., BOTHWELL, T. H., FINCH, C. A.: Iron absorption. II. The absorption of radioiron administered with a standard meal in man. J. Lab. clin. Med. *51,* 37 (1958).
REYNAFARJE, C., LOZANO, R., VALDIVIESO, J.: The polycythemia of high altitudes. Iron metabolism and related aspects. Blood *14,* 433 (1959).
RIEBER, E. E., CONRAD, M. E., CROSBY, W. H.: Gastrectomy and iron absorption. Effects of bleeding, iron loading and ascorbic acid in rats. Proc. Soc. exp. Biol. (N. Y.) *124,* 577 (1967).
RUDZKI, Z., DELLER, D. J.: The iron-binding glycoprotein of human gastric juice. I. Isolation and characterisation. Digestion (Basel) *8,* 31 (1973).
SAYLOR, L., FINCH, C. A.: Determination of iron absorption using two isotopes of iron. Amer. J. Physiol. *172,* 372 (1953).
SCHULZ, J., SMITH, N. J.: A quantitative study of the absorption of iron salts in infants and children. Amer. J. Dis. Child. *95,* 120 (1958).
SMITH, J. A., DRYSDALE, J. W., MUNRO, H. N., GOLDBERG, A.: The effect of enteral and parenteral iron on the synthesis of ferritin in the intestinal mucosa of the rat. Brit. J. Haemat. *14,* 79 (1968).
SMITH, M. D., PANNACCIULLI, I. M.: Absorption of inorganic iron from graded doses; its significance in relation to iron absorption tests and the mucosal block theory. Brit. J. Haemat. *4,* 428 (1958).
SMITH, P. M., STUDLEY, F., WILLIAMS, R.: Postulated gastric factor enhancing iron absorption in haemochromatosis. Brit. J. Haemat. *16,* 443 (1969).
SOBEL, H. D., WAYE, J. D.: Pancreatic changes in various types of cirrhosis in alcoholics. Gastroenterology *45,* 341 (1963).
SÖLVELL, L.: Effect of iron and transferrin intravenously on iron absorption and turnover in man. Acta med. scand. *168* (Suppl. 358), 71 (1960).
STEVENS, A. R., PIRZIO-BIROLI, G., HARKINS, H. N., NYHUS, L. M., FINCH, C. A.: Iron metabolism in patients after partial gastrectomy. Ann. Surg. *149,* 534 (1959).
STEWART, W. B., VASSAR, P. S., STONE, R. S.: Iron absorption in dogs during anemia due to acetylphenylhydrazine. J. clin. Invest. *32,* 1225 (1953).
STEWART, W. B., YUILE, C. L., CLAIBORNE, H. A., SNOWMAN, R. T., WHIPPLE, G. H.: Radioiron absorption in anemic dogs. Fluctuations in the mucosal block and evidence for a gradient of absorption in the gastrointestinal tract. J. Exp. Med. *92,* 375 (1950).
STROHMEYER, G.: Messung der Körperradioaktivität zur Bestimmung der Eisenresorption beim Menschen. In: KEIDERLING, W., HOFFMANN, G. (eds.): Radioisotope in der Hämatologie, p. 69. Schattauer, Stuttgart 1963.

TAYLOR, J., STIVEN, D., REID, E. W.: Haemochromatosis in a depancreatized cat. J. Path. Bact. *34*, 793 (1931).
TAYLOR, J., STIVEN, D., REID, E. W.: Experimental and idiopathic siderosis in cats. J. Path. Bact. *41*, 397 (1935).
TURNBERG, L. A.: Iron absorption in acute hepatitis. Amer. J. dig. Dis. *11*, 20 (1966).
TURNBULL, A.: The absorption of radioiron given with a standard meal after Pólya partial gastrectomy. Clin. Sci. *28*, 499 (1965).
TURNBULL, A., FINCH, C. A. (1961), cit. BOTHWELL, FINCH, 1962.
TURNBULL, A., CLETON, F., FINCH, C. A.: Iron absorption. IV. The absorption of hemoglobin iron. J. clin. Invest. *41*, 1897 (1962).
VAN DYKE, D., LAYRISSE, M., LAWRENCE, J. H., GARCIA, J. F., POLLYCOVE, M.: Relation between severity of anemia and erythropoietin titre in human beings. Blood *18*, 187 (1961).
VAN HOEK, R., CONRAD, M. E., JR.: Iron absorption. Measurement of ingested iron^{59} by a human whole body liquid scintillation counter. J. clin. Invest. *40*, 1153 (1961).
WACK, J. P., WYATT, J. P.: Studies on ferrodynamics. I. Gastrointestinal absorption of Fe^{59} in the rat under differing dietary states. Arch. Path. *67*, 237 (1959).
WALTERS, G. O., JACOBS, A., WORWOOD, M., TREVETT, D.: Iron absorption in normal subjects and patients with idiopathic haemochromatosis: relationship with serum ferritin concentration. Gut *16*, 188 (1975).
WEINFELD, A.: Iron stores. In: HALLBERG, L., HARWERTH, H. G., VANNOTTI, A. (eds.): Iron Deficiency. Academic Press, London–New York 1970.
WEINTRAUB, L. R., WEINSTEIN, M. B., HUSER, H. J., RAFEL, S.: Absorption of hemoglobin iron; the role of a heme-splitting substance in the intestinal mucosa. J. clin. Invest. *47*, 531 (1968).
WHITEHEAD, J. S. W., BANNERMAN, R. M.: Absorption of iron by gastrectomized rats. Gut *5*, 38 (1964).
WÖHLER, F., HEILMEYER, L., EMRICH, D., KANG, S. H.: Zur Funktion des Ferritins bei der Eisenresorption. Arch. exp. Path. Pharmakol. *230*, 107 (1957).
WYNTER, C. V. A., WILLIAMS, R.: Iron binding properties of gastric juice in idiopathic haemochromatosis. Lancet *2*, 534 (1968a).
WYNTER, C. V. A., WILLIAMS, R.: Gastric iron binding in haemochromatosis. Lancet *2*, 1243 (1968b).

CHAPTER 7

IRON TRANSPORT

PLASMA IRON

Iron circulating in the plasma is bound to transferrin. Most of this iron originates from the breakdown of effete red cells in the reticuloendothelial system (RES) and it is transported to the bone marrow for new hemoglobin synthesis. A smaller proportion of iron enters the transport pool from the gastrointestinal tract, from storage iron, and from catabolism of tissue hemins.

In the healthy adult, the total amount of the circulating iron is 3–4 mg, and although this represents less than 1‰ of the iron content of the body, it is functionally clearly important because the entire iron metabolism takes place through this plasma pool.

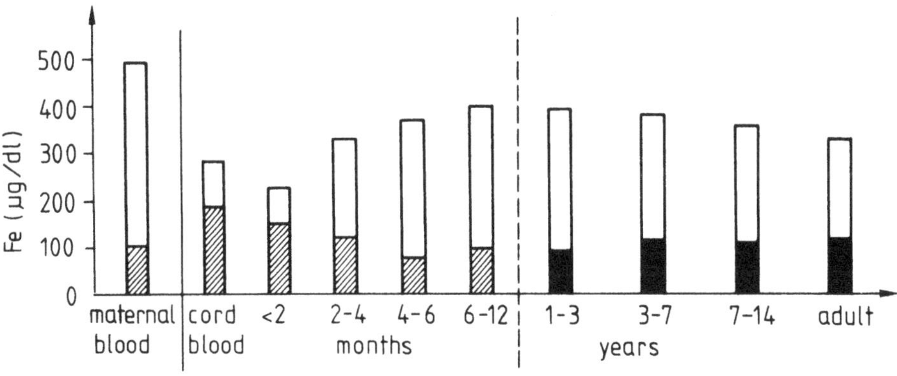

Fig. 7/1. Serum iron and iron-binding capacity from birth until maturity (after Hagberg, B.: Acta pediat. [Uppsala] *42*, 589, 1953). □ free transferrin; ▨ serum iron; □ + ▨ TIBC (total iron-binding capacity)

The plasma iron concentration displays characteristic changes throughout life. In the newborn the mean value is 193 μg/dl (Fig. 7/1), but in the first few hours of life there is a rapid decline to a mean value of 46 μg/dl (Fig. 7/2). This is followed by a slow rise, and by the end of the second week it reaches 125 μg/dl. There is then another slow decrease, reaching the lowest point of about 50 μg/dl variously reported as being between the fourth and sixth months (Hagberg, 1953) or the sixth to twelfth month (Waldenström, 1959). Later the level rises slowly, and at the age of 2 years the mean value is 102 μg/dl, and the adult values develop around puberty. From then on there is a pronounced sex difference, the average level in adult men

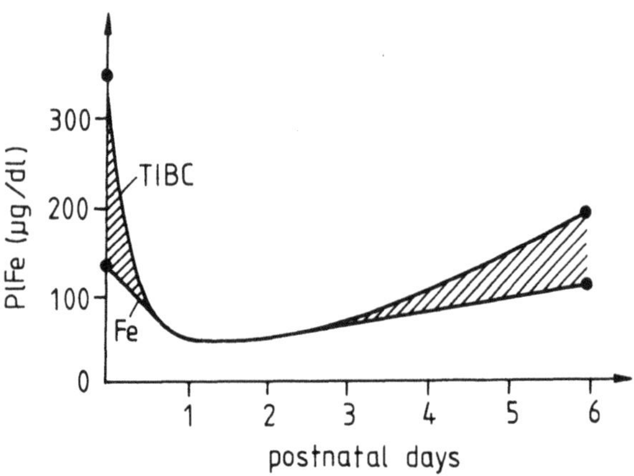

Fig. 7/2. Serum iron and iron-binding capacity in the first days after birth (after Sturgeon, P.: Pediatrics *13*, 107, 1954)

being 125 μg/dl, while that of adult women is 110 μg/dl, with a range of 80–150 μg/dl and 70–130 μg/dl in men and women, respectively. During pregnancy the level diminishes due to a relative iron deficiency, which can be prevented by the administration of iron (Göltner, 1975). With advancing years the mean plasma iron level tends to decrease in both sexes, this being most evident after 40 years of age. The sex difference, present during the reproductive period, disappears after the menopause (Fig. 7/3).

The plasma iron shows a diurnal variation in healthy humans (Vahlquist, 1941; Hemmeler, 1944; Høyer, 1944; Bothwell and Mallett, 1955); it is highest in the morning and lowest toward evening (Fig. 7/4). The variation depends on the morning level. Hamilton et al. (1950) found a 57% drop from an average morning value of 155 μg/dl. Vahlquist found a 27% drop at an initial value of 135 μg/dl, and Johnston (1947) a 24% drop from 96 μg/dl. The diurnal variation is related to sleep (Hemmeler, 1944; Høyer, 1944; Schäfer and Boenecke, 1949), and the direction of change is reversed in night workers who sleep during the day (Sinniah et al., 1969). It has been suggested that the decrease during waking hours is related to a higher iron uptake elicited by neurogenic and humoral factors (Schäfer, 1964), or alternatively

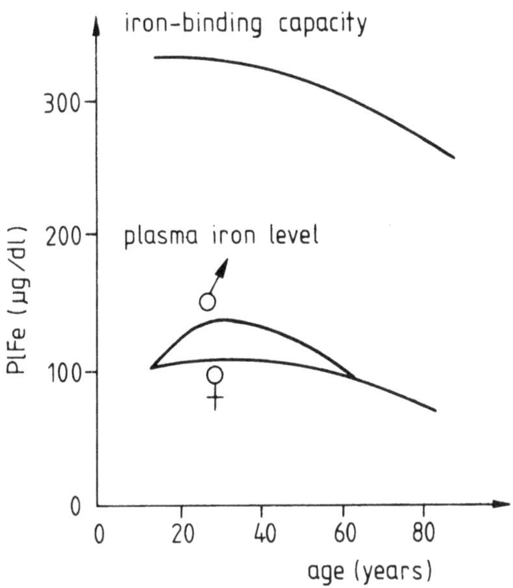

Fig. 7/3. Plasma iron and iron-binding capacity in healthy adult humans. Mean values between 15 and 80 years of age

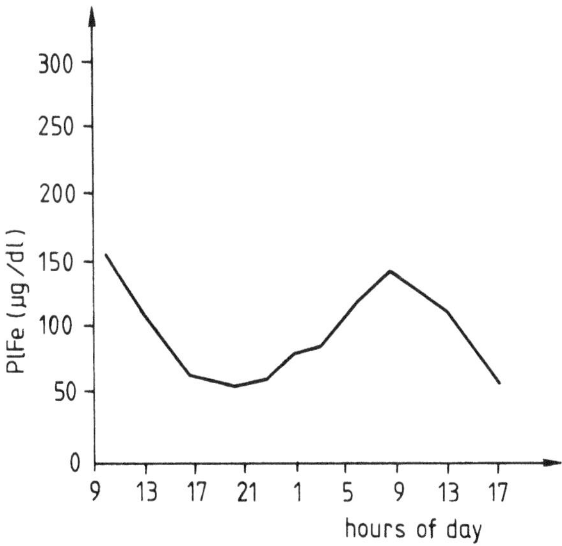

Fig. 7/4. Diurnal variation in plasma iron level; mean values in 19 humans (after Hamilton, L. D. et al.: Proc. Soc. exp. Biol. Med. *75*, 65, 1950)

that the variation is the consequence of cyclic changes in bone marrow function (Paterson, 1957). The diurnal variations only become evident in children over 20 months of age (Schwartz and Baehner, 1968), and they tend to disappear in old age. The variation also decreases or disappears in pathological conditions such as iron deficiency, aplastic anemia, pernicious anemia, and idiopathic hemochromatosis (Paterson et al., 1953; Finch and Finch, 1955).

Stress is accompanied by a decrease in plasma iron level (Schäfer, 1964). This decrease can be produced experimentally by estrogens (Palmer, 1952), ACTH, adrenocortical extract, cortisone, desoxycorticosterone acetate, and histamine (Hamilton et al., 1951). It can also be provoked by injection of toxins, vaccines, and protein decomposition products. Muscular work (Dreyfus and Schapira, 1949) and food intake (Tötterman, 1949; Bernát and Kovács, 1956) do not influence the plasma iron concentration (see Fig. 6/27).

The plasma iron concentration is determined by the amount of iron per unit time that flows into and out of the plasma and is influenced by the rate of hemoglobin production and degradation, by the avidity of the RES for iron, and by the rate of liberation of storage iron.

HYPERSIDEREMIA

Plasma iron concentration increases (Table 7/1)

(1) if erythropoiesis is diminished,

(2) if red cell destruction is increased,

(3) if the release of iron from the RE system is increased.

Thus the plasma iron level is high in aregenerative anemia (Laurell, 1952; Cartwright et al., 1946). Ionizing irradiation in rats (Chanutin and Ludewig, 1951)

Table 7/1

Plasma iron level in various pathological conditions

Hypersideremia	Hyposideremia
Hemosiderosis, hemochromatosis	Iron deficiency
Hypersideremic hypochromic anemia	RHS hyperfunction (infection, thermal injury, tissue destruction, stress)
Acute hepatitis	Hypo- and atransferrinemia
Hemolytic anemias (in several phases of the disease)	Hemolytic anemias (when erythropoiesis preponderates over red cell destruction)
Addison–Biermer's disease (in the stage of decompensation)	Addison–Biermer's disease (following effect of specific therapy)
Shahidi anemia	Hypovitaminosis-C
Porphyria	

and in monkeys (Hartwig et al., 1959) produced increases in plasma iron, but Melville et al. (1957) found this increase not to be significant, and the change was biphasic. Nitrogen mustard (Bertenchamps et al., 1958) and chloramphenicol (Rubin et al., 1960) both induce a rise in plasma iron concentration, and Rubin et al. (1960) believe that the determination of plasma iron may aid early recognition of drug-induced erythropoietic disturbances.

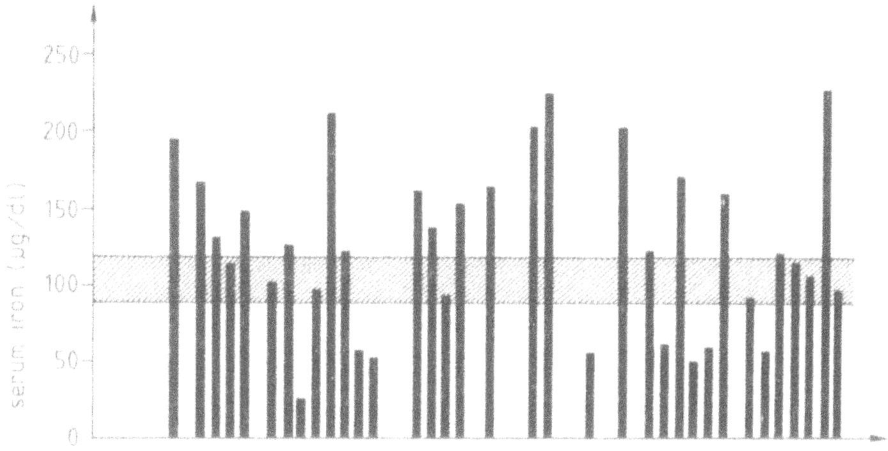

Fig. 7/5. Serum iron level in 35 patients with hemolytic anemia. Shaded area represents normal values (after Heilmeyer, L., in: Gross, F.: Iron Metabolism. Springer, Berlin–Göttingen–Heidelberg 1964)

There are exceptions to the increase in plasma iron associated with diminishing erythropoiesis, as, for example, in subjects returning to the plains from a high altitude who may show a drop rather than a rise in plasma iron (Reynafarje et al., 1959).

The rise in plasma iron level associated with acute destruction of red cells can be shown experimentally following the injection of hemolytic compounds (Laurell, 1958). In chronic hemolytic states the situation is more complex because the plasma iron level is determined not only by the rate of iron released from destroyed red cells but also by the speed at which the bone marrow re-utilizes the iron in the course of enhanced erythropoiesis (Fig. 7/5). The RE cells play the key role in the maintenance of the equilibrium.

In some hemolytic anemias, notably those of childhood, the iron concentration of the plasma is usually higher than normal (Laurell, 1952). The level is particularly elevated in those hemolytic anemias associated with ineffective erythropoiesis (de Raadt, 1942; Laurell, 1958; Rath and Finch, 1949; Finch, 1961) or where the incorporation of iron is inhibited, as in lead poisoning (Josephs, 1954; Saita et al., 1955). The plasma iron is also elevated in pyridoxine deficiency (Harris, 1958), in Cooley's anemia (Erlandson et al., 1958), and in Di Guglielmo's disease (Baldini et al., 1959). The rise in plasma iron concentration is limited by the iron-binding capacity (Erlandson et al., 1958).

In hemolytic anemias in which the rate of erythropoiesis exceeds that of erythrocyte destruction the plasma iron may be normal or low.

If the iron requirement of the bone marrow increases, for example, following an acute hemorrhage, more iron flows out of the iron stores. This gives rise to a transient rise in the plasma iron level (Hagberg et al., 1958), which lasts until the amount of mobilized iron exceeds the rate of entry of iron into normoblasts. Later, hyposideremia usually develops. Such a transient rise in plasma iron concentration has been shown experimentally in hemorrhagic shock by Mazur et al. (1955).

In acute hepatitis, as in other infections, at the beginning of the febrile period there is a drop in plasma iron concentration, but in this type of infection the iron value rises above normal during the second to third week, reaching a peak value between days 12–31 of the jaundice (Hemmeler, 1943; Ducci et al., 1952; Rechenberger, 1955;

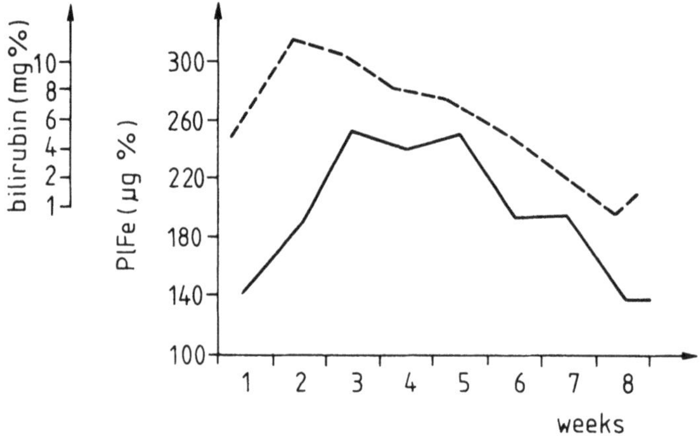

Fig. 7/6. Bilirubin (– – –) and iron level (——) of the plasma in acute hepatitis (mean values of determinations in 30 patients) (after Rechenberger, J.: Dtsch. Z. Verdau.- u. Stoffwechselkr. *15*, *70*, 1955)

Rumball et al., 1959) (Fig. 7/6). The rise in plasma iron has been correlated with the degree of necrosis of the hepatic cells (Reissman et al., 1954). In addition to the transferrin-bound iron, ferritin iron may also be released from damaged hepatic cells (Reissman and Dietrich, 1956). In subacute or chronic hepatitis the plasma iron may be normal or reduced. In cirrhosis the plasma iron may be elevated, normal, or low. Where low, it may be the result of hemorrhage.

The level of plasma iron is always high in iron overload and is usually above 200 μg/dl in established hemochromatosis. A high plasma iron may be an indication that a relative carries the gene even at a stage before the accumulation of excess iron stores (Bothwell et al., 1959). When an intercurrent infection or complicating malignant neoplasm occurs in iron overload, the plasma iron may not be elevated (Finch and Finch, 1955).

The interpretation of hypersideremia in transfusional siderosis is complicated by the fact that such patients frequently have a high plasma iron even before transfusions are instituted. The level therefore gives little indication of increased iron stores.

In Bantu siderosis there is some correlation between the level of plasma iron and the degree of hepatic siderosis (Higginson et al., 1957; Mathorn et al., 1960), but this is not a constant finding.

HYPOSIDEREMIA

The level of iron in the plasma decreases (see Table 7/1)

(1) when the inflow of iron from the RES into the plasma is slow,

(2) when the plasma iron clearance is rapid,

(3) when the iron content of the body diminishes.

The first two conditions prevail in infections and tissue damage (Cartwright et al., 1946, 1951; Cartwright and Wintrobe, 1949; Jasinski, 1950). In such cases iron release from the RES is impeded (Freireich et al., 1957a). In addition, the iron leaves the plasma more rapidly than normal (Keiderling, 1959). Similar hypoferremia develops in malignant tumors (Keiderling and Scharpf, 1953; Miller et al., 1956), in thermal injuries (Bernát et al., 1965), in rheumatoid arthritis (Brendstrup, 1953b; Freireich et al., 1957b), in chronic renal disease (Rath and Finch, 1949), in myocardial infarction (Myhrman and Wilander, 1955), in scurvy (Bronte-Stewart, 1953; Greenberg and Rinehart, 1955), and after bone fractures or surgery (Nylander, 1955).

In iron-deficiency anemia and latent iron deficiency the plasma iron level is always low (Ramsay, 1957; Beutler et al., 1958; Hallgren, 1953; Coleman et al., 1955). A drop in plasma iron occurs when the iron stores are depleted and bears no correlation with the hemoglobin level of the blood (Waldenström, 1959; Jasinski and Roth, 1954).

After acute hemorrhage, as already discussed, there may be a transient rise in plasma iron concentration followed by a decline, and if the hemorrhage has exhausted the iron reserves, the level of plasma iron remains low.

Hypoxia, by its effect of enhancing erythropoiesis, may result in a decrease in plasma iron (Schade et al., 1954). Such changes can be observed in subjects going from sea level to a high altitude, since even if there is adequate storage iron the mobilization from the reticuloendothelial system cannot usually keep pace with the increased requirement for erythropoiesis (Huff et al., 1951).

TRANSFERRIN

It has long been recognized that plasma iron is protein-bound, and at physiological pH it cannot be dialyzed or ultrafiltered (Barkan, 1927; Warburg and Krebs, 1927). Starkenstein and Harvalik (1933) showed by ammonium sulfate precipitation that the iron-binding protein is a globulin. In 1947 this specific protein was named transferrin, but it was not isolated until 1949. Transferrin has also been called the metal-binding globulin, iron-binding protein, and siderophilin.

Transferrin is a glycoprotein with the electrophoretic mobility of a β_1-globulin (Wallenius, 1952; Neale, 1955) and can be separated from the IV/7 Cohn fraction (Surgenor et al., 1949). Several genetic variants of the iron-binding protein can be separated by starch-gel electrophoresis (Smithies, 1957, 1959a, b; Horsfall and

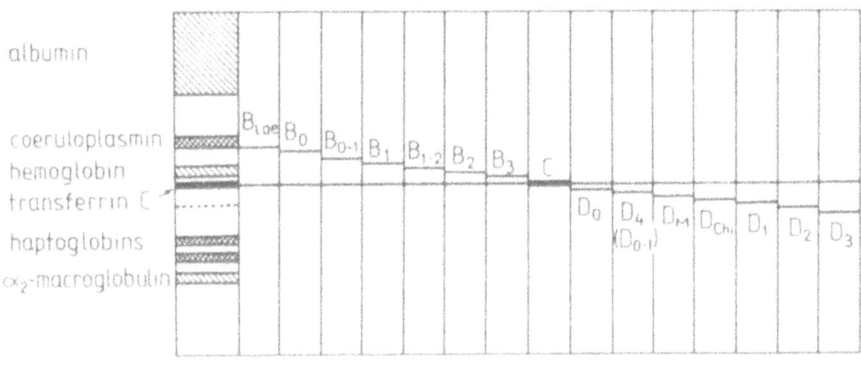

Fig. 7/7. Genetic variants of human transferrin (after Bearn, A. G. and Parker, W. C., in: Gross, F.: Iron Metabolism. Springer, Berlin–Göttingen–Heidelberg 1964)

Smithies, 1958; Giblett et al., 1959; Harris et al., 1958; Smithies and Hiller, 1959; Turnbull and Giblett, 1961). The variants differ from one another in amino acid sequence. In order of decreasing electrophoretic mobility, variants have been named B_{Lae}, B_0, B_1, B_2, B_3, C, D_0, D_4, $D_{Montreal}$, D_{Chi}, D_1, D_2, D_3 (Bearn and Parker, 1964) (Fig. 7/7). At present, at least 19 transferrin variants are known (Giblett, 1960). Only transferrin C can be found in the plasma of the majority of humans. These individuals are homozygotes from the point of view of the transferrin C gene (Tf^C). About 1% of the Caucasian race has B_2C genotype. In 10–12% of the American Negroes CD_1 is found, and in 8% of the Navajo Indians B_{0-1} variations can be detected. Bearn and Parker (1964) showed that 6% of the Chinese had the D_{Chi} type. The remaining transferrin variants are very rare. Functionally the transferrins appear to be the same (Turnbull and Giblett, 1961).

The molecular weight of transferrin has been variously estimated from 66,000 to 95,000 Daltons, the earlier estimations tending to be higher than those most recently obtained. Probably 76,000–80,000 Daltons is the most acceptable estimate for human transferrin (Surgenor et al., 1949; Koechlin, 1952; Katz, 1961, Allerton, 1962; Jandl and Katz, 1963).

The transferrin molecule is able to bind two atoms of ferric iron. The iron–transferrin complex is more resistant to heat and enzymes than the pure protein (Azari and Feeney, 1958, 1961). The link between iron and transferrin is strictly pH-dependent. *In vitro* the binding of iron is complete only above pH 7, and dissociation begins below pH 6.5 (Fig. 7/8). The reaction is reversible, and the equilibrium

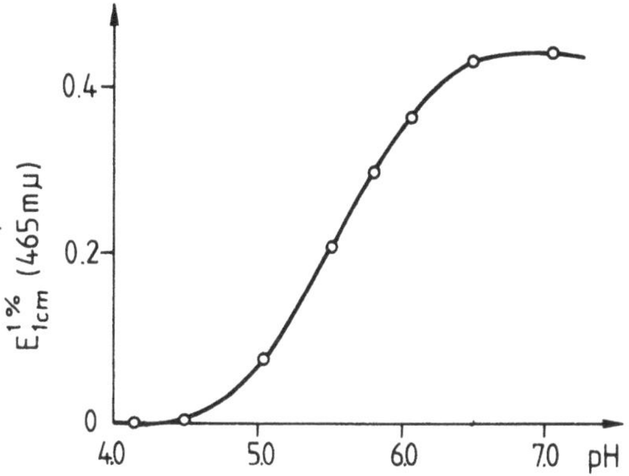

Fig. 7/8. Dissociation curve of the transferrin–iron complex (after Surgenor, D. M. et al.: J. clin. Invest. *28*, 73, 1949)

constant of iron transferrin is 10^{26} to 10^{30} (Saltman and Charley, 1960). Transferrin is a true carrier protein. It delivers iron to the erythroid precursors and is transiently bound to specific receptors on the cell membrane of the immature red cells (Jandl et al., 1959). Only those cells that are actively synthesizing hemoglobin, erythroblasts and reticulocytes, take up the iron (Morgan, 1964). The iron-carrying molecules replace iron-free molecules at the surface of the cell (Jandl et al., 1959). The biological half-life of transferrin has been variously reported as 10–12 days (Gitlin et al., 1956) or 6–8 days (Katz, 1961, 1970) (Fig. 7/9). The transferrin molecule can be divided into two subunits by 8 M urea. The subunits are not alike. Although the mechanism of the link between iron and transferrin has not been clarified, it is known that three tyrosine and two histidine side chains participate in it (Aasa et al., 1963), and that the tertiary configuration of the protein is essential to optimal binding (Katz, 1961).

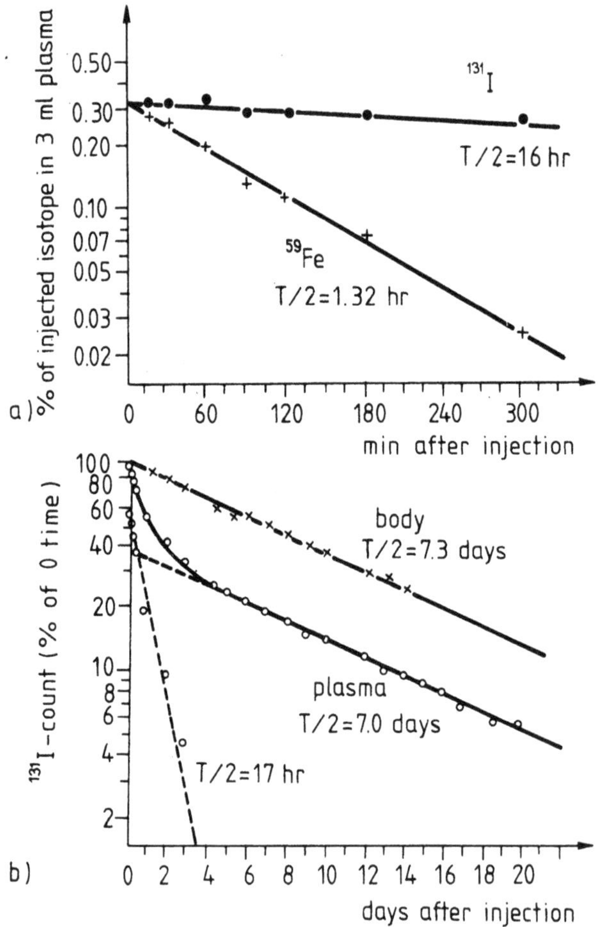

Fig. 7/9. Disappearance rate of ^{59}Fe and ^{131}I from the plasma after the injection of double-labeled iron–transferrin complex. ^{131}I marks the protein fraction of the complex. a) Disappearance rate during the first 5 hours; b) in the course of the first 20 days. The initial rapid decline in activity is attributed to the dilution of transferrin in the extravascular space and to its binding to the cell membrane. The second component of the curve is the result of the degradation of protein (after Katz, J. H.: J. clin. Invest. *40,* 2143, 1961)

Fletcher and Huehns (1968) suggested that either the two iron atoms bind separately to the two subunits of transferrin or the fine structures that surround the two binding sites differ from one another. They also suggested that functional differences exist between the two iron-binding sites, site A being mainly involved in the delivery of iron to erythroid cells and site B donating its iron preferentially to cells involved in the absorption and storage of iron (Fletcher and Huehns, 1967, 1968; Fletcher and Suter, 1969). Four transferrin forms can exist, apotransferrin, A and B site monoferric-transferrin, and diferric-transferrin, in which both A and B sites are occupied. When the iron saturation of transferrin is high, diferric-transferrin will predominate; where the transferrin saturation is low, most of the protein molecules will be free of iron and the number of molecules transporting one or two atoms of iron is low (Fletcher, 1970). The B site appears to be the weaker binding site.

The erythroblasts take up less iron from transferrin than would be expected on the basis of the transferrin turnover (Jandl and Katz, 1963), and some of the protein molecules may leave the surface of the erythroblasts without releasing their iron (Morgan et al., 1966). Fletcher and Huehns (1968) believe that erythroblasts and reticulocytes take up iron more easily from diferric-transferrin, and *in vitro* experiments suggest that the rate of uptake of iron from the diferric molecules is 10 times faster than from molecules with only one iron atom (Fletcher and Huehns, 1967). Verhoef et al. (1978) have failed to confirm this in experiments using iron–transferrin mixtures in which radioiron was present at low or high iron saturation. They could find no difference in the iron uptake of rat bone marrow cells exposed to monoferric as opposed to diferric-transferrin.

Fletcher (1970) has also suggested that the mucosal cells of the small intestine have a greater affinity for transferrin molecules that carry two iron atoms as opposed to one. The state of the iron stores affects the ratio of the different forms of transferrin molecules; hence this ratio could be the "messenger" that informs the intestinal mucosa about the state of the iron stores and erythropoietic activity of the bone marrow (Fletcher, 1970).

The amount of transferrin in the plasma can be determined directly using immunochemical methods (Jager, 1949; Jager and Gubler, 1952; Goodman et al., 1958), but the procedure is not suitable for a routine estimation. Instead, the amount of iron that is required to saturate the transferrin is estimated, i.e., the free iron-binding capacity of the plasma, and this, together with the plasma iron, is taken to represent the total iron-binding capacity of the plasma. Normally only about one-third of the transferrin is saturated (Cartwright and Wintrobe, 1949) (see Figs. 7/11 and 20/4).

The transferrin level of the newborn is lower than its mother's (Hagberg, 1953). During the first day of life the iron-binding capacity shows a marked and sudden decline (45 μg/dl), but then starts to rise slowly, and by days 5–6 it reaches 230–260 μg/dl (Hagberg, 1953; Dreyfus and Schapira, 1955) (see Fig. 7/2). In infancy the concentration of transferrin increases to a level higher than that found in the adult (Hagberg, 1953) (see Fig. 7/1).

In the healthy adult human the total iron-binding capacity of the plasma ranges between 300 and 340 μg/dl, the extreme values being 250–400 μg/dl (Holmberg and Laurell, 1945; Cartwright and Wintrobe, 1949; Brendstrup, 1953a; Dreyfus and Schapira, 1955; Bothwell et al., 1959). The iron-binding protein represents approximately 3% of the plasma protein. If the iron-binding capacity is tested after treatment with oral iron so that the presence of latent iron deficiency is excluded, the iron-binding capacity of the plasma shows less variability: 285–330 μg/dl, with a mean value of 305 μg/dl (Bernát, 1971).

In contrast to the plasma iron level, there is no sex difference in the transferrin concentration of the plasma in normal subjects. There is also no diurnal variation. There is a tendency for the level to fall with age. Over 60 years of age it is about 280 μg/dl, and at about 80 years of age it is 250 μg/dl (see Fig. 7/3).

The iron-binding capacity increases during pregnancy. In the fifth month the value is about 400 μg/dl, and the maximum value is reached during the eighth month (about 500 μg/dl). There may be a slight drop immediately before delivery (Ventura and Klopper, 1951), and there is a return to normal following delivery.

HYPERTRANSFERRINEMIA

The most common cause of a rise in transferrin level is iron deficiency (Cartwright and Wintrobe, 1949; Laurell, 1952; Kind, 1954). There is some disagreement as to whether the transferrin level rises when the iron stores are exhausted or only with the onset of iron-deficiency anemia (Hagberg et al., 1958; Pirzio-Biroli and Finch, 1960; Beutler et al., 1958).

HYPO- AND ATRANSFERRINEMIA

Infection is the most common cause of decrease in the transferrin level (Cartwright and Wintrobe, 1949; Rechenberger, 1956b; Bernát, 1971). In animals with sterile abscesses, or following the injection of toxin, the transferrin level drops by 40% within 48 hours (Cartwright and Wintrobe, 1949).

Thermal injury will produce a moderate fall in the iron-binding capacity to levels of between 100 and 275 μg/dl, with a mean of 212 μg/dl (Bernát et al., 1966; Bernát, 1971). Rheumatoid arthritis (Brendstrup, 1953a) and malignant tumors (Simpson et al., 1959) also produce moderate reduction in TIBC. Further causes of reduction in the iron-binding capacity are hemolytic anemia, pernicious anemia, disturbances of plasma protein synthesis (e.g., chronic liver disease), and increased protein loss (e.g., nephrosis) (Rechenberger, 1956a, c; Lange and Oberhoffer, 1958; Horst and Schäfer, 1953; Cartwright et al., 1954; Neale, 1955; Rifkind et al., 1961).

The total amount of transferrin is distributed between the intravascular and extravascular spaces, and some of the extravascular transferrin in the course of iron exchange binds transitorily to the surface of cells (Mitchell et al., 1960; Jandl, 1960). Transferrin is fixed particularly on the membrane of the reticuloendothelial cells when iron stores are accumulating, as is the case in infections and thermal injuries, and this may account, at least in part, for the change in plasma transferrin concentration (Fig. 7/10; see also Fig. 20/4).

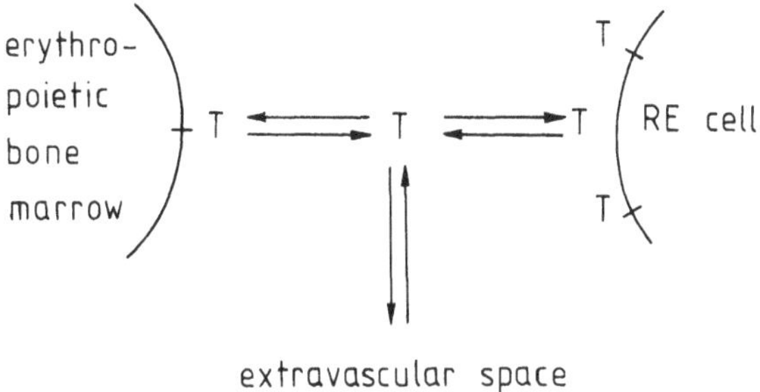

Fig. 7/10. Distribution of transferrin (T) within the organism. Part of the transferrin is exchanged between the plasma and the extravascular space, some binds to the membranes of the erythroid precursors, and a larger proportion to the surface of the RE cells. In infections there is an increase in the latter fraction (after Cartwright, G. E. and Wintrobe, M. M.: J. clin. Invest. *28*, 86, 1949)

Heilmeyer et al. (1961) described a single case of congenital atransferrinemia in a 7-year-old girl. Both parents showed a reduction in their transferrin level (see Chapter 27).

The physiological and pathological conditions that influence the plasma transferrin level are summarized in Table 7/2.

Table 7/2

Factors influencing the plasma transferrin level

Hypertransferrinemia	Hypo- and atransferrinemia
Iron deficiency	Infections
Pregnancy	Collagen diseases
Infancy	Malignant tumors
	Thermal injuries
	Hemolytic anemias
	Hypoproteinemia
	Congenital atransferrinemia

Because the iron level and the iron-binding capacity of the plasma both undergo changes in pathological conditions, the relationship between these two values, i.e., the degree of saturation of transferrin, is a better indicator of the state of iron metabolism than the two values separately. The saturation is usually expressed as a percentage, and the normal value is 30–40% (Cartwright and Wintrobe, 1949; Laurell, 1952; Beutler et al., 1958).

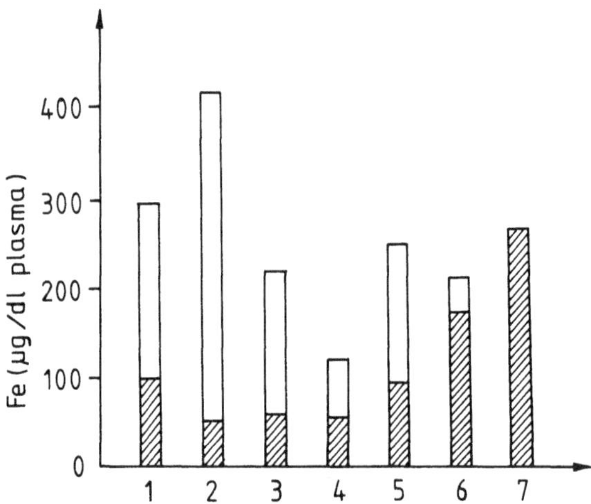

Fig. 7/11. Iron level and iron-binding capacity of the plasma in healthy humans and in various pathological processes. □ unsaturated iron-binding capacity; ▨ plasma iron level. 1 = normal; 2 = iron deficiency; 3 = chronic diseases; 4 = protein deficiency; 5 = hemolytic anemia (effective erythropoiesis); 6 = hemolytic anemia (ineffective erythropoiesis); 7 = hemochromatosis

The transferrin saturation is increased (Fig. 7/11):

1. in pathological conditions associated with increased iron storage, as in hemochromatosis or transfusion siderosis;
2. where there is ineffective erythropoiesis, as in untreated pernicious anemia and the thalassemias;
3. in disturbances of hemoglobin synthesis, as in lead poisoning, sideroblastic anemias, pyridoxine deficiency.

The transferrin saturation is reduced (Fig. 7/11):

1. in iron-deficiency anemia where the plasma iron is low and the TIBC high;
2. in pregnancy where the plasma iron may be normal or low but the TIBC high;
3. in infections, in malignant tumors, and after thermal injury if the plasma iron is reduced to a greater extent than the iron-binding capacity (see Fig. 20/4).

Iron is most readily taken up by erythroblasts when the saturation of transferrin ranges between 30 and 60% (Jandl et al., 1959). If the saturation is in excess of 60%, the iron is incorporated both in the RE system and in other tissues.

BIBLIOGRAPHY

AASA, R., MALMSTRÖM, B. G., SALTMAN, P., VÄRMGÅRD, T.: The specific binding of iron(III) and copper(II) to transferrin and conalbumin. Biochim. biophys. Acta *75*, 203 (1963).

ALLERTON, S. E.: Doctoral thesis. Division of Medical Sciences, Harvard University 1962.

AZARI, P. R., FEENEY, R. E.: Resistance of metal complexes of conalbumin and transferrin to proteolysis and to thermal denaturation. J. biol. Chem. *232*, 293 (1958).

AZARI, P. R., FEENEY, R. E.: The resistance of conalbumin and its iron complex to physical and chemical treatments. Arch. Biochem. *92*, 44 (1961).

BALDINI, M., FUDENBERG, H. H., FUKUTAKE, K., DAMESHEK, W.: The anemia of the Di Guglielmo syndrome. Blood *14*, 334 (1959).

BARKAN, G.: Eisenstudien. Die Verteilung des leicht abspaltbaren Eisens zwischen Blutkörperchen und Plasma und sein Verhalten unter experimentellen Bedingungen. Z. physiol. Chem. *171*, 194 (1927).

BEARN, A. G., PARKER, W. C.: Some observations on transferrin. In: GROSS, F. (ed.): Iron Metabolism, p. 60. Springer, Berlin–Göttingen–Heidelberg 1964.

BERNÁT, I.: Az égési anaemia pathogenesise (The Pathogenesis of Anemia after Thermal Injury). Akadémiai Kiadó, Budapest 1971.

BERNÁT, I., DÓZSÁN, G., NOVÁK, J., ELEK, S.: Anaemia after thermal injury. II. Acta med. Acad. Sci. hung. *22*, 253 (1966).

BERNÁT, I., KOVÁCS, E.: Vizsgálatok a normális és kóros vasanyagcsere köréből. III. A vasfelszívódást és a vaskezelés hatásosságát befolyásoló egyes tényezőkről (Investigations on the normal and pathological iron metabolism. III. Factors affecting the iron absorption and the effectivity of iron treatment). Katonaorvosi Szle *8*, 882 (1956).

BERNÁT, I., NOVÁK, J., FÁBER, V., DÓZSÁN, G., ELEK, S.: Újabb adatok az égési anaemia pathogenesiséhez (Some new data on the pathogenesis of anemia after thermal injury). Haemat. hung. *5*, 9 (1965).

BERTENCHAMPS, A., KENIS, Y., TAGNON, H. J.: The effect of the administration of HN_2 on the utilization of plasma iron. Cancer *11*, 117 (1958).

BEUTLER, E., ROBSON, M. J., BUTTENWIESER, E. A.: Comparison of the plasma iron, iron-binding capacity, sternal marrow iron and other methods in the clinical evaluation of iron stores. Ann. intern. Med. *48*, 60 (1958).

BOTHWELL, T. H., FINCH, C. A.: Iron Metabolism. Little, Brown and Co., Boston 1962.

BOTHWELL, T. H., MALLETT, B.: Diurnal variation in the turnover of iron through the plasma. Clin. Sci. *14*, 235 (1955).

BOTHWELL, T. H., COHEN, I., ABRAHAMS, O. L., PEROLD, S. M.: A familial study in idiopathic hemochromatosis. Amer. J. Med. *27*, 730 (1959).

BOTHWELL, T. H., JACOBS, P., KAMENER, R.: The determination of the unsaturated iron-binding capacity of serum using radioactive iron. S. Afr. J. med. Sci. *24*, 93 (1959a).

BRENDSTRUP, P.: Serum iron, total iron-binding capacity of serum, and serum copper in normals. Scand. J. clin. Lab. Invest. *5*, 312 (1953a).

BRENDSTRUP, P.: Serum copper, serum iron and total iron-binding capacity of serum in patients with chronic rheumatoid arthritis. Acta med. scand. *146*, 384 (1953b).

BRONTE-STEWART, B.: The anaemia of adult scurvy. Quart. J. Med. *22*, 309 (1953).

CARTWRIGHT, G. E., WINTROBE, M. M.: Chemical, clinical and immunological studies on the products of human plasma fractionation. XXXIX. The anemia of infection. Studies on the iron-binding capacity of serum. J. clin. Invest. *28*, 86 (1949).

CARTWRIGHT, G. E., GUBLER, C. J., WINTROBE, M. M.: The anaemia of infection. XII. The effect of turpentine and colloidal thorium dioxide on the plasma iron and plasma copper of dogs. J. biol. Chem. *184*, 579 (1950).
CARTWRIGHT, G. E., GUBLER, C. J., WINTROBE, M. M.: Studies on copper metabolism. XI. Copper and iron metabolism in the nephrotic syndrome. J. clin. Invest. *33*, 685 (1954).
CARTWRIGHT, G. E., HAMILTON, L. D., GUBLER, C. J., FELLOWS, N. M., ASHENBRUCKER, H., WINTROBE, M. M.: The anaemia of infection. XIII. Studies on experimentally produced acute hypoferremia in dogs and the relationship of the adrenal cortex to hypoferremia. J. clin. Invest. *30*, 161 (1951).
CARTWRIGHT, G. E., LAURITSEN, M., JONES, P., MERRILL, I., WINTROBE, M. M.: The anemia of infection. II. Experimental production of hypoferremia and anemia in dogs. J. clin. Invest. *25*, 81 (1946).
CHANUTIN, A., LUDEWIG, S.: Effect of whole body X-irradiation on serum iron concentration of rats. Amer. J. Physiol. *166*, 380 (1951).
COHEN, I., BOTHWELL, T. H.: Hemochromatosis in a young female. S. Afr. med. J. *32*, 629 (1958).
COLEMAN, D. H., STEVENS, A. R., FINCH, C. A.: The treatment of iron deficiency anemia. Blood *10*, 567 (1955).
DE RAADT, M. F.: The iron metabolism in pernicious anemia. Acta med. scand. *110*, 376 (1942).
DREYFUS, J. C., SCHAPIRA, G.: Le fer sérique. Sang *20*, 347 (1949).
DREYFUS, J. C., SCHAPIRA, G.: Siderophiline. Bull. Soc. Chim. biol. (Paris) *37*, 541 (1955).
DUCCI, H., SPOERER, A., KATZ, R.: Serum iron in liver disease. Gastroenterology *22*, 52 (1952).
EDOZIEN, J. C., UDEOZO, I. O. K.: Serum copper, iron and iron-binding capacity in kwashiorkor. J. trop. Pediat. *6*, 60 (1960).
ERLANDSON, M. E., SCHULMAN, I., STERN, E., SMITH, C. H.: Studies of congenital hemolytic syndromes. I. Rates of destruction and production of erythrocytes in thalassemia. Pediatrics *22*, 910 (1958).
FINCH, C. A. (1961), cit. BOTHWELL, FINCH, 1962.
FINCH, S. C., FINCH, C. A.: Idiopathic hemochromatosis, an iron-storage disease. Medicine (Baltimore) *34*, 381 (1955).
FLETCHER, J.: Variation in the availability of transferrin-bound iron for uptake by immature red cells. Clin. Sci. *37*, 273 (1969).
FLETCHER, J.: Iron transport in the blood. Proc. roy. Soc. Med. *63*, 1216 (1970).
FLETCHER, J., HUEHNS, E. R.: Significance of the binding of iron by transferrin. Nature *215*, 584 (1967).
FLETCHER, J., HUEHNS, E. R.: Function of transferrin. Nature *218*, 1211 (1968).
FLETCHER, J., SUTER, P. E. N.: The transport of iron by the human placenta. Clin. Sci. *36*, 209 (1969).
FOY, H. (1962), cit. BOTHWELL, FINCH, 1962.
FREIREICH, E. J., MILLER, A., EMERSON, C. P., ROSS, J. F.: The effect of inflammation on the utilisation of erythrocyte and transferrin bound radioiron for red cell production. Blood *12*, 972 (1957a).
FREIREICH, E. J., ROSS, J. F., BAYLES, T. B., EMERSON, C. P., FINCH, S. C.: Radioactive iron metabolism and erythrocyte survival studies of the mechanism of the anemia associated with rheumatoid arthritis. J. clin. Invest. *36*, 1043 (1957b).
GIBLETT, E. R.: Genetic markers in human blood. Davis, Philadelphia 1960.
GIBLETT, E. R., HICKMAN, C. G., SMITHIES, O.: Serum transferrins. Nature *183*, 1589 (1959).
GITLIN, D., JANEWAY, C. A., FARR, L. E.: Studies on metabolism of plasma proteins in nephrotic syndrome. Albumin, gamma-globulin and iron-binding globulin. J. clin. Invest. *35*, 44 (1956).
GITLIN, D., KUMATE, J., URRUSTI, J., MORALES, C.: The selectivity of the human placenta in the transfer of plasma proteins from mother to fetus. J. clin. Invest. *43*, 1938 (1964).
GÖLTNER, E.: Iron requirement and deficiency in menstruating and pregnant women. In: KIEF, H. (ed.): Iron Metabolism and its Disorders. Excerpta Medica, Amsterdam–Oxford; American Elsevier, New York 1975.
GOODMAN, M., NEWMAN, H. S., RAMSAY, D. S.: The use of chicken antiserum for the rapid determination of plasma protein components. III. The assay of human serum transferrin. J. Lab. clin. Med. *51*, 814 (1958).
GREENBERG, L. D., RINEHART, J. F.: Serum iron levels in rhesus monkeys with chronic vitamin C deficiency. Proc. Soc. exp. Biol. Med. *88*, 325 (1955).

GREENBERG, M. S., WONG, H., MILLER, S. A., SCARLATA, R. W., CHALMERS, T. C.: Iron absorption and turnover in hypoxia. J. clin. Invest. *39,* 992 (1960).
HAGBERG, B.: The iron-binding capacity of serum in infants and children. Acta paediat. (Uppsala) *42,* 589 (1953).
HAGBERG, B., WALLENIUS, G., WRANNE, L.: Latent iron deficiency after repeated removal of blood in blood donors. Scand. J. clin. Lab. Invest. *10,* 63 (1958).
HALLGREN, B.: Haemoglobin formation and iron storage in protein deficiency. Acta Soc. Med. upsalien. *59,* 79 (1953).
HAMILTON, L. D., GUBLER, C. J., ASHENBRUCKER, H., CARTWRIGHT, G. E., WINTROBE, M. M.: Studies on the relationship of the adrenal cortex to the experimental production of hypoferremia in rats. Endocrinology *48,* 44 (1951).
HAMILTON, L. D., GUBLER, C. J., CARTWRIGHT, G. E., WINTROBE, M. M.: Diurnal variation in the plasma iron level of man. Proc. Soc. exp. Biol. Med. *75,* 65 (1950).
HANSEN, N. E. et al.: Plasma myeloperoxidase and lactoferrin measured by radioimmunoassay: relations to neutrophil kinetics. Acta. med. scand. *198,* 437 (1975).
HARRIS, H., ROBSON, E. B., SINISEALCO, M.: Beta-globulin variants in man. Nature *182,* 452 (1958).
HARRIS, J. W.: Alterations in iron metabolism secondary to vitamin-B_6 and vitamin-B_{12} abnormalities. In: WALLERSTEIN, R. O., METTIER, S. R. (eds.): Iron in Clinical Medicine, pp. 227–250. University of California Press, Berkeley 1958.
HARTWIG, Q. L., MELVILLE, G. S., LEFFINGWELL, T. P., YOUNG, R. J.: Iron^{59} metabolism as an index of erythropoietic damage and recovery in monkeys exposed to nuclear radiations. Amer. J. Physiol. *196,* 156 (1959).
HAWKINS, C. F.: Value of serum iron levels in assessing effects of haematinics in the macrocytic anaemias. Brit. med. J. *1,* 383 (1955).
HEILMEYER, L.: Human hyposideraemia. In: GROSS, F. (ed.): Iron Metabolism, p. 201. Springer, Berlin–Göttingen–Heidelberg 1964.
HEILMEYER, L., KELLER, W., VIVELL, O., KEIDERLING, W., BETKE, K., WÖHLER, F., SCHULTZE, H. E.: Kongenitale Atransferrinaemie bei einem sieben Jahre alten Kind. Dtsch. med. Wschr. *86,* 1745 (1961).
HEMMELER, G.: Le fer sérique dans les ictères parenchymateux et par rétention. Schweiz. med. Wschr. *73,* 1056 (1943).
HEMMELER, G.: Nouvelles recherches sur le métabolisme du fer; les oscillations du fer sérique dans la journée. Helv. med. Acta *11,* 201 (1944).
HIGGINSON, J., KEELLEY, K. J., ANDERSSON, M., WALKER, A. R. P.: Serum iron levels in siderosis due to habitually excessive iron intake. J. clin. Invest. *36,* 1723 (1957).
HOLMBERG, C. G., LAURELL, C. B.: Studies on the capacity of serum to bind iron. A contribution to our knowledge of the regulation of serum iron. Acta physiol. scand. *10,* 307 (1945).
HORSFALL, W. R., SMITHIES, O.: Genetic control of some human serum beta-globulins. Science *128,* 35 (1958).
HORST, W., SCHÄFER, K. H.: Die Eisenbindung im Serum und in weiteren biologischen Flüssigkeiten, untersucht mit Papierelektrophorese und Radioeisen (Fe^{59} und Fe^{55}). Klin. Wschr. *31,* 791 (1953).
HØYER, K.: Physiologic variations in the iron content of human blood serum from week to week, from day to day and through 24 hours. Acta med. scand. *119,* 562 (1944); *119,* 577 (1944).
HUFF, R. L., LAWRENCE, J. H., SIRI, W. E., WASSERMAN, L. R., HENNESSY, T. G.: Effects of changes in altitude on hematopoietic activity. Medicine (Baltimore) *30,* 197 (1951).
JAGER, B. V.: Immunological studies of an iron-binding protein in human serum. J. clin. Invest. *28,* 792 (1949).
JAGER, B. V., GUBLER, C. J.: An immunological study of the iron-binding protein. J. Immunol. *69,* 311 (1952).
JANDL, J. H.: The agglutination and sequestration of immature red cells. J. Lab. clin. Med. *55,* 663 (1960).
JANDL, J. H., KATZ, J. H.: The plasma-to-cell cycle of transferrin. J. clin. Invest. *42,* 314 (1963).

JANDL, J. H., INMAN, J. K., SIMMONS, R. L., ALLEN, D. W.: Transfer of iron from serum iron-binding protein to human reticulocytes. J. clin. Invest. *38*, 161 (1959).
JASINSKI, B.: Eisenresorption und Infektion... Pathogenese der Infektanämie. Acta haemat. (Basel) *3*, 17 (1950).
JASINSKI, B., ROTH, O.: Die larvierte Eisenmangelkrankheit. Schwabe, Basel 1954.
JEPPSON, J. O., SJÖQUIST, J.: Proc. 6th Int. Congr. Biochem. Abstract II, 1964.
JOHNSTON, F. A.: Serum iron levels in adolescent girls. Amer. J. Dis. Child. *74*, 716 (1947).
JOSEPHS, H. W.: The determination of iron in small amounts of serum and whole blood with use of thiocyanate. J. Lab. clin. Med. *44*, 63 (1954).
KATZ, J. H.: Iron and protein kinetics studied by means of doubly labeled human crystalline transferrin. J. clin. Invest. *40*, 2143 (1961).
KATZ, J. H.: Transferrin and its functions in the regulation of iron metabolism. In: GORDON, A. S. (ed.): Regulation of Hematopoiesis. Appleton–Century–Crofts, New York 1970.
KEIDERLING, W.: Eisenstoffwechsel. Thieme, Stuttgart 1959.
KEIDERLING, W., SCHARPF, H.: Über die klinische Bedeutung der Serumkupfer- und Serumeisenbestimmung bei neoplastischen Krankheitszuständen. Münch. med. Wschr. *95*, 437 (1953).
KERR, L. M. H., RAMSAY, W. N. M.: Plasma iron in a terminal case of hemochromatosis. Biochem. J. *57*, 17 (1954).
KIND, A.: Eisenresorptionsversuch und Eisenbindungskapazität im Serum bei der Diagnose des Eisenmangels. Helv. med. Acta *21*, 402 (1954).
KOECHLIN, B.: Preparation and properties of serum and plasma proteins. XXVIII. The $beta_1$-metal combining protein of human plasma. J. Amer. chem. Soc. *74*, 2649 (1952).
LANGE, J., OBERHOFFER, G.: Die Eisenbindungskapazität des Blutserums bei akuten und chronischen Erkrankungen der Leber und der Gallenwege. Dtsch. Arch. klin. Med. *204*, 697 (1958).
LAURELL, C. B.: Plasma iron and the transport of iron in the organism. Pharmacol. Rev. *4*, 371 (1952).
LAURELL, C. B.: Iron transportation. In: WALLERSTEIN, R. O., METTIER, S. R. (eds.): Iron in Clinical Medicine. University of California Press, Berkeley 1958.
MALMQUIST, J. et al.: Lactoferrin in haematology. Scand. J. Haemat. *21*, 5 (1978).
MATHORN, M., GILLMAN, T., CANHAM, P. A., LAMONT, N. M.: Plasma iron and iron-binding capacity in African males with siderosis. Clin. Sci. *19*, 35 (1960).
MAZUR, A., BAEZ, S., SHORR, E.: The mechanism of iron release from ferritin as related to its biological properties. J. biol. Chem. *213*, 147 (1955).
MELVILLE, G. S., CONTE, F. P., UPTON, A. C.: Biphasic hyperferremia induced in rats by X-irradiation. Amer. J. Physiol. *190*, 17 (1957).
MILLER, A., CHODOS, R. B., EMERSON, C. P., ROSS, J. F.: Studies of the anemia and iron metabolism in cancer. J. clin. Invest. *35*, 1248 (1956).
MITCHELL, J., HALDEN, E. R., JONES, F., BRYAN, S., STIRMAN, J. A., MUIRHEAD, E. E.: Lowering of transferrin during iron absorption in iron deficiency. J. Lab. clin. Med. *56*, 555 (1960).
MOORE, C. V., DUBACH, R.: Iron. In: Mineral Metabolism. Academic Press, New York 1962.
MORGAN, E. H.: The interaction between rabbit, human and rat transferrin and reticulocytes. Brit. J. Haemat. *10*, 442 (1964).
MORGAN, E. H., HUEHNS, E. R., FINCH, C. A.: Iron reflux from reticulocytes and bone marrow cells in vitro. Amer. J. Physiol. *210*, 579 (1966).
MYHRMAN, G., WILANDER, O.: Inflammatory anemia and serum iron changes in myocardial infarction. Acta med. scand. *151*, 407 (1955).
NEALE, F. C.: The demonstration of the iron-binding globulin (transferrin) in serum and urine proteins by the use of Fe^{59} combined with paper electrophoresis. J. clin. Path. *8*, 334 (1955).
NYLANDER, G.: Iron metabolism in fractures. Acta endocr. (Kbh.) *20*, 148 (1955).
OLOFSSON, T. et al.: Myeloperoxidase and lactoferrin of blood neutrophils and plasma in chronic granulocytic leukaemia. Scand. J. Haemat. *18*, 113 (1977).
PALMER, H.: The influence of sex hormones on plasma iron. Scand. J. clin. Lab. Invest. *4*, 81 (1952).

PAOLETTI, C.: Rôle des beta-globulines plasmatiques dans le transport du fer utilisé par les cellules érythroformatrices. C. R. Acad. Sci. (Paris) *245*, 377 (1957).
PATERSON, J. C. S.: Disappearance of radioactive iron from plasma by day and by night. Proc. Soc. exp. Biol. Med. *96*, 97 (1957).
PATERSON, J. C. S., MARRACK, D., WIGGINS, H. S.: The diurnal variations of serum iron level in erythropoietic disorders. J. clin. Path. *6*, 105 (1953).
PIRZIO-BIROLI, G., FINCH, C. A.: Iron absorption. III. The influence of iron stores on iron absorption in the normal subject. J. Lab. clin. Med. *55*, 216 (1960).
RAMSAY, W. N. M.: The determination of iron in blood plasma or serum. Clin. chim. Acta *2*, 214 (1957).
RATH, C. E., FINCH, C. A.: Chemical, clinical and immunological studies on the products of human plasma fraction. XXXVIII. Serum iron transport. Measurement of iron-binding capacity of serum in man. J. clin. Invest. *28*, 79 (1949).
RECHENBERGER, J.: Die Bestimmung des Serumeisens und Serumkupfers in der Diagnostik der Erkrankungen des Leberparenchyms und der Gallenwege; Serumeisen und Serumkupfer bei der Hepatitis epidemica und beim Verschlußikterus. Dtsch. Z. Verdau.- u. Stoffwechselkr. *15*, 70 (1955).
RECHENBERGER, J.: Über die Eisenbindungskapazität des Blutserums; das Verhalten der Eisenbindungskapazität bei Leberzirrhosen und Hämochromatosen. Dtsch. Z. Verdau.- u. Stoffwechselkr. *16*, 155 (1956a).
RECHENBERGER, J.: Über die Eisenbindungskapazität des Blutserums; die Eisenbindungkapazität bei Blutungsanämien und chronischen Infekten. Dtsch. Z. Verdau.- u. Stoffwechselkr. *16*, 162 (1956b).
RECHENBERGER, J.: Über die Eisenbindungskapazität des Blutserums; das Verhalten der Eisenbindungskapazität und des Serumeisens bei der unbehandelten und regenerierenden perniziösen Anämie. Dtsch. Z. Verdau.- u. Stoffwechselkr. *16*, 223 (1956c).
REISSMAN, K. R., BOLEY, J., CHRISTIANSON, J. F., DELP, M. H.: The serum iron in experimental hepatocellular necrosis. J. Lab. clin. Med. *43*, 572 (1954).
REISSMAN, K. R., DIETRICH, M. R.: On the presence of ferritin in the peripheral blood of patients with hepatocellular disease. J. clin. Invest. *35*, 588 (1956).
REYNAFARJE, C., LOZANO, R., VALDIVIESO, J.: The polycythemia of high altitudes: Iron metabolism and related aspects. Blood *14*, 433 (1959).
RIFKIND, D., KRAVETZ, H. M., KNIGHT, V., SCHADE, A. L.: Urinary excretion of iron-binding protein in the nephrotic syndrome. New Engl. J. Med. *265*, 115 (1961).
RUBIN, D., WEISBERGER, A. S., CLARK, D. R.: Early detection of drug induced erythropoietic depression. J. Lab. clin. Med. *56*, 453 (1960).
RUMBALL, J. M., STONE, C. M., HASSETT, C.: The behaviour of serum iron in acute hepatitis. Gastroenterology *36*, 219 (1959).
SAITA, G., MOREO, L., PETROCCHI, V.: Sideremia e transferrinemia nel saturnismo professionale; il quadro biochemico dell'anemia saturnia. Med. d. Lavoro *46*, 463 (1955).
SALTMAN, P., CHARLEY, P. J.: The regulation of iron metabolism by equilibrium binding and chelation. In: SEVEN, M. J. (ed.): Metal-binding in Medicine, pp. 241–244. Lippincott, Philadelphia 1960.
SCHADE, A. L., OYAMA, J., REINHART, R. W., MILLER, J. R.: Bound iron and unsaturated iron-binding capacity of serum, a rapid and reliable quantitative determination. Proc. Soc. exp. Biol. Med. *87*, 443 (1954).
SCHÄFER, K. H.: Neuro-endocrine control of iron metabolism. In: GROSS, F. (ed.): Iron Metabolism, p. 289. Springer, Berlin–Göttingen–Heidelberg 1964.
SCHÄFER, K. H., BOENECKE, I.: Die neurovegetative Lenkung des Eisenstoffwechsels. Arch. exp. Path. Pharmak. *207*, 666 (1949).
SCHWARTZ, E., BAEHNER, R. L.: Diurnal variation of serum iron in infants and children. Acta paediat. scand. *57*, 533 (1968).
SIMPSON, W. L., BRENNAN, M. J., NEWMAN, H. S.: Serial study of transferrin levels in cancer patients. Proc. Amer. Ass. Cancer Res. *3*, 64 (1959).
SINNIAH, R., DOGGART, J. R., NEILL, D. W.: Diurnal variations of the serum iron in normal subjects and in patients with haemochromatosis. Brit. J. Haemat. *17*, 351 (1969).

SMITHIES, O.: Variations in human serum beta-globulins. Nature *180*, 1482 (1957).
SMITHIES, O.: An improved procedure for starch-gel electrophoresis: Further variations in the serum proteins of normal individuals. Biochem. J. *71*, 585 (1959a).
SMITHIES, O.: Zone electrophoresis in starch gels and its application to studies of serum proteins. In: ANFINSEN, C. B., BAILEY, K., LANSON, M. J., EDSALL, J. T. (eds.): Advances in Protein Chemistry, Vol. 14, pp. 65–113. Academic Press, New York 1959b.
SMITHIES, O., HILLER, O.: The genetic control of transferrins in humans. Biochem. J. *72*, 121 (1959).
SPITZNAGEL, J. K. et al.: Character of azurophil and specific granules purified from human polymorphonuclear leukocytes. Lab. Invest. *30*, 774 (1974).
STARKENSTEIN, E., HARVALIK, Z.: Über eine im intermediären Eisenstoffwechsel entstehende Ferriglobulinverbindung. Arch. exp. Path. Pharmak. *172*, 75 (1933).
STURGEON, P.: Studies of iron requirements in infants and children; normal values for serum iron, copper and free erythrocyte protoporphyrin. Pediatrics *13*, 107 (1954).
SURGENOR, D. M. et al.: Chemical, clinical and immunological studies on the products of human plasma fraction. XXXVII. J. clin. Invest. *28*, 73 (1949).
TÖTTERMAN, L.: Vitamin C and iron metabolism. Acta med. scand. *230*, 10 (1949).
TURNBULL, A., GIBLETT, E. R.: The binding and transport of iron by transferrin variants. J. Lab. clin. Med. *57*, 450 (1961).
VAHLQUIST, B. C.: Das Serumeisen. Acta pediat. (Uppsala) *28* (Suppl. 5), 1 (1941).
VAN SNICK, J. L., MASSON, P. L.: The binding of human lactoferrin to mouse peritoneal cells. J. exp. Med. *144*, 1568 (1976).
VENTURA, S., KLOPPER, A.: Iron metabolism in pregnancy. J. Obstet. Gynaec. Brit. Emp. *58*, 173 (1951).
VERHOEF, N. J. et al.: Functional heterogeneity of transferrin-bound iron. Acta haemat. (Basel) *60*, 210 (1978).
WALDENSTRÖM, J.: Eisenmangelkrankheit. In: KEIDERLING, W. (ed.): Eisenstoffwechsel. Thieme, Stuttgart 1959.
WALLENIUS, G.: A note on serum iron transportation. Scand. J. clin. Lab. Invest. *4*, 24 (1952).
WARBURG, O., KREBS, H. A.: Über locker gebundenes Kupfer und Eisen im Blutserum. Biochem. Z. *190*, 143 (1927).

CHAPTER 8

STORAGE IRON

In the normal adult male about 25–30% of the total iron content is present in iron stores, the amount depending upon the balance between iron absorption from the gut and iron loss. Because of the greater physiological requirements of women during reproductive life, their stores are usually smaller than those of men (Weinfeld, 1964).

FERRITIN AND HEMOSIDERIN

The bulk of the iron is stored in the liver, spleen, and bone marrow, although small amounts may be found in other organs.

The storage iron is present in two forms:

(1) ferritin, and

(2) hemosiderin.

Ferritin is water-soluble and not visible by the light microscope, whereas hemosiderin is insoluble and stainable by the Prussian blue reaction (Heilmeyer, 1958). In both forms the iron is bound to apoferritin (Wöhler, 1960, 1964), but the composition of hemosiderin is more variable than that of ferritin, and it contains other substances.

Electron microscopic examination shows that within the ferritin molecule the iron is situated in the six apices of an octahedron (Bessis, 1958) (see Chapter 17). The six ferruginous micelles are surrounded by a shell of apoferritin (Kerr and Muir, 1960). The structure of apoferritin has been described by Harrison (1963) (Fig. 8/1).

The molecule consists of 24 subunits (polypeptide chains) and forms a shell within which ferric ions, hydroxyl ions, and oxygen are dispersed in a latticelike relationship (Harrison et al., 1967, 1980; Brady et al., 1968). This permits a semicrystalline structure to form, probably due to close stacking of octahedral and tetrahedral aggregates (Fig. 8/2). Each apoferritin molecule can accommodate about 2000–4000 iron atoms. The diameter of the protein shell in the native state is 122 Å; that of the inorganic nucleus is 74 Å. Electron microscopic investigations have given values of 105 and 55 Å, respectively. The molecular weight of apoferritin is about 445,000–465,000 (Bryce and Crichton, 1971) or, according to more recent studies, somewhat more, 480,000 (Harrison, 1980). . .

The iron–apoferritin complex (ferritin) is not entirely homogeneous. It contains 20–26% of iron (Harrison, 1964), which can be liberated by reducing substances such as ascorbic acid, glutathione, cysteine. Mazur et al. (1958) have shown that it can be released *in vivo* by xanthinoxidase. It can also be mobilized by low molecular weight chelating agents without the necessity for reduction to the ferrous form.

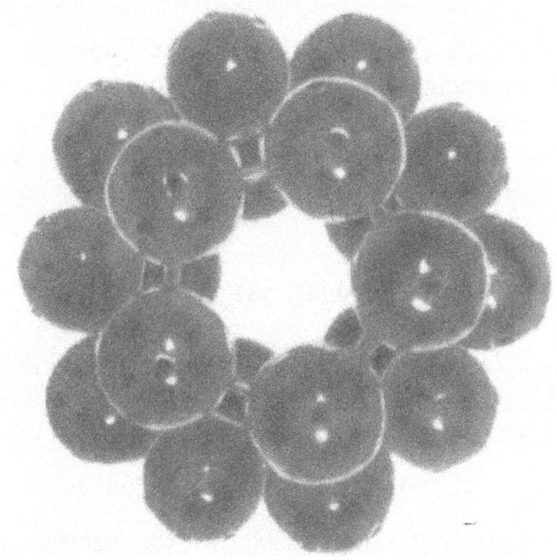

Fig. 8/1. Model of the quaternary structure of apoferritin (after Harrison, P. M.: J. Med. Biol. [G. B.] *6*, 404, 1963)

The proportion of free protein and apoferritin–iron complex varies. From ultracentrifugation studies it has been shown that following cadmium sulfate precipitation about 25% of the crystals are apoferritin and the remainder the iron–protein complex ferritin (Rothen, 1944). Within cells the ferritin molecules can be observed either in dispersed form or in aggregates (siderosomes). Ferritin has been isolated by chemical methods from the tissues of many vertebrate animals such as horse, guinea pig, mouse, rat, rabbit, and cat. In horses it has been found in the spleen, bone marrow, liver, testes, kidney, pancreas, ovary, and lymph nodes (Fig. 8/2). The ferritins isolated from human and horse spleen and liver have been

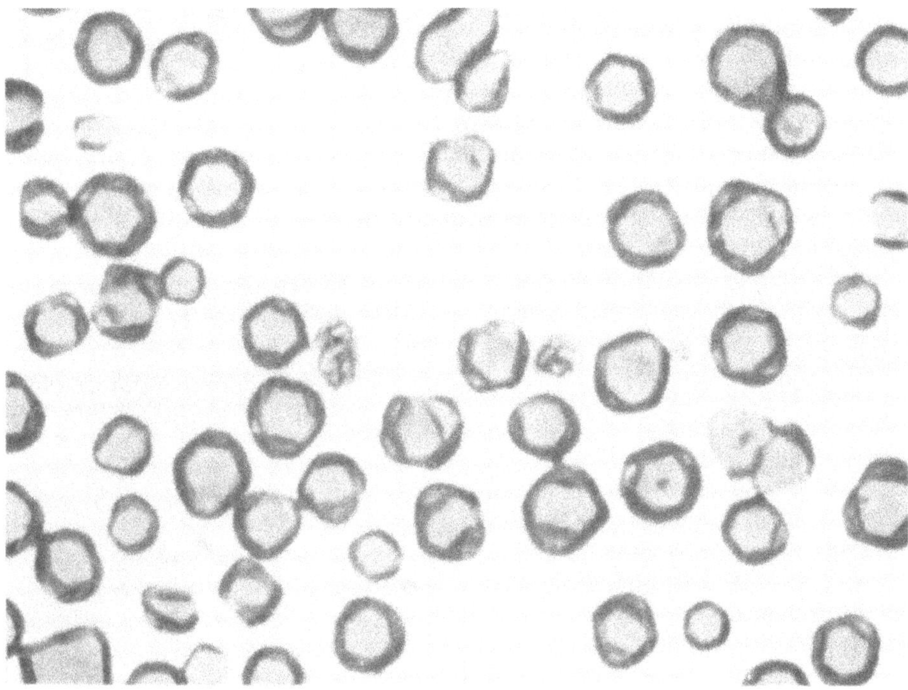

Fig. 8/2. Ferritin crystals, magnification: × 100 (after Sturgeon, P. and Shoden, A., in: Gross, F.: Iron Metabolism. Springer, Berlin–Göttingen–Heidelberg 1964)

compared. Homologous ferritins share reactions of identity in immunodiffusion experiments, whereas heterogeneous ferritins show only partial identity. The amino acid compositions of the proteins reveal distinct differences, between both organs and species (Crichton et al., 1973). van Kreel et al. (1972) have shown microheterogeneity of normal rabbit ferritin using isoelectric focusing. They found a variant with an isoelectric point of 4.46.

By definition, *isoferritins* differ from one another in the molecular structure of the protein shell. On the basis of electrophoretic separation of amino acid composition and sequence, and of antibody reactivity, it has been securely established that ferritins vary in structure in both normal and malignant cells (Bomford and Munro, 1980).

In man, ferritin has been isolated from the spleen, placenta, and intestinal mucosa (Granick, 1946; Shoden et al., 1953; Richter, 1957; Heilmeyer, 1958; Wöhler, 1959) and from human red cells (Agner, 1943). It is also present in serum and can be determined by an immunoradiometric method (Addison et al., 1972).

Ferritin, as well as transferrin and lactoferrin, can be demonstrated in histological sections by the use of an immunoperoxidase technique (Mason and Taylor, 1978).

Bessis and Breton-Gorius (1959, 1962), using electron microscopy, demonstrated that the siderocyte granules are aggregates of ferritin molecules. Reticulum cells (Bessis and Breton-Gorius, 1959) and cells of the renal tubules have also been shown to contain ferritin (Richter, 1957). Most human tissues probably contain some ferritin but often in insufficient quantity to be demonstrated by routine procedures (Granick, 1951).

The synthesis of apoferritin is stimulated by exposure to iron, both *in vitro* and *in vivo*. This phenomenon is blocked by inhibitors of protein synthesis but not by inhibitors of RNA synthesis (Crichton, 1975).

Hemosiderin contains iron, protein, and lipoids (Bessis and Breton-Gorius, 1962) and occasionally porphyrin and other pigments (Shoden and Sturgeon, 1960). The

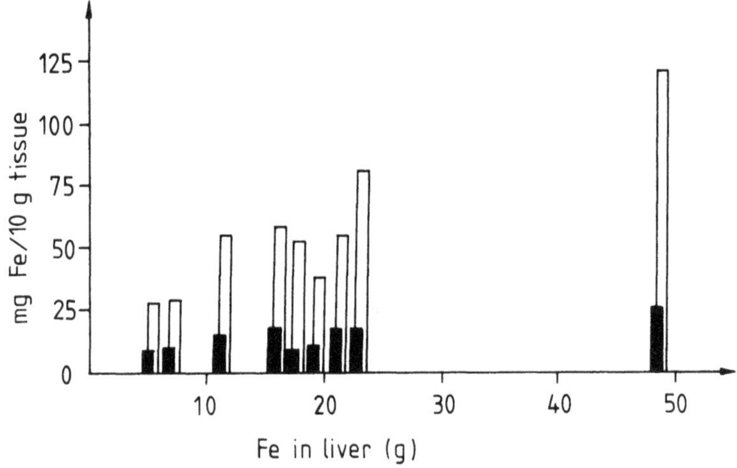

Fig. 8/3. The ratio of ferritin (■) and hemosiderin (□) in various sideroses (idiopathic hemochromatosis, transfusional siderosis and Laënnec cirrhosis) (after Finch, S. C. and Finch, C. A.: Medicine [Baltimore] *34,* 381, 1955)

iron content of hemosiderin varies between 30 and 37%. Normally hemosiderin is confined to the liver, spleen, and bone marrow (Rath and Finch, 1948; Shoden et al., 1953), but in hypersiderosis it can be demonstrated in almost every tissue (Finch and Finch, 1955).

Normally there is more ferritin than hemosiderin in the tissues (Shoden et al., 1953; Kaldor, 1958), the ratio being about 60:40 (Weinfeld, 1964). In iron-storage diseases the amount of both ferritin and hemosiderin increases, but the bulk of the storage iron in these circumstances is in hemosiderin (Finch and Finch, 1955; Kaldor, 1958; Bothwell et al., 1965) (Fig. 8/3).

DEPOSITION AND MOBILIZATION OF IRON

Ferritin and hemosiderin are in close functional interrelationship. If the iron supply is abundant, iron appears first in the form of hemosiderin and later, depending on the rate of apoferritin synthesis, is transformed into ferritin (Wöhler, 1955, 1959). Ferritin iron can be rapidly mobilized from storage cells (Weinfeld, 1964); hemosiderin iron is liberated more slowly and to a varying degree. In general, iron that has recently been deposited in the stores is more easily mobilizable than that which has been there longer (Hampton, 1954). Merker (1968) found that the iron granules stored in the sideromacrophages showed certain morphological

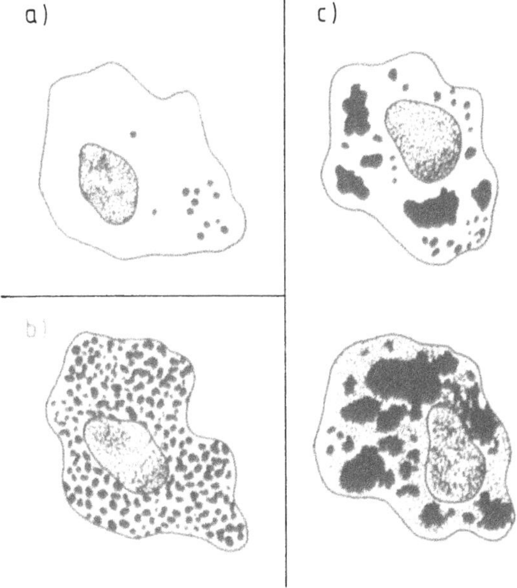

Fig. 8/4. Various forms of iron storage in the reticulum cells of the bone marrow. a) Finely granulated iron granules in the macrophage of a healthy human; b) relatively small iron lumps of even distribution and sharp margin in hemolytic anemia (accelerated iron turnover, freshly deposited reserve); c) large lumps of iron in uneven distribution in chronic infectious diseases and in malignancy. The earlier deposited iron is mobilized with difficulty (after Merker, H., in: Heilmeyer, L.: Blut und Blutkrankheiten. Springer, Berlin–Heidelberg–New York 1968)

features indicating the mobility of the iron (Fig. 8/4). The small granules (0.5–2.0 μm in diameter) or the relatively evenly distributed iron in the cytoplasm appeared to correspond to fresh reserves (Fig. 8/4b). In contrast, the iron found in large hemosiderin agglomerates, e.g., in the anemia of infections (Fig. 8/4c), is less easily mobilized and presumably represents earlier depositions in the cells (Beutler, 1958).

In the course of hemoglobin synthesis, iron released from decomposing erythrocytes is used in preference to storage iron (Noyes et al., 1960), and there is apparently little mixing between this labile iron pool and the storage compartment. Tracer doses of radioactive iron given intravenously also appear to mix with the labile pool rather than with iron deposited earlier in the RE cells (Finch, 1959).

Little is known about the mechanisms of iron deposition and mobilization. Cells normally contain little ferritin in the preformed state (Gabrio and Salomon, 1950), but the specific protein is synthesized when iron enters the cells. Bothwell and Finch (1962) draw a parallel between the iron induction of apoferritin synthesis and antigen stimulation of antibody production. Apoferritin synthesis in response to iron has been shown in the mucosal cells of the gastrointestinal tract after both oral and parenteral administration of iron (Granick, 1946; Gabrio and Salomon, 1950; Heilmeyer, 1958) and in the liver after intramuscular injection of iron compounds (Hahn et al., 1943; Loftfield and Eigner, 1958).

Ferritin molecules can be observed around the membrane of phagocytosed red cells in the cytoplasm of reticulum cells (Bessis and Breton-Gorius, 1957; Hampton, 1960). Ferritin from the reticulum cells can, by rhopheocytosis, be transferred to the developing erythroblast (Bessis, 1958). It has also been suggested that ferritin can behave as an intermediate in iron transport between transferrin and heme. There is, however, some disagreement about this.

Recently, emphasis has been placed on the heterogeneity of the ferritin molecule. Isoferritins variously distributed in different tissues have been described (Drysdale et al., 1977). Ferritin of varying iron content differs in the ease with which iron is taken up or mobilized (Harrison et al., 1974). Yamada and Gabuzda (1974a, b) have shown that ferritin of low iron content in red cell precursors is metabolically highly active.

Speyer and Fielding (1979) have described techniques for the fractionation of whole reticulocytes after incubation with ^{59}Fe–^{125}I-transferrin. Membrane ^{59}Fe was separated into two components, A and B, and the latter further separated into two subcomponents, B_1 and B_2. B_2 has been recognized as the complex of transferrin and its membrane receptor (Speyer and Fielding, 1979; Fielding and Speyer, 1974; van der Heul et al., 1978). Two labeled components were found in the cytosol: hemoglobin and component C (Fielding et al., 1977). Component C was found to behave as an intermediate in the iron transport pathway to heme and was identified as ferritin, and it was concluded that cytosol ferritin is an obligatory intermediate in reticulocyte iron transport (Speyer and Fielding, 1979).

QUANTITATIVE ASPECTS OF IRON STORES

The amount of storage iron has been assessed by mobilization following phlebotomy, repeated until evidence of iron deficiency appeared. Using this method, the storage iron was assessed in normal males as about 1000–1500 mg. The iron stores have also been assessed by isotope dilution, which indicates normal iron stores in the male of about 1200 mg and in the female of about 800 mg (Finch, 1959). The male and female differences are largely determined by physiological differences in iron requirements. There are also considerable individual variations due to the numerous factors controlling iron supply and iron loss (Table 8/1).

Table 8/1
Factors influencing the extent of iron storage (from Weinfeld, 1970)

Factors	Effect*
Effective erythropoiesis (hemoglobin synthesis)	−
Expanding body mass	−
Hemorrhage	−
Iron-deficient diet	−
Disturbance in iron absorption	−
Pregnancy	−
Iron-rich diet	+
Blood transfusion	+
Parenteral iron injections	+
Increased iron absorption	+

* −: decreases the reserve
+: increases the reserve

The liver is an important organ of iron storage. Weinfeld (1964) estimated the non-heme iron content of the liver at 400 mg in males, 235 mg in postmenopausal females, and 130 mg in menstruating females. The non-heme iron concentration of the liver in different parts of the world is remarkably uniform (Meier et al., 1959; Mayet and Bothwell, 1964; Weinfeld, 1964; Hofvander, 1968) (Table 8/2).

About 1/3rd to 1/4th of the storage iron is in the liver, about 1/6th is in the striated muscles, and the rest is mainly in the spleen and bone marrow, the bone marrow amounting to about 300 mg on an average (Bothwell and Gale, 1962). The isotope dilution method has given estimates of 288 mg of iron in the bone marrow of males and 99 mg of iron in females (Gale et al., 1963). A fairly good correlation can be

Table 8/2

The non-heme iron concentration of the liver according to literary data

Authors	Country	No. of males examined	Non-heme Fe in the liver (mg/100 g substance)	Cause of death
Schairer and Rechenberger (1948)	Germany	18	95 ± 12.5	Accident
Meier et al. (1959)	Germany	136	95	Accident or suicide
Morgan and Walters (1963)	Australia	18	81.5 ± 10.7	Accident
Mayet and Bothwell (1964)	South Africa	85	89	Accident
Hofvander (1968)	Ethiopia	65	87.8 ± 7.3	Accident or suicide
Weinfeld (1964)	Sweden	29	80.5 ± 7.6	(Biopsy examinations)

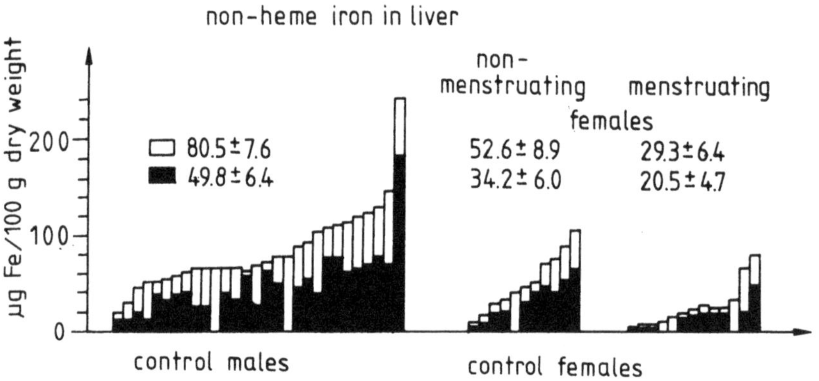

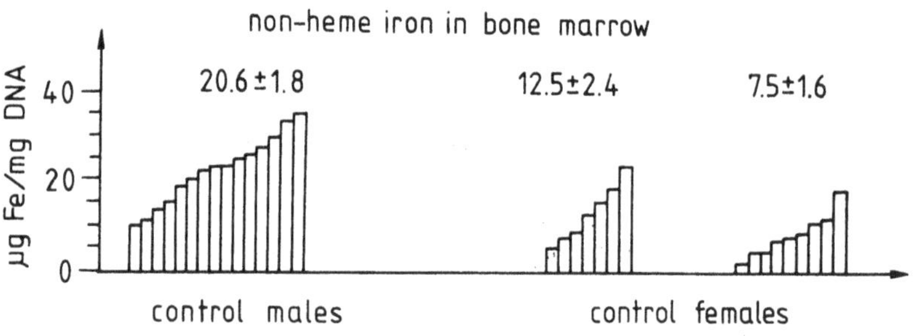

Fig. 8/5. Non-heme iron content of the liver of males, nonmenstruating and menstruating females (results of biopsy examinations). The height of the columns represents the total amount of iron, the black area the amount of ferritin iron (after Weinfeld, A.: Acta med. scand. Suppl. 427, 1964)

established between the iron content of the liver and the bone marrow (Weinfeld, 1964) (Figs. 8/5 and 8/6).

Serum ferritin has been used as an indirect measure of iron stores (Addison et al., 1972). The mean level is higher in males than in females; the levels are lower in iron deficiency than normal and are high in patients with iron overload.

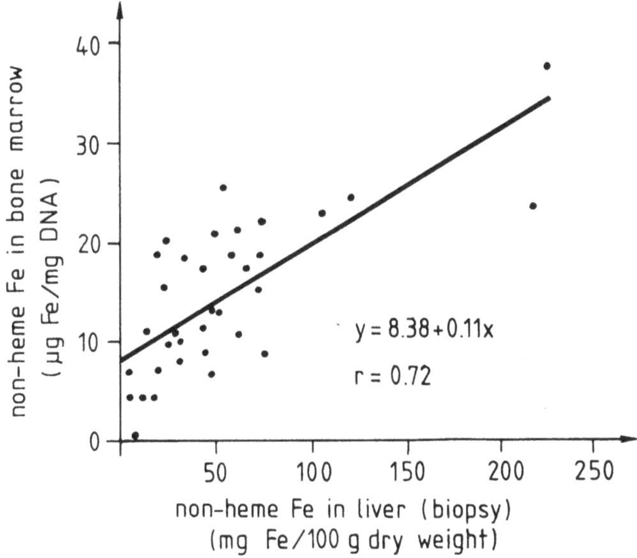

Fig. 8/6. Correlation between the non-heme iron content of the liver and bone marrow (after Weinfeld, A.: Acta med. scand. Suppl. 427, 1964)

Iron stores decrease gradually during infancy, reaching their lowest level between 6 and 24 months (Roth et al., 1951; Smith et al., 1955; Erlandson, 1962) (Fig. 8/7). Subsequently, if the diet is adequate in respect to iron, the reserves increase gradually until 20–22 years of age. The minimal obligatory loss of iron can normally be replaced from the diet, and an equilibrium is maintained.

In the presence of a negative iron balance the iron concentration of the liver and bone marrow is reduced. Weinfeld (1964) demonstrated this both in chronic iron-deficiency anemia and in nonanemic blood donors (Fig. 8/8). He also showed that following partial gastrectomy the iron content of the bone marrow declined with time since operation (Fig. 8/9), while the total iron-binding capacity of the plasma tended to rise (Fig. 8/10).

Olsson (1972), by quantitative phlebotomy, showed that the mobilizable iron stores in normal men averaged 750 mg with a range of 180–1350 mg, whereas

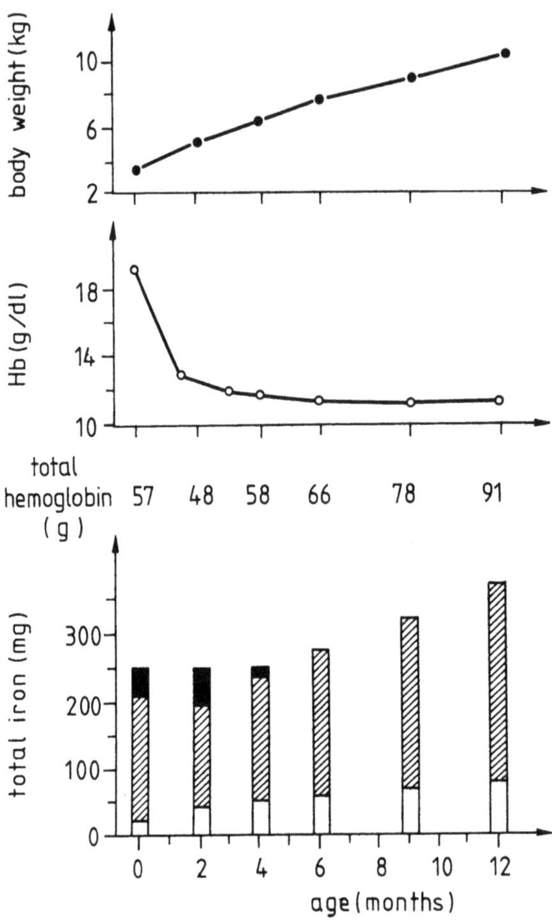

Fig. 8/7. Correlation between the total amount of iron reserve and blood pigment. During a rise in hemoglobin the reserve iron is mobilized (after Erlandson, M. E.: Pediat. Clin. N. Amer. *9*, 672, 1962). ■ reserve Fe; ▨ Hb Fe; □ parenchymal Fe

regular blood donors had significantly lower values with a mean of 110 mg and a range of 0–250 mg.

Excessive iron stores are found with a long continued positive iron balance such as may occur in hemochromatosis, where an excess of iron is absorbed from a normal diet, or in Bantu siderosis, where the iron content of the diet is excessively high and in an easily absorbable form. The excess iron is retained in cells of the parenchymatous organs as well as in the RE cells. In idiopathic hemochromatosis the iron-binding capacity of the RE system may be relatively reduced (Bothwell and Finch, 1962), and the deposition is primarily in the parenchymatous organs.

Excessive iron storage is also found in association with transfusion siderosis and in the anemias associated with ineffective erythropoiesis. In transfusion siderosis the iron released from the donors' blood is deposited first in the RE cells, and there is a great accumulation of iron in the Kupffer cells and macrophages of the liver, in the spleen and lymph nodes, and in the bone marrow (Cappell et al., 1957). Later, part of the iron is liberated from the RE system and enters the plasma and is transported to the parenchymal cells, where it is deposited in large amounts (Oliver, 1959). This redistribution takes some time, and the factors determining it are not known (Finch, 1958).

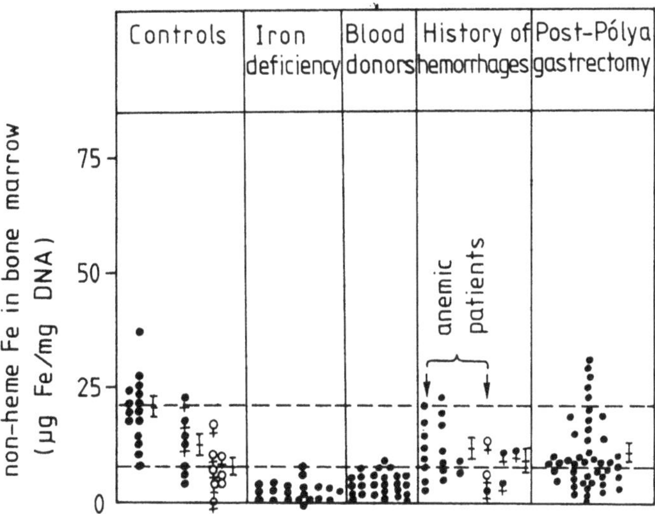

Fig. 8/8. Non-heme iron concentration in the bone marrow of males (●), of nonmenstruating females (♀), and of menstruating females (♀) (after Weinfeld, A.: Acta med. scand. Suppl. 427, 1964) ‡ = mean value ± S. D. Upper broken line = mean value of males; lower broken line = upper limit in iron-deficiency anemia

With ineffective erythropoiesis the iron overload depends not only on transfusion but also on excessive iron absorption, and the pattern of iron distribution is somewhat different from that found with transfusion siderosis alone. There is considerable iron deposition in the parenchymal cells in agreement with Fletcher's (1970) hypothesis that the cells have a higher affinity for saturated transferrin than for half-saturated transferrin molecules.

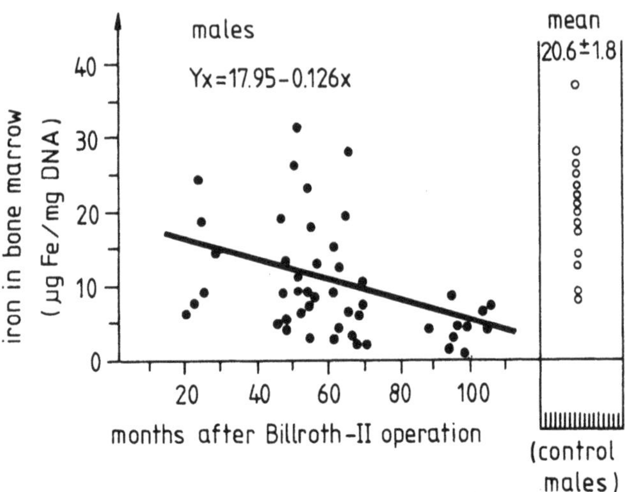

Fig. 8/9. Changes in the non-heme iron content of the bone marrow after Billroth II operation as a function of time (after Weinfeld. A.: Acta med. scand. Suppl. 427, 1964)

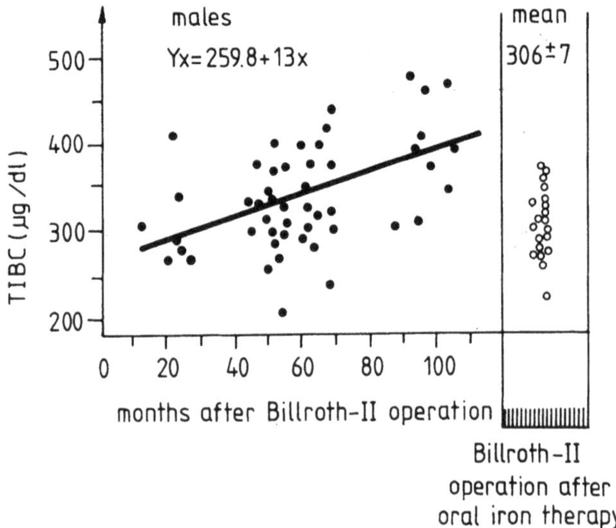

Fig. 8/10. Changes in the total iron-binding capacity of the plasma after Billroth II operation as a function of time (after Weinfeld, A.: Acta med. scand. Suppl. 427, 1964)

CLINICAL METHODS USED FOR THE ESTIMATION OF AVAILABLE IRON STORES

1. CYTOCHEMICAL OR HISTOCHEMICAL EXAMINATION OF THE BONE MARROW

This is a simple and direct method for estimating the amount of storage iron and can be used on smears and sections of bone marrow. The Prussian blue reaction is usually used to stain the iron (Rath and Finch, 1948; Beutler et al., 1954; Coleman et al., 1955; Vries and Izak, 1955; Finch and Finch, 1955; McCrea, 1958). The results are in good agreement with the data obtained by the chemical determination of non-heme iron of the bone marrow (Kerr, 1957; Bothwell and Gale, 1962; Weinfeld, 1964) (Fig. 8/11). The method is particularly useful for distinguishing between iron deficiency and the hypochromic anemia of chronic disorders (Fig. 8/12), but the findings can possibly be misinterpreted if an iron-deficient patient has received

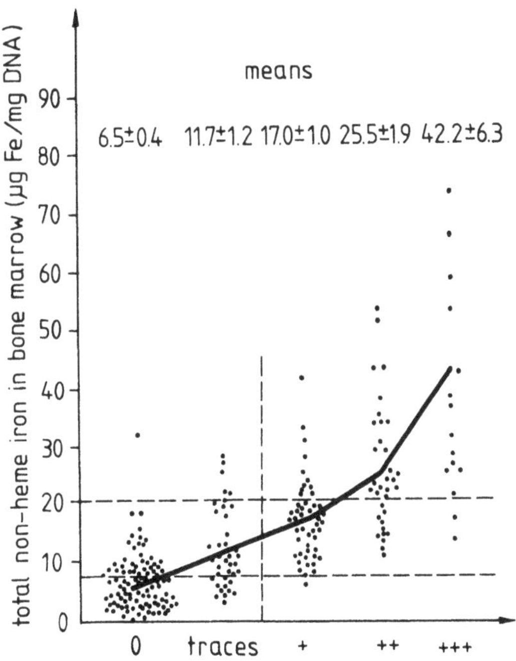

Fig. 8/11. Relationship between the non-heme iron concentration and the histochemically demonstrable iron content of the bone marrow. Black dots represent the non-heme iron concentration corresponding to the different degrees of the Prussian blue reaction. The upper broken line indicates the normal values, the lower broken line the upper limit of patients with iron deficiency (after Weinfeld, A.: Acta med. scand. Suppl. 427, 1964)

parenteral iron that has not yet been utilized (Bothwell and Finch, 1962). In general, lack of hemosiderin in the reticulum cells of the bone marrow indicates an iron-deficiency state. In most healthy subjects some iron can be demonstrated in the bone marrow, which can be graded quantitatively. Excess iron is seen in hemolytic anemias, long-standing infections (Freireich et al., 1957), and thermal injuries

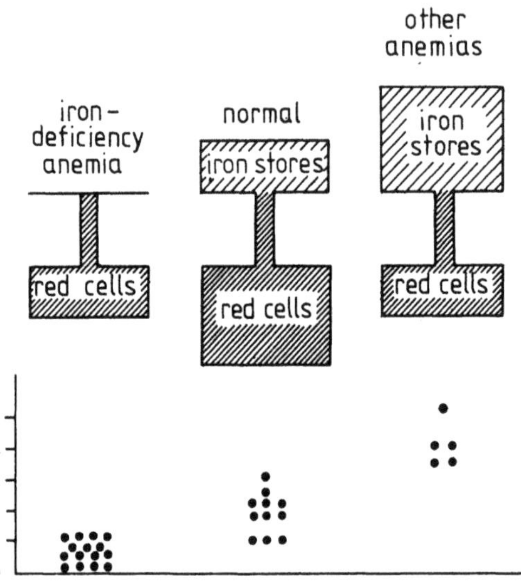

Fig. 8/12. Significance of bone marrow hemosiderin in the differential diagnosis of iron-deficiency anemia. The amount of iron stored in the RE cells is indicated by dots in the lower part of the figure. In iron-deficiency anemia the iron is absent from the bone marrow. In other types of anemia, e.g., infectious, hemolytic, sideroblastic, the iron stored in the RE cells is increased. The upper part of the figure demonstrates the relationship between the total volume of red cells and the degree of saturation of the iron stores (after Bothwell, T. H. and Finch, C. A.: Iron Metabolism. Little, Brown and Co., Boston 1962)

Table 8/3

Iron content of the sideromacrophages of the bone marrow in iron deficiency in healthy individuals, in thermal injury, and in sideroblastic anemia (Bernát, 1971)

	The amount of iron in the cell			
	0	1 (slight)	2 (moderate)	3 (much)
Iron-deficiency anemia	*92*	8	0	0
Healthy individuals	15	*73*	12	0
Thermal injury	7	21	*54*	18
Björkman's anemia	0	6	33	*61*

(Bernát, 1971). Heavy deposition is seen in various conditions associated with iron overload, i.e., idiopathic hemochromatosis, transfusion siderosis, etc.

Quantitative analysis should include the examination of 50–100 reticulum cells, the iron content being classified as no iron (0), very little iron (1), a medium amount of iron (2), and large amounts of iron (3) (Table 8/3). It is useful also to note whether the iron is evenly distributed in the form of fine granules in the cytoplasm or in big clusters, the latter being less easily mobilizable than the former (Wallerstein and Pollycove, 1958; Beutler, 1958; Merker, 1968).

2. ESTIMATION OF IRON STORES USING THE DESFERRIOXAMINE IRON EXCRETION TEST

Following the intramuscular injection of 500 mg of desferrioxamine B (Desferal®, Ciba), healthy humans excrete an average of 0.4–0.8 mg of iron in the urine in the next 24 hours. There is a reasonable correlation between the storage iron and the effect of Desferal® on iron excretion. The excretion is reduced in iron deficiency (Dagg et al., 1966; Hallberg and Hedenberg, 1967) (Fig. 8/13) and increased in iron overload (Wöhler, 1961, 1964; Hallberg et al., 1966; Balcerzak et al., 1968; Ploem et al., 1966; Olsson, 1972) (Fig. 8/14).

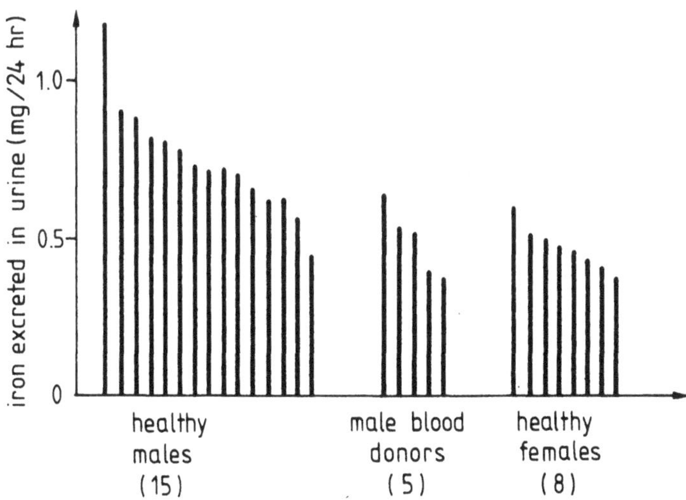

Fig. 8/13. The effect of desferrioxamine-B (10 mg/kg i.v.) upon the amount of iron excreted with the urine in healthy males, male blood donors, and healthy menstruating females. Iron excretion is reduced if the iron stores are lower than normal (after Hallberg, L., in: Gross, F.: Iron Metabolism. Springer, Berlin–Göttingen–Heidelberg 1964)

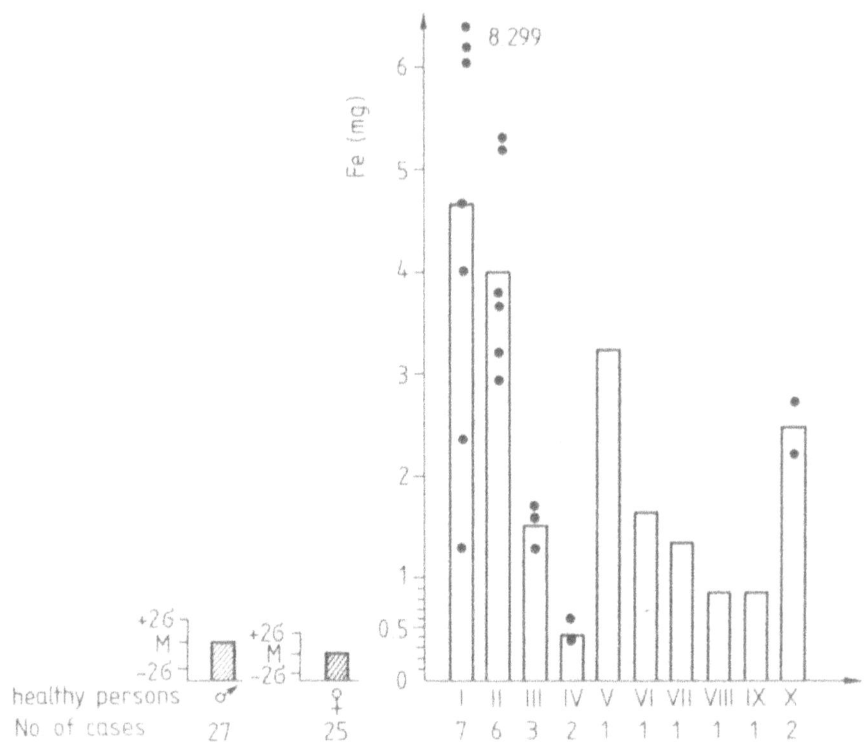

Fig. 8/14. In diseases associated with enhanced iron storage (I–X) the DFO-induced iron excretion increases as compared to the normal, depending on the amount of storage iron (after Wöhler, F.: Proc. 8th Congr. Europ. Soc. Haemat. Karger, Basel–New York 1961)

3. MEASUREMENT OF SERUM FERRITIN

In normal subjects the serum ferritin is proportional to the size of body iron stores (Saarinen and Siimes, 1979; Worwood, 1980). A highly significant inverse correlation has been demonstrated in normal subjects between serum ferritin level and radioiron absorption, the latter being a sensitive indirect measure of body iron stores (Cook et al., 1974; Walters et al., 1975; Heinrich et al., 1977; Kaltwasser and Werner, 1977; Kaltwasser et al., 1977; Worwood, 1980). Bezwoda et al. (1979) have also shown an inverse relationship between plasma ferritin and marrow iron in addition to iron absorption (Bezwoda et al., 1979). Oertel et al. (1978) examined levels of non-heme iron and ferritin in the bone marrow in patients with iron deficiency or iron overload, as well as in patients with no abnormality of iron metabolism. They found an excellent overall correlation between non-heme iron and ferritin in the bone marrow and also between bone marrow and serum ferritin. Lipschitz et al. (1974) have found a mean value of 59 ng/ml in normal controls with a 95% confidence limit of 12–300 ng/ml, and in patients with iron overload they found the serum ferritin to be invariably elevated.

The measurement of the serum ferritin allows a distinction between iron-deficiency anemia and the anemia of chronic disorders, it being decreased in the former but increased in the latter (Fig. 8/15). Serum ferritin concentrations of up to 510 μg/l have been found in children with various common, acute infections (Siimes

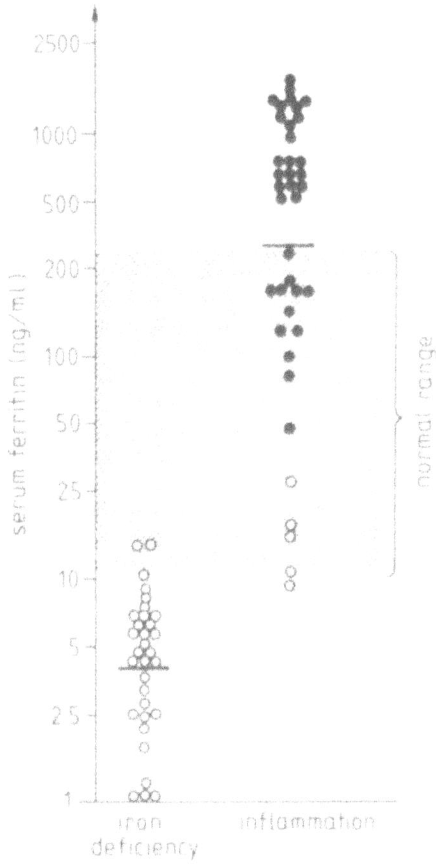

Fig. 8/15. Serum ferritin level in iron deficiency and in various inflammatory processes (after Lipschitz, D. A. et al.: New Engl. J. Med. *290*, 1213, 1974). ○ = patients with iron deficiency (TIBC > 400 μg/dl; transferrin saturation < 16% or no iron in bone marrow)

et al., 1974), and in adult patients with inflammation the mean serum ferritin level was 305 μg/l with a range of 10–1650 μg/l (Lipschitz et al., 1974).

Myocardial infarction causes an inflammatory reaction and the changes in ferritin concentration follow the pattern described for induced fever (Birgegård et al., 1979). The finding of an elevated serum ferritin level in an anemic subject with inflammation would exclude inadequate iron stores and might obviate the need for bone marrow examination (Lipschitz et al., 1974).

BIBLIOGRAPHY

Addison, G. M., Beamish, M. R., Hales, C. N. et al.: An immunoradiometric assay for ferritin in the serum of normal subjects and patients with iron deficiency and iron overload. J. clin. Path. *25*, 326 (1972).

Agner, K.: Erythrocyte catalase. Arkiv. Kem. Mineral. Geol. *17*, 1 (1943).

Balcerzak, S. P., Westerman, M. P., Heinle, E. W., Taylor, F. H.: Measurement of iron stores using desferrioxamine. Ann. intern. Med. *68*, 518 (1968).

Bernát, I.: Az égési anaemia pathogenesise (The Pathogenesis of Anemia after Thermal Injury). Akadémiai Kiadó, Budapest 1971.

Bessis, M.: Étude au microscope électronique de la destinée d'une molécule dans l'organisme: la ferritine et le cycle hémoglobinique du fer. Bull. Acad. nat. Méd. (Paris) *142*, 629 (1958).

Bessis, M. C., Breton-Gorius, J.: Iron particles in normal erythroblasts and normal and pathological erythrocytes. J. biophys. biochem. Cytol. *3*, 503 (1957).

Bessis, M. C., Breton-Gorius, J.: Ferritin and ferruginous micelles in normal erythroblasts and hypochromic hypersideremic anemias. Blood *14*, 423 (1959).

Bessis, M. C., Breton-Gorius, J.: Iron metabolism in the bone marrow as seen by electron microscopy: A critical review. Blood *19*, 635 (1962).

Beutler, E.: The utilisation of saccharated Fe^{59} oxide in red cell formation. J. Lab. clin. Med. *51*, 415 (1958).

Beutler, E., Drennan, W., Block, M.: The bone marrow and liver in iron-deficiency anemia, a histopathologic study of sections with special reference to stainable iron content. J. Lab. clin. Med. *43*, 427 (1954).

Bezwoda, W. R., Bothwell, T. H., Torrance, J. D., MacFail, A. P., Charlton, R. W.: The relationship between marrow iron stores, plasma ferritin concentration and iron absorption. Scand. J. Haemat. *22*, 113 (1979).

Birgegård, G., Hallgren, R., Venge, P., Wide, I.: Acta med. scand. *206*, 361 (1979).

Bomford, A. B., Munro, H. N.: Biosynthesis of ferritin and isoferritins. In: Jacobs, A., Worwood, M. (eds.): Iron in Biochemistry and Medicine, Vol. 2. Academic Press, London–New York 1980.

Bothwell, T. H., Finch, C. A.: Iron Metabolism. Little, Brown and Co., Boston 1962.

Bothwell, T. H., Gale, G. E. G. (1962), cit. Bothwell, Finch, 1962.

Bothwell, T. H., Abrahams, C., Bradlow, B. A., Charlton, R. W.: Idiopathic and Bantu haemochromatosis. Comparative histological study. Arch. Pathol. *79*, 163 (1965).

Bothwell, T. H., Hurtado, A. V., Donohue, D. M., Finch, C. A.: Erythrokinetics. IV. The plasma iron turnover as a measure of erythropoiesis. Blood *12*, 409 (1957).

Bothwell, T. H., Pirzio-Biroli, G., Finch, C. A.: Iron absorption. I. Factors influencing absorption. J. Lab. clin. Med. *51*, 24 (1958).

Brady, G. W., Kurjian, C. R., Lyden, E. F. X., Robin, M. B., Saltman, P., Spiro, T., Terzis, A.: The structure of an iron core analog of ferritin. Biochemistry *7*, 2185 (1968).

Brown, E. B., Dubach, R., Smith, D. E., Reynafarje, C., Moore, C. V.: Studies in iron transportation and metabolism. X. Long term iron overload in dogs. J. Lab. clin. Med. *50*, 862 (1957).

BRYCE, C. F. A., CRICHTON, R. R.: J. Biol. Chem. *246*, 4198 (1971).
CAPPELL, D. F., HUTCHISON, H. E., JOWETT, M.: Transfusional siderosis: the effects of excessive iron deposits on the tissues. J. Path. Bact. *74*, 245 (1957).
COLEMAN, D. H., STEVENS, A. R., FINCH, C. A.: The treatment of iron deficiency anemia. Blood *10*, 567 (1955).
COOK, J. D., LIPSCHITZ, D. A., MILES, L. E. M., FINCH, C. A.: Amer. J. Clin. Nutr. *27*, 681 (1974).
CRICHTON, R. R.: Ferritin: structure, function and role in intracellular iron metabolism. In: KIEF, H. (ed.): Iron Metabolism and its Disorders, pp. 81–89. Excerpta Medica, Amsterdam–Oxford; American Elsevier, New York 1975.
CRICHTON, R. R., MILLAR, J. A., CUMMING, R. L. C., BRYCE, C. F. A.: The organ specificity of ferritin in human and horse liver and spleen. Biochem. J. *131*, 51 (1973).
CROSBY, W. H.: The metabolism of hemoglobin and bile pigment in hemolytic disease. Amer. J. Med. *18*, 112 (1955).
DACIE, J. V., SMITH, M. D., WHITE, J. C., MOLLIN, D. L.: Refractory normoblastic anemia. A clinical and hematological study of seven cases. Brit. J. Haemat. *5*, 56 (1959).
DAGG, J. H., SMITH, J. A., GOLDBERG, A.: Urinary excretion of iron. Clin. Sci. *30*, 495 (1966).
DOUGLAS, A. S., DACIE, J. V.: The incidence and significance of iron-containing granules in human erythrocytes and their precursors. J. clin. Path. *6*, 307 (1953).
DRYSDALE, J. U., ADELMAN, T. G., ARIOSO, P.: Human isoferritins in normal and disease states. Semin. Haemat. *14*, 71 (1977).
DUGGAN, D. E., STREETER, K. B. (1973), cit. CRICHTON, 1975.
ELLIS, J. T., SCHULMAN, I., SMITH, C. H.: Generalized siderosis with fibrosis of liver and pancreas in Cooley's (Mediterranean) anemia. Amer. J. Path. *30*, 287 (1954).
ERLANDSON, M. E.: Iron metabolism and iron deficiency anemia. Pediat. Clin. N. Amer. *9*, 672 (1962).
FIELDING, J., SPEYER, B. E.: Iron transport intermediates in human reticulocytes and the membrane binding site of iron-transferrin. Biochim. Biophys. Acta *363*, 387 (1974).
FIELDING, J., DONEY, V. J., SPEYER, B. E.: Patterns of plasma iron fractions after intramuscular Ferastral. Scand. J. Haemat. *32*, 228 (1977).
FINCH, C. A.: Discussion of hemosiderosis and hemochromatosis. In: WALLERSTEIN, R. O., METTIER, S. R. (eds.): Iron in Clinical Medicine, pp. 131–132. University of California Press, Berkeley 1958.
FINCH, C. A.: Body iron exchange in man. J. clin. Invest. *38*, 392 (1959).
FINCH, S. C., FINCH, C. A.: Idiopathic hemochromatosis, an iron storage disease. Medicine (Baltimore) *34*, 381 (1955).
FINCH, S. C., HASKINS, D., FINCH, C. A.: Iron metabolism. Hematopoiësis following phlebotomy. Iron as a limiting factor. J. clin. Invest. *29*, 1078 (1950).
FLETCHER, J.: Iron transport in the blood. Proc. roy. Soc. Med. *63*, 1216 (1970).
FREIREICH, E. J., MILLER, A., EMERSON, C. P., ROSS, J. F.: The effect of inflammation on the utilization of erythrocyte and transferrin bound radioiron for red cell production. Blood *12*, 972 (1957).
FRUMIN, A. M., MILLER, E. E.: Exogenous hemochromatosis in sickle cell anemia. Gastroenterology *24*, 130 (1953).
GABRIO, B. W., SALOMON, K.: Distribution of total ferritin in intestine and mesenteric lymph nodes of horses after iron feeding. Proc. Soc. exp. Biol. Med. *75*, 124 (1950).
GALE, E., TORRANCE, J., BOTHWELL, T.: The quantitative estimation of total iron stores in human bone marrow. J. clin. Invest. *42*, 1076 (1963).
GELPI, A. P., ENDE, N.: An hereditary anemia with hemochromatosis. Studies of an unusual hemopathic syndrome resembling thalassaemia. Amer. J. Med. *25*, 303 (1958).
GOLDISH, R. J., AUFDERHEIDE, A. C.: Secondary hemochromatosis. II. Report of a case not attributable to blood transfusions. Blood *8*, 837 (1953).
GRACE, N. D. et al. (1970), cit. CRICHTON, 1975.
GRANICK, S.: Ferritin. IX. Increase of protein apoferritin in the gastrointestinal mucosa as a direct response to iron feeding. The function of ferritin in the regulation of iron absorption. J. biol. Chem. *164*, 737 (1946).

Granick, S.: Structure and physiological functions of ferritin. Physiol. Rev. *31*, 489 (1951).
Hahn, P. F., Granick, S., Bale, W. F., Michaelis, L.: Ferritin; conversion of inorganic and hemoglobin iron into ferritin iron in animal body. Storage function of ferritin iron as shown by radioactive and magnetic measurements. J. biol. Chem. *156*, 407 (1943).
Hallberg, L.: In: Gross, F. (ed.): Iron Metabolism. Springer, Berlin–Göttingen–Heidelberg 1964.
Hallberg, L., Hedenberg, L.: Desferrioxamine-induced urinary iron excretion in normal and iron-deficient subjects. Scand. J. Haemat. *4*, 11 (1967).
Hallberg, L., Hedenberg, L., Weinfeld, A.: Liver iron and desferrioxamine-induced urinary iron excretion. Scand. J. Haemat. *3*, 85 (1966).
Hampton, J. C.: An electron microscope study of the source and distribution of ferritin in hepatic parenchymal cells of the newborn rabbit. Blood *15*, 480 (1960).
Hampton, J. K., Jr.: Uptake of radioiron in tissue storage compounds in normal and hemosiderotic mice and its utilisation for erythropoiesis. Amer. J. Physiol. *176*, 20 (1954).
Harrison, P. M.: J. Med. Biol. (G. B.) *6*, 404 (1963), cit. Harrison, 1964.
Harrison, P. M.: Ferritin and haemosiderin. In: Gross, F. (ed.): Iron Metabolism. Springer, Berlin–Göttingen–Heidelberg 1964.
Harrison, P. M., Clegg, G. A., May, Keith: Ferritin structure and function. In: Jacobs, A., Worwood, M. (eds.): Iron in Biochemistry and Medicine, Vol. 2. Academic Press, London–New York 1980.
Harrison, P. M., Fischbach, F. A., Hoy, T. G., Haggis, G. H.: Ferric oxyhydroxide core of ferritin. Nature (Lond.) *216*, 1188 (1967).
Harrison, P. R., Conkie, D., Afforo, N.: In situ localization of globin messenger RNA formation. J. Cell. Biol. *63*, 402 (1974).
Heffernan, C. K., Jaswon, N.: A case of paroxysmal nocturnal haemoglobinuria associated with secondary haemochromatosis, a lower nephron nephrosis and a megaloblastic anaemia. J. clin. Path. *8*, 211 (1955).
Heilmeyer, L.: Ferritin. In: Wallerstein, R. O., Mettier, S. R. (eds.): Iron in Clinical Medicine, pp. 24–42. University of California Press, Berkeley 1958.
Heinrich, H. C., Brüggemann, J., Cable, E. E., Gläser, M.: Z. Naturforsch. *32*, 1023 (1977).
Hillman, R. S., Henderson, P. A.: Control of marrow production by the level of iron supply. J. clin. Invest. *48*, 454 (1969).
Higginson, J., Gerritsen, T., Walker, A. R. P.: Siderosis in the Bantu of Southern Africa. Amer. J. Path. *29*, 779 (1953).
Hofvander, Y.: Hematological investigations in Ethiopia with special reference to a high iron intake. Acta med. scand. Suppl. 494 (1968).
Howell, J., Wyatt, J. P.: Development of pigmentary cirrhosis in Cooley's anemia. Arch. Path. *55*, 423 (1953).
Jacobs, A., Miller, F., Worwood, M.: Ferritin in the serum of normal subjects and patients with iron deficiency and iron overload. Brit. med. J. *4*, 206 (1972).
Kaldor, I.: Studies on intermediary iron metabolism. XII. Measurement of the iron derived from water soluble and water insoluble non-haem compounds (ferritin and haemosiderin iron) in liver and spleen. Aust. J. exp. Biol. med. Sci. *36*, 173 (1958).
Kaltwasser, J. P., Werner, E.: Die radio-immunologische Messung von Ferritin in Serum und ihre klinische Bedeutung. Klin. Wschr. *55*, 1103 (1977).
Kent, G., Popper, H.: Secondary hemochromatosis. Its association with anemia. Arch. Path. *70*, 623 (1960).
Kerr, D. N. S., Muir, A. R.: A demonstration of the structure and disposition of ferritin in the human liver cell. J. Ultrastruct. Res. *3*, 313 (1960).
Kerr, L. M. H.: A method for the determination of non-haem iron in bone marrow. Biochem. J. *67*, 627 (1957).
Kleckner, M. S., Baggenstoss, A. H., Weir, J. F.: Iron-storage diseases. Amer. J. clin. Path. *25*, 915 (1955).

Kubanek, B.: Konservative Behandlung der Infekt- und Tumoranämie. Dtsch. med. Wschr. *104,* 238 (1979).

Lipschitz, D. A., Cook, J. D., Finch, C. A.: A clinical evaluation of serum ferritin as an index of iron stores. New Engl. J. Med. *290,* 1213 (1974).

Loftfield, R. B., Eigner, E. A.: The time required for the synthesis of ferritin molecule in rat liver. J. biol. Chem. *231,* 925 (1958).

Lukl, P., Wiedermann, B., Barborik, M.: Hereditäre Leptocytenanämie bei Männern mit Hämochromatose. Folia haemat. (Lpz.) *3,* 17 (1958).

Mason, D. Y., Taylor, C. R.: Distribution of transferrin, ferritin, and lactoferrin in human tissues. J. clin. Pathol. *31,* 316 (1978).

Mayet, F. G. H., Bothwell, T. H.: Hepatic iron stores in three different racial groups. S. Afr. J. med. Sci. *29,* 55 (1964).

Mazur, A., Green, S., Saha, A., Carleton, A.: Mechanism of release of ferritin iron in vivo by xanthine oxidase. J. clin. Invest. *37,* 1809 (1958).

McCrea, P. C.: Marrow iron examination in the diagnosis of iron deficiency in rheumatoid arthritis. Ann. rheum. Dis. *17,* 89 (1958).

Meier, W., Beneke, G., Ahlert, G.: Die Veränderungen von Ferritin in Relation zum Hämosiderin und Gesamteisen der Leber beim Mann in den verschiedenen Altersstufen. Z. ges. inn. Med. *14,* 1065 (1959).

Merker, H.: Cytochemie der Blutzellen. In: Heilmeyer, L. (ed.): Blut und Blutkrankheiten, Part 1, p. 130. Springer, Berlin–Heidelberg–New York 1968.

Millar, J. A., Goldberg, A., Cumming, R. L. C.: Studies in ferritin synthesis in relation to iron absorption. In: Hallberg, L., Harwerth, H. G., Vannotti, A. (eds.): Iron Deficiency. Academic Press, London–New York 1970.

Moeschlin, S.: Klinik und Therapie der Vergiftungen, pp. 378–401. 3rd ed. Thieme, Stuttgart 1959.

Morgan, E. H., Walters, M. N. I.: Iron storage in human disease. J. clin. Path. *16,* 101 (1963).

Noyes, W. D., Bothwell, T. H., Finch, C. A.: The role of the reticuloendothelial cell in iron metabolism. Brit. J. Haemat. *6,* 43 (1960).

Oertel, J., Bombik, B. M., Stephan, M., Gerhartz, H.: Blut *37,* 113 (1978).

Oliver, R. A.: Siderosis following transfusions of blood. J. Path. Bact. *77,* 171 (1959).

Olsson, K. S.: Iron stores in normal man and male blood donors, as measured by desferrioxamine and quantitative phlebotomy. Acta med. scand. *192,* 401 (1972).

Pirzio-Biroli, G., Finch, C. A.: Iron absorption. III. The influence of the iron stores on iron absorption in the normal subject. J. Lab. clin. Med. *55,* 216 (1960).

Pirzio-Biroli, G., Bothwell, T. H., Finch, C. A.: Iron absorption. II. The absorption of radioiron administered with a standard meal in man. J. Lab. clin. Med. *51,* 37 (1958).

Ploem, J. E., de Wall, J., Verloop, M. C., Punt, K.: Sideruria following a single dose of desferrioxamine-B as a diagnostic test in iron overload. Brit. J. Haemat. *12,* 396 (1966).

Rath, C. E., Finch, C. A.: Sternal marrow hemosiderin. A method for the determination of available iron stores in man. J. Lab. clin. Med. *33,* 81 (1948).

Richter, G. W.: A study of hemosiderosis with the aid of electron microscopy. J. Exp. Med. *106,* 203 (1957).

Richter, G. W.: The nature of storage iron in idiopathic hemochromatosis and in hemosiderosis. Electron optical, chemical and serologic studies on isolated hemosiderin granules. J. Exp. Med. *112,* 551 (1960).

Roth, O., Jasinski, B., Bidder, H.: Das Gewebeeisen beim Menschen bei normalen und pathologischen Zuständen. Helv. med. Acta *18,* 159 (1951).

Saarinen, U. M., Siimes, M. A.: Pediat. Res. *13,* 143 (1979).

Schairer, E., Rechenberger, J.: Das Leber- und Milzeisen bei Mann und Frau in verschiedenen Lebensaltern. Virchows Arch. path. Anat. *315,* 309 (1948).

Shoden, A., Sturgeon, P.: Hemosiderin. I. A physico-chemical study. Acta haemat. (Basel) *23,* 376 (1960).

SHODEN, A., GABRIO, B. W., FINCH, C. A.: The relationship between ferritin and hemosiderin in rabbits and man. J. biol. Chem. *204*, 823 (1953).

SIIMES, M. A., ADDIEGO, J. E., DALLMAN, P. R.: Brit. J. Haemat. *28*, 7 (1974).

SMITH, N. J., ROSELLO, S., SAY, M. B., YEYA, K.: Iron storage in the first five years of life. Pediatrics *16*, 166 (1955).

SPEYER, B. E., FIELDING, J.: Ferritin as a cytosol iron transport intermediate in human reticulocytes. Brit. J. Haemat. *42*, 255 (1979).

STURGEON, P., SHODEN, A.: Mechanism of iron storage. In: GROSS, F. (ed.): Iron Metabolism, p. 121. Springer, Berlin–Göttingen–Heidelberg 1964.

VÁCHA, J., DUNGEL, J., KLEINWÄCHTER, V.: Determination of heme and non-heme iron content of mouse erythropoietic organs. Exp. Hemat. *6*, 718 (1978).

VAN DER HEUL, C., KROOS, M. J., VAN EIJK, H. G.: Binding sites of iron transferrin on rat reticulocytes. Biochim. Biophys. Acta *511*, 430 (1978).

VAN KREEL, B. K., VAN EIJK, H. G., LEIJNSE, B.: The isoelectric fractionation of rabbit ferritin. Acta haemat. (Basel) *47*, 59 (1972).

VRIES, A., IZAK, G.: Variations du taux de l'hémosidérine dans la moelle osseuse dans différentes conditions hématologiques. Rev. hémat. *10*, 657 (1955).

WALLERSTEIN, R. O., POLLYCOVE, M.: Bone marrow hemosiderin and ferrokinetic patterns in anemia. Arch. Intern. Med. *101*, 418 (1958).

WALTERS, G. O., JACOBS, A., WORWOOD, M., TREVETT, D., THOMSON, W.: Gut *16*, 188 (1975).

WEINFELD, A.: Storage iron in man. Acta med. scand. Suppl. 427 (1964).

WEINFELD, A.: Iron stores. In: HALLBERG, L., HARWERTH, H. G., VANNOTTI, A. (eds.): Iron Deficiency. Academic Press, London–New York 1970.

WHIPPLE, G. H., BRADFORD, W. L.: Mediterranean disease — thalassemia (erythroblastic anemia of Cooley); associated pigment abnormalities simulating hemochromatosis. J. Pediat. *9*, 279 (1936).

WÖHLER, F.: Zur Physiologie und Pathologie des Speichereisens. III. Dtsch. med. Wschr. *80*, 30 (1955).

WÖHLER, F.: Über das Depoteisen, Ferritin. In: KEIDERLING, W. (ed.): Eisenstoffwechsel. Thieme, Stuttgart 1959.

WÖHLER, F.: Über die Natur des Hämosiderins. Acta haemat. (Basel) *23*, 342 (1960).

WÖHLER, F.: Über die Freisetzung von Eisen aus dem menschlichen Organismus durch eine Hydroxamsäureverbindung. Proc. 8th Congr. Europ. Soc. Haemat., p. 244. Karger, Basel–New York 1961.

WÖHLER, F.: Ferritin and haemosiderin. Germ. med. Mth. *9*, 377 (1964).

WORWOOD, M.: Serum ferritin. In: JACOBS, A., WORWOOD, M. (eds.): Iron in Biochemistry and Medicine, Vol. 2. Academic Press, London–New York 1980.

WYATT, J. P.: Patterns of pathological iron storage. II. Exogenic siderosis in chronic anemia due to prolonged oral iron medication. Arch. Path. *61*, 56 (1956).

YAMADA, H., GABUZDA, T. G.: Interaction between rabbit erythroblast ferritin and normal plasma proteins. Blood *43*, 875 (1974a).

YAMADA, H., GABUZDA, T. G.: Erythroblast ferritin: synthesis, structure, and function in developing erythroid cells. J. Lab. clin. Med. *83*, 478 (1974b).

CHAPTER 9

IRON LOSS AND IRON REQUIREMENT

Measurement of iron loss is technically difficult. Chemical iron balances require meticulous technique to avoid contamination, and they do not allow the differentiation between nonabsorbed iron and iron excreted through the gastrointestinal tract. Isotope techniques require knowledge of the distribution of the radioiron within the organism and the extent of the various iron pools.

The daily iron loss has been variously estimated as between 0.5 and 1.3 mg (Dubach et al., 1955; Finch, 1959; Saito et al., 1964). The most comprehensive studies of iron loss are those of Green et al. (1968) using the long-lived isotope ^{55}Fe. These workers gave an intravenous dose of ^{55}Fe to various groups of subjects and measured the radioactivity in the blood at regular intervals thereafter (Fig. 9/1).

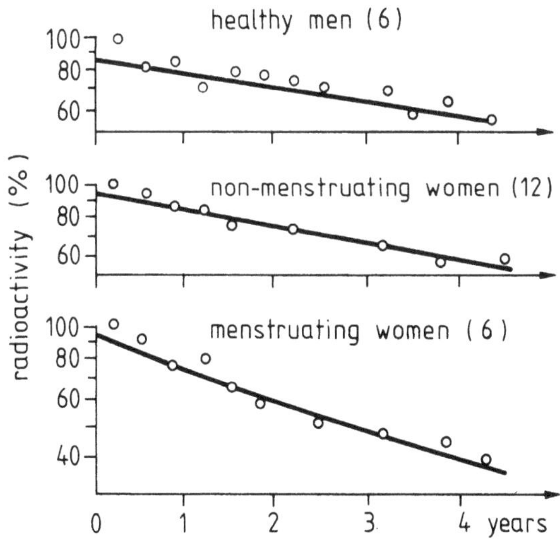

Fig. 9/1. Iron loss of the organism. Decline in red cell radioactivity after the intravenous injection of ^{55}Fe. The annual decline in the activity is about 8% in males, 11% in non-menstruating females, and 20% in menstruating females (after Finch, C. A.: J. clin. Invest. *38*, 392, 1959)

During the first year the level of radioactivity decreased as a consequence of the mixing of the radioactive isotope with the slowly exchanging iron pool of the body. Later the activity decreased more slowly at an exponential rate, and from this the daily iron loss was calculated (Fig. 9/2). The groups studied were from various parts of the world, living in different climatic and working conditions. From Seattle, in the United States, a temperate zone, there was a group of men with sedentary

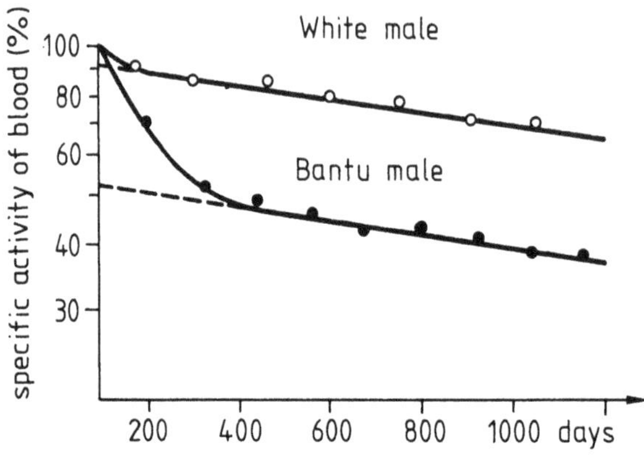

Fig. 9/2. Decline in the specific activity of the blood as a function of time in a white male with normal iron stores and in a Bantu male with hypersiderosis (after Green, R. et al: Amer. J. Med. *45*, 336, 1968)

occupations; from Venezuela, men working in a more humid climate; and from South Africa, Indians working in a very humid environment. These three groups all showed very similar daily iron losses, 0.95, 0.9, and 1.02 mg, respectively (Fig. 9/3). The fourth group consisted of South African Bantus in whom hemosiderosis is common; the average daily iron loss in these was 2.2 mg, and in some individuals as high as 3–4 mg daily (see Fig. 9/3). The iron loss therefore is related to the extent of iron stores rather than to any environmental or physical condition (Fig. 9/4). The earlier studies of Dubach et al. (1955) are in keeping with this, as these workers found iron loss to be reduced in iron-deficiency states (Figs. 9/5 and 9/6).

About 0.25 mg of the iron loss is in the bile, but only about 40 μg of this amount originates directly from the blood serum. The iron content of the bile is in proportion to the amount of iron in the liver (Bothwell, 1970), and the parenchymal or reticuloendothelial cells of the liver are probably the source of most of the iron lost in the bile.

Some of the fecal iron loss is from desquamated intestinal mucosal cells. This can be demonstrated following the intravenous injection of radioiron. During the first few days after the radioiron administration, the radioactivity in the feces reaches a

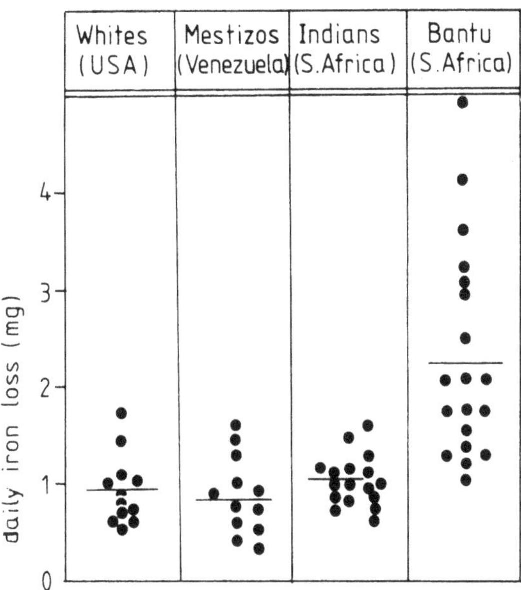

Fig. 9/3. Average daily iron loss in different population groups (after Green, R. et al.: Amer. J. Med. *45*, 336, 1968)

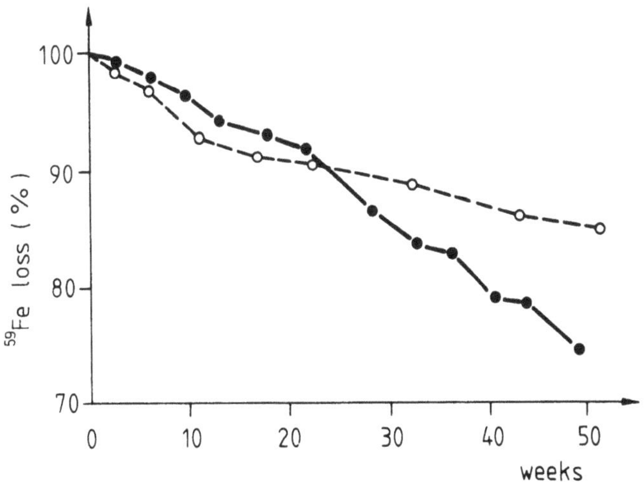

Fig. 9/4. Radioactive iron loss measured in a total body counter following the intravenous injection of ^{59}Fe in patients with hemochromatosis and healthy controls. The iron loss increased after week 22 in the patients with hemochromatosis (after Dymock, I. W. et al.: Proc. roy. Soc. Med. *63*, 1227, 1970). ●–●–● hemochromatosis (n = 7); ○– –○– –○ controls (n = 6)

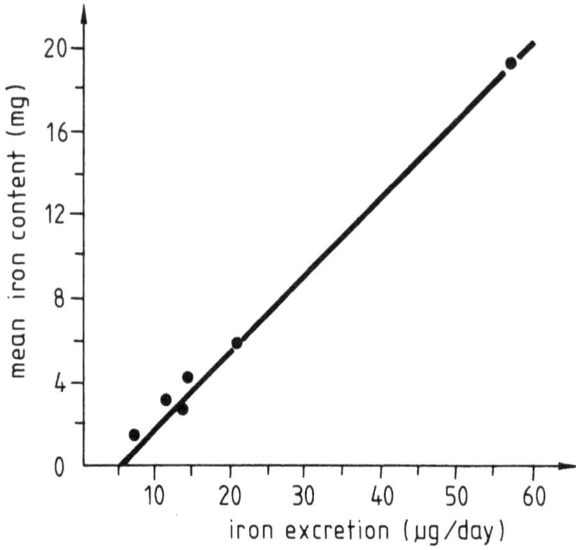

Fig. 9/5. Correlation between the mean iron content of the organism and iron excretion (after Chappelle, E. et al.: Amer. J. Physiol. *182*, 390, 1955)

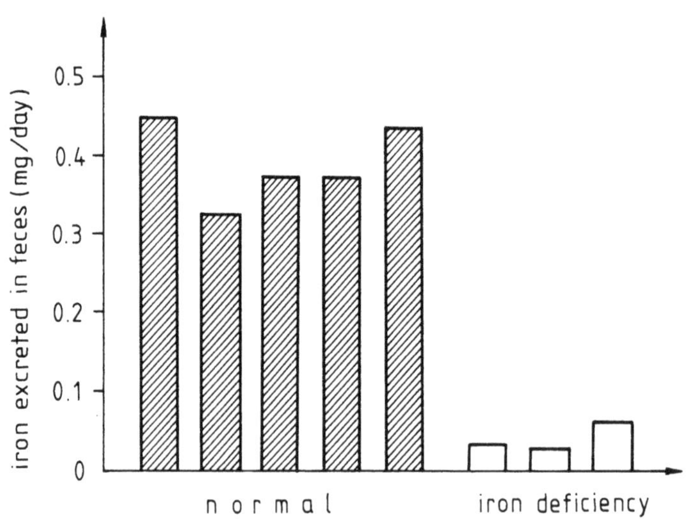

Fig. 9/6. Excretion of iron in the feces in healthy subjects and in iron-deficiency disease (after Dubach, R. et al.: J. Lab. clin. Med. *45*, 599, 1955)

peak and then declines rapidly (Fig. 9/7). The earlier peak correlates with the time during which the radioiron is incorporated into the epithelial cells in the crypts of Lieberkühn. These cells migrate up to be desquamated from the tips of the villi (Conrad et al., 1964). It has been estimated that about 0.1 mg of iron is excreted daily with the desquamated cells, and a further 0.4 mg is lost from traces of blood in the

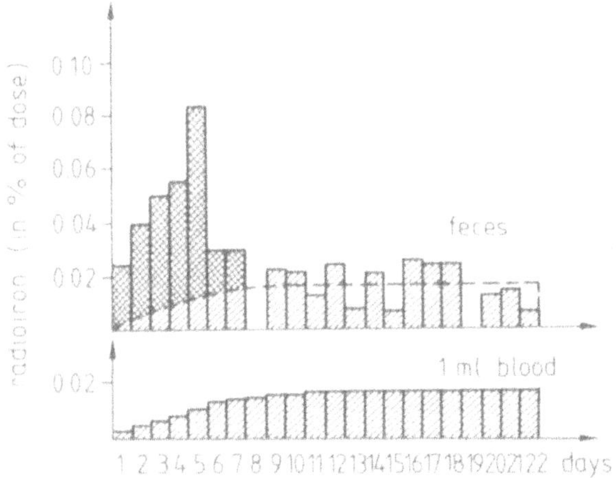

Fig. 9/7. Radioactivity of the feces after the intravenous administration of labeled iron. The cross-hatched parts of the columns indicate the iron excreted with the desquamated epithelial cells, the rest of the activity can be ascribed to blood loss (upper part of the figure). The lower part of the figure serves for comparison. The individual columns indicate the radioactivity present in 1 ml of blood (after Green, R. et al.: Amer. J. Med. *45*, 336, 1968)

feces (Bothwell, 1970). Part of the increased loss in iron overload is due to the epithelial cells being richer in iron.

Normally the urinary excretion of iron is less than 0.1 mg (Dagg et al., 1966; Bothwell, 1970; Bernát, 1971). In iron overload this amount may be increased. Bernát has noted an increase in urinary iron excretion in association with thermal injury, the mean daily urinary iron being 328 ± 152 μg in the patients, as compared with normal values of 60 ± 22 μg.

Urinary iron losses may be significantly increased in patients with a nephrotic syndrome, although not in excess of 0.5 mg/day (Wilting et al., 1972).

Some of the highest urinary losses are found in conditions such as paroxysmal nocturnal hemoglobinuria, where one of the diagnostic features of the condition is a deposit of hemosiderin in the urine.

There has been some debate as to the significance of iron loss from the skin. Some workers have claimed losses of milligrams of iron per day, particularly under conditions of excessive perspiration (Foy and Kondi, 1957; Hussain et al., 1960). These high figures probably are attributable to contamination of the skin, and most

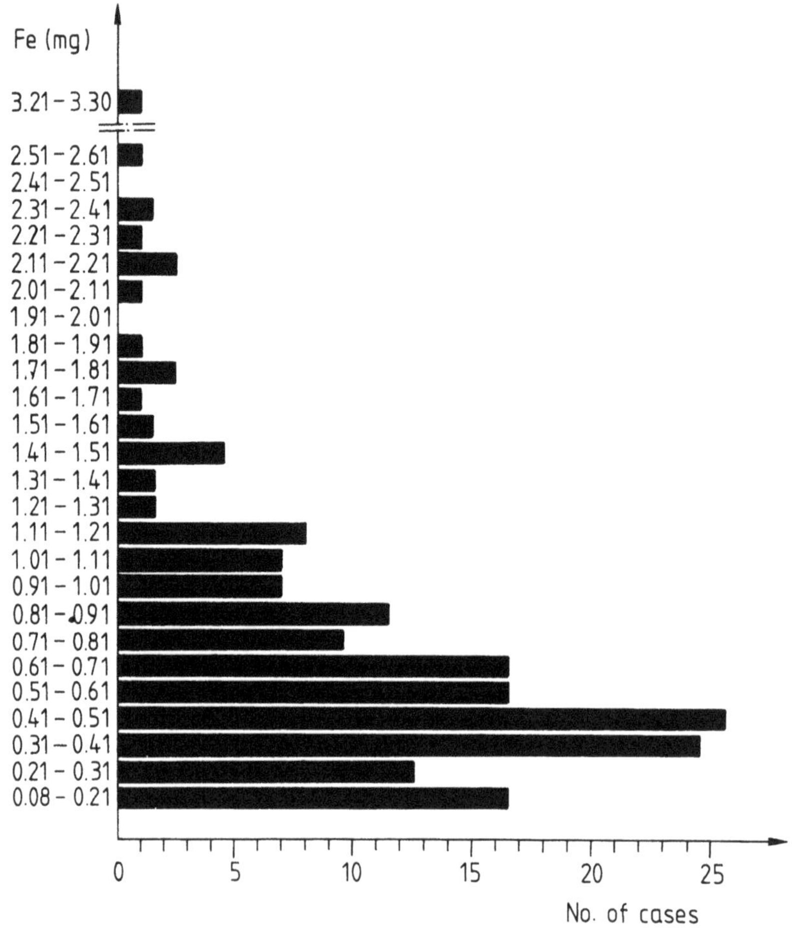

Fig. 9/8. Menstrual loss of iron expressed as daily iron loss (after Frenchman, R. and Johnston, F. A.: J. Amer. diet. Ass. *25,* 217, 1949)

workers agree that there is only minimal iron excretion in sweat (Dubach et al., 1955; Green et al., 1968), and any iron loss that does occur is due to desquamation of the skin and amounts to 0.1–0.6 mg/day (Bothwell, 1970).

IRON LOSS IN WOMEN

The minimal obligatory losses in adult males and females are the same, but women during reproductive years have additional physiological iron losses associated with menstruation, pregnancy, delivery, and breast-feeding.

MENSTRUAL IRON LOSS

Measurement of menstrual blood loss in a random population of women showed that the average loss was 34 ml per period, corresponding to a daily iron loss of 0.5 mg (Hallberg and Nilsson, 1964). Many women, however, lose more (Fig. 9/8), and these heavy losses are a significant factor in the development of iron deficiency (Rybo, 1970).

Later studies by Hallberg et al. (1966a) in 476 women indicated a higher mean blood loss (43.4 ± 2.3 ml) corresponding to a daily loss of 0.8 mg of iron (Fig. 9/9). These authors also measured the hemoglobin concentration, the plasma iron, and the total iron-binding capacity and concluded that the normal upper limit of menstrual blood loss was between 60 and 80 ml. A loss of more than 80 ml was associated with a significant increase in anemia, a lowering of plasma iron (Fig. 9/10), an increase in TIBC (Fig. 9/11), and a loss of 80 ml was therefore considered the limit of tolerance. Losses of this degree were found in 11% of the women examined.

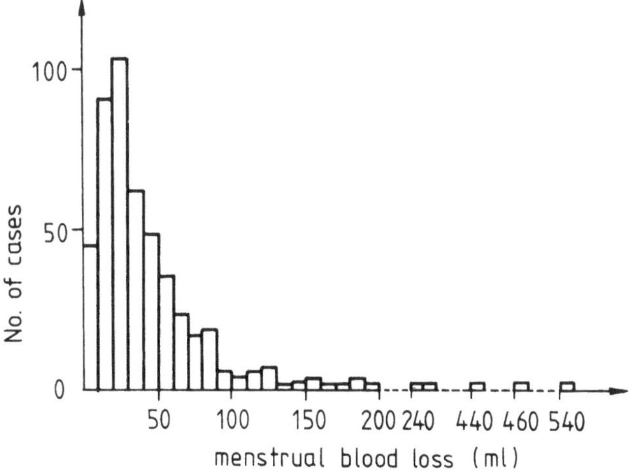

Fig. 9/9. Distribution of menstrual blood loss in a population of women. Mean loss 43.4 ± 2.3 ml (after Hallberg, L. et al.: Int. Soc. Haemat. Sydney, 1966a)

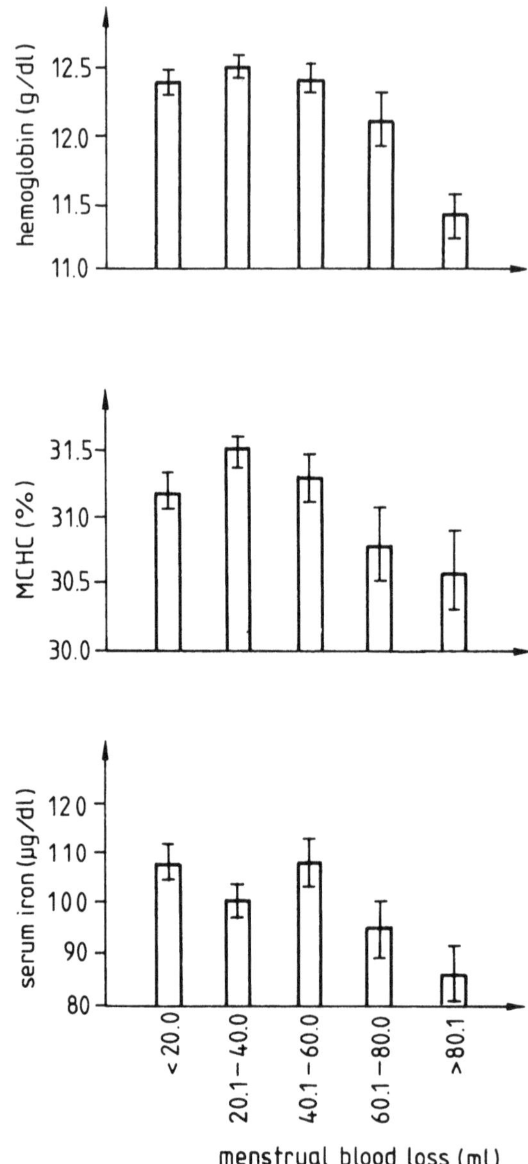

Fig. 9/10. Correlation of menstrual blood loss, hemoglobin level, mean hemoglobin concentration of red cells (MCHC) and plasma iron level (after Rybo, G., in: Hallberg, L. et al.: Iron Deficiency. Academic Press, London–New York 1970)

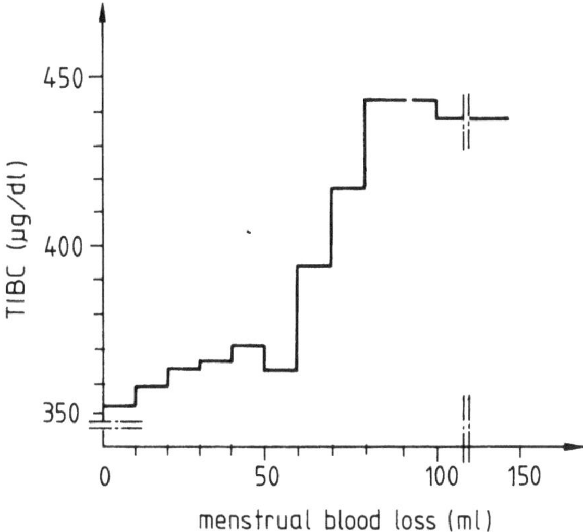

Fig. 9/11. Correlation between menstrual blood loss and total iron-binding capacity of the plasma. Data on 420 individuals (after Rybo, G., in: Hallberg, L. et al.: Iron Deficiency. Academic Press, London–New York 1970)

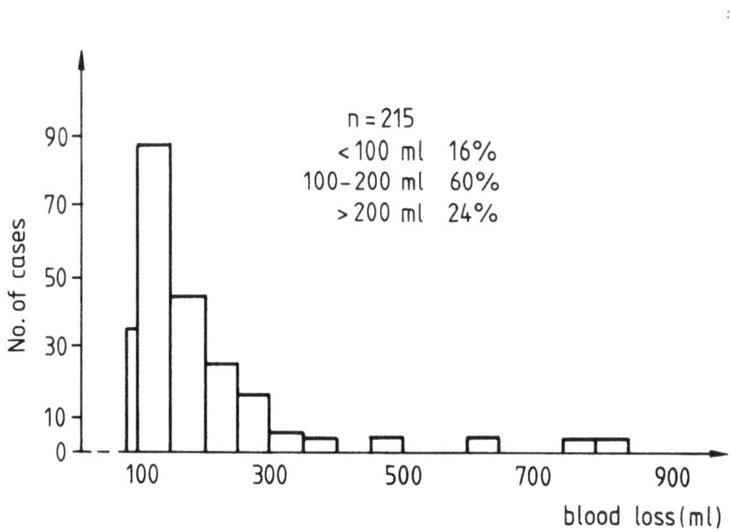

Fig. 9/12. Distribution of the volume of menstrual blood loss in 215 menorrhagic women (after Rybo, G., in: Hallberg, L. et al.: Iron Deficiency. Academic Press, London–New York 1970)

In a study of women with menorrhagia, 215 of the 268 lost more than 80 ml per period, and in a quarter of them the loss exceeded 200 ml (Fig. 9/12) (Rybo, 1970). This again emphasizes the significance of menorrhagia in the development of iron deficiency in women (Göltner, 1975).

Oral contraceptives can reduce the menstrual blood loss by 50–60%, occasionally even more (Heinrich, 1970; Rybo, 1970).

THE IRON REQUIREMENT OF PREGNANCY

During pregnancy, iron is required for the fetus and the placenta, and for the increase in the total red cell mass that occurs in the mother. The fetus and the placenta between them contain 270–300 mg of iron (Sturgeon, 1959, 1966; Widdowson, 1968; Pritchard and Scott, 1970), this being accumulated largely during the last trimester (Bothwell et al., 1958; Widdowson, 1968; Pritchard and Scott, 1970) (Figs. 9/13 and 9/14).

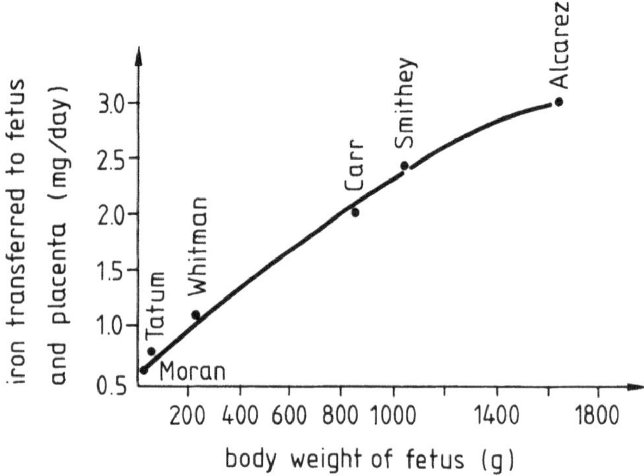

Fig. 9/13. The daily amount of iron reaching the fetus and placenta from the maternal organism as a function of fetal body weight (after Pritchard, J. A. and Scott, D. E., in: Hallberg, L. et al.: Iron Deficiency. Academic Press, London–New York 1970)

During pregnancy, the mother's blood volume begins to rise even in the first trimester, but the increase is accelerated in the second trimester and the beginning of the third trimester. The increase in total red cell mass may be as much as 32% with an additional requirement for iron of about 500 mg (Pritchard, 1965; de Leeuw et al., 1966).

The total iron requirement for a pregnancy has been estimated as between 0.8 and 1.0 gram, and if the blood loss during delivery is also included, the total demand increases to 1.275 grams (Rybo, 1973). The increased maternal red cell mass should, however, not be taken into this calculation, and the obligatory daily loss could also be excluded if one is to get a realistic assessment of the net cost of pregnancy (Fig. 9/14).

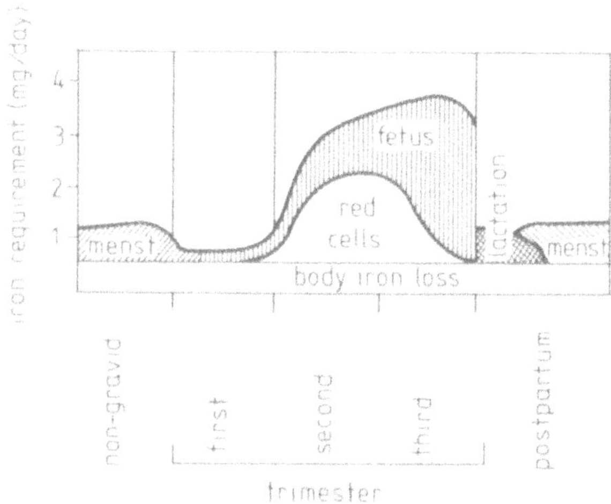

Fig. 9/14. Iron requirement of adult nonpregnant and pregnant women. During pregnancy additional iron is needed for the supply to the fetus and placenta and for the increase in maternal red cell volume (after Bothwell, T. H. and Finch, C. A.: Iron Metabolism. Little, Brown and Co., Boston 1962)

The daily iron supply from food often makes the iron balance during pregnancy very precarious. Studies of dietary iron intake in pregnant women suggest an intake of about 13–14 mg/day (Darby et al., 1955; Chanarin et al., 1968), whereas the National Academy of Sciences, Washington, 1968, has recommended an intake of 18 mg of iron per day for pregnant women.

Blood loss at parturition has been estimated variously as 500–600 ml, corresponding to the loss of about 250 mg of iron (Newton et al., 1961; Newton, 1966; Pritchard et al., 1962; de Leeuw et al., 1968).

The iron in milk accounts for about 0.5–1.0 mg of iron per day (Wintrobe, 1975), but this is compensated for by the temporary absence of menstrual loss during lactation.

In normal pregnancy the hemoglobin level and hematocrit show a gradual decline up to the 16th to 22nd week, after which the value becomes stabilized at approximately 11 g/dl or a hematocrit of 32–34%. At about 32–39 weeks the values may even rise a little. After delivery the values revert to normal.

Plasma iron concentration decreases in a high proportion of pregnant women, and the iron-binding capacity increases (Verloop et al., 1958, 1959). A proportion of pregnant women develop a hypochromic microcytic anemia, and in the majority of cases this is the consequence of iron deficiency. It is noteworthy that the plasma iron level of Bantu women does not decrease significantly during pregnancy, and they usually do not develop anemia (Gerritsen and Walker, 1954), presumably because of the high iron content of the Bantu diet. During pregnancy serum ferritin concentrations steadily decline (Fenton et al., 1977; van Eijk et al., 1978; Taft et al. 1978; Német et al., 1981), and during the later stages of pregnancy ferritin concentrations are significantly lower in women who have not received iron therapy.

IRON REQUIREMENTS FOR NEWBORNS, INFANTS, AND CHILDREN

The mean red blood cell count of the umbilical blood at birth is 4.5 million/μl and the hemoglobin 15.6 g/dl (Künzer, 1962). About 12 hours after birth the count of the infant is about 5.5 million and the hemoglobin concentration 20 g/dl (Künzer, 1957; Betke, 1958) (Fig. 9/15). The level of the count, however, depends upon how soon the umbilical cord is clamped. If it is not clamped too early, at the beginning of the pulmonary circulation the newborn "absorbs" a considerable quantity of blood from the placenta, as much as 40 ml being pumped over within 10–15 seconds. Some of the rise in red cell count and hemoglobin concentration is, however, due to hemoconcentration as the result of plasma flow into the tissues after birth. The hemoglobin level gradually drops from the second week of birth, the value reaching its lowest level at about the tenth postnatal week (Guest and Brown, 1957). On the average, in about the third month the red blood cell count is 3.8 million/μl and the hemoglobin 11.5 g/dl. This postnatal decline is not affected by altitude (2400 meters), although at 4 months of age the values are higher than those normally found at sea level. It appears therefore that the stimulus from increased erythropoietin production at high altitudes starts to act some time between the third and fourth months of age (Tafari and Habte, 1972).

The hemoglobin level for premature infants and small-for-dates babies may go as low as 8–10 g/dl (Betke, 1958, 1964).

By the end of the first year of life the average red blood cell count is 4.5–4.7 million, and the hemoglobin 12–12.5 g/dl (Guest and Brown, 1957) (Fig. 9/16). Throughout childhood the values tend to remain low as compared with the adult values, which are reached only at puberty (Peter, 1964) (Fig. 9/17). Until puberty there is no sex difference in the red cell count and hemoglobin, but from then on the hemoglobin levels of males are higher (mean 15.5 g/dl) than those of females (mean 13.5 g/dl).

There are certain morphological differences between the red blood cells of neonates and adults. The erythrocytes are larger, their mean diameter being 8.1–8.7 μ (Weicker et al., 1953; Betke, 1964; Wintrobe, 1975) (Fig. 9/18), the MCV is 105 μ^3, the mean hemoglobin content 37.6 pg (Fig. 9/19), and the mean hemoglobin concentration 35.7%. Anisocytosis is more pronounced, and the polychromasia may be conspicuous (Künzer and Jacobi, 1970).

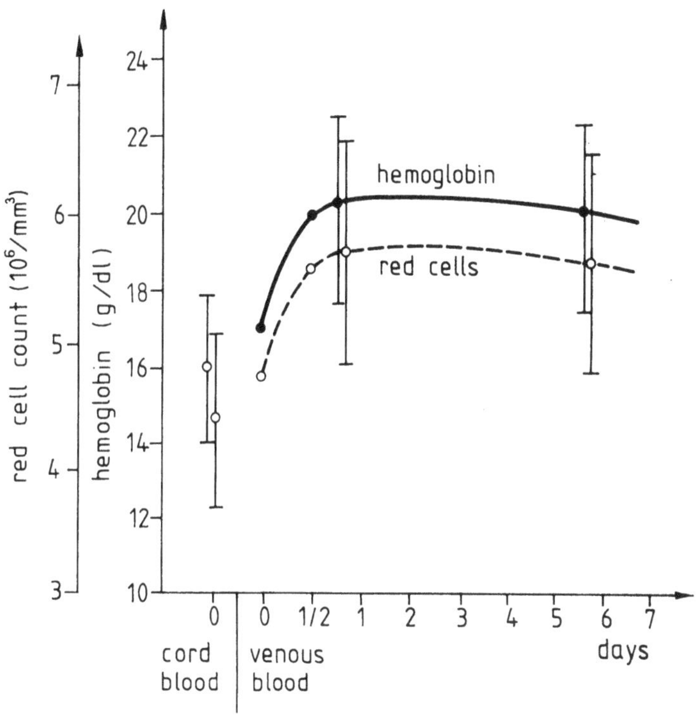

Fig. 9/15. Changes in the hemoglobin level and red cell count at birth and in the first days of life (after Künzer, W. and Jacobi, H., in: Heilmeyer, L.: Blut und Blutkrankheiten. Springer, Berlin–Heidelberg–New York 1970)

The life-span of the newborn's red cells may be reduced somewhat, the average for mature newborns being estimated at 70–100 days (Vest, 1962) and for premature infants 70–80 days (Garby et al., 1964) (Fig. 9/20). There are also serological (Seelemann, 1952), electron microscopic (Dervichian et al., 1952), and electrophoretic (Sachtleben et al., 1961) differences between neonatal and adult red blood

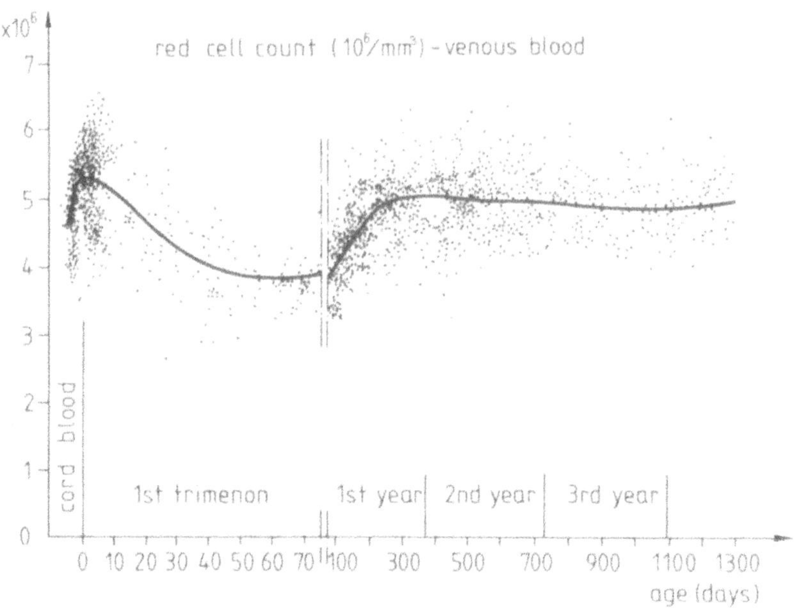

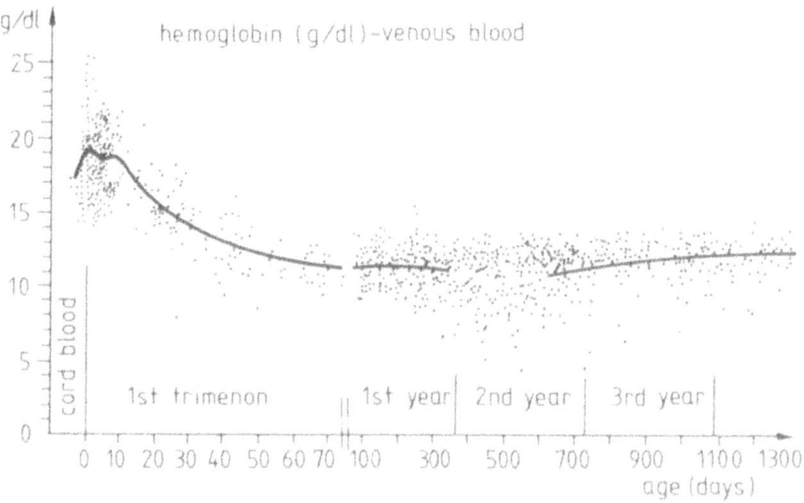

Fig. 9/16. Changes in the red cell count and hemoglobin level during the first 3 years (after Guest, G. and Brown, E.: Amer. J. Dis. Child. *93*, 486, 1957)

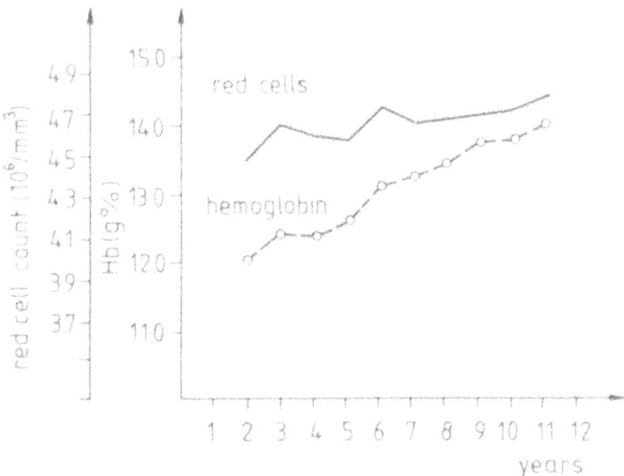

Fig. 9/17. Changes in the red cell count and hemoglobin level between 2 and 11 years of age (mean values of capillary blood of 496 healthy children) (after Peter, H., 1964)

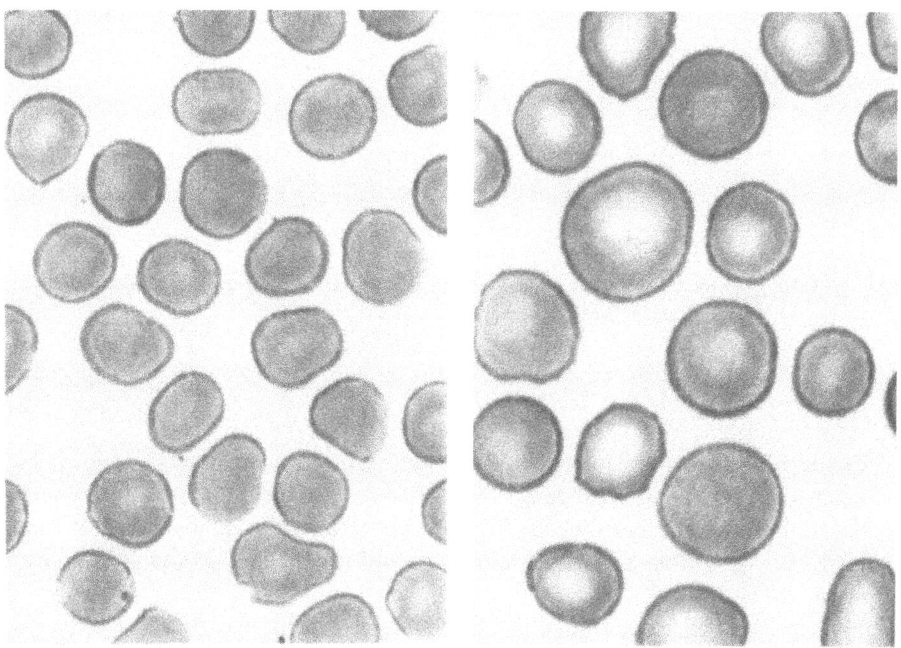

Fig. 9/18. Red cells of adult man (left) and neonate (right) (after Künzer, W. and Jacobi, H., in: Heilmeyer, L.: Blut und Blutkrankheiten. Springer, Berlin–Heidelberg–New York 1970)

corpuscles. Other differences are found in their spontaneous hemolysis (Sjölin, 1954; Zipursky et al., 1960), osmotic resistance (Kleihauer, 1965), mechanical resistance (Sjölin, 1954; Aalam, 1959; Schubothe et al., 1961; Kleihauer, 1965), in their nonelectrolytic permeability (Hollán and Szelényi, 1967), and in the activity of several red cell enzymes (Künzer and Jacobi, 1970).

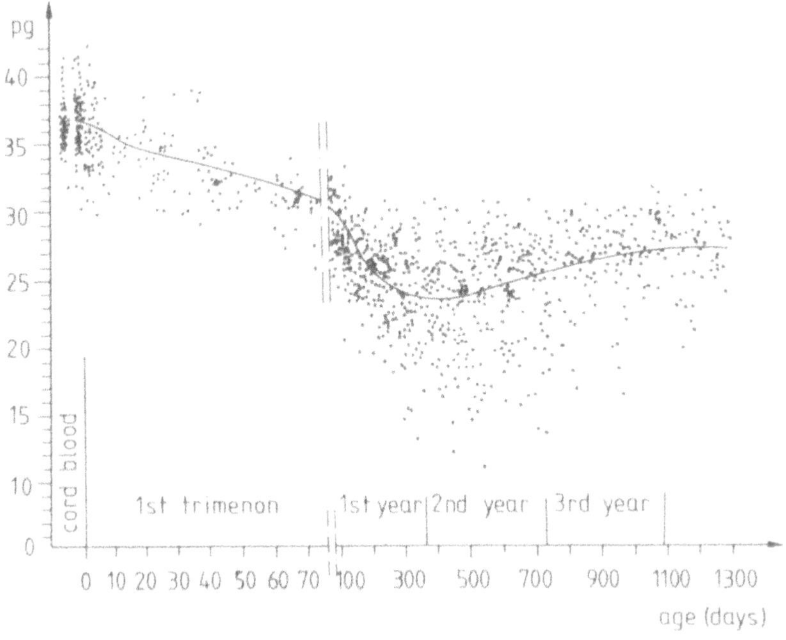

Fig. 9/19. Mean hemoglobin content (pg) of the red cells in the first 3 years of life (after Guest, G. and Brown, E.: Amer. J. Dis. Child. *93*, 486, 1957)

The total iron content of the healthy mature newborn is about 250–280 mg (Gubler, 1956; Saddi and Schapira, 1970). This is about 6% of the total iron content of the adult, and in the infant about 75% of the iron can be found in the hemoglobin, 10% in the myoglobin and tissue enzymes, and about 15% in the storage iron (Josephs, 1953).

The iron of the fetus is mainly accumulated during the last few weeks of intrauterine life (Fig. 9/21), hence premature birth results in a reduction of the iron stores (Schulman, 1961) (Fig. 9/22). Severe maternal iron deficiency may also affect the newborn (Woodruff and Bridgeforth, 1953; Sisson and Lund, 1958).

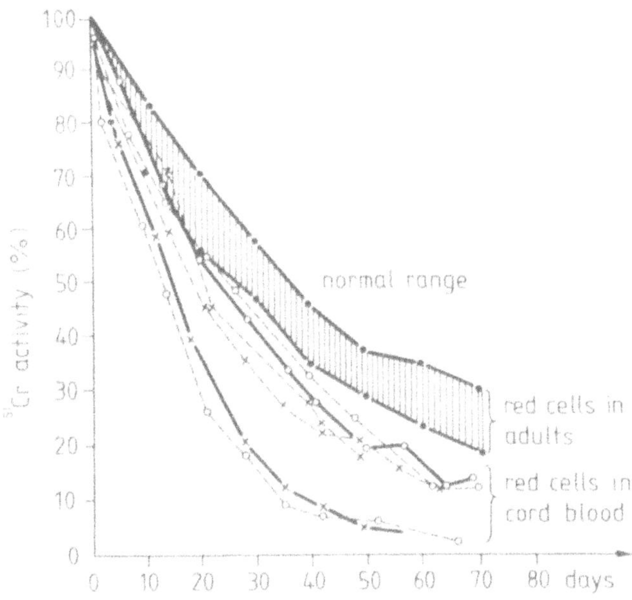

Fig. 9/20. Life-span of red cells in the neonate (after Hollingsworth, J.: J. Lab. clin. Med. *45*, 469, 1955)

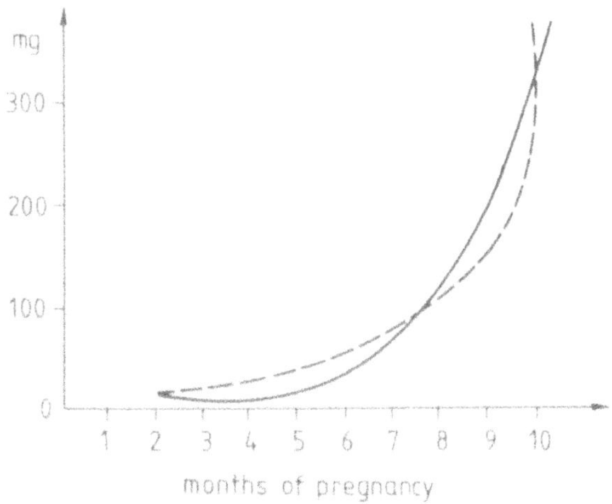

Fig. 9/21. Increase in the iron content of the fetal organism. – – – Jerlow (1934); ——— Iob and Swanson (1938)

Fenton et al. (1977) and Kelly et al. (1978) found significantly lower ferritin concentrations in sera from babies born to mothers with pre-delivery ferritin concentrations of less than 12 μg/l. Rios et al. (1975) did not find lower cord blood concentrations in babies born to mothers with pre-delivery ferritin concentrations of less than 9 μg/l but only 6 babies were studied. However, fetal iron stores may be reduced when maternal iron stores are absent (Worwood, 1980).

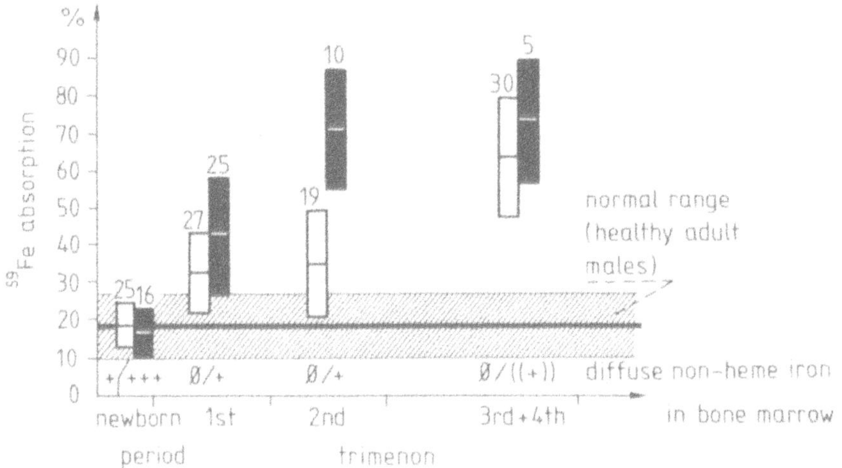

Fig. 9/22. Iron absorption in infants. The results reflect relative iron deficiency, which is more marked in premature than in full-term infants (after Schäfer, K. H., in: Hallberg, L. et al.: Iron Deficiency. Academic Press, London–New York 1970). Birth weight: □ > 2500 g; ■ < 2500 g

The iron requirement for the first year of life is in the region of 160–280 mg (Erlandson, 1962; Saddi and Schapira, 1970). The high hemoglobin at birth, and early destruction of excess red cells supplement the supply of iron for the growing infant for about the first 4–5 months. Beyond this the supply may become critical, especially where feeding with milk alone is continued. In the first postpartum week the iron concentration of the mother's milk ranges from about 130 to 150 μg/dl, but this level declines gradually during lactation, and by 4–6 months it is reduced to about 90 μg/dl. The iron content of cow's milk is lower (about 70 μg/dl) and that of goat's milk only about 25 μg/dl (Schäfer, 1953). Only about 10% of the iron in milk is absorbed (Moore, 1955; Schulz and Smith, 1958), which means that only 0.15 mg iron is absorbed from a liter of mother's milk per day. This means that by the end of the fifth or sixth month the iron stores of most infants are exhausted (Fig. 9/23), and subsequently there is a tendency to develop iron-deficiency anemia. The anemia of infancy is therefore almost exclusively a true nutritional iron deficiency. It can be prevented by the introduction of iron-supplemented cereals.

Further factors concerned in the production of iron deficiency of infants are multiple birth, peri- and postnatal bleeding such as fetomaternal bleeding (Chown,

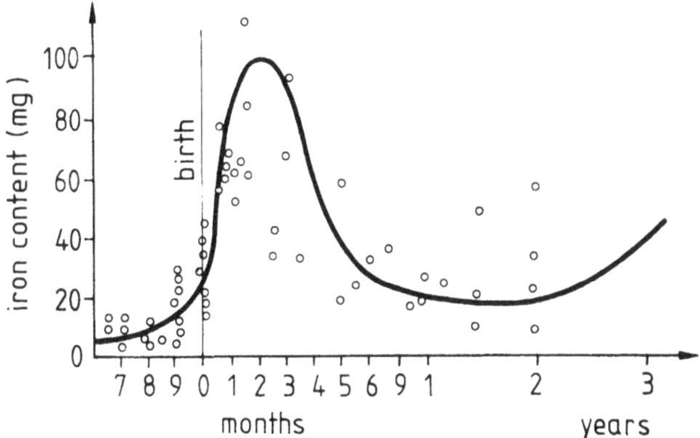

Fig. 9/23. Iron reserves of infants in the last trimester of pregnancy and in the first 2–3 years after delivery estimated from the liver iron content. During the first 3 months of extrauterine life the hemoglobin concentration of the blood falls and the iron content of the liver increases. From the fifth or sixth month up to the second year of life the iron reserve is greatly reduced (after Lintzel, W. et al.: Z. ges. exp. Med. *113*, 591, 1944)

1955; Kirkman and Riley, 1959), hemophilia and occult intestinal hemorrhage (Hoag et al., 1961), or early ligation of the umbilical cord (Wilson et al., 1941). A delay in clamping the cord for 3 minutes leads to a 58% increase in the total red cell volume (Yao et al., 1969); early ligation deprives the infant of this valuable iron supply.

IRON REQUIREMENTS FOR GROWTH

From birth to maturity the iron content of the body increases from 250–280 mg to about 3–4 g. This averages out at a yearly increase of 200 mg, or 0.6 mg of iron daily. However, the rate of body growth is uneven; it is most rapid during the first year, then slows down and is relatively steady between the second and eleventh years of life, and is then followed by an adolescent growth spurt that is particularly rapid between 11 and 14 years in the case of girls and 14–17 years in boys (Fig. 9/24) (Jackson et al., 1945).

In the first year of life the body weight increases from an average of 3.5 kg to 10.5 kg. One kg of weight increase is associated with 35–45 mg of iron requirement; therefore, a 7 kg weight increase requires about 280 mg of iron, i.e., a daily requirement of 0.7–0.8 mg.

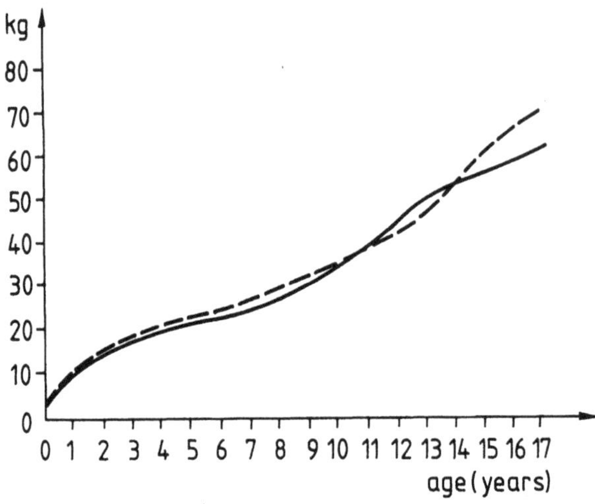

Fig. 9/24. Rate of growth from birth until the 17th year of age (after Jackson, R. L. et al.: J. Pediat. *27*, 215, 1945). ——— girls; – – – boys

During the slower growth rate of 2–11 years there is an accumulation of about 120 mg of iron yearly, i.e., an average of about 0.3 mg of iron daily, but during the adolescent growth spurt the requirement may go up to 180 mg yearly, or 0.5 mg daily. These are the additional requirements that are needed over and above that of the minimal obligatory loss, and in girls the additional loss occurs at the onset of menstruation. The balance is most precarious in infancy and during the adolescent growth spurt (Fig. 9/25).

Although the hypochromic anemia of infancy has been considered as physiological by some authors, it occurs at a time when the iron stores are exhausted, the plasma iron concentration is usually low, and the iron-binding capacity high, and it can be reversed by iron therapy. There is a rapid change in ferritin concentration during the first few months after birth, and the ferritin levels remain low from the age of 6 months to about 15 years (Worwood, 1980) (see Table 9/1). The hypochromic anemia of infancy should probably therefore be regarded as pathological rather than physiological. The various measurements reflecting the iron status at different ages are shown in Table 9/2.

Table 9/1
Serum ferritin concentrations in children (male and female)
(from Jacobs and Worwood, 1980)

Age	Mean ferritin concentration (μg l)	95% confidence range	References
0.5 month	238	90–628	Saarinen and Siimes (1978)
1 month	240	144–399	
2 months	194	87–430	
4 months	91	37–223	
6 months	51	19–142	
9 months	39	14–103	
12 months	31	1– 99	
6 months to 15 years	30	7–142	Siimes et al. (1974)

Table 9/2
Non-heme (storage) iron content of the liver, hemoglobin concentration of the blood, iron level and iron-binding capacity of the serum in the various age groups (Dreyfus and Schapira, 1958; Hagberg, 1953)

Age	Fe content of liver (mg/kg dry matter)	Hb (g/dl)	Plasma-Fe (μg/dl)	Iron-binding capacity (μg/dl)
Newborn*	1800	17–18	170	—
Newborn**	—	—	173 ± 6.9	259 ± 10.5
0.5–2 months	—	—	142 ± 7.1	212 ± 6.6
3 months	700	11–12	110	—
2–4 months**	—	—	113 ± 5.0	308 ± 11.3
4–6 months**	—	—	78 ± 6.1	360 ± 12.3
6–12 months**	—	—	93 ± 6.5	394 ± 13.4
1 year*	700	12–13	60	—
1–3 years**	—	—	99 ± 5.9	387 ± 9.7
2 years*	170	12–13	90	—
3–7 years**	—	—	124 ± 8.7	368 ± 9.6
10 years*	350	13–14	90	—
7–14 years**	—	—	119 ± 6.7	353 ± 7.6
10–15 years*	700	14–15	80 ± 120	—
Adult*	1000	15–16	Male: 135	—
			Female: 120	—
Adult**	—	—	130 ± 5.2	330 ± 4.9

* Dreyfus and Schapira, 1958.
** Hagberg, 1953.

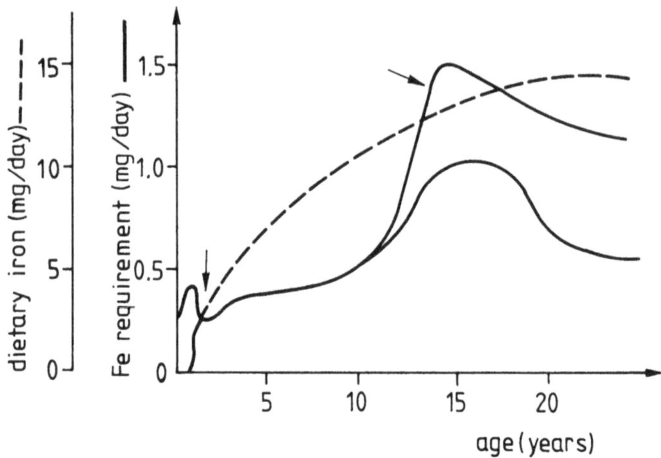

Fig. 9/25. Changes in iron requirement from birth until the completion of growth (continuous black line) in girls (upper curve) and in boys (lower curve). The broken line indicates the iron content of the diet. Arrows point to the two critical periods of life (after Bothwell, T. H. and Finch, C. A.: Iron Metabolism. Little, Brown and Co., Boston 1962)

BIBLIOGRAPHY

Aalam, F.: Vergleichende Untersuchungen der mechanischen Resistenz von Erwachsenen- und Neugeborenenerythrozyten, geprüft im Homogenisator. Diss., Freiburg 1959.

Bernát, I.: Az égési anaemia pathogenesise (The Pathogenesis of Anemia after Thermal Injury). Akadémiai Kiadó, Budapest 1971.

Betke, K.: Hämatologie der ersten Lebenszeit. Ergebn. inn. Med. Kinderheilk. *9,* 437 (1958).

Betke, K.: Blutfarbstoffanomalien: Diagnostik und pathogenetische Gesichtspunkte. Schweiz. med. Wschr. *91,* 1069 (1961).

Betke, K.: Haemoglobin und Erythrocyten. In: Wiesener, H. (ed.): Entwicklungsphysiologie des Kindes: Springer, Berlin–Göttingen–Heidelberg 1964.

Betke, K., Kleihauer, E.: Fetaler und bleibender Blutfarbstoff in Erythrocyten und Erythroblasten von menschlichen Feten und Neugeborenen. Blut *4,* 241 (1958).

Bloom, W., Bartelmez, G.: Hematopoiesis in young human embryos. Amer. J. Anat. *67,* 21 (1940).

Boroviczeny, K., Balló, T.: Normoblasten im Blutbild des Neugeborenen. Wien. Z. inn. Med. *5,* 196 (1957).

Bothwell, T. H.: Total iron loss and relative importance of different sources. In: Hallberg, L., Harwerth, H. G., Vannotti, A. (eds.): Iron Deficiency, p. 151. Academic Press, London–New York 1970.

Bothwell, T. H., Finch, C. A.: Iron Metabolism. Little, Brown and Co., Boston 1962.

Bothwell, T. H., Pribilla, W. F., Mebust, W., Finch, C. A.: Iron metabolism in the pregnant rabbit. Iron transport across the placenta. Amer. J. Physiol. *193,* 615 (1958).

Burman, D.: Iron deficiency in infants and children. In: Callender, S. T. (ed.): Clinics in Haematology, Vol. 2/2. Saunders, London–Philadelphia–Toronto 1973.

Chanarin, I., Rothman, D., Perry, J., Stratfull, D.: Normal dietary folate, iron and protein intake, with particular reference to pregnancy. Brit. Med. J. *2,* 394 (1968).

Chandra, R. K., Saraya, A. K.: Impaired immunocompetence associated with iron deficiency. J. Pediat. *86,* 899 (1975).

Chappelle, E. et al.: Regulation of body iron content through excretion in the mouse. Amer. J. Physiol. *182,* 390 (1955).

Chisholm, M.: A controlled clinical trial of prophylactic folic acid and iron in pregnancy. J. Obstet. Gynaec. Brit. Cwlth *73,* 191 (1966).

Chown, B.: The fetus can bleed: three clinico-pathological pictures. Amer. J. Obstet. Gynec. *70,* 1298 (1955).

Conrad, M. E., Weintraub, L. R., Crosby, W. H.: The role of the intestine in iron kinetics. J. clin. Invest. *43,* 963 (1964).

Cook, Ch. et al.: Measurement of fetal hemoglobin in newborn infants. Pediatrics *20,* 272 (1957).

Crosby, W. H., Conrad, M. E., Wheby, M. S.: The rate of iron accumulation in iron storage disease. Blood *22,* 429 (1963).

Dagg, J. H., Smith, J. A., Goldberg, A.: Urinary excretion of iron. Clin. Sci. *30,* 495 (1966).

Darby, W. J., Bridgeforth, E., Martini, M. P., McGanity, W. J.: The Vanderbilt cooperative study of maternal and infant nutrition. IX. Some obstetric implications. Obstet. and Gynec. *5,* 528 (1955).

de Gruchy, G. C.: Clinical Haematology in Medical Practice, 4th ed., Blackwell, Oxford–Edinburgh 1978.

de Leeuw, N. H. W., Lowenstein, L., Hsieh, Y.: Iron deficiency and hydremia in normal pregnancy. Medicine (Baltimore) *45,* 291 (1966).

de Leeuw, N. H. W., Lowenstein, L., Tucker, E. C., Dayal, S.: Correlation of red-cell loss at delivery with changes in red-cell mass. Amer. J. Obstet. Gynec. *100,* 1092 (1968).

Dervichian, D., Fournet, G., Guinier, A., Ponder, E.: Structure submicroscopique des globules rouges contenant des hémoglobines anormales. Rev. Hémat. *7,* 567 (1952).

Drescher, H., Künzer, W.: Der Blutfarbstoff des menschlichen Feten. Klin. Wschr. *32,* 92 (1954).

Dreyfus, J. C., Schapira, G.: Le fer. L'Expansion Scientifique Française, Paris 1958.

Dubach, R., Moore, C. V., Callender, S.: Studies on iron transportation and metabolism. IX. The excretion of iron as measured by the isotope technique. J. Lab. clin. Med. *45,* 599 (1955).

Dymock, I. W. et al.: Proc. roy. Soc. Med. *63,* 1227 (1970).

Erlandson, M.: Iron metabolism and iron deficiency anemia. Paediat. Clin. N. Amer. *9,* 673 (1962).

Fenton, V., Cavill, I., Fisher, J.: Iron stores in pregnancy. Brit. J. Haemat. *37,* 145 (1977).

Finch, C. A.: Body iron exchange in man. J. clin. Invest. *38,* 392 (1959).

Foy, H., Kondi, A.: Anaemias of the tropics. Relation to iron intake, absorption and losses during growth, pregnancy, and lactation. J. trop. Med. Hyg. *60,* 105 (1957).

Frenchman, R., Johnston, F. A.: Relation of menstrual losses to iron requirement. J. Amer. diet. Ass. *25,* 217 (1949).

Garby, L., Sjölin, S., Vuille, J.: Studies on erythrokinetics in infancy. V. Estimation of life span of red cells in the newborn. Acta paediat. (Uppsala) *53,* 165 (1964).

Gerritsen, T., Walker, A. R. P.: The effect of habitually high intake on certain blood values in pregnant Bantu women. J. clin. Invest. *33,* 23 (1954).

Gilardi, A., Miescher, P.: Die Lebensdauer von autologen und homologen Erythrocyten bei Frühgeborenen und älteren Kindern. Schweiz. med. Wschr. 1957, 1456.

Göltner, E.: Iron requirement and deficiency in menstruating and pregnant women. In: Kief, H. (ed.): Iron Metabolism and its Disorders. Excerpta Medica, Amsterdam–Oxford; American Elsevier, New York 1975.

Green, R., Charlton, R., Seftel, H., Bothwell, T. H., Mayet, F., Adams, B., Finch, C. A., Layrisse, M.: Body iron excretion in man. Amer. J. Med. *45,* 336 (1968).

Gubler, C.: Absorption and metabolism of iron. Science *123,* 87 (1956).

Guest, G., Brown, E.: Erythrocytes and hemoglobin of the blood in infancy and childhood. Amer. J. Dis. Child. *93,* 486 (1957).

HAGBERG, G.: The iron-binding capacity of serum in infants and children. Acta paediat. (Uppsala) *42*, Suppl. 93 (1953).
HALLBERG, L., NILSSON, L.: Determination of menstrual blood loss. Scand. J. clin. Lab. Invest. *16*, 244 (1964).
HALLBERG, L., HÖGDAHL, A. M., NILSSON, L., RYBO, G.: Frequency and pathogenesis of iron deficiency in a female Swedish population sample. XIth Congr. Int. Soc. Haemat. Sydney 1966a.
HALLBERG, L., HÖGDAHL, A. M., NILSSON, L., RYBO, G.: Constancy of individual menstrual blood loss. Acta obstet. gynec. scand. *45*, 352 (1966b).
HEILMEYER, L.: Die Störungen der Bluthämsynthese. Thieme, Stuttgart 1964.
HEINRICH, H. C.: In: HALLBERG, L., HARWERTH, H. G., VANNOTTI, A. (eds.): Iron Deficiency, p. 170. Academic Press, London–New York, 1970.
HEINRICH, H. C., BRÜGGEMANN, J., CABLE, E. E., GLÄSER, M.: Z. Naturforsch. *32*, 1023 (1977).
HEINRICH, H. C. et al. (1968), cit. GÖLTNER, 1975.
HOAG, M., WALLERSTINE, R., POLLYCOVE, M.: Occult blood loss in iron deficiency anemia of infancy. Pediatrics *27*, 199 (1961).
HOLLÁN, S. R., SZELÉNYI, J. G.: Glycerol and thiourea permeability of normal and abnormal human erythrocytes. In: Protides of the Biological Fluids. Vol. 15, p. 377. Elsevier, Amsterdam 1967.
HOLLINGSWORTH, J.: Lifespan of fetal erythrocytes. J. Lab. clin. Med. *45*, 469 (1955).
HUEHNS, E., DANCE, N., BEAVEN, G., KEIL, J., HECHT, F., MOTULSKY, A.: Human embryonic haemoglobins. Nature *201*, 1095 (1964).
HUSSAIN, R., PATWARDHAN, V. N., SRIRAMACHARI, S.: Dermal loss of iron in healthy Indian men. Indian J. med. Res. *48*, 235 (1960).
JACKSON, R. L. et al.: Growth charts. J. Pediat. *27*, 215 (1945).
JACOBI, H.: Gestörte Hämoglobinsynthese. In: Handbuch der Kinderheilkunde, Vol. 6, p. 883. Springer, Berlin–Heidelberg–New York 1967.
JACOBS, A., BUTLER, E. B.: Menstrual blood loss in iron-deficiency anaemia. Lancet *2*, 407 (1965).
JACOBS, A., WORWOOD, M.: Iron in Biochemistry and Medicine. Vol. 2. Academic Press, London–New York 1980.
JOSEPHS, H.: Iron metabolism and the hypochromic anemia of infancy. Medicine (Baltimore) *32*, 125 (1953).
JOUNSON, D. H. M., JACOBS, A. et al.: Defect of cell-mediated immunity in patients with iron deficiency anaemia. Lancet *2*, 1058 (1972).
KALTWASSER, J. P., WERNER, E., BECKER, H.: Dtsch. med. Wschr. *102*, 1150 (1977).
KELLY, A. M., MACDONALD, D. J., MCDOUGALL, A. N.: Brit. J. Obstet. Gynaec. *85*, 338 (1978).
KIMBER, C. et al.: Malabsorption of iron secondary to iron deficiency. New Engl. J. Med. *279*, 453 (1968).
KIRKMAN, H., RILEY, H.: Posthaemorrhagic anemia and shock in the newborn. Pediatrics *24*, 92, 97 (1959).
KLEIHAUER, E.: Fetales Haemoglobin und fetale Erythrocyten. Habil.-Schr. Tübingen 1965.
KNOLL, W.: Die embryonale Blutbildung beim Menschen. Zollikofer, St. Gallen 1950.
KNOLL, W.: Die Entwicklung des blutbildenden Gewebes und des Blutes beim Menschen. In: HEILMEYER, L., HITTMAIR, A. (eds.): Handbuch der gesamten Hämatologie, Vol. 1/1, p. 36. Urban und Schwarzenberg, München–Berlin–Wien 1957.
KÜNZER, W.: Über den Blutfarbstoffwechsel gesunder Säuglinge und Kinder. Mit besonderer Berücksichtigung der Anaemisierungsvorgänge im Verlauf des ersten Trimenons. Karger, Basel 1951.
KÜNZER, W.: Untersuchungen zur Reifungszeit der Retikulozyten von Neugeborenen, jungen Säuglingen und Frühgeburten. Z. Kinderheilk. *77*, 249 (1955).
KÜNZER, W.: Das Blut des normalen Säuglings und Kindes. In: HEILMEYER, L., HITTMAIR, A. (eds.): Handbuch der gesamten Hämatologie, Vol. 1/1, p. 60. Urban und Schwarzenberg, München–Berlin–Wien 1957.
KÜNZER, W.: Rotes Blutzellsystem bei Feten und Neugeborenen. In: KEPP, R., OEHLERT, G. (eds.): Blutbildung und Blutumsatz beim Feten und Neugeborenen. Enke, Stuttgart 1962.

KÜNZER, W., JACOBI, H.: Anämien des Kindes. In: HEILMEYER, L. (ed.): Blut und Blutkrankheiten, Vol. 2/2, Springer, Berlin–Heidelberg–New York 1970.
LAHEY, E.: Iron deficiency anemia. Pediat. Clin. N. Amer. 481 (1957).
LANGLEY, F.: Haemopoiesis and siderosis in the foetus and newborn. Arch. Dis. Childh. *26*, 64 (1951).
LINTZEL, W. et al.: Über den Eisenstoffwechsel des Neugeborenen und des Säuglings. Z. ges. exp. Med. *113*, 591 (1944).
LOW, J. A., JOHNSTON, E. E., MCBRIDE, R. L.: Blood volume adjustments in the normal obstetric patient with particular reference to the third trimester of pregnancy. Amer. J. Obstet. Gynec. *91*, 356 (1965).
MACDOUGALL, L. G. et al.: The immune response in iron deficiency children: impaired cellular defense mechanism with altered humoral components. J. Pediat. *86*, 833 (1975).
MAJEWSKI, A.: Über Blutbildung und Blutumsatz beim Feten. Arch. Gynäk. *191*, 88 (1958).
MARTI, H.: Normale und anormale menschliche Hämoglobine. Springer, Berlin–Göttingen–Heidelberg 1963.
MAXIMOW, A.: Relation of blood cells to connective tissues and endothelium. Physiol. Rev. *4*, 533 (1924).
MAXIMOW, A.: Bindegewebe und blutbildende Gewebe. In: MÖLLENDORF, W. (ed.): Handbuch der mikroskopischen Anatomie des Menschen, Vol. II/1. Springer, Berlin 1927.
MERKER, H., CARSTEN, P.: Elektronenmikroskopische Untersuchungen über die Erythropoese in der Leber menschlicher Embryonen. Blut *9*, 329 (1963).
MOE, P. J.: The diagnosis of iron deficiency anemia in children. Acta paediat. scand. *58*, 141 (1969).
MONCRIEFF, A., KOUMIDES, O., CLAYTON, B., PATRICK, A., REINWICK, A., ROBERTS, G.: Lead poisoning in children. Arch. Dis. Childh. *39*, 1 (1964).
MONSEN, E. R., KUHN, I. N., FINCH, C. A.: Iron status of menstruating women. Amer. J. clin. Nutr. *20*, 842 (1967).
MOORE, C. V.: Importance of nutritional factors in the pathogenesis of iron deficiency anemias. Amer. J. clin. Nutr. *3*, 3 (1955); in: GROSS, F. (ed.): Iron Metabolism. Springer, Berlin–Göttingen–Heidelberg 1964.
NAIMAN, J., OSKI, F., DIAMOND, L., VAWTER, G., SCHWACHMAN, H.: The gastrointestinal effects of iron deficiency anemia. Pediatrics *33*, 83 (1964).
National Academy of Sciences, Washington, D. C.: Recommended Dietary Allowances, 7th ed., No. 1694, 1968.
NÉMET, K., ANDRÁSSY, K., MANN, V., FÜZES, E., KOCSÁR, L., SIMONOVITS, I.: Iron stores in healthy pregnant women as determined by immunoradiometry of their serum ferritin levels. Haematologia *14*, 307 (1981).
NEWTON, M.: Postpartum hemorrhage. Amer. J. Obstet. Gynec. *94* 711 (1966).
NEWTON, M., MOSEY, L. M., EGLY, G. E., GIFFORD, W. B., HULL, C. T.: Blood loss during and immediately after delivery. Obstet. and Gynec. *17*, 9 (1961).
PAINTIN, D. B., THOMSON, A. M., HYTTEN, F. E.: Iron and the haemoglobin level in pregnancy. J. Obstet. Gynec. Brit. Cwlth *73*, 181 (1966).
PEARSON, H. A.: J. Pediat. *69*, 473 (1966).
PETER, H.: Vergleich der roten Blutwerte von gesunden Kindern aus Freiburger Kinder- und Waisenheimen mit denen gleichaltriger Kinder aus Familien. Inaug.-Diss., Freiburg 1964.
PRITCHARD, J. A.: Changes in the blood volume during pregnancy and delivery. Anesthesiology *26*, 393 (1965).
PRITCHARD, J. A., HUNT, C. F.: A comparison of the haematologic responses following the routine prenatal administration of intramuscular and oral iron. Surg. Gynec. Obstet. *166*, 516 (1958).
PRITCHARD, J. A., MASON, R. A.: Iron stores of normal adults and replenishment with oral iron therapy. J. Amer. med. Ass. *190*, 897 (1964).
PRITCHARD, J. A., SCOTT, D. E.: Iron demands during pregnancy. In: HALLBERG, L., HARWERTH, H. G., VANNOTTI, A. (eds.): Iron Deficiency, p. 173. Academic Press, London–New York 1970.
PRITCHARD, J. A., BALDWIN, R. M., DICKEY, J. C., WIPPINS, K. M.: Blood volume changes in pregnancy and the puerperium. II. Red blood cell loss and changes in apparent blood volume during and following vaginal delivery, Cesarean section, and Cesarean section plus total hysterectomy. Amer. J. Obstet. Gynec. *84*, 1271 (1962).

Reimann, F.: Wachstumsanomalien und Mißbildungen bei Eisenmangelzuständen (Asiderosen). 5. Kongreß der europäischen Gesellschaft für Hämatologie, Freiburg/Br. 1955, p. 546. Springer, Berlin–Göttingen–Heidelberg 1956.

Report of the Food and Nutrition Board. National Research Council Publication, 1146. Washington, D. C. 1964.

Rios, E., Lipschitz, D. A., Cook, J. D., Smith, N. J.: Relationship of maternal and infant iron stores as assessed by determination of plasma ferritin. Pediatrics *55,* 694 (1975).

Rohr, K.: Das menschliche Knochenmark. 3rd ed. Thieme, Stuttgart 1960.

Rosenberg, M.: Fetal hematopoiesis. Blood *33,* 66 (1969).

Rybo, G.: Menstrual loss of iron. In: Hallberg, L., Harwerth, H. G., Vannotti, A. (eds.): Iron Deficiency, p. 163. Academic Press, London–New York 1970.

Rybo, G.: Physiological causes of iron deficiency in women: menstruation and pregnancy. In: Callender, S. T. (ed.): Clinics in Haematology, Vol. 2/2. Saunders, London–Philadelphia–Toronto 1973.

Saarinen, U. M., Siimes, M. A.: Acta Paediat. Scand. *67,* 741 (1978).

Sachtleben, P., Lehmann, H., Ruhentroth-Bauer, G.: Zur Frage der Strukturspezifizität der Membranen von Neugeborenen-Erythrozyten. Blut *7,* 369 (1961).

Saddi, R., Schapira, G.: Iron requirements during growth. In: Hallberg, L., Harwerth, H. G., Vannotti, A. (eds.): Iron Deficiency, p. 183. Academic Press, London–New York 1970.

Saito, H., Sargent, T., Parker, H. G., Lawrence, J. H.: Whole-body iron loss in normal man measured with a gamma spectrometer. J. nucl. Med. *5,* 571 (1964).

Schäfer, K.: Der Eisenstoffwechsel des wachsenden Organismus. Ergebn. inn. Med. Kinderheilk. N. F. *4,* 706 (1953).

Schäfer, K. H.: In: Hallberg, L., Harwerth, H. G., Vannotti, A. (eds.): Iron Deficiency, pp. 524–526. Academic Press, London–New York (1970).

Schmidt, D., Dech, G., Fikentscher, R., Stich, W.: Über die fetale Hämsynthese in verschiedenen Blutbildungsphasen. Blut *25,* 85 (1972).

Schubothe, H., Middendorf, K., Behrends, D.: Vergleichende Untersuchungen über die mechanische Haemolyse beim Menschen und bei verschiedenen Tierarten. Klin. Wschr. *39,* 875 (1961).

Schulman, J.: Iron requirements in infancy. J. Amer. med. Ass. *175,* 118 (1961).

Schulz, J., Smith, N.: A quantitative study of the absorption of food iron in infants and children. Amer. J. Dis. Child. *95,* 109 (1958).

Scott, D. E., Saltin, A. S., Pritchard, J. A., cit. Pritchard, Scott, 1970.

Seelemann, K.: Zur Differentialdiagnose der Erythroblastosis foetalis. Z. Kinderheilk. *71,* 61 (1952).

Shahidi, N. T., Diamond, L. K.: Skull changes in infants with chronic iron deficiency anaemia. New Engl. J. Med. *262,* 137 (1960).

Siimes, M. A., Addiego, J. E., Dallman, P. R.: Blood *43,* 581 (1974).

Singh, A. K., Loehry, C. A.: Mechanism of excretion of iron by the small intestine; an experimental study in rats with normal and abnormal mucosae. Brit. J. Haemat. *15,* 495 (1968).

Sisson, T., Lund, C.: Influence of maternal iron deficiency on the newborn. Amer. J. clin. Nutr. *6,* 376 (1958).

Sjölin, S.: The resistance of red cells in vitro. A study of the osmotic properties, the mechanical resistance and the storage behaviour of red cells of fetuses, children and adults. Acta paediat. *43,* Suppl. 98 (1954).

Smith, C., Schulman, J., Morgenthau, J.: Iron metabolism in infants and children. Serum iron and iron-binding protein: Diagnostic and therapeutic implications. In: Levine, S. Z. (ed.): Advances in Paediatrics, Vol. 5, p. 195. Yearbook Publishers, Chicago 1952.

Smith, N. J. et al.: Pediatrics *16,* 166 (1955).

Sturgeon, P.: Studies of iron requirements in infants. III. Influence of supplemental iron during normal pregnancy on mother and infant. A. The mother. Brit. J. Haemat. *5,* 31 (1959).

Sturgeon, P.: Iron metabolism. A review with special consideration of iron requirement during normal infancy. Pediatrics *18,* 267 (1966).

TAFARI, N., HABTE, D.: Physiologic anaemia of infancy at high altitude. Acta Paediat. Scand. *61*, 706 (1972).

TAFT, L. I., HALLIDAY, J. W., RUSSO, A. M., FRANCIS, B. H.: Aust. N. Z. J. Obst. Gynaec. *18*, 226 (1978).

VAN EIJK, H. G., KROOS, M. J., HOOGENDOORN, G. A., WALLENBURG, H. C. S.: Clin. Chim. Acta *83*, 81 (1978).

VERLOOP, M. C., BLOKHUIS, E. W. M., BOS, C. C.: Über die Ursachen der (physiologischen) Schwangerschaftsanämie. Schweiz. med. Wschr. *88*, 1051 (1958) und Acta haemat. (Basel) *22*, 158 (1959).

VEST, M.: Physiologie und Pathologie des Neugeborenen. Bibl. paediat. Suppl. Ann. paediat. *69* (1959).

VEST, M.: Lebensdauer fetaler und kindlicher Erythrocyten. In: KEPP, R., OEHLERT, G. (eds.): Blutbildung und Blutumsatz beim Feten und Neugeborenen, p. 108. Enke, Stuttgart 1962.

WALKER, J., TURNBULL, E.: Haemoglobin and red cells in the human foetus. Lancet *2*, 312 (1953).

WALTERS, G. O., JACOBS, A., WORWOOD, M., TREVETT, D., THOMSON, W.: Gut *16*, 188 (1975).

WARNINGHOFF, G., HAUSMANN, K.: Die Morphologie der embryonalen Hämatopoese des Menschen im Vergleich zu postfetalen Blutbildungsstörungen. Acta haemat. (Basel) *14*, 273 (1955).

WEICKER, H., WAGNER, I., GUTTMANN, A., KRIEGER, F., LOREY, H., ZIMMERMANN, H.: Der Erythrocytendurchmesser des Kindes. Acta haemat. (Basel) *10*, 50 (1953).

WHITE, H. S.: Iron nutriture of girls and women—a review. J. Amer. diet. Ass. *53*, 563 (1968).

WIDDOWSON, E. M.: In: ASSALI, N. S. (ed.): Biology of Gestation, Vol. II. The fetus and neonate. Academic Press, London–New York 1968.

WILSON, E., WINDLE, W., ALT, H.: Deprivation of placental blood as a cause of iron deficiency in infants. Amer. J. Dis. Child. *62*, 320 (1941).

WILTING, W. F., VAN EYK, H. G., BOBECK RUTSAERT, M. M.: Urinary iron excretion in nephrotic syndrome. Acta haemat. (Basel) *47*, 269 (1972).

WINTROBE, M. M.: Clinical Haematology. Lea and Febiger, Philadelphia 1967, 1975.

WOODRUFF, C., BRIDGEFORTH, F.: Relationship between the hemogram of the infant to that of the mother during pregnancy. Pediatrics *12*, 681 (1953).

WORWOOD, M.: Serum ferritin. In: JACOBS, A., WORWOOD, M. (eds.): Iron in Biochemistry and Medicine, Vol. 2. Academic Press, London–New York 1980.

YAO, A. C., MOINIAN, M., LIND, J.: Distribution of blood between infant and placenta after birth. Lancet *2*, 871 (1969).

YETGIN et al.: Myeloperoxidase activity and bactericidal function of PMN in iron deficiency. Acta haemat. (Basel) *61*, 10 (1979).

ZIPURSKY, A., LARUE, T., ISRAELS, L.: The in vitro metabolism of erythrocytes from newborn infants. Canad. J. Biochem. *38*, 727 (1960).

CHAPTER 10

ERYTHROPOIESIS

In the course of cell maturation the basic biochemical processes consist of the synthesis of various proteins (nucleoproteins, enzyme proteins, and structural proteins). The basic information for the synthesis of specific proteins is stored in the DNA, and the mRNA conveys the message of the code to the ribosomes. The latter adhere to the long chain of the mRNA and scan the information to select in a given order the amino acids that are required for the synthesis of polypeptide chains.

The amino acids must first become activated in order to reach the ribosomes, and the energy for this process is gained from adenosine triphosphate (ATP). The transfer or soluble ribonucleic acids (sRNA) recognize the activated amino acids and bind to them via specific enzymes (Fig. 10/1).

Cell protein synthesis is most active in the earliest phases of cell development. With the disappearance of the nucleoli, the amounts of DNA and of RNA begin to diminish. Hemoglobin synthesis starts relatively late, at the maturation stage of the basophilic erythroblasts, and simultaneously with the reduction of RNA synthesis it increases rapidly (Fig. 10/2).

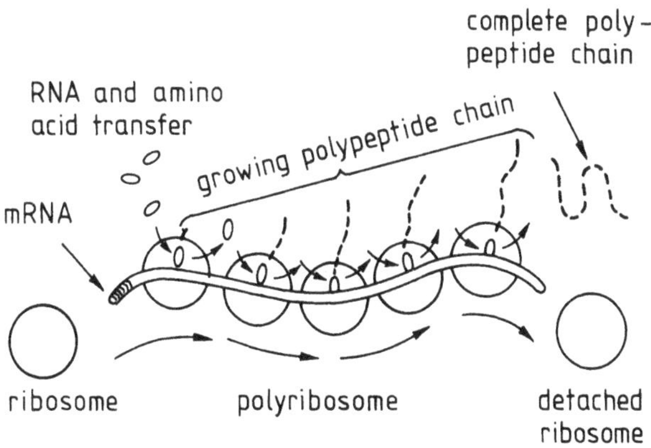

Fig. 10/1. Development of the polypeptide chain in the polyribosomes (after Rich, A.: Scientific American *209,* 44, 1963; Science *138,* 1399, 1962)

Although the RNA synthesis takes place in general only in the nucleated cells, ribonucleic acids can also be demonstrated in the reticulocytes as a result of the RNA stability after the expulsion of the normoblast nuclei. The reticulocytes also contain ribosomes and mitochondria, and hemoglobin synthesis continues to take place in these cells (Fig. 10/3).

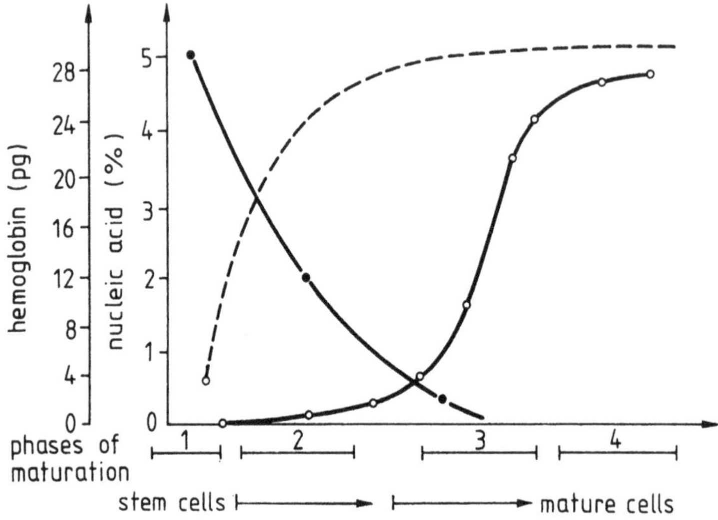

Fig. 10/2. Changes in nucleic acid, globin and heme concentration in the course of red cell maturation (after Thorell, B.: Acta med. scand. *129* [Suppl. 200], 1, 1947)

Cell production in the bone marrow is controlled by fine regulating mechanisms that not only secure the relative stability of the number of peripheral blood cells but also enable the adjustment of cell production to changing physiological and pathophysiological conditions. The rate of erythropoiesis is controlled by the tissue tension of oxygen (Erslev, 1969, 1977); this in turn depends on the relation between oxygen supply and oxygen demand. The mechanism is mediated by a renal erythropoietic hormone, erythropoietin. When oxygen supply exceeds the demand, erythropoietin is reduced and red cell production decreased. Conversely, when oxygen demand exceeds the supply, erythropoietin is increased and erythropoiesis is also increased. Hypothalamic stimulation and various pituitary hormones may induce the release of erythropoietin (Mirand, 1968; Meincke and Crafts, 1968), and feedback control of red cell production may be mediated by the hypothalamus and pituitary or pituitary-dependent hormones. Products released through the

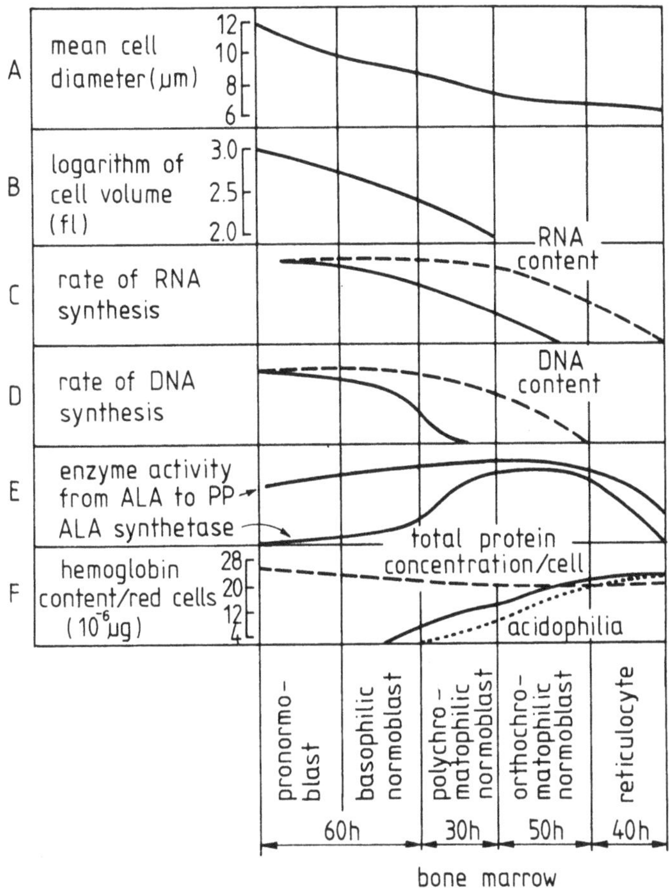

Fig. 10/3. Changes in the diameter and volume of erythroid precursors, in the rate of the RNA and DNA synthesis in the cells, in the activity of enzymes participating in heme synthesis, and in the mean protein concentration and hemoglobin level of the cells during cell differentiation and maturation (after Granick, S. and Levere, R. D., in: Moore, C. V. and Brown, E. B.: Progress in Hematology. Grune and Stratton, New York 1964)

destruction of red cells also have a positive feedback effect on the rate of erythropoiesis mediated via the release of erythropoietin (Erslev, 1971).

Humoral inhibitors of erythropoiesis have been found in the serum IgG fraction from occasional patients with pure red cell aplasia. Zalusky et al. (1973) studied the mechanism of action of such an inhibitor obtained from a female patient. Both the whole serum and its IgG fraction suppressed erythropoiesis in normal mice and heme synthesis in rat and human bone marrow *in vitro*. The addition of excess erythropoietin failed to overcome this inhibition. The inhibitor, however, did not behave like an antibody to erythropoietin. The authors suggested that the inhibitor interfered with erythropoiesis by acting directly on the early erythroid cell.

The red blood cells are formed from stem cells via mitosis and cell differentiation. Various kinetic models have been constructed to explain the mechanism of cell maturation and mitosis (Osgood, 1959, 1963; Lajtha et al., 1962; Lajtha, 1965; Boggs, 1966). Most of the models are based on autoradiographic studies using ^{59}Fe and ^{3}H-thymidine, a DNA label.

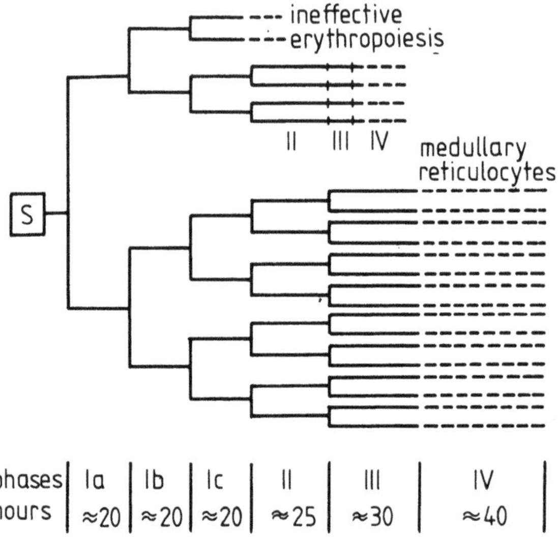

Fig. 10/4. Model of erythropoiesis (after Lajtha, L. G.: Series Haematologica *2*, 26, 1965)

The generation time or cell cycle is composed of four phases:

(1) the mitosis, which lasts for 1–1.5 hours
(2) the postmitotic rest period
(3) the period of DNA synthesis, and
(4) the premitotic rest period,

the last three phases being the intermitotic period.

The first three cycles, in the course of which the early pronormoblasts differentiate into late basophilic normoblasts, last for about 20 hours each (Fig. 10/4). At this stage of development some of the cells are destroyed (ineffective erythropoiesis). The next cycle, during which the basophilic normoblasts are transformed into polychromatophilic normoblasts, requires about 25 hours. This is followed by the last cell mitosis, after which the cells differentiate into late polychromatophilic normoblasts in about 30 hours. They then lose their nuclei and transform into reticulocytes.

Thus the proerythroblasts develop into mature acidophilic normoblasts in about 5 days. This estimate is in good agreement with the results of radioactive iron kinetic studies (Pollycove and Mortimer, 1961). The reticulocytes remain in the bone marrow for another 1.5–2 days, which means that the complete maturation time of erythroid cells from birth to delivery into the bloodstream is about 7 days. The erythroid cells of smaller animals require less time for maturation (Alpen and Cranmore, 1959; Alpen et al., 1962; Bond et al., 1962). The red cells enter the blood stream in the form of reticulocytes and lose their reticular network in about 24 hours. The average life-span of the intact red cell is 120 days.

BIBLIOGRAPHY

ALPEN, E. L., CRANMORE, D.: Observations on the regulation of erythropoiesis and on cellular dynamics by Fe^{59} autoradiography. In: STOHLMAN, F. (ed.): The Kinetics of Cellular Proliferation. Grune and Stratton, New York–London 1959.

ALPEN, E. L., CRANMORE, D., JOHNSTON, E. In: Erythropoiesis. Grune and Stratton, New York 1962.

BOGGS, D. R.: Homeostatic regulatory mechanisms of hematopoiesis. Ann. Rev. Physiol. *28*, 39 (1966); Amer. J. Physiol. *211*, 51 (1966).

BOND, V. P., ODARTCHENKO, N., COTTIER, H., FEINDEGEN, L. E., CRONKITE, E. P.: The kinetics of the more mature erythrocytic precursors studied with tritiated thymidine. In: Erythropoiesis. Grune and Stratton, New York 1962.

ERSLEV, A. J.: Control of red cell production. Ann. Rev. Med. 315 (1969).

ERSLEV, A. J.: The effect of hemolysates on red cell production and erythropoietin release. J. Lab. clin. Med. *78*, 1 (1971).

ERSLEV, A. J.: Production of erythrocytes. In: Williams, W. J., BENTLER, E., ERSLEV, A. J., RUNDLES, R. W. (eds.): Hematology, p. 203. McGraw-Hill, New York 1977.

GRANICK, S., LEVERE, R. D.: Heme synthesis in erythroid cells. In: MOORE, C. V., BROWN, E. B. (eds.): Progress in Hematology, Vol. 4, p. 1. Grune and Stratton, New York 1964.

LAJTHA, L. G.: Cellular mechanism of red cell production. Series Haematologica *2*, 26 (1965).

LAJTHA, L. G., OLIVER, R.: Studies on the kinetics of erythropoiesis: A model of the erythron. In: Haemopoiesis, cell production and its regulation. Churchill, London 1960.

LAJTHA, L. G., OLIVER, R., GURNEY, C. W.: Kinetic model of a bone marrow stem-cell population. Brit. J. Haemat. *8*, 442 (1962).

MEINCKE, H. A., CRAFTS, R. C.: Further observations on the mechanisms by which androgens and growth hormone influence erythropoiesis. Ann. N. Y. Acad. Sci. *149*, 298 (1968).

MIRAND, E. A.: Extrarenal and renal control of erythropoietin production. Ann. N. Y. Acad. Sci. *149*, 94 (1968).

OSGOOD, E. E.: In.: STOHLMAN, F. (ed.): The Kinetics of Cellular Proliferation. Grune and Stratton, New York–London 1959.

POLLYCOVE, M., MORTIMER, R.: The quantitative determination of iron kinetics in hemoglobin synthesis in human subjects. J. clin. Invest. *40*, 753 (1961).

RICH, A.: Polyribosomes. Scientific American *209*, 44 (1963); Science *138*, 1399 (1962).

STOHLMAN, F.: Observations on the kinetics of red cell proliferation. In: STOHLMAN, F. (ed.): The Kinetics of Cellular Proliferation. Grune and Stratton, New York–London 1959.

STOHLMAN, F.: Erythropoiesis. New Engl. J. Med. *267*, 342 (1962).

THORELL, B.: Studies on the formation of cellular substances during blood cell production. Acta med. scand. *129*, Suppl. 200, 1 (1947).

ZALUSKY, R., ZANJANI, E. D., GIDARI, A. S., ROSS, J.: Site of action of a serum inhibitor of erythropoiesis. J. Lab. clin. Med. *81*, 867 (1973).

CHAPTER 11

HEMOGLOBIN SYNTHESIS

Hemoglobin is a complex chromoprotein that is composed of four heme groups and four polypeptide globin chains. In adult hemoglobin, hemoglobin A, the globin consists of two alpha and two beta chains. The alpha chains comprise 141 and the beta chains 146 amino acids. The molecular weight of hemoglobin is 64,458. Under biological conditions the molecular weight is somewhat higher, 66,000 according to Drabkin.

Each heme group is situated in a well-defined crevice of the polypeptide chain to which it belongs, between the helical segments E and F near the surface of the

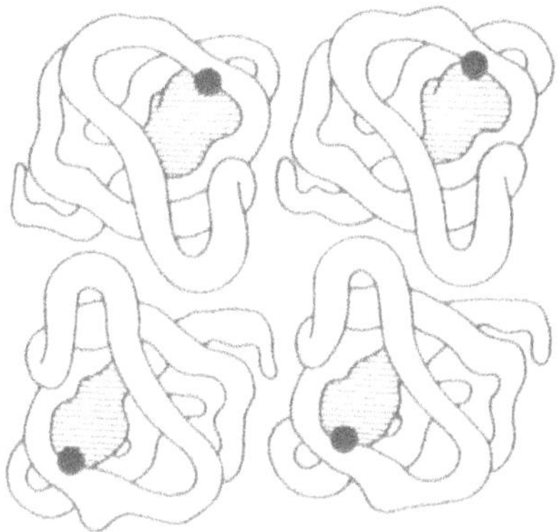

Fig. 11/1. Quaternary structure of the hemoglobin molecule

molecule (Perutz et al., 1960, 1963a, b). The forces that hold the individual chains together secure at the same time the spheroidal shape of the molecule (Fig. 11/1).

The heme amounts to only 4% of the hemoglobin, and the iron 0.3466%. The iron is situated in the center of the flat heme molecule; with its 5th coordinate valency it is linked to the 87th histidine residue (proximal histidine) of the alpha chain and to the

92nd histidine residue of the beta chain, while with its 6th coordinate valency, mediated either by the oxygen molecule (O_2-Hb) or by the water molecule (reduced Hb), it is linked to the opposite 58 alpha and 63 beta histidine radical (distal histidine) (Fig. 11/2).

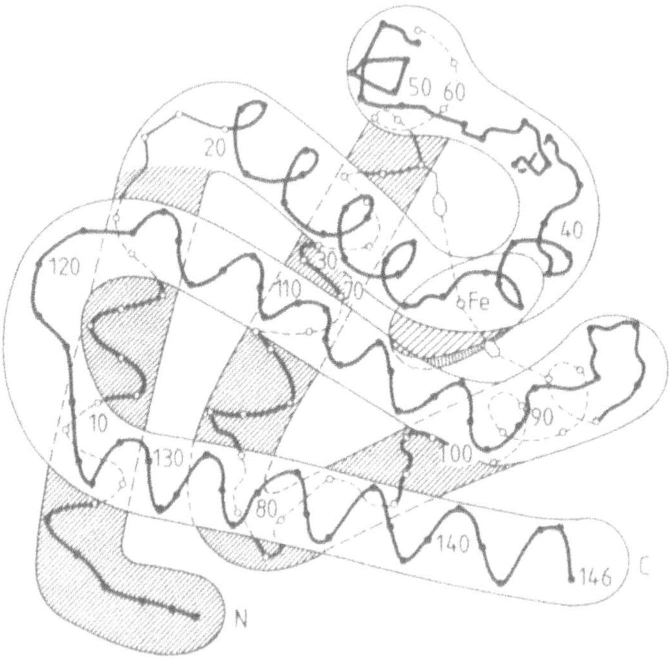

Fig. 11/2. Stereoscopic structure of the beta chain of human hemoglobin (schematic drawing). The sequence of amino acids is indicated by numerals; the heme molecule is represented by a flat disk. The iron, with its two coordinate valences, binds to the 92nd and 63rd histidine radicals (after Perutz, M. F.: Scientific American *211,* 64, 1964)

Haurowitz was the first to state that the hemoglobin molecule changes its crystalline structure when it takes up oxygen. Muirhead and Perutz (1963) clarified the basis of this phenomenon by demonstrating that the distance between the beta chains in the reduced hemoglobin is greater than in oxyhemoglobin; i.e., the blood pigment expands when it gives off oxygen and contracts when it takes up oxygen (Perutz, 1964).

Hemoglobin is a widespread compound in the living world; it occurs not only in higher animals but also in certain protozoa, plants, and even fungi. Although these hemoglobins differ considerably from each other in their biological, chemical, and physical qualities, they are identical in relation to their reversible oxygen-binding capacity and in the structure of their heme complexes.

There are, however, differences in the composition of the globin even within a single species. In the healthy human, in addition to hemoglobin A, which constitutes 95–98% of the adult blood pigment, there are small amounts of hemoglobin F (less than 1%) and hemoglobin A_2 (about 2.5%). Hemoglobin F consists of two alpha and two gamma chains, and hemoglobin A_2 of two alpha and two delta chains. The alpha chains of hemoglobin A, hemoglobin F, and hemoglobin A_2 are identical. Research into the composition and sequence of the individual amino acids has revealed that there are about 80 deviations between the alpha and beta chains, about 40 between the beta and gamma chains, somewhat more than 80 between the alpha and gamma chains, and, finally, 8 between the beta and delta chains. The similarities and differences that exist in the primary structure of individual polypeptide chains indicate that the genes of these chains originate from a mutual gene, so that the alpha, gamma, beta, and delta chains develop from this common gene in the course of genetic development.

BIOSYNTHESIS OF HEME

The process of heme synthesis (Fig. 11/3) has been followed by *in vivo* isotope studies (Shemin and Rittenberg, 1946a; Wittenberg and Shemin, 1949) and by the incubation of labeled precursors with bird red cells (Shemin et al., 1948, 1950, 1951, 1952, 1953, 1954) or mammalian reticulocytes (Kruh et al., 1961, 1962, 1964). Shemin and Rittenberg (1946a) demonstrated that shortly after the administration of ^{15}N-labeled glycine, heavy nitrogen became detectable in the heme. By the use of the stable and radioactive isotopes (^{15}N and ^{14}C) they succeeded in proving *in vitro* that only glycine and succinic acid furnish the nitrogen and carbon atoms necessary for the construction of the porphyrin molecule. Delta-aminolevulinic acid is the condensation product of these two components in the biological system. Succinyl coenzyme A proved to be one of the active precursors (Brown, 1958a, b), whereas glycine, the other component, is activated by pyridoxal-5-phosphate in the presence of Fe^{2+} (Laver et al., 1958; Gibson et al., 1958; Kikuchi et al., 1958, 1959). Delta-ALA synthetase is the catalyzer ferment of the reaction. First, a labile compound (alpha-amino-beta-keto-adipinic acid) develops that quickly undergoes decarboxylation and leads to delta-ALA formation. The enzyme reaction is bound to mitochondria.

Dresel and Falk (1953) demonstrated that two molecules of delta-ALA are converted to porphobilinogen (PBG). PBG itself was demonstrated by Waldenström and Vahlquist (1939) in the urine of a patient suffering from acute porphyria. Westall (1952) crystallized the compound, and its structure was

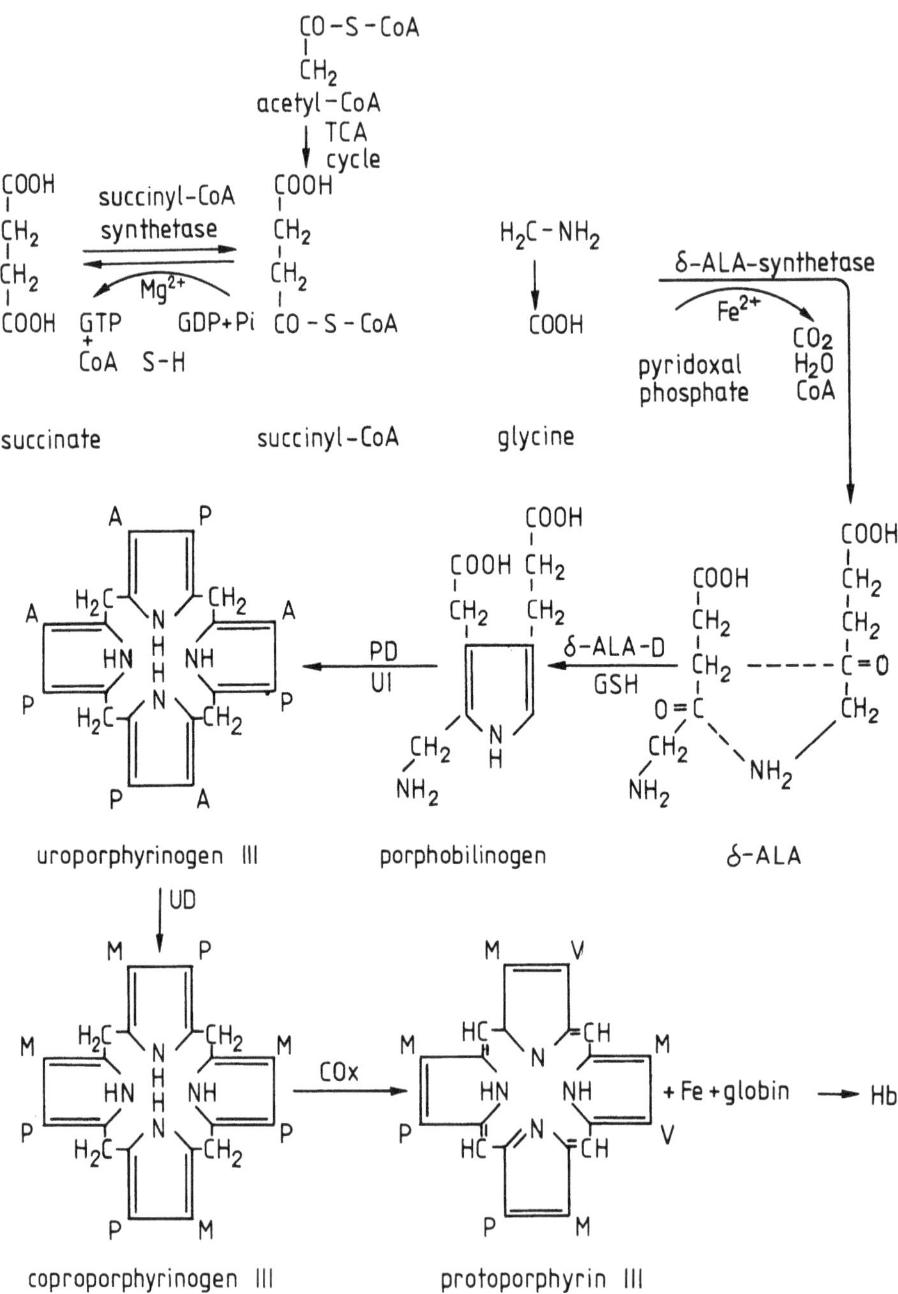

Fig. 11/3. Scheme of hemoglobin biosynthesis (after Plaut, G. W. E. and Cartwright, G. E., cit. Wintrobe, M. M.: Clinical Hematology. Lea and Febiger, Philadelphia 1967). Abbreviations: A = acetyl; M = methyl; P = propionic acid; V = vinyl (radicals)

clarified by Cookson and Rimington in 1953. PBG is a pyrrol that contains one 3-propionate, one 4-acetate, and one 5-aminomethylene side chain (Granick et al., 1956, 1958). In 1953, Falk and his collaborators proved that PBG was a true intermediary product of biosynthesis.

The enzyme of the delta-ALA-PBG conversion is delta-ALA-dehydrase, which contains SH groups. The presence of potassium is required for the full activity of the enzyme: the optimal pH of the reaction varies between 7.8 and 8.0 (Burnham and Lascelles, 1963).

Part of the delta-ALA is reconverted to succinate and participates again in PBG synthesis (Shemin cycle).

Uroporphyrinogen (UPG-III), the first reduced tetramer, is formed from the PBG molecule in the presence of PBG desaminase and UPG isomerase, which possess SH groups and are present in the cytoplasm of the cells. Biosynthesis takes place through this reduced form (Neve et al., 1956; Granick, cited by Heilmeyer, 1964), and uroporphyrin (UP) is a by-product of the biosynthesis. Falk et al. (1953) demonstrated that there is little or no enzymatic transformation of UP and CP to heme.

Coproporphyrinogen (CPG-III) is synthesized from uroporphyrinogen (UPG-III). The conversion is catalyzed by the UPG decarboxylase, which supplies SH groups. In the next step, protoporphyrinogen (PPG-9) develops in the mitochondria from CPG-III in the presence of CPG decarboxylase and a dehydrogenase.

Bogorad (1958) demonstrated the *in vitro* enzymatic synthesis of UPG to coproporphyrinogen and protoporphyrinogen, and the enzymatic oxidation of individual porphyrinogens to porphyrins. Sano and Granick (1961) found that the enzymes that induced the CPG-PPG transformation were demonstrable in the mitochondria of liver cells. Rat liver cell mitochondria were also found to induce transformation of delta-ALA to CPG through PBG and UPG. PPG-9 was not synthesized in this system.

ALA-S, CPG-D, and heme synthetase are all found in mitochondria, while the other enzymes occur in other parts of the cytoplasm (Table 11/1). Mitochondrial damage will reduce the activity of the mitochondrial enzymes.

The last phase of heme biosynthesis is the incorporation of iron into the PP-9 molecule. Transferrin carries the iron to the erythroblasts, and the membrane of the developing erythroblasts and reticulocytes possesses specific sites for the reversible binding of the transferrin (Jandl et al., 1959, 1960, 1963, 1964).

As the iron is taken over by the appropriate receptors, the transferrin is liberated (Fig. 11/4). The iron passes through the cell membrane, probably by an active process that requires the presence of ATP and SH groups (Wiese and Archdeacon, 1965). After entry into the cell the iron goes to the mitochondria, probably by means of an iron transport protein, although the nature of this is not clear. The protoporphyrin–iron–globin link is formed subsequently. Bessis and Breton-Gorius (1959, 1962) have suggested a second mode of iron transfer from the

Table 11/1

The pathway of heme synthesis (from Heilmeyer, 1964)

Substrate	Enzyme
1. Glycine + succinyl coenzyme A	
↓ ←	Delta-ALA synthetase (in the mitochondria); coenzyme: pyridoxal-5-phosphate
2. Delta-aminolevulinic acid	
↓ ←	Delta-ALA dehydratase (outside the mitochondria)
3. Porphobilinogen	
↓ ←	PBG desaminase + UPG isomerase (steric configuration!) (outside the mitochondria)
4. Uroporphyrinogen III	
←	Specific decarboxylase (outside the mitochondria)
Porphyrinogens (7,6,5-carboxyl groups) ↓	
5. Coproporphyrinogen III	
↓ ←	Specific decarboxylase, oxidase (mainly in the mitochondria)
6. Protoporphyrinogen 9	
↓ ←	Dehydrating system (in the mitochondria)
7. Protoporphyrin 9	
←	Heme synthetase (only in the mitochondria)
↓ + Fe	
8. Heme	

reticuloendothelial cells to erythroblasts, i.e., rhopheocytosis, but it is not known how much of the iron reaches the erythroblasts in this way under normal circumstances.

Ferrous iron is regularly incorporated into the PP-9 molecule *in vitro*. For the incorporation of ferric iron, Goldberg et al. (1956) showed that enzyme participation was necessary, the enzyme being obtained from a hemolysate of hen blood. The enzyme was activated by ascorbic acid, by glutathione-SH, and by ergothioneine, which reduced the ferric iron (Goldberg, 1959). The peak of enzyme activity was reached when the concentration of reducing agents corresponded to their natural intracellular concentration, and at pH 4 (Neve, 1961).

Measurement of the iron and protoporphyrin content of the cells can serve as the basis for the clarification of disturbances in heme synthesis. Concentration in the amount of other precursors in the peripheral red cells and in the urine can also be helpful (Fig. 11/5). It is not necessary to determine the porphyrin precursors in the bone marrow because their concentration in peripheral red cells reflects that in the bone marrow (Heilmeyer, 1964). The radioisotopes of iron can be used in addition to study the distribution of iron throughout the body and iron incorporation into red cells, and to give additional information about disturbances of heme synthesis (Gardner and Nathan, 1962).

Concentrations of the free PP and CP in the red cells normally range within relatively narrow limits and vary only with disturbances of heme synthesis. On the other hand, the delta-ALA, PBG, UP, and CP content of the urine is affected by

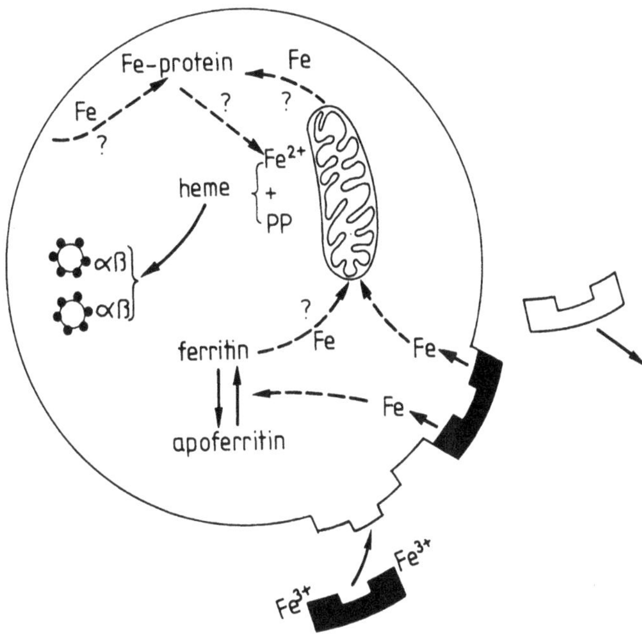

Fig. 11/4. Schematic representation of iron incorporation and heme and globin synthesis (after London, I. M.: Series Haematologica *2*, 1, 1965)

disturbances in other tissues, particularly the liver, and this must be kept in mind when interpreting abnormalities of excretion.

Since the enzymatic activity of reticulocytes is higher than that of mature erythrocytes, the proportion of these cells must be taken into account when assessing the free PP and CP concentrations in red cells (Watson, 1950; Prato et al., 1961).

If a rise in the PP level is associated with a drop in the hemoglobin concentration without an increase in the reticulocytes, there may be a block in the last step of heme synthesis.

Enzyme blockade may occur at an earlier stage of heme synthesis giving accumulation of some of the intermediary products.

In considering the autoregulation of heme synthesis, the function of the feedback mechanisms has also to be taken into account. Burnham and Lascelles (1963) showed that the final formation of the heme molecule had a specific reducing effect on the activity of delta-aminolevulinic acid-synthetase, which catalyzes the first step in the process of heme synthesis. PP-9, as the second-last link in the chain of heme

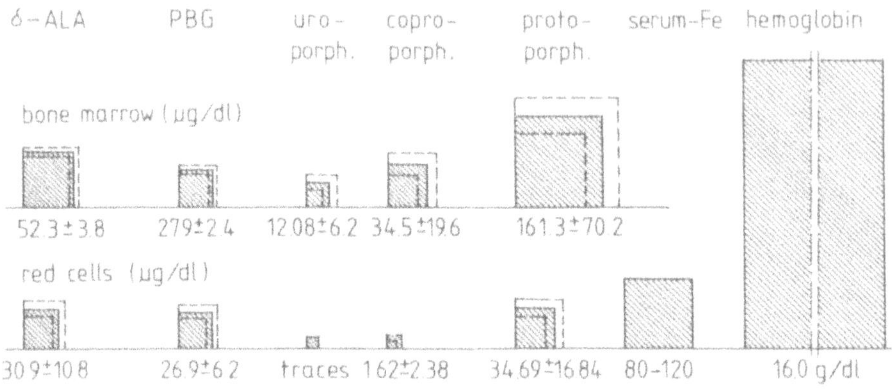

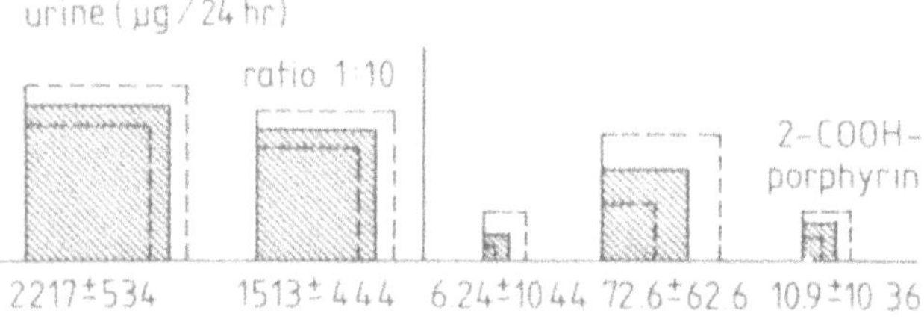

Fig. 11/5. Concentration and quantity of heme precursors in the bone marrow, red blood cells, and urine of healthy man (after Heilmeyer, L.: Die Störungen der Bluthämsynthese. Thieme, Stuttgart 1964). —— normal mean; - - - $\pm 2\sigma$

synthesis, inhibits the activity of delta-aminolevulinic acid-dehydrase, which catalyzes the second step of the biosynthesis (Gibson et al., 1961; Heilmeyer, 1964).

Sano and Granick (1961) suggest that the cell itself participates in autoregulation by virtue of the fact that the enzymes that catalyze the first and last stages of the synthesis are situated in the mitochondria, and the intermediary steps take place elsewhere in the cytoplasm. During the synthesis delta-ALA must leave the mitochondria and CPG-III must return to the mitochondria, and the intermediary products must overcome either actively or passively the resistance of the mitochondrial membrane. These membranes are vulnerable to damage, which will affect the biosynthesis of heme.

The hematological picture that results from disturbed heme synthesis depends on whether the disturbance is in heme production, in which case a hypochromic anemia will result without necessarily any diminution in the red cell count, or whether there is some disorder in erythropoiesis associated with the disturbance of heme synthesis, in which case hypochromia may not be present and the blood picture may be normochromic or even hyperchromic (Heilmeyer, 1964).

Normally there is very little free iron or free protoporphyrin in the red cells of the healthy individual. If there is a lack of supply of iron, the free protoporphyrin concentration is increased, but where protoporphyrin synthesis is inhibited, as for example in pyridoxine deficiency or sideroblastic anemia, free non-heme iron is increased in the red cells. An increase in both iron and protoporphyrin in the red cells indicates either reduced activity of heme synthetase (Goldberg enzyme), some disturbance of intracellular iron metabolism, or damaged globin production.

BIOSYNTHESIS OF GLOBIN

The synthesis of heme in the mitochondria and that of the globin in the ribosomes take place separately, but the two processes are nevertheless in close correlation with one another (London, 1961, 1966; Kruh and Borsook, 1956). The development of the polypeptide chain starts out from the N terminal and proceeds toward the C terminal (Naughton and Dintzis, 1962; Schweet et al., 1961a, b, 1962), and the complete alpha-beta subunits leave the ribosomes (Huehns and Shooter, 1965). The development of the link between the two subunits is promoted by the heme molecule (Gribble and Schwartz, 1965; Winterhalter and Huehns, 1964).

The production of protoporphyrin-9 and the incorporation of iron into the molecule are the factors that determine the rate of hemoglobin formation (Neve, 1961). If reticulocytes are incubated with ^{59}Fe together with cycloheximide, which inhibits protein synthesis, the incorporation of iron into the reticulocytes proceeds normally (Felicetti et al., 1966). If, however, the heme precursors or the iron are lacking, globin production is diminished (Morell et al., 1958). The addition of heme to iron-deficient reticulocyte preparations stimulates globin synthesis, but accumulation of globin inhibits further synthesis (Karibian and London, 1965; Felicetti et al., 1966). The further addition of iron, however, may result in renewal of globin production (Waxman and Rabinowitz, 1965).

BIBLIOGRAPHY

Bessis, M. C., Breton-Gorius, J.: Ferritin and ferruginous micelles in normal erythroblasts and hypochromic hypersideremic anemias. Blood *14*, 423 (1959); Rev. Hémat. *14*, 165 (1959).

Bessis, M. C., Breton-Gorius, J.: Iron metabolism in the bone marrow as seen by electron microscopy: A critical review. Blood *9*, 635 (1962).

Bogorad, L.: The enzymatic synthesis of porphyrins from porphobilinogen. J. biol. Chem. *233*, 501, 510, 516 (1958).
Brown, E. G.: The relationship of the tricarboxylic acid cycle to the synthesis of delta-aminolaevulinic acid in avian erythrocyte preparations. Biochem. J. *70*, 313 (1958a).
Brown, E. G.: Evidence for the involvement of ferrous iron in the biosynthesis of delta-ALA by chicken erythrocyte preparations. Nature *182*, 313 (1958b).
Burnham, B. F., Lascelles, J.: Control of porphyrin biosynthesis through a negative feedback mechanism. Biochem. J. *87*, 462 (1963).
Cookson, G. H., Rimington, C.: Porphobilinogen, chemical constitution. Nature *171*, 875 (1953).
Dresel, E. I. B., Falk, J. E.: Conversion of delta-amino-laevulinic acid to porphobilinogen in a tissue system. Nature *172*, 1185 (1953).
Falk, J. E., Dresel, E. I. B., Rimington, C.: Porphobilinogen as a porphyrin precursor and interconversion of porphyrin in a tissue system. Nature *172*, 292 (1953).
Felicetti, L., Colombo, B., Baglioni, C.: Assembly of hemoglobin. Biochim. biophys. Acta *129*, 380 (1966).
Gardner, F. H., Nathan, D. G.: Hypochromic anemia and hemochromatosis. Response to combined testosterone, pyridoxine and liver extract therapy. Amer. J. med. Sci. *243*, 81 (1962).
Gibson, K. D., Laver, W. G., Neuberger, A.: Initial stages in the biosynthesis of porphyrins. Biochem. J. *70*, 71 (1958).
Gibson, K. D., Mathew, M., Neuberger, A., Tait, F. R. S., Tait, G. H.: Biosynthesis of porphyrins and chlorophylls. Nature *192*, 204 (1961).
Goldberg, A.: The enzymatic formation of haem by the incorporation of iron into protoporphyrin, importance of ascorbic acid, ergothionine and glutathionine. Brit. J. Haemat. *5*, 150 (1959).
Goldberg, A., Ashenbrucker, H., Cartwright, G. E., Wintrobe, M. M.: Studies on the biosynthesis of heme in vitro by avian erythrocytes. Blood *11*, 821 (1956).
Granick, S. et al.: The occurrence and determination of delta-aminolaevulinic acid and porphobilinogen in urine. J. biol. Chem. *219*, 435 (1956); *232*, 1101, 1119, 1141 (1958).
Gribble, T. J., Schwartz, H. C.: Effect of protoporphyrin on hemoglobin synthesis. Biochim. biophys. Acta *103*, 333 (1965).
Heilmeyer, L.: Die Störungen der Bluthämsynthese. Thieme, Stuttgart 1964.
Huehns, E. R., Shooter, E. M.: Human haemoglobins. J. Med. Genet. *2*, 1 (1965).
Jandl, J. H., Inman, J. K., Simmons, R. L., Allen, D. W.: Transfer of iron from serum iron-binding protein to human reticulocytes. J. clin. Invest. *38*, 161 (1959); J. Lab. clin. Med. *55*, 663 (1960); J. clin. Invest. *42*, 314 (1963); International Symposium on Iron Metabolism, p. 103. Springer, Berlin 1964.
Karibian, D., London, I. M.: Control of heme synthesis by feedback inhibition. Biochem. biophys. Res. Commun. *18*, 243 (1965).
Kikuchi, G., Kumar, A., Shemin, D.: Mechanism of enzymatic synthesis of delta-amino-laevulinic acid. Fed. Proc. *18*, 259 (1959).
Kikuchi, G., Kumar, A., Talmage, P., Shemin, D.: The enzymatic synthesis of delta-amino-laevulinic acid. J. biol. Chem. *233*, 1214 (1958).
Kruh, J., Borsook, H.: Hemoglobin synthesis in rabbit reticulocytes in vitro. J. biol. Chem. *220*, 905 (1956).
Kruh, J. et al.: Synthèse d'hémoglobines par des systèmes acellulaires des réticulocytes. Biochim. biophys. Acta *49*, 509 (1961); *55*, 690 (1962); *87*, 253 (1964); *87*, 669 (1964).
Lascelles, J.: Adaptation to form bacteriochlorophylls in Rhodopseudomonas spheroides. Biochem. J. *72*, 508 (1959).
Laver, W. G., Neuberger, A., Udenfriend, S.: Initial stages in the biosynthesis of porphyrins. Biochem. J. *70*, 4 (1958).
London, I. M.: The metabolism of the erythrocyte. Harvey Lect. *56*, 151 (1961).
London, I. M.: The biosynthesis of hemoglobin and its control in relation to some hypochromic anemias in man. Series Haematologica *2*, 1 (1965).
London, I. M.: The metabolism of the erythrocyte. Proc. nat. Acad. Sci. (Wash.) *55*, 650 (1966).

MORELL, H., SAVOIE, J. C., LONDON, I. M.: The biosynthesis of heme and the incorporation of glycine into globin in rabbit bone marrow in vitro. J. biol. Chem. *233*, 923 (1958).
MUIRHEAD, H., PERUTZ, M. F.: Structure of haemoglobin, a three-dimensional Fourier synthesis of reduced human haemoglobin at 5–5 A resolution. Nature (Lond.) *199*, 633 (1963).
NAUGHTON, M. A., DINTZIS, H. M.: Sequential biosynthesis of the peptide chains of hemoglobin. Proc. nat. Acad. Sci. (Wash.) *48*, 1822 (1962).
NEVE, R. A.: The enzymic incorporation of iron into protoporphyrin. In: Haematin Enzymes, p. 207. Pergamon Press, Oxford–London 1961.
NEVE, R. A., LABBE, R. F., ALDRICH, R. A.: Reduced uroporphyrin-III in the biosynthesis of heme. J. Amer. chem. Soc. *78*, 691 (1956).
PERUTZ, M. F.: The hemoglobin molecule. Scientific Amer. *211*, 64 (1964).
PERUTZ, M. F.: Hemoglobin structure and respiratory transport. Scientific Amer. *239*, 68 (1978).
PERUTZ, M. F., ROSSMANN, M. G., CULLIS ANN, F., MUIRHEAD, H., WILL, G., NORTH, A. C. T.: Structure of haemoglobin. Nature *185*, 416 (1960); Science *140*, 863 (1963a); Nature *199*, 633 (1963b).
PLAUT, G. W. E., CARTWRIGHT, G. E., cit. WINTROBE, M. M.: Clinical Hematology, p. 142. Lea and Febiger, Philadelphia 1967.
PRATO, V., MAZZA, U.: Some aspects of porphyrin-metabolism in thalassemia. Symposion on the Normal and Pathological Porphyrin Metabolism, p. 44. Panminerva Medica, Torino 1961.
PRATO, V., RUBINO, G. F., MAZZA, U., RASETTI, L.: Some aspects of porphyrin metabolism in hereditary spherocytosis and in acquired haemolytic anaemias. Symposion on the Normal and Pathological Porphyrin Metabolism, p. 76. Panminerva Medica, Torino 1961.
SANO, S., GRANICK, S.: Mitochondrial coproporphyrinogen oxidase and protoporphyrin formation. J. biol. Chem. *236*, 1173 (1961).
SCHWEET, R., BISHOP, J., MORRIS, A.: Protein synthesis with particular reference to hemoglobin synthesis—a review. Lab. Invest. *10*, 992 (1961a); Nature *191*, 1365 (1961b); J. biol. Chem. *237*, 760 (1962).
SHEMIN, D., RITTENBERG, D.: The biological utilization of glycine for the synthesis of the protoporphyrin of hemoglobin. J. biol. Chem. *166*, 621 (1946a).
SHEMIN, D., RITTENBERG, D.: The life span of the human red-blood cell. J. biol. Chem. *166*, 627 (1946b).
SHEMIN, D., LONDON, I. M., RITTENBERG, D.: The in vitro synthesis of heme from glycine by the nucleated red blood cell. J. biol. Chem. *173*, 799 (1948); *183*, 749, 757 (1950); *184*, 745, 755 (1950); *185*, 103 (1950); *192*, 315 (1951); *198*, 827 (1952); J. Amer. chem. Soc. *75*, 4873 (1953); *76*, 1204 (1954).
WALDENSTRÖM, J., VAHLQUIST, B.: Studien über die Entstehung der roten Harnpigmente (Uroporphyrin und Porphobilin) bei der akuten Porphyrie aus ihrer farblosen Vorstufe (Porphobilinogen). Z. physiol. Chem. *260*, 189 (1939).
WATSON, C. J.: The erythrocyte coproporphyrin and its variations in respect to protoporphyrin and reticulocytes in certain types of the anemias. Arch. Intern. Med. *86*, 797 (1950).
WAXMAN, H. S., RABINOWITZ, M.: Iron supplementation in vitro and the state of aggregation and function of reticulocyte ribosomes in hemoglobin synthesis. Biochem. biophys. Res. Commun. *19*, 538 (1965).
WESTALL, R. G.: Isolation of porphobilinogen from the urine of a patient with acute porphyria. Nature *170*, 614 (1952).
WIESE, W. C., ARCHDEACON, J. W.: Sodium- and potassium-activated ATP-ase in uptake of Fe^{59} by the rabbit reticulocyte. Proc. Soc. exp. Biol. Med. *118*, 653 (1965).
WINTERHALTER, K. H., HUEHNS, E. R.: Preparation, properties and specific recombination of alpha-, beta-globin subunits. J. biol. Chem. *239*, 3699 (1964).
WINTROBE, M. M.: Clinical Hematology. Lea and Febiger, Philadelphia 1967, 1975.
WITTENBERG, J., SHEMIN, D.: The utilization of glycine for the biosynthesis of both types of pyrrol in protoporphyrin. J. biol. Chem. *178*, 47 (1949).

CHAPTER 12

RED CELL DESTRUCTION

Normal red cells survive in man for about 120 days, and random destruction is minimal. At the end of their life-span the red cells are taken up by phagocytes. Presumably these cells can sense the biophysical and biochemical changes that occur during the life cycle of the red cells and can recognize and remove the aging cells. The changes that occur with aging may be structural or changes in the chemical composition of the cells or in their metabolism. The normal intact cells possess considerable reduction potentials. Glutathione is essential to the integrity of the cells, and it is reduced by glutathione-reductase. Most of the NADPH that is formed in the course of the hexose-monophosphate cycle is utilized in the reduction of glutathione, and the reduced glutathione protects the cells against the oxidative processes. The stability particularly of the SH-radical-containing enzymes, of the SH groups of hemoglobin, and of the structural proteins is related to the presence of GSH (Beutler, 1956; Allen and Jandl, 1961; Rapoport, 1962; Scheuch and Wagenknecht, 1962; Waller and Löhr, 1963).

During the aging process of the cells, the glycolytic enzymes diminish at an exponential rate; the half-time of the activity of glyceraldehyde-3-phosphate-dehydrogenase is 34 days, that of glucose-6-phosphate-dehydrogenase 53 days (Löhr et al., 1959). Hexokinase, the rate-limiting enzyme of glycolysis, is very slowly inactivated (Löhr and Schabert, 1962) and only starts to diminish after the GSH/GSSG ratio and the ATP content have decreased. The ATP/ADP ratio drops to one-fifth of the original value between 60 and 90 days, and as the ATP content diminishes, the stability of the cell structure alters and the shape of the cells changes (Waller, 1965). The alterations in membrane structure render the cells vulnerable to environmental effects (Bessis and Burté, 1965), but intracellular changes also play an important role in the deterioration of cell function (London and Schwartz, 1953; Harris, 1963).

The mechanism whereby the phagocytes recognize the aging red cells is not clear. The red cells are presumed to adhere to the surface of the macrophages during stasis in the sinusoids of the spleen and bone marrow, and this adhesion stimulates the process of phagocytosis. The possibility of a chemical stimulus originating from the unviable red cells that would attract the phagocytes has been suggested by Bessis

(1964) and Bessis and Burté (1965) and has been termed "necrotaxis." These authors suggest that the stimulating substance attracts neighboring phagocytes, which ingest part or the whole of the effete cells. If only part of the cell is ingested, the remainder (a schistocyte) has a very short survival and will also be ingested later. Bessis and Breton-Gorius (1959) studied the phases of ingestion by electron microscopy and found that vacuoles develop around the ingested erythrocytes and also within the ingested red cells, and that actual hemolysis starts only after this process. Ferritin molecules appear on the external surface of the vacuoles but occasionally also in the vacuoles themselves. The stroma of the cells is split into small fragments in which myelin formation can be observed (Policard et al., 1957; Stoeckenius, 1957; Bessis, 1961). Finally, hemosiderin develops around the debris of the stroma, this hemosiderin consisting of ferritin molecules, lipids (myelin), proteins, and other constituents of the stroma.

Normally the erythrophagocytosis takes place primarily in the bone marrow, but in pathological conditions red cells are destroyed also in the spleen and liver.

BIBLIOGRAPHY

ALLEN, D. W., JANDL, J. H.: Oxidative hemolysis and precipitation of hemoglobin. II. Role of thiols in oxidant drug action. J. clin. Invest. *40*, 454 (1961).

BESSIS, M.: Certains aspects morphologiques de l'hémolyse. Les figures myéliniques des globules rouges. In: Hämolyse und hämolytische Erkrankungen, pp. 3–21. Springer, Berlin 1961.

BESSIS, M.: Studies on cell agony and death. Attempt at classification. CIBA Symposium of Cell Injury, pp. 287–316. 1964.

BESSIS, M., BRETON-GORIUS, J.: Différents aspects du fer dans l'organisme. II. J. biophys. biochem. Cytol. *6*, 237 (1959).

BESSIS, M., BURTÉ, B.: Positive and negative chemotaxis as observed after the destruction of a cell by UV or laser microbeams, cit. BESSIS, M.: Destruction of erythrocytes. Series Haematologica *2*, 59 (1965).

BEUTLER, E.: In vitro studies of the stability of red cell glutathione. A new test for drug sensitivity. J. clin. Invest. *35*, 690 (1956).

ESSNER, E.: An electron microscopy study of erythrophagocytosis. J. biophys. biochem. Cytol. *7*, 329 (1960).

HARRIS, J. W.: The Red Cell, Vol. I. Harvard University Press, Cambridge, Mass. 1963.

LÖHR, G. W., SCHABERT. J. (1962), cit. WALLER, 1965.

LÖHR, G. W., WALLER, H. D., KARGES, O., SCHLEGEL, B., MÜLLER, A. A.: Zur Biochemie der Alterung menschlicher Erythrocyten. Klin. Wschr. *37*, 833 (1959).

LONDON, I. M., SCHWARTZ, H.: Erythrocyte metabolism. The metabolic behaviour of the cholesterol of human erythrocytes. J. clin. Invest. *32*, 1248 (1953).

POLICARD, A., BESSIS, M., BRETON-GORIUS, J.: Structures myéliniques observées au M. E. sur des coupes de globules rouges en voie de lyse. Exp. Cell. Res. *13*, 184 (1957).

RAPOPORT, S.: Reifung und Alterungsvorgänge in Erythrocyten. Folia haemat. (Lpz.) *78*, 364 (1962).

SCHEUCH, D., WAGENKNECHT, C.: Über die Bedeutung des Glutathions für den Stoffwechsel intakter Zellen am Beispiel der roten Blutkörperchen. Acta biol. med. germ. Suppl. II, 199 (1962).

STOECKENIUS, W.: Morphologische Beobachtungen beim intrazellulären Erythrocytenabbau und der Eisenspeicherung in der Milz des Kaninchens. Klin. Wschr. *35*, 700 (1957).

WALLER, H. D.: Biochemical determinants of red cell life span. Series Haematologica *2*, 34 (1965).

WALLER, H. D., LÖHR, G. W.: Neue Ergebnisse zum Mechanismus der Heinz-Körperchenbildung in Erythrocyten. Folia haemat. (Frankfurt) N. F. *8*, 360 (1963).

CHAPTER 13

HEMOGLOBIN CATABOLISM

During the breakdown of hemoglobin virtually all the iron is conserved and re-utilized for the synthesis of the various iron-containing compounds (Moore and Dubach, 1956). The globin decomposes to amino acids, and the porphyrin ring is split and converted to bilirubin (Fig. 13/1).

The protoporphyrin ring opens along the alpha-methene bridge, and the carbon atom of the bridge is oxidized to CO, which enters the bloodstream in the form of CO-hemoglobin (Sjöstrand, 1951). The CO-hemoglobin concentration of the circulating blood gives an indirect measure of the rate of hemoglobin destruction (Coburn et al., 1964).

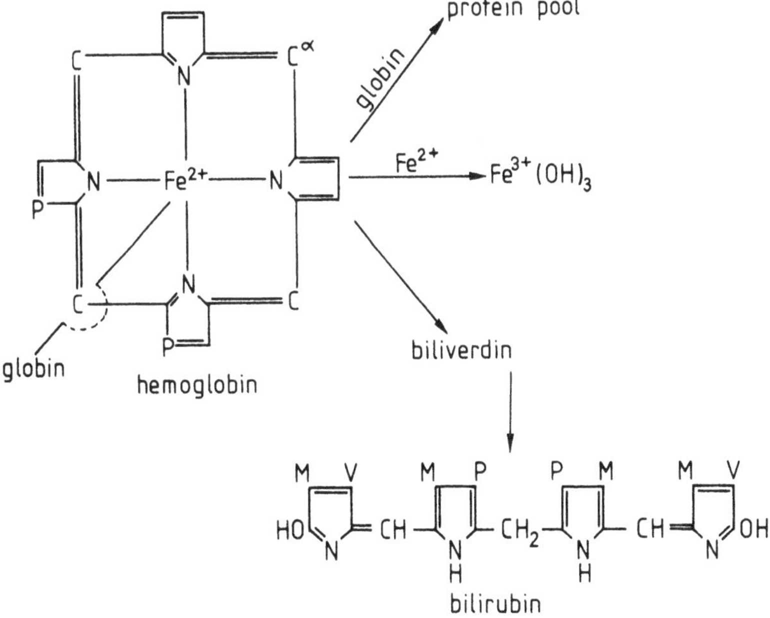

Fig. 13/1. Schematic representation of the decomposition of hemoglobin (after Schmid, R.: Series Haematologica *2*, 69, 1965). For abbreviations see Fig. 11/3

The first stage of hemoglobin decomposition is the separation of heme and globin, and the next step is the opening of the heme ring and the splitting of the iron from the molecule. The oxidized form of heme ("hemin") loses the iron atom after the CO has split off (Fig. 13/2), and it is converted to biliverdin, which is then further reduced to bilirubin.

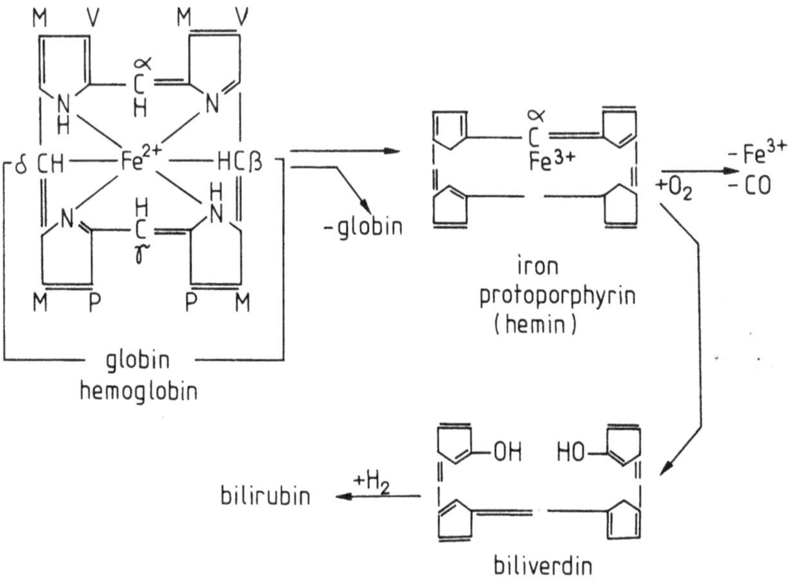

Fig. 13/2. The iron–protoporphyrin complex is the obligatory intermediary product of the transformation of heme to bilirubin (after Schmid, R.: Series Haematologica *2*, 69, 1965)

Biliverdin can be demonstrated in the bile of birds and some amphibia, but in men and mammals it is excreted only in its reduced form. Thus, when ^{14}C-biliverdin was injected into rats, the isotope was recovered quantitatively from the ^{14}C-bilirubin in the bile (Goldstein and Lester, 1964). The conversion of hematin to bile pigment has been shown both *in vivo* and *in vitro* (Pass et al., 1945; London, 1950; Kench et al., 1950; Snyder and Schmid, 1964).

In vivo a small proportion of isotopically labeled protoporphyrin is converted to stercobilin (London et al., 1951), but this stercobilin is probably derived from the small proportion of protoporphyrin that is converted to heme (Mauzerall and Granick, 1958).

Heme-alpha-methenyl oxidase, which can be obtained from the soluble fraction of liver and kidney homogenates, catalyzes hemoglobin to a biliverdin precursor, which is rapidly oxidized to biliverdin (Nakajima, 1963; Nakajima and associates,

1963). The decomposition of hemoglobin to bile pigment is relatively rapid, as shown, for example, in the rat, where ^{14}C-labeled blood pigment appears in the bile on average in 3 hours (Fig. 13/3) (Ostrow et al., 1962).

About 80–90% of the bile pigment excreted in normal man originates from red cell destruction, the rest is from other sources. Using ^{15}N-glycine, London et al. (1950)

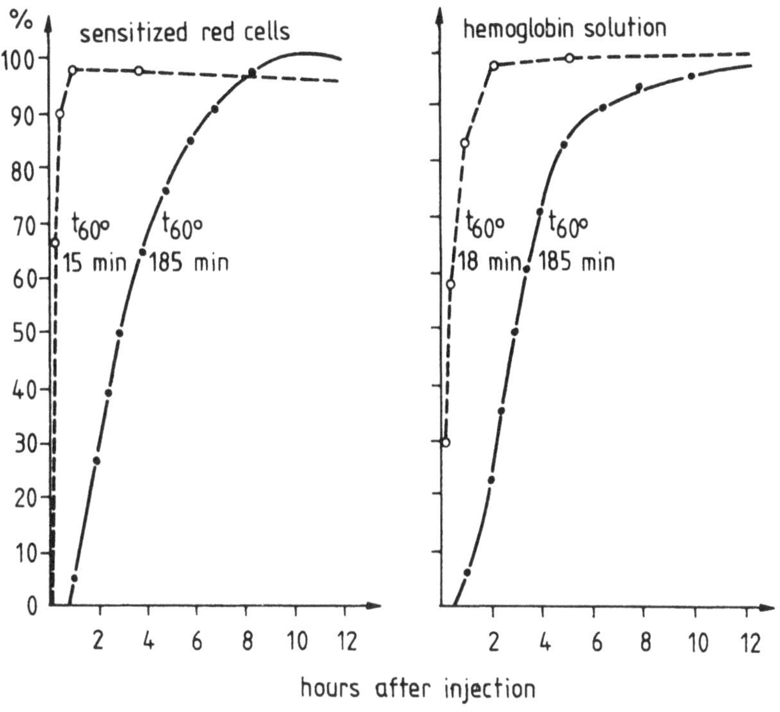

Fig. 13/3. The time taken for bilirubin to appear in bile following the administration of labeled hemoglobin solution or sensitized red cells; 50% of the bilirubin excretion takes place within 3 hours (after Ostrow, J. D. et al.: J. clin. Invest. *41*, 1628, 1962). ○– –○– –○ ^{59}Fe outflow; ●–●–● ^{14}C-bilirubin excretion

and Gray et al. (1950) have shown that 10–20% of the ^{15}N appears in the feces soon after administration and at a time when the isotope concentration in the heme is increasing. This early labeled bile pigment has been attributed to various sources. There is probably always some degree of ineffective erythropoiesis with destruction of red cell precursors before they enter the circulation. Furthermore, heme may be formed in unnecessary quantities as compared with globin, and part of the blood pigment may be degraded in the red cell precursors. Sources other than the hemoglobin, such as myoglobin, catalase, peroxidase, and cytochrome, may be responsible for a proportion of the early labeled bile pigment, and possibly bilirubin may be formed directly from the porphyrins or pyrrol precursors.

From their electron microscopic examinations Bessis et al. (1961) have suggested that during the phase of erythropoiesis in which the nuclei are expelled from the normoblasts, the expelled nuclei may be surrounded by a thin cytoplasmic border and that the early bile pigment could originate from the hemoglobin that is present in these fragments of cytoplasm.

In some observations in pathological conditions associated with marked ineffective erythropoiesis, there is support for the view that the bulk of the early labeled bile pigment is derived from this ineffective erythropoiesis; for example, in thalassemia the early fraction amounts to 40–80% (Grinstein et al., 1960; Robinson et al., 1962). High values are also found in pernicious anemia (London and West, 1950), in refractory normoblastic anemia (Barrett et al., 1964), in erythropoietic porphyria (London et al., 1950; Gray et al., 1950), and in some atypical forms of hereditary spherocytosis (Israels and Zipursky, 1962). Under physiological conditions there is more doubt about the origin of the early labeled bile pigment, and it has been suggested that some of it comes from the liver (Israels et al., 1963a, b; Gray et al., 1964; Schmid, 1965).

BIBLIOGRAPHY

BARRETT, P. V. D., CLINE, M. J., BERLIN, M. I.: The association of the urobilin "early peak" and erythrocytosis. Clin. Res. *12*, 108 (1964).

BESSIS, M., BRETON-GORIUS, J., THIERY, J. P.: Rôle possible de l'hémoglobine accompagnant le noyau des érythroblastes, dans l'origine de la stercobiline éliminée précocement. C. R. Acad. Sci. (Paris) *252*, 2300 (1961).

COBURN, R. F., WILLIAMS, W. J., FORSTER, R. E.: Effect of erythrocyte destruction on carbon monoxide production in man. J. clin. Invest. *43*, 1098 (1964).

GOLDSTEIN, G. W., LESTER, R.: Reduction of biliverdin-C^{14} to bilirubin-C^{14} in vivo (1964), cit. SCHMID, 1965.

GRAY, C. H., KULCZYCKA, A., NICHOLSON, D. C., MAGNUS, I. A., RIMINGTON, C.: Isotopic studies on a case of erythropoietic protoporphyria. Clin. Sci. *26*, 7 (1964).

GRAY, C. H., NEUBERGER, A., SNEATH, P. H. A.: Studies in congenital porphyria: incorporation of ^{15}N in stercobilin in normal and porphyric patients. Biochem. J. *47*, 87 (1950).

GRINSTEIN, M., BANNERMAN, R. N., VARRA, J. D., MOORE, C. V.: Hemoglobin metabolism in thalassemia: in vivo studies. Amer. J. Med. *29*, 18 (1960).

ISRAELS, L. G., ZIPURSKY, A.: Primary shunt hyperbilirubinaemia. Nature *193*, 73 (1962).

ISRAELS, L. G., SKANDERBEG, J., GUYDA, H., ZINGG, W., ZIPURSKY, A.: Study of the early-labeled fraction of bile pigment: the effect of altering erythropoiesis on the incorporation of 2-^{14}C-glycine into haem and bilirubin. Brit. J. Haemat. *9*, 50 (1963b).

ISRAELS, L. G., YAMAMOTO, T., SKANDERBEG, J., ZIPURSKY, A.: Shunt bilirubin: evidence for two components. Science *139*, 1054 (1963a).

KENCH, J. E., GARDIKAS, C., WILKINSON, J. F.: Bile pigment formation in vitro from haematin and other haem derivates. Biochem. J. *47*, 129 (1950).

LONDON, I. M.: Conversion of hematin to bile pigment. J. biol. Chem. *184*, 373 (1950).

London, I. M., West, R.: Formation of bile pigment in pernicious anemia. J. biol. Chem. *184*, 359 (1950).
London, I. M., West, R., Shemin, D., Rittenberg, D.: On the origin of bile pigment in normal man. J. biol. Chem. *184*, 351 (1950).
London, I. M., Yamasaki, M., Sabella, A. G.: Conversion of protoporphyrin to bile pigment. Fed. Proc. *10*, 217 (1951).
Mauzerall, D., Granick, S.: Porphyrin biosynthesis in erythrocytes. III. Uroporphyrinogen and its decarboxylase. J. biol. Chem. *232*, 1141 (1958).
Moore, C. V., Dubach, R.: Metabolism and requirements of iron in man. J. Amer. med. Ass. *162*, 197 (1956).
Nakajima, H.: Studies on heme alfa-methenyl oxygenase. II. Isolation and characterization of final reaction product, possible precursor of bilirubin. J. biol. Chem. *238*, 3797 (1963).
Nakajima, H., Takemura, T., Nakajima, O., Yamaoka, K.: Studies of heme alfa-methenyl oxygenase. I. Enzymatic conversion of pyridine hemichromogen and hemoglobin-haptoglobin into possible precursor of biliverdin. J. biol. Chem. *238*, 3784 (1963).
Ostrow, J. D., Jandl, J. H., Schmid, R.: Formation of bilirubin from hemoglobin in vivo. J. clin. Invest. *41*, 1628 (1962).
Pass, I. J., Schwartz, S., Watson, C. J.: The conversion of hematin to bilirubin following intravenous administration in human subjects. J. clin. Invest. *24*, 283 (1945).
Robinson, S., Vanier, T., Desforges, J. F., Schmid, R.: Jaundice in thalassemia minor: a consequence of ineffective erythropoiesis. New Engl. J. Med. *267*, 523 (1962).
Schmid, R.: Kinetics and mechanisms of hemoglobin catabolism. Series Haematologica *2*, 69 (1965).
Sjöstrand, T.: Endogenous formation of carbon monoxide. Acta physiol. scand. *22*, 137 (1951).
Snyder, A. L., Schmid, R. (1964), cit. Schmid, 1965.

CHAPTER 14

FERROKINETICS

SURVEY

The levels of plasma iron and iron-binding capacity are static indices only of iron metabolism. They do not yield information about ferrokinetics, i.e., the flow and distribution of iron within the organism.

Early attempts to assess iron transport depended on the estimation of the rate at which the plasma iron concentration diminished after the saturation of transferrin with iron (Waldenström, 1941; Holmberg and Laurell 1945; Brendstrup, 1949). From this the rate of outflow of plasma iron was inferred. Tötterman (1949) and Brendstrup (1949) found that the plasma iron level of healthy men dropped by about 40 μg/dl or 58 μg/dl, respectively, within 2 hours of the saturation of transferrin. Bernát et al. (1965, 1966) found that the rate of reduction of plasma iron was not linear; in the first hour the level decreased by 113 μg/dl but in the subsequent 2 hours by only 14 μg/dl, and the curve then flattened out for several hours (see Fig. 20/4).

The rate of iron outflow is accelerated in infections (Cartwright et al., 1946; Brendstrup, 1949; Goldeck and Remy, 1953), in hemorrhagic anemia (Brøchner-Mortensen, 1943; Goldeck and Remy, 1953), in malignancy (Tötterman, 1951; Goldeck and Remy, 1953), and in thermal injury (Bernát et al., 1965). In the latter, even the peak values 5 minutes after the injection of iron are lower than in healthy individuals. The plasma iron continues to decline over some hours and is exponential in character.

The fall in plasma iron is determined by the capacity of tissue iron receptors and by the avidity of the erythroblasts (Bothwell and Finch, 1962) and of the RE system (Hoppe et al., 1961). Hoppe et al. (1961) have used an intravenous iron load as a test of function of the reticuloendothelial system. In posthemorrhagic anemia and in iron deficiency, iron flows mainly from plasma to the erythroid cells of the bone marrow, while in infections and in anemia of thermal injury, a greater proportion than normal is directed to the cells of the reticuloendothelial system (Bernát et al., 1965, 1966).

The first reliable quantitative kinetic studies followed the use of radioactive iron by Huff et al. in 1950. These workers measured the rate of decline of plasma activity after the intravenous injection of radioiron, and from the time taken for the radioactivity to drop by 50% (T/2) (Fig. 14/1) they calculated the percentage of iron that leaves the plasma in unit time. The product of the so-called outflow constant $\left(\frac{\log 2}{T/2}\right)$ and the total circulating iron $\left(\frac{\text{Pl. Fe}}{100} \times \text{Pl. vol.}\right)$ expresses the absolute value of the iron that leaves or enters the circulation during a given time (e.g., 1 or 24 hours). Huff et al. (1950, 1951, 1952), and later other investigators, called this value *plasma iron turnover rate*, but more recently the less ambiguous *plasma iron transport rate* is preferred (Lajtha, 1961).

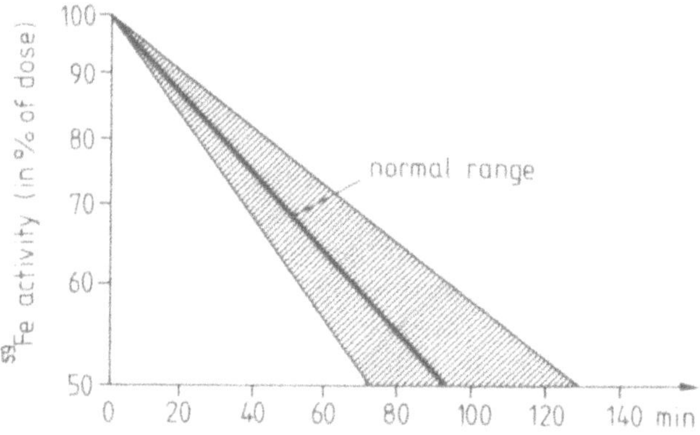

Fig. 14/1. Decline in the radioactivity of plasma after the i.v. injection of ^{59}Fe. In healthy man T/2 is 70–130 min. The rate of decline is exponential (after Bernát, I. et al., 1968)

Under normal conditions most of the radioactive iron cleared from the plasma is incorporated into hemoglobin in the bone marrow and subsequently appears in the circulating erythrocytes. The activity of the blood of the healthy individual increases in proportion to the newly formed red cells that reach the periphery, and within 10–14 days about 70–90% of the radioactive iron appears in the red cells (Fig. 14/2). The shape of the incorporation curve indicates not only the extent of the effective erythropoiesis but also the duration of maturation of the erythroblasts ("mean marrow duration time", "mean transit time") (Dubach et al., 1946; Finch et al., 1949a, b; Loeb et al., 1953).

If the daily plasma iron transport rate is multiplied by the proportion that is incorporated into the erythroid cells of the bone marrow, the amount of iron utilized in 24 hours for red cell production can be determined. This parameter (*red cell iron turnover rate*) is the quantitative indicator of effective erythropoiesis (see Fig. 20/9).

The absolute life-span of the red cells can also be calculated from the further development of the activity of the blood (see Fig. 15/2).

^{59}Fe emits both β particles and γ rays, the latter of sufficient penetration to allow for *in vivo* surface counting measurements (Huff et al., 1951; Elmlinger et al., 1953; Pollycove, 1959). Measurements are made over the sacrum, liver, spleen, and heart (Fig. 14/3), appropriate corrections being made for the activity of the blood flowing through the various organs (Veall and Vetter, 1958).

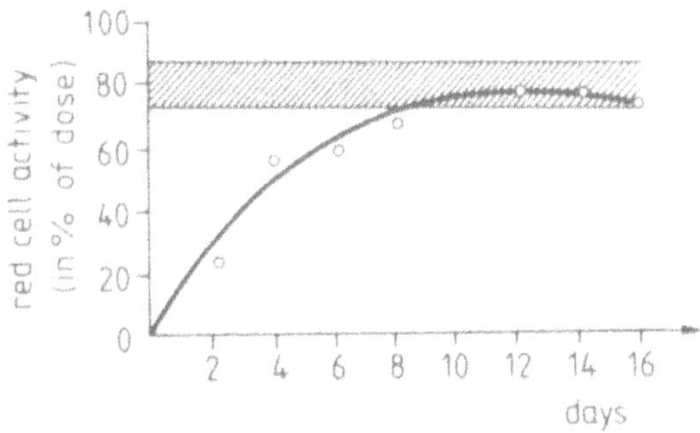

Fig. 14/2. Red cell activity in healthy man after the i.v. injection of ^{59}Fe as a function of time ("incorporation curve") (after Bernát, I., 1971)

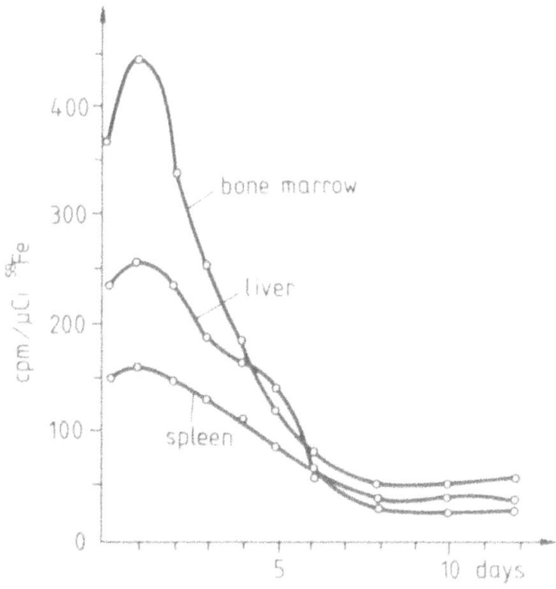

Fig. 14/3. Radioactivity over the bone marrow (os sacrum), liver and spleen after the i.v. injection of ^{59}Fe in healthy man (after Bernát, I., 1971)

The ferrokinetic studies have extended our knowledge concerning the quantitative aspect of iron metabolism, erythropoiesis, and hemoglobin synthesis, although the kinetic laws are much more complex than originally thought. Various mathematical models have been elaborated (Pollycove and Mortimer, 1961; Sharney et al., 1965; Najean et al., 1969) for the better understanding of the flow and distribution of iron in the organism, but such analyses are of doubtful value in clinical practice.

PLASMA IRON CLEARANCE AND PLASMA IRON TRANSPORT RATE

After an intravenous injection of a tracer dose of radioactive iron the plasma radioactivity declines exponentially over the first 2–4 hours, and when plotted on semilogarithmic paper a more or less straight line is obtained, the slope of which characterizes the *iron clearance rate* (Fig. 14/4). In the healthy subject the T/2 is about 70–130 minutes, which means that under normal conditions about 30–35% of the plasma iron content is exchanged during 1 hour. If, for example, the T/2 is 2 hours, the proportion of the plasma iron content that is exchanged in 1 hour equals 0.346, or 34.6% (log 2 : T/2). The total iron content of the plasma therefore is exchanged in 3 hours, or 8 times a day.

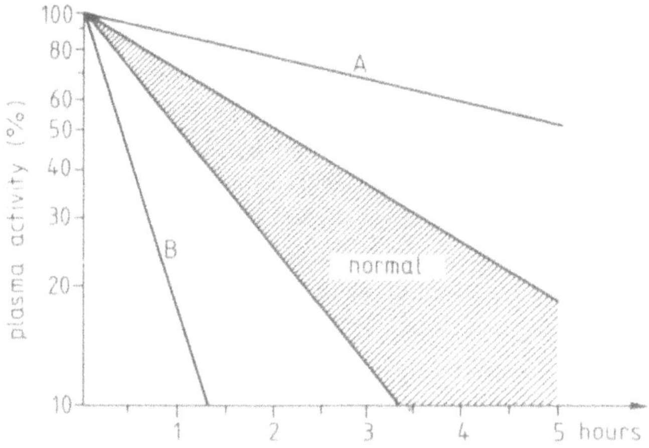

Fig. 14/4. Plasma iron clearance in healthy man and in various pathological processes (after Bothwell, T. H. et al.: Brit. J. Haemat. *2*, 1, 1956). A: Aplastic anemia; B: iron deficiency, infectious and hemolytic anemia

The rate of iron outflow does not indicate, however, the extent of iron exchange. Bothwell et al. (1957) demonstrated that in healthy subjects and nonhematological patients there is a correlation between plasma clearance T/2 and plasma iron concentration (Fig. 14/5); e.g., if the plasma iron level is half the normal value, the rate of iron outflow doubles. Thus the absolute value of the iron leaving the plasma

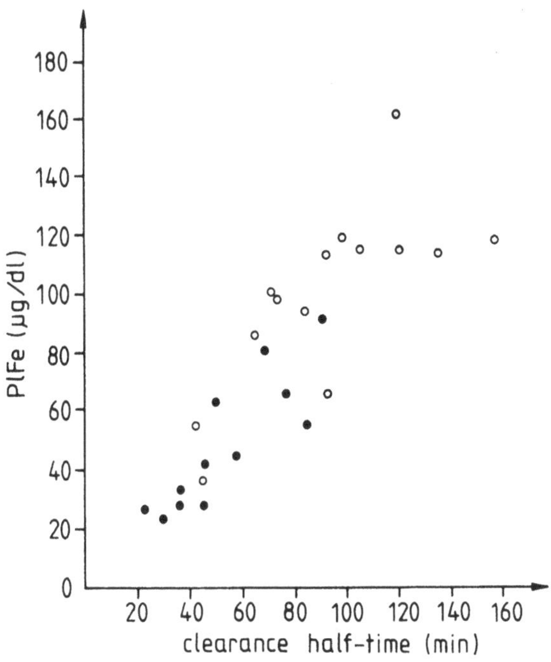

Fig. 14/5. Correlation between plasma iron level and the rate of iron clearance from the plasma in normal subjects (○) and patients with nonhematological disorders (●). The plasma iron turnover is normally maintained at a relatively constant value in spite of variations in the plasma iron level (after Bothwell, T. H. et al.: Blood, *12*, 409, 1957)

remains unchanged in spite of the accelerated clearance. The transport rate is not affected by the diurnal variation of the plasma iron (Bothwell and Mallett, 1955), but it is reduced during sleep, which may be related to diurnal variations of erythropoiesis (Paterson, 1957). *The extent of iron exchange can therefore be reliably assessed only from the absolute amount of iron leaving the plasma.*

The transport rate varies normally between 20 and 42 mg/24 hours (Huff et al., 1950; Hyman and Harvey, 1955; Bothwell et al., 1955, 1957; Freireich et al., 1957; Giannopoulos and Bergsagel, 1959). The amount varies with body weight. Bernát et al. (1968) found the transport rate to vary between 25 and 34 mg/24 hours with a mean value of 30 mg/24 hours in individuals of 55–70 kg body weight.

In general, the plasma iron transport rate of females is lower than that of males, but this sex difference disappears if the values are related to body weight (Bothwell et al., 1955).

For comparative measurements the following units have been used

mg/kg/24 hr
mg/liter RBC/hr
mg/dl whole blood/24 hr.

Bernát et al. (1968) found that the normal plasma iron transport rate varied from 0.41 to 0.52 mg/kg/24 hr with a mean of 0.47. Bothwell et al. (1957) and Freireich et al. (1957) have recommended the use of mg/dl whole blood/24 hr; the normal mean value is approximately 0.6 mg ± 25%. About 0.42 mg of iron/dl whole blood is utilized for hemoglobin production daily. Up to 10–20% more iron may reach the bone marrow, but as a result of intramedullary hemolysis it is returned to the plasma through the RHS within 3–5 days. A further 5–10% of the transport iron goes to other tissues, notably the liver, and 5–10% is exchanged in the extravascular space.

The early exponential decline of plasma radioactivity indicates that there is little early reflux of iron into the plasma, but more detailed analyses of the later stages of the curve have shown that there may be significant feedback from the labile iron pool. This is indicated by one or more slower exponential curves (Fig. 14/6), from

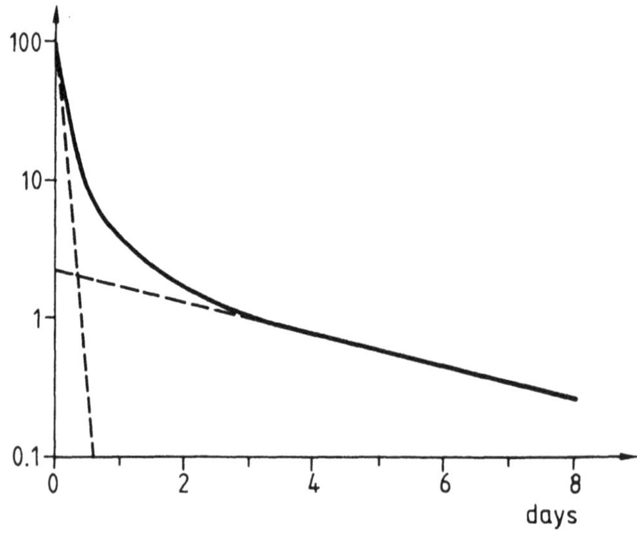

Fig. 14/6. Plasma radioiron clearance in the course of the first 8 days after the i.v. injection of ^{59}Fe. The broken lines represent the extrapolations of the curve's two components (after Lajtha, L. G.: The Use of Isotopes in Haematology. Blackwell, Oxford 1961)

which it has been estimated that the labile iron pool contains 80–90 mg of iron in normal individuals (Fairbanks and Beutler, 1977).

The amount of iron that flows through the plasma during the unit time is influenced largely by the rate of erythropoiesis; iron exchange between the plasma and the iron stores is a less significant factor. For this reason the plasma iron transport rate varies with the total erythropoiesis. It shows a 1.5- to 2-fold rise if erythropoiesis is enhanced as a consequence of hemorrhage (Bothwell et al., 1957) or of relative hypoxia, e.g., when moving from sea level to a high altitude (Lawrence et al., 1952; Reynafarje et al., 1959). The rate drops by 70–90% on return to sea level with a consequent decrease in erythropoiesis (Lawrence et al., 1952; Reynafarje et al., 1959). A similar reduction takes place in response to ionizing irradiation, to cytotoxic treatment (nitrogen mustard) (Keiderling et al., 1957; Bothwell et al., 1957; Bertenchamps et al., 1958), or to hypertransfusion (Bothwell et al., 1957).

In hypoplastic and aplastic anemias the transport rate may remain more or less unchanged in spite of the reduced erythropoiesis because of increased iron flow to tissues other than the bone marrow (Huff et al., 1950; Bothwell et al., 1956, 1957; Bernát, 1971).

Normally the capacity of the other tissues for the uptake of iron is limited. The only exception is the placenta, whose iron avidity nearing term exceeds that of the maternal erythroid tissue (Bothwell et al., 1958; Davies et al., 1959).

In pathological conditions the proportion of iron taken up by other tissues may be increased, e.g., in hypoplastic anemias or idiopathic hemochromatosis (Elmlinger et al., 1953; Pollycove, 1959). When erythropoiesis is enhanced, less iron is directed to the tissues and more to the bone marrow, and occasionally this may be as much as 100%.

The plasma iron transport rate primarily reflects changes in bone marrow function and gives an indication of the total erythropoiesis. In pathological conditions where there is ineffective erythropoiesis, disturbance of heme synthesis, or increased iron stores, it gives no indication of the effective red cell production.

INCORPORATION OF RADIOIRON INTO THE ERYTHROBLASTS AND RETICULOCYTES. IRON UTILIZATION IN THE COURSE OF RED CELL PRODUCTION. EFFECTIVE ERYTHROPOIESIS

If the activity of the red cells is determined daily or every other day for about 2 weeks after the injection of radioiron and the values are plotted in a coordinate system, normally 70–90% of the radioactive dose will be incorporated into the red cells within 10–14 days, i.e., about 4/5th (24 mg iron) of the amount of iron that passes through the plasma daily. This fraction of the transport iron indicates effective erythropoiesis. Under pathological conditions the iron incorporation may be either enhanced or diminished.

Iron incorporation increases in iron deficiency, probably because, as a consequence of the low iron saturation of transferrin, other tissues are less able to

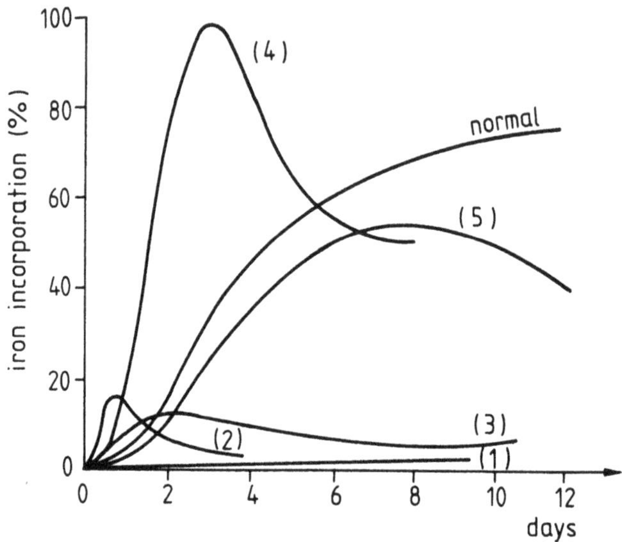

Fig. 14/7. Curves of radioiron incorporation in healthy man and in various pathological processes. The activity of the blood is influenced by the rate of both red cell production and destruction (after Giblett, E. R. et al.: Blood, *11*, 291, 1956). (1) and (2) aplastic anemia; (3) Cooley's anemia; (4) acquired hemolytic anemia; (5) hereditary spherocytic anemia

bind the iron (Bothwell and Finch, 1962). However, in the presence of iron deficiency where the plasma iron pool is much reduced, the specific activity of the plasma iron following the radioiron injection is higher than in normal subjects. It is possible, therefore, that the concentration and not the absolute amount of incorporated iron in erythroblasts is higher than normal. An appropriate correction could be made taking into consideration the reduced plasma iron pool.

There may be an unusually steep iron incorporation curve in some hemolytic anemias with the entry of many immature reticulocytes into the peripheral blood, but because of the hemolysis and random shortening of the life-span of the red cells the activity of the blood begins to decline within a few days (Fig. 14/7).

Iron incorporation is lower in hemochromatosis, where the unsaturated iron-binding capacity is slight or nonexistent (Bothwell et al., 1955), and also in bone marrow hypoplasia or aplasia (e.g., in radiation injury or drug-induced aplasia), where the diminished iron incorporation is a sensitive and direct indicator of disturbed bone marrow function (Belcher et al., 1954) (Fig. 14/8).

In ineffective erythropoiesis, i.e., when some of the red cell precursors are destroyed in the bone marrow and only a proportion reach the peripheral blood,

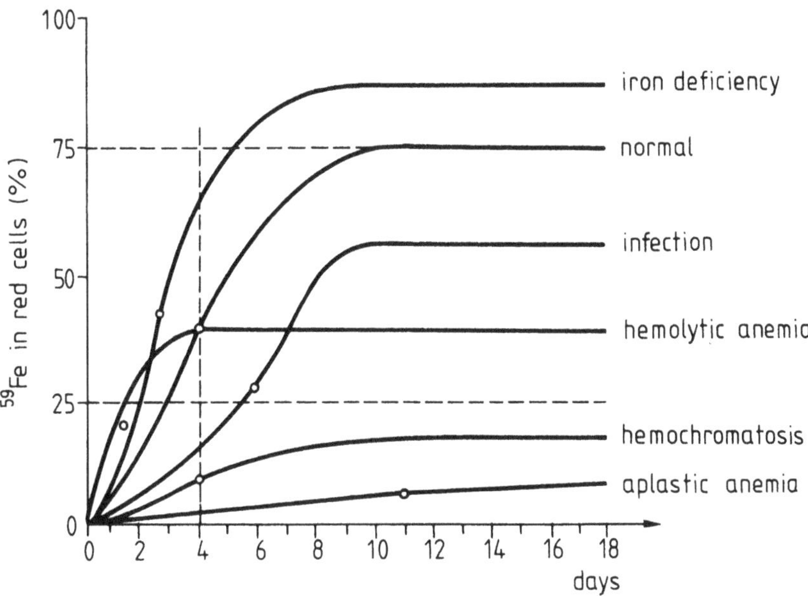

Fig. 14/8. Curves of iron incorporation in healthy man and in various pathological processes. (after Dreyfus, J. C. and Schapira, G., in: Gross, F.: Iron Metabolism. Springer Berlin–Göttingen–Heidelberg 1964)

iron incorporation is also reduced, as it may be in those cases of disturbed heme synthesis where non-heme iron is accumulated intracellularly and is removed by the spleen without necessarily destroying the red cells themselves (Crosby, 1957, 1959). Reduced iron incorporation is also seen in infection (Freireich et al., 1957) and thermal injuries (Bernát, 1971), which are examples of conditions in which there is an increased avidity of the RE system for iron and a diversion from the erythropoietic tissue.

Reduction in iron incorporation may be only partly apparent in those cases, in which the plasma iron pool is greatly increased and the specific activity of the injected radioiron is consequently decreased.

DISTRIBUTION OF IRON AMONG THE BONE MARROW, LIVER, AND SPLEEN

In vivo surface measurements of radioactivity may give some indication of the flow and distribution of iron within the organism. Measurements are usually taken over the heart for assessment of blood activity, over the sacrum for assessment of the bone marrow uptake, and over the liver and spleen. Normally the activity increases rapidly over the sacrum as the plasma activity decreases. The peak over the sacrum is usually reached within 20–24 hours and gradually drops as the result of the outflow of newly formed red cells. The increase in activity of the blood mirrors the decrease in bone marrow activity (Fig. 14/9).

Activity over the liver and spleen increases slightly with the early rise in activity over the sacrum, and it falls off over these organs as the bone marrow activity starts to decline (see Fig. 14/3).

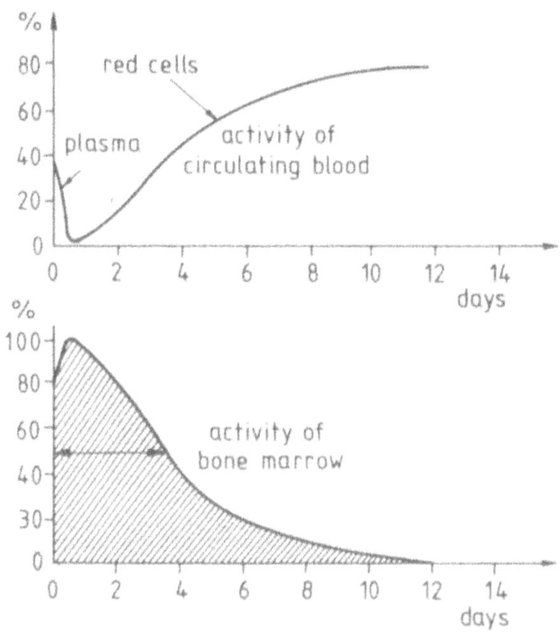

Fig. 14/9. Correlation between the radioactivity of the circulating blood and bone marrow after the i.v. injection of ^{59}Fe. The upper part of the figure indicates the decline in plasma activity and the subsequent rise in the activity of the red blood cells. The lower part shows the change in activity over the bone marrow. The latter curve corresponds to the reciprocal of the former. The horizontal arrow points to the mean of the bone marrow turnover time (after Bothwell, T. H. and Finch, C. A.: Iron Metabolism. Little, Brown and Co., Boston 1962)

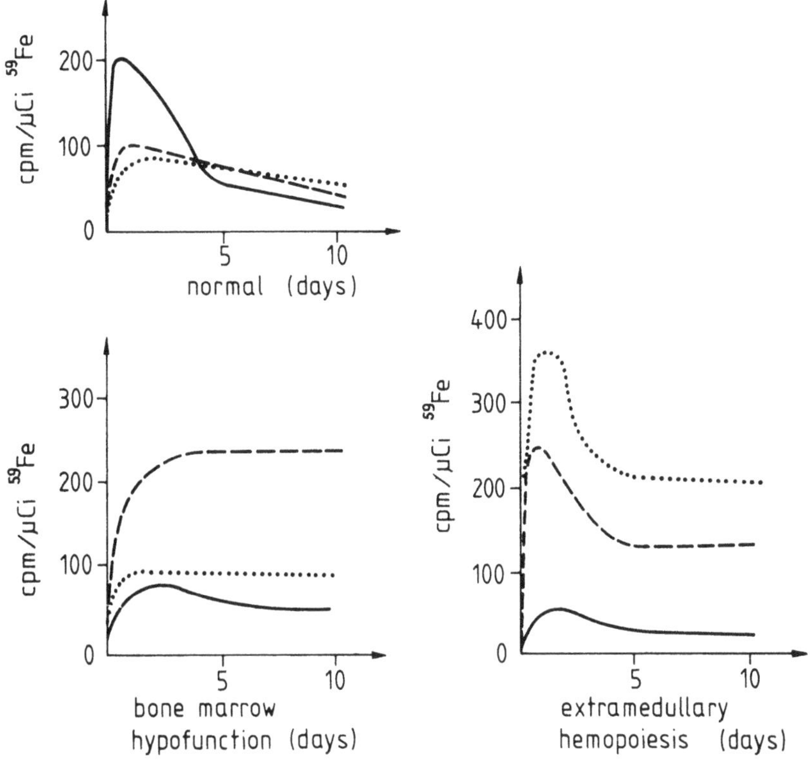

Fig. 14/10. Changes in the radioactivity over the bone marrow (———), liver (– – –) and spleen (····) after the i.v. injection of ^{59}Fe in healthy man, in bone marrow hypofunction, and in the presence of extramedullary hemopoiesis

In hemolytic anemias there is a more rapid rise in the bone marrow activity than normal, and as the activity of the sacrum starts to decline there is a rise over the spleen, where the red cells are destroyed, and sometimes also over the liver.

If the bone marrow is hypoplastic, little or no activity is found over the sacrum, and more radioiron is accumulated in the liver and spleen (Fig. 14/10). If there is extramedullary hemopoiesis in the liver and spleen, the initial activity may gradually decline if there is any degree of effective erythropoiesis in these organs.

In thermal injuries radioactivity over the liver and spleen decreases at a slower rate, and the values are stabilized at a higher level than under normal conditions (Bernát et al., 1968). This can be ascribed to accumulation of the radioiron in the reticuloendothelial system as the result of increased avidity of the reticulum cells and histiocytes for iron.

Table 14/1

The norms of ferrokinetic investigations

Parameter	Norms and literature		
Plasma iron clearance half-time (min)	97	(85–138)	[1]*
		(70–140)	[2]
Plasma iron transport rate (mg/24 hr)	30	(20–42)	[3–6]
	30	(25–34)	[1]
(mg/dl blood/24 hr)	0.61 ± 25%		[5]
(mg/kg/24 hr)	0.41 ± 0.52		[1]
^{59}Fe incorporation (%)	83	(75–94)	[1]
Amount of iron utilized for effective erythropoiesis ("RBC-iron-turnover rate") (mg/24 hr)	24	(20–27)	[1]
In vivo (body surface) measurements (activity ratios)	Bone marrow ≫ liver > spleen		

* [1] Bernát et al., 1968; [2] Keiderling et al., 1957; [3] Huff et al., 1950; [4] Hyman and Harvey, 1955; [5] Bothwell et al., 1955, 1956, 1957; [6] Freireich et al., 1957.

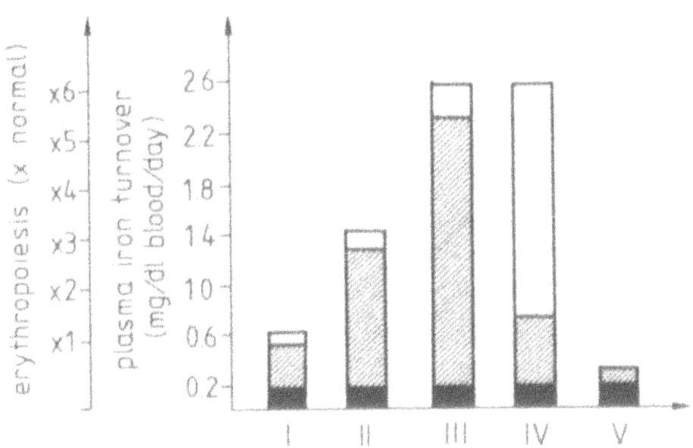

Fig. 14/11. Ferrokinetic classification of anemias on the basis of plasma iron transport rate and iron utilization. Full height of the columns (= plasma iron transport rate) reflects total erythropoiesis. Shaded areas correspond to effective and blank areas to ineffective erythropoiesis. The black areas represent iron exchange not related to hemopoiesis (after Bothwell, T. H. and Finch, C. A.: Iron Metabolism. Little, Brown and Co., Boston 1962). I, normal; II, optimal response in acute anemia; III, optimal response in chronic anemia; IV, ineffective erythropoiesis; V, bone marrow aplasia

Surface counting alone cannot give an accurate quantitative assessment of iron kinetics, but when it is combined with the measurements of the plasma iron transport rate and iron incorporation, they can supplement the information about the flow and distribution of iron within the organism.

The normal values of ferrokinetic measurements are given in Table 14/1, and the ferrokinetic classification of anemias in Fig. 14/11.

BIBLIOGRAPHY

BELCHER, E. H., GILBERT, I. G. F., LAMERTON, L. F.: Experimental studies with radioactive iron. Brit. J. Radiol. *27*, 387 (1954).

BERLIN, N. I.: Techniques for the study of erythrokinetics. In: WILLIAMS, W. J. et al. (eds.): Hematology pp. 152–162. McGraw-Hill Inc., New York 1972, 1977.

BERNÁT, I.: Az égési anaemia pathogenesise (The Pathogenesis of Anemia after Thermal Injury). Akadémiai Kiadó, Budapest 1971.

BERNÁT, I., DÓZSÁN, G., MAGYARI, J., NOVÁK, J.: Anaemia after thermal injury. IV. Iron kinetics in burned patients. Haematologia *2*, 279 (1968).

BERNÁT, I., NOVÁK, J., ELEK, S., DÓZSÁN, G.: Anemia after thermal injury. I. Acta med. Acad. Sci. hung. *21*, 121 (1965).

BERNÁT, I., NOVÁK, J., FÁBER, V., DÓZSÁN, G., ELEK, S.: Neue Beiträge zur Pathogenese der Verbrennungsanämie. Folia haemat. (Lpz.) *86*, 85 (1966).

BERTENCHAMPS, A., KENIS, Y., TAGNON, H. J.: The effect of the administrations of NH_2 on the utilization of plasma iron. Cancer *11*, 117 (1958).

BERZINI, C. et al.: Iron kinetics: modelling and parameter estimation in normal and anemic states. Computers and Biomedical Research *11*, 209 (1978).

BOTHWELL, T. H., FINCH, C. A.: Iron Metabolism. Little, Brown and Co., Boston 1962.

BOTHWELL, T. H., MALLETT, B.: Diurnal variation in the turnover of iron through the plasma. Clin. Sci. *14*, 235 (1955).

BOTHWELL, T. H., CALLENDER, S., MALLETT, B., WITTS, L. J.: The study of erythropoiesis using tracer quantities of radioactive iron. Brit. J. Haemat. *2*, 1 (1956).

BOTHWELL, T. H., ELLIS, B. C., VAN DOORN-WITTKAMPF, H. W., ABRAHAMS, O. L.: Radioiron studies in haemochromatosis. The effects of repeated phlebotomies. J. Lab. clin. Med. *45*, 167 (1955).

BOTHWELL, T. H., HURTADO, A. V., DONOHUE, D. M., FINCH, C. A.: Erythrokinetics. IV. The plasma iron turnover as a measure of erythropoiesis. Blood *12*, 409 (1957).

BOTHWELL, T. H., PRIBILLA, W. F., MEBUST, W., FINCH, C. A.: Iron metabolism in the pregnant rabbit. Iron transport across the placenta. Amer. J. Physiol. *193*, 615 (1958).

BRENDSTRUP, P.: Intravenous iron loadings of patients with infections. Proc. 21st Scand. Congr. Intern. Med., p. 1, 1949.

BRØCHNER-MORTENSEN, K.: Iron content of the serum in patients with hemorrhagic anemia. Acta med. scand. *113*, 345 (1943).

CARTWRIGHT, G. E., LAURITSEN, M. A., JONES, P. J., MERRILL, I. M., WINTROBE, M. M.: Anemia of infection; hypoferremia, hypercupremia, and alterations in porphyrin metabolism in patients. J. clin. Invest. *25*, 65 (1946).

CROSBY, W. H.: Siderocytes and the spleen. Blood *12*, 165 (1957).

CROSBY, W. H.: Evidence of recycling of siderocyte iron. J. clin. Invest. *38*, 997 (1959).

DAVIES, J., BROWN, E. B., STEWART, D., TERRY, C. W., SISSON, J.: Transfer of radioactive iron via the placenta and accessory fetal membranes in the rabbit. Amer. J. Physiol. *197*, 87 (1959).

DREYFUS, J. C., SCHAPIRA, G.: The metabolism of iron in haemochromatosis. In: GROSS, F. (ed.): Iron Metabolism. Springer, Berlin–Göttingen–Heidelberg 1964.

DUBACH, R., MOORE, C. V., MINNICH, V.: Studies in iron utilization and metabolism. V. Utilization of intravenously injected radioactive iron for Hb synthesis and evaluation of the radioactive iron method for studying iron absorption. J. Lab. clin. Med. *31,* 1201 (1946).

ELMLINGER, P. J., HUFF, R. L., TOBIAS, C. A., LAWRENCE, J. H.: Iron turnover abnormalities in patients having anemia; serial blood and in vivo tissue studies with Fe^{59}. Acta haemat. (Basel) *9,* 73 (1953).

FAIRBANKS, V. F., BEUTLER, E.: Iron Metabolism. In: WILLIAMS, W. J. et al. (eds.): Hematology. McGraw-Hill Inc., New York 1977.

FINCH, C. A. (1958), cit. BOTHWELL, FINCH, 1962.

FINCH, C. A., GIBSON, J. G., PEACOCK, W. C., FLUHARTY, R. G.: Iron metabolism. Utilization of intravenous radioactive iron. Blood *4,* 905 (1949a).

FINCH, C. A., WOLFF, J. A., RATH, C. E., FLUHARTY, R. G.: Iron metabolism. Erythrocyte iron turnover. J. Lab. clin. Med. *34,* 1480 (1949b).

FREIREICH, E. J., MILLER, A., EMERSON, C. P., ROSS, J. F.: The effect of inflammation on the utilization of erythrocyte and transferrin-bound radioiron for red cell production. Blood *12,* 972 (1957).

FREIREICH, E. J., ROSS, J. F., BAYLES, T. B., EMERSON, C. P., FINCH, S. C.: Radioactive iron metabolism and erythrocyte survival studies of the mechanism of the anemia associated with rheumatoid arthritis. J. clin. Invest. *36,* 1043 (1957).

GIANNOPOULOS, P. P., BERGSAGEL, D. E.: The mechanism of the anemia associated with Hodgkin's disease. Blood *14,* 856 (1959).

GIBLETT, E. R. et al.: Erythrokinetics: Quantitative measurement of red cell production and destruction in normal subjects and patients with anemia. Blood *11,* 291 (1956).

GOLDECK, H., REMY, D.: Über die Abwanderungsgeschwindigkeit des Eisens nach intravenösen Eisen-III-Gaben. Klin. Wschr. *31,* 608 (1953).

HOLMBERG, C. G., LAURELL, C. B.: Studies on the capacity of serum to bind iron. Contribution to our knowledge of the regulation mechanism of serum iron. Acta physiol. scand. *10,* 307 (1945).

HOPPE, I., GÖTTE, W., DEMKE, C., JAHN, E., SCHWEIKART, R., OVERKAMP, H.: Die intravenöse Eisenbelastung als Funktionsprobe des RES. Blut *7,* 385 (1961).

HUFF, R. L., JUDD, O. J.: Kinetics of iron metabolism. In: LAWRENCE, J. H., TOBIAS, C. A. (eds.): Advances in Biological and Medical Physics, Vol. 4, pp. 223–237. Academic Press, New York 1956.

HUFF, R. L., ELMLINGER, P. J., GARCIA, J. F., ODA, J. M., COCKRELL, M. C., LAWRENCE, J. H.: Ferrokinetics in normal persons and in patients having various erythropoietic disorders. J. clin. Invest. *30,* 1512 (1951).

HUFF, R. L., HENNESSY, T. G., AUSTIN, R. E., GARCIA, J. F., ROBERTS, B. M., LAWRENCE, J. H.: Plasma and red cell iron turnover in normal subjects and in patients having various hematopoietic disorders. J. clin. Invest. *29,* 1041 (1950).

HUFF, R. L., TOBIAS, C. A., LAWRENCE, J. H.: A test for red cell production. Acta haemat. (Basel) *7,* 129 (1952).

HYMAN, G. A., HARVEY, J. E.: The pathogenesis of anemia in patients with carcinoma. Amer. J. Med. *19,* 350 (1955).

KEIDERLING, W., SCHMIDT, H. A. E., LEE, M.: Radioisotope in Klinik und Forschung, Vol. 2. Urban und Schwarzenberg, München–Berlin 1956.

KEIDERLING, W., SCHMIDT, H. A. E., LEE, M.: Radioaktive Isotope in Klinik und Forschung. Sonderbände für Strahlentherapie *36,* 24 (1957).

LAJTHA, L. G.: The Use of Isotopes in Haematology. Blackwell, Oxford 1961.

LAJTHA, L. G. (1967): Personal communication.

LAWRENCE, J. H., HUFF, R. L., SIRI, W., WASSERMAN, L. R., HENNESSY, T. G.: A physiological study in the Peruvian Andes. Acta med. scand. *142,* 117 (1952).

LOEB, V., MOORE, C. V., DUBACH, R.: The physiologic evaluation and management of chronic bone marrow failure. Amer. J. Med. *15,* 499 (1953).

NAJEAN, Y., ARDAILLON, N., DRESCH, C.: Utilisation des techniques isotopiques en haematologie. Baillières et Fils, Paris 1969.

Orr, J. S., Horton, P. W., Hutcheon, A. W., Dagg, J. H.: Erythropoiesis and iron kinetics. Brit. J. Haemat. *42,* 155 (1979).
Paterson, J. C.: Disappearance of radioactive iron from plasma by day and by night. Proc. Soc. exp. Biol. Med. *96,* 97 (1957).
Pollycove, M.: Iron kinetics. In: Wallerstein, R. O., Mettier, S. R. (eds.): Iron in Clinical Medicine, pp. 43–57. University of California Press, Berkeley 1958.
Pollycove, M.: Ferrokinetics: Techniques. In: Keiderling, W. (ed.): Eisenstoffwechsel, pp. 20–40. Thieme, Stuttgart 1959.
Pollycove, M., Mortimer, R.: The quantitative determination of iron kinetics and hemoglobin synthesis in human subjects. J. clin. Invest. *40,* 753 (1961).
Reynafarje, C., Lozano, R., Valdivieso, J.: The polycythaemia of high altitudes: Iron metabolism and related aspects. Blood *14,* 433 (1959).
Ricketts, C., Jacobs, A., Cavill, J.: Ferrokinetics and erythropoiesis, ineffective erythropoiesis and red cell lifespan using ^{59}Fe. Brit. J. Haemat. *31,* 65 (1975).
Sharney, L. et al.: Multiple-pool analysis in tracer studies of metabolic kinetics. J. Mt. Sinai Hosp. *32,* 201 (1965).
Tötterman, L. E.: Vitamin C and iron metabolism. Acta med. scand. *230* (Suppl.), 10 (1949).
Tötterman, L. E.: Intravenous iron tolerance tests in malignant neoplasms. Acta med. scand. *140,* 265 (1951).
Veall, N., Vetter, H.: Radioactive Techniques in Clinical Research and Diagnosis. Butterworth, London 1958.
Waldenström, J.: Serumjärnbestämmingen och des praktiska betydelse. Nord. med. *11,* 234 (1941).

CHAPTER 15

ERYTHROKINETICS

A deeper insight into the quantitative aspects of red cell production and destruction has added to the better understanding of the pathomechanisms underlying the various forms of anemia. In clinical practice an estimate of the total erythropoiesis is obtained by measuring the erythroid/myeloid (E/M) ratio in the bone marrow and the plasma iron transport rate, whereas effective erythropoiesis can be estimated from the iron utilization and corrected reticulocyte count.

Red cell destruction can be determined from the urobilinogen content of the feces and the life-span of the red cells (Fig. 15/1).

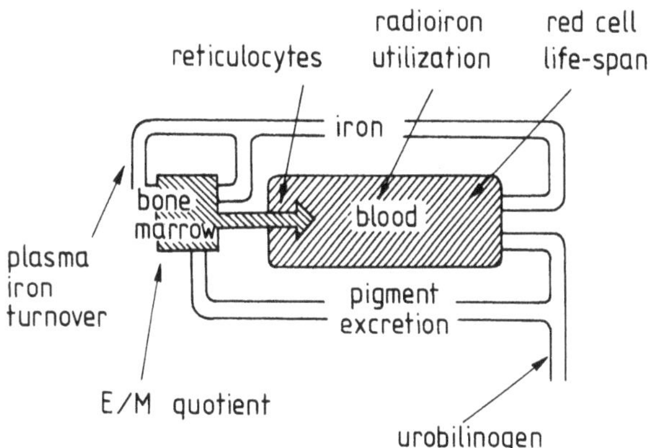

Fig. 15/1. Procedures for the quantitative determination of red cell production and destruction. Arrows mark the site of determinations within the system (after Bothwell, T. H. and Finch, C. A.: Iron Metabolism. Little, Brown and Co., Boston 1962)

The E/M ratio can assess erythropoiesis only when the myeloid cells have not appreciably altered in number and when there is minimal contamination with peripheral blood. The standard error is considerable even when the examination is carried out with utmost care; in practice it is sufficient to estimate whether or not the ratio is (a) diminished, (b) normal, (c) moderately increased (doubled), or (d) markedly elevated (more than threefold) (Bothwell and Finch, 1962). Thus the bone marrow that is not capable of effective reaction can be distinguished from the bone marrow that is capable of optimal response.

At least 1000 myeloid cells (granulocytes and granulocyte precursors) should be counted, and the ratio of nucleated red cells to the myeloid cells ascertained. Giblett et al. (1956) and Reiff et al. (1958) have reported that the normal value of the E/M ratio is approximately 0.40–0.45. Bernát (1971) found a wider variation in healthy individuals, 0.19–0.46 with a mean of 0.32.

The plasma iron transport rate is a more precise indicator of the extent of *total erythropoiesis* than the E/M ratio, but as already discussed, the iron flowing through the plasma reaches other tissues besides the bone marrow. Normally about 75% goes to the bone marrow and 25% to other tissues, which means that if the plasma iron transport rate is 0.6 mg/dl blood/24 hr, about 0.45 mg enters the erythroid cells of the bone marrow. This corresponds to 22.5 mg of iron daily used for erythropoiesis, or to nearly 6.5 grams of hemoglobin (3.47 mg of iron per gram of hemoglobin).

The distribution of iron between the bone marrow and other tissues depends on the functional state of the bone marrow and the iron-binding capacity of other tissues. In hyperplasia of the erythroid marrow a larger proportion of the iron that flows through the plasma reaches the bone marrow, and a rise in iron transport rate is related to the enhanced red cell production. With hypoplasia of the bone marrow, however, the plasma iron turnover rate is quantitatively less informative because it does not tell precisely the proportion of transport iron that reaches the bone marrow in relation to other tissues. The extent of *effective erythropoiesis* is reflected by the fraction of iron utilized in the course of viable mature red cell production during unit time (red cell iron turnover rate) (see Chapter 14, Ferrokinetics). This value is calculated from the plasma iron transport rate and iron incorporation.

The absolute or corrected reticulocyte count also indicates effective erythropoiesis. Its reliability is based on the not necessarily valid assumptions that the red cells all enter the circulation as reticulocytes (Nizet, 1946; Seip, 1953) and that the reticulocytes have a more or less constant maturation time.

It is not enough to determine only the percent of reticulocytes; the absolute value must be calculated from the red cell count. For example, if the reticulocyte count is 8% with a red cell count of $2.5 \times 10^6/\mu l$, this corresponds to an absolute count of reticulocytes of $200{,}000/\mu l$, i.e., 4% of a normal red cell count of $5.0 \times 10^6/\mu l$. If the normal reticulocyte count is taken as 1%, the rate of red cell production in the given example is four times and not eight times the normal.

The corrected reticulocytes may also be calculated from a normal hematocrit value of 45%. For example, if the reticulocytes are 18% and the hematocrit value is 30%, the corrected reticulocyte count will be 12% (30/45 × 18).

The use of the reticulocyte count as a quantitative indicator of hemopoiesis is limited in conditions where the maturation time of the reticulocytes is greater or less than normal. In acquired hemolytic anemia (Young and Lawrence, 1945) or pernicious anemia responding to treatment (Finch et al., 1956), less mature reticulocytes are released into the blood (Reiff et al., 1958) and the maturation time of the circulating reticulocytes is increased. Although the proportion of the circulating reticulocytes released early cannot be ascertained, the arbitrary practice is to take the peripheral maturation time as double and to divide the corrected reticulocyte count by 2. Thus, in the example quoted above, the final estimate of effective erythropoiesis would be 6 times normal and not 12 times normal.

In some conditions, for example, in untreated megaloblastic anemias (vitamin B_{12} or folic acid deficiency), most of the red cells have already lost their reticular network before entering the circulation (Finch et al., 1956), and an assessment of erythropoiesis from the corrected reticulocyte count will be inaccurate.

In spite of these limitations, the rate of erythropoiesis estimated on the basis of the corrected reticulocyte count is certainly better than using the uncorrected percent of reticulocytes.

RED CELL DESTRUCTION

An elevated plasma hemoglobin level, a rise in the nonconjugated bilirubin concentration, and a drop in the plasma haptoglobin level are all evidence of increased red cell destruction, although they do not give a quantitative estimate of hemolysis, with the possible exception of the rise in the serum bilirubin level of neonates, which does correlate with the extent of red cell destruction (Sjöstrand, 1948, 1949a, b, 1951; Brown et al., 1958).

Hallberg (1955) has shown that hemolysis can be measured from the CO content of the expired air. As the major proportion of the endogenous carbon monoxide originates from the alpha carbon atom of the pyrrol ring, the rise in CO concentration in the alveolar air correlates with the increasing red cell destruction. A more widespread use of this method is hindered by the complexity of the procedure and by the influence of other factors, for example, smoking.

ASSESSMENT OF RED CELL DESTRUCTION BASED ON THE UROBILINOGEN CONTENT OF THE FECES

Protoporphyrin, liberated in the course of the decomposition of the blood pigment, is not re-utilized in hemoglobin synthesis but is decomposed to be excreted with the feces as urobilinogen. The urobilinogen content of the feces is therefore an indicator of the extent of blood pigment destruction. On average 148–180 mg of urobilinogen (extreme values: 40–280 mg) are excreted daily in the feces of the healthy individual (Heilmeyer and Oetzel, 1931; Giblett et al., 1956), which is slightly less than hypothetically expected (193–234 mg/24 hr) (Giblett et al., 1956). The cause for this difference is not known, but the decomposition in the gut may be incomplete, and differences in the intestinal flora may play a part. If the bacterial flora is altered by antibiotics, for example, the amount of the urobilinogen excreted in the feces is considerably diminished (Dearing et al., 1958). This must be borne in mind when evaluating the results.

DETERMINATION OF THE LIFE-SPAN OF THE RED CELLS

This is the most direct method for assessment of red cell destruction (Berlin et al., 1959). Two principles can be used for measuring red cell life-span: (a) cohort labeling, (b) random labeling.

(a) COHORT LABELING

The erythrocytes can be labeled with a stable or radioactive isotope, which is incorporated into the red cell precursors only; this enables the selective labeling of a relatively homogeneous cohort of cells formed within a definite, short time. The labels that have been used are the hemoglobin precursors ^{15}N-glycine (Shemin and Rittenberg, 1946a), ^{14}C-glycine (Berlin, 1964), ^{3}H-glycine, and, more recently, ^{75}Se-methionine (Penner, 1966). Labeling with radioactive iron (^{59}Fe, ^{55}Fe) also belongs to this group.

(b) RANDOM LABELING

In the second group of methods the survival of red cells of different ages, i.e., a mixed red cell population, is measured. The first to be introduced was the Ashby differential agglutination method (Ashby, 1919). The recipient is transfused with red cells from a donor with compatible but immunologically recognizable and differentiable red cells, and at intervals subsequent samples are taken, the patient's own red cells are agglutinated or lysed by appropriate serum (Hurley and Weisman, 1954), and the donor red cells are counted.

This method has been superseded by *in vitro* labeling with ^{51}Cr or DF^{32}P of the subject's own red cells. The DF^{32}P can also be used for *in vivo* labeling.

***

The determination of the life-span of red cells with ^{59}Fe has the advantage that red cell destruction can be examined within the ferrokinetic studies. Its disadvantage is that re-utilization of iron renders evaluation difficult. To some extent this can be prevented by the administration of therapeutic doses of stable iron, which prevents some of the re-utilization of the isotopic iron, so that a distinct decrease in red cell radioactive iron may occur at the end of the red cell life-span.

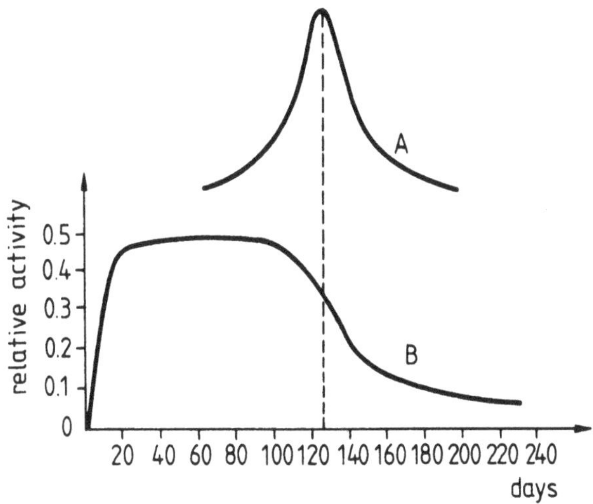

Fig. 15/2. Radioactive labeling of a cohort of red cells (B), the curve (A) being derived to show the maximal rate of destruction at about 120 days

Three phases can be distinguished in the course of the iron incorporation and iron disappearance curves: the ascending line is followed by a plateau and then the curve starts to decline, following an S-shaped curve, and stabilizing at a lower value (Fig. 15/2). The steepest segment of the descending line of the curve (the maximal rate of red cell destruction) represents the average life-span of the red cells.

In the blood of healthy man the radioactivity remains practically at the same level for about 80 days after the initial rapid rise; then between days 80 and 130 there is a decline, reaching the lowest value between days 130 and 140; then there is again a slow rise (Fig. 15/3). Theoretically, the curve ought to reach the abscissa (Fig. 15/4) but this does not take place because of the re-utilization of the iron liberated from effete red cells. The degree of iron re-utilization varies with the rate of erythropoiesis and the saturation of the iron stores. If the rate of erythropoiesis is diminished with bone marrow hypofunction or if the storage iron is considerable, re-utilization is slight (Fig. 15/5).

Determination of the life-span of red cells using ^{14}C-glycine has the advantage that glycine is not re-utilized after the catabolism of the hemoglobin, but the long half-life of ^{14}C (5568 years) is a disadvantage. The half-life of tritium (12.3 years) also makes ^{3}H-glycine relatively unsuitable.

Labeling with the stable isotope ^{15}N-glycine has several advantages (Shemin and Rittenberg, 1946a, b; London et al., 1949; Box et al., 1953). The glycine that is incorporated into the heme molecule is not re-utilized; some is incorporated into the

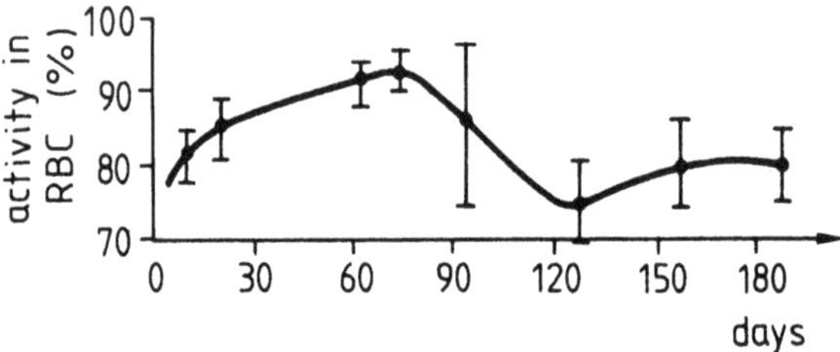

Fig. 15/3. Red cell activity after the i.v. injection of ^{59}Fe. The curve indicates the mean values of 5 healthy individuals. The perpendicular lines represent extreme values. Deviation of the curve from its theoretically expected course (see Fig. 15/4) is explained by the re-utilization of iron (after Bothwell, T. H. and Finch, C. A.: Iron Metabolism. Little, Brown and Co., Boston 1962)

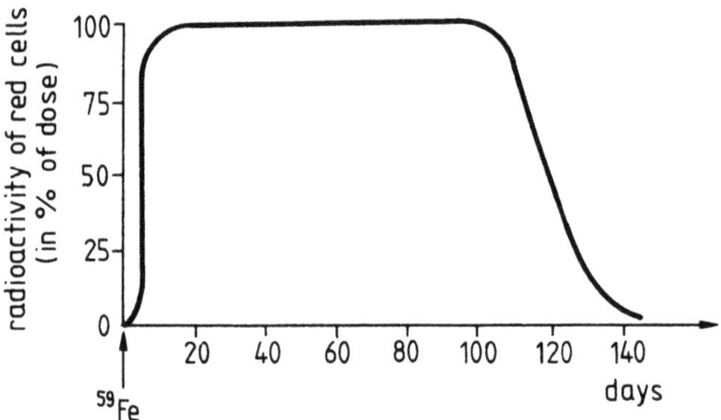

Fig. 15/4. The theoretical curve expected from labeling a cohort of red cells with radioactive iron where there is neither a shortened life-span nor re-utilization of radioactive iron

globin and is re-utilized, but is diluted in the amino acid pool and does not interfere with the evaluation of the examination (Cooper and Jandl, 1972). ^{15}N-glycine, unfortunately, is expensive, and the measurements require access to a mass spectrometer (Strumia, 1956).

The labeled glycine is given orally and is incorporated into a cohort of red cells in the bone marrow. In the blood of healthy man the ^{15}N atom percentage excess in the heme increases rapidly and reaches its peak between days 12 and 14 (Fig. 15/6), and the values remain practically at the same level until the 105th day. In the next few days the values decrease slowly, and then more rapidly, to day 146, where the value is quite low.

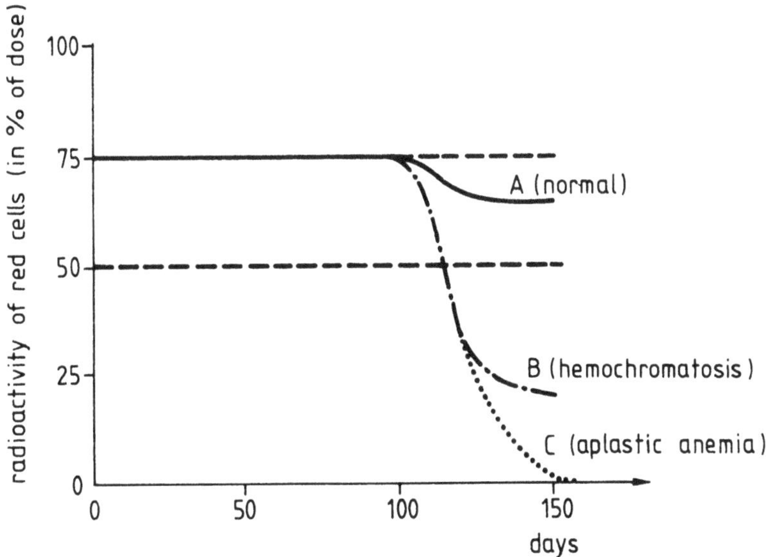

Fig. 15/5. The disappearance curves at the end of the normal life-span of labeled red cells in (A) healthy man and in (B) hemochromatosis and (C) aplastic anemia (after Dreyfus, J. C. and Schapira, G.: Le fer. L'Expansion, Paris 1958)

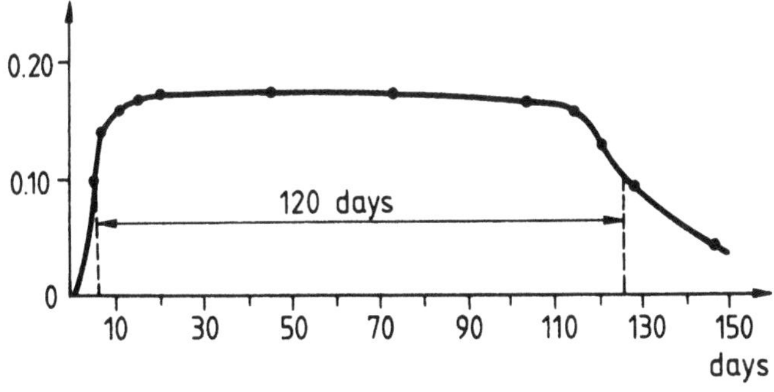

Fig. 15/6. The life-span of red cells labeled with ^{15}N after the oral intake of ^{15}N-glycine in a healthy man (after Bernát, I. and Cornides, I., 1970)

Bernát and Cornides (1970) found the rate of incorporation to be fastest between days 3 and 7 and the rate of decline maximal between days 121 and 126. The life-span is taken as the length of time between the centers of these two phases; thus the average life-span of the red cells is 118–119 days.

Increased red cell production during the course of the study will result in dilution of the isotope and reduction of the percent incorporation. Reduction in hemopoiesis will have the reverse effect. Such variations have to be taken into consideration, and the results will be more accurate if the life-span of the red cells is determined on the basis of the total amount of the circulating ^{15}N instead of the atom percentage excess of the heavy nitrogen incorporated in the heme (Berlin, 1964) (Fig. 15/7).

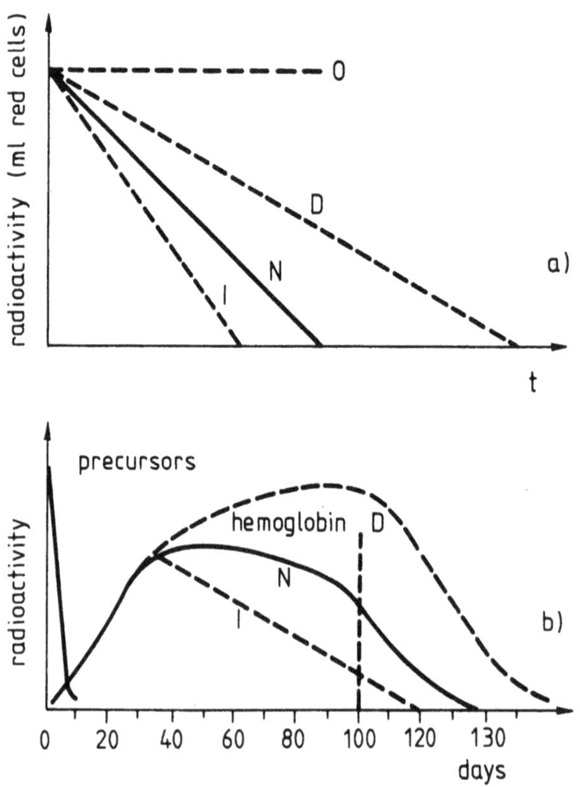

Fig. 15/7. Red cell survival curves of a mixed cell population (a). Life-span curves obtained with labeling a cohort of new red cells (b). The interpretation may be complicated by the effect of a changing total red cell volume over the period of study, which will alter the concentration of isotope-labeled cells. (I) Result of enhancement of erythropoiesis and increase in the total amount of circulating heme that dilutes the isotope. (D) Reduction of erythropoiesis, which increases the concentration of the isotope. N: normal. O: aplasia (after Berlin, N. I., in: Bishop, C. and Surgenor, D. M.: The Red Blood Cell. Academic Press, New York 1964)

The curve constructed on the basis of the total amount of ^{15}N reaches its peak on the 14th day. The plateau lasts until day 105, and the rate of destruction is fastest between days 121 and 129. From this curve the average life-span of the red cells of the healthy human is 120 days.

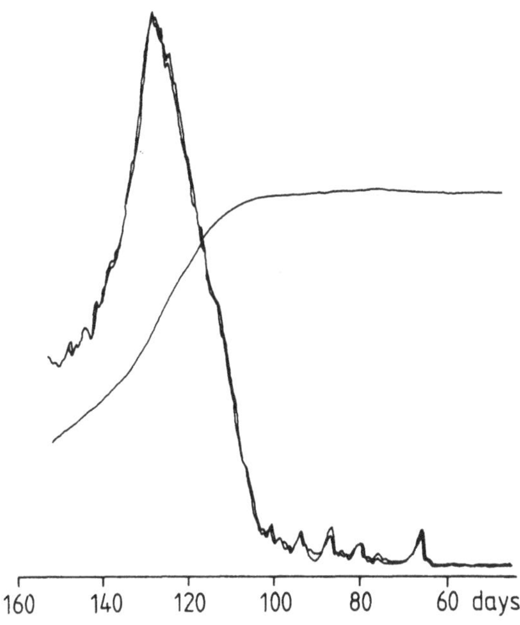

Fig. 15/8. Isotope disappearance curve and its derived (differential quotient) curve calculated with analogue computer method (after Bernát, I., 1971)

According to Shemin and Rittenberg (1946), the average life-span of the red cells is marked by the moment in which the rate of destruction is fastest. Mathematically, this moment is defined as the differentiation of the descending line of the curve. Instead of the cumbersome graphic-numerical procedure, the determination of the differential quotient curve (derived curve) can be carried out by an electronic analogous method (Bernát, 1971). The empirical "disappearance curve" is "read" by a moving photoelectric head, the differentiation is carried out electronically by the machine, and the derived curve together with the original one is drawn by it with the help of an X–Y coordinate writer (Fig. 15/8). The derived curves obtained by the electronic analogue method not only indicate the average life-span of the red cells but yield very precise information about the time course of the red cell destruction as well. The normal curve (see Fig. 15/8) shows little difference in the life-span of the red cells. The curve is nearly symmetrical in relation to the perpendicular axis drawn through the peak of the curve, and random destruction is minimal. Bernát (1971) found that the majority of the red cells survived between 99 and 140 days, the average life-span being 123 days.

If some of the red cells are randomly destroyed, the derived curve around the perpendicular axis will not have a symmetrical course. Where there is great variation in life-span, the curve has a broad base (Fig. 15/9).

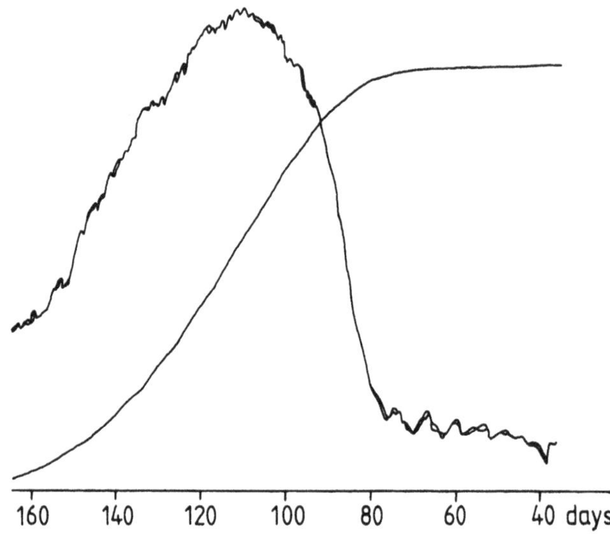

Fig. 15/9. Isotope disappearance curve and its derived curve calculated with analogue computer method in thermal injury (after Bernát, I., 1971)

According to the literature, the average life-span of the red cells of the healthy human varies between 109 and 127 days; i.e., about 0.83% of the red cells are destroyed and replaced daily. London et al. (1949) found that the average life-span of erythrocytes was 85 days in untreated pernicious anemia and 129 days after treatment, 42 days in sickle cell anemia, and 131 days in polycythemia rubra vera.

The use of these blood pigment precursors is not practical in a clinical situation because the results take some time to obtain and the stable isotope is expensive. They do, however, probably give the most accurate estimation of red cell survival.

Labeling with ^{51}Cr and $DF^{32}P$. In clinical practice these two isotopes have proved to be the most useful. With the former, the red cells are labeled *in vitro* with $Na_2{}^{51}CrO_4$.

Ninety percent of the sodium chromate penetrates the red cells in 45–60 minutes and is fixed intracellularly. Further chromium incorporation is hindered by reduction with ascorbic acid, or the surplus isotope is removed by washing. The labeled cells are injected intravenously, and serial blood samples are taken subsequently to plot the disappearance of radioactivity (Figs. 15/7 and 15/10). Since cells of all ages are labeled, the type of curve is entirely different from that obtained with cohort labeling. There should be a straight line as obtained with the differential

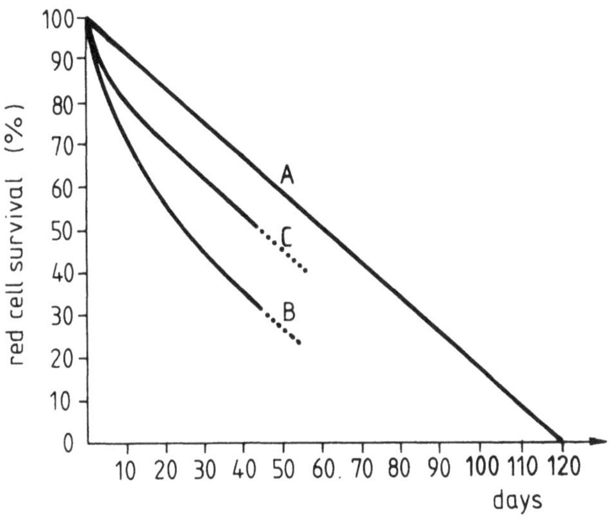

Fig. 15/10. Survival curves produced by the Ashby or $DF^{32}P$ methods (A), ^{51}Cr labeling (B), and the corrected ^{51}Cr curve (C) (after Lajtha, L. G.: The Use of Isotopes in Haematology: Blackwell, Oxford 1961)

agglutination technique of Ashby, but elution of the isotope, particularly in the early stages of the curve, results in an exponential type of curve, which produces a more or less straight line when plotted on semilog paper (Wintrobe, 1974). The T/2 shows considerable variations (25–35 days) even in healthy individuals.

Attempts have been made to correct the error that arises from elution, but there may be considerable differences in the rate of elution in pathological conditions (Rigas, 1961, cited by Bothwell and Finch, 1962; Donohue et al., 1955). Even if the corrected ^{51}Cr isotope disappearance curve is plotted, the T/2 is not as long as that derived from Ashby survival data or from cohort labeling (see Fig. 15/10 C and A), and in general, attempts to correct the survival have in practice been abandoned.

The degree of hemolysis is roughly estimated from the T/2 compared with a normal of 25–35 days.

Donaldson et al. (1968) described the use of stable chromium (^{50}Cr) to measure the red cell survival using activation analysis. This method may be useful in those subjects where the use of radioactive isotopes is contraindicated, for example, in normal children and pregnant women.

^{32}P-tagged diisopropyl fluorophosphate (DFP) interferes with the cholinesterase activity of red cells, and the radiophosphorus is fixed in the cells. The labeling can be done *in vivo* by the intramuscular or intravenous injection of $DF^{32}P$ or *in vitro* by incubation of the blood with the isotope. There is slight elution during the first few days (Cline and Berlin, 1963) but subsequently the decline in radioactivity is linear, and the daily destruction in normals, 0.57–1.47%, with a mean 0.86% (Garby, 1962), is in good agreement with the results obtained with other procedures (Cline and Berlin, 1963).

Details of the methods recommended by the International Committee for the Standardization in Haematology for red cell survival studies with ^{51}Cr or $DF^{32}P$ are described in Blut (1972), *25*, 108–115.

Barrett et al. (1966a, b, c) and Berk et al. (1969) have suggested a method for the indirect determination of the life-span of the red cells that will reduce the necessary observation period. The procedure is based on the calculation of bilirubin production. The method requires the knowledge of the following parameters:

1. The rate of outflow from the plasma of an i.v. injection of ^{14}C-labeled bilirubin (the clearance can be expressed by the sum of three exponential functions).
2. The nonconjugated bilirubin concentration of the serum.
3. The total red cell volume.

The labeled bilirubin is biosynthetically produced with ^{14}C or ^{3}H-delta-aminolevulinic acid as precursor. The isotope is incorporated in the course of heme synthesis, and the labeled hemoglobin is transformed to bilirubin (Barrett et al., 1968; Howe et al., 1970).

One milliliter of red cells contains 330 mg of hemoglobin, the decomposition of which results in 12 mg of bilirubin, therefore

$$\frac{(1-x)\times(\text{BRP})}{12} = \text{the rate of cell destruction (in ml/day)}$$

where BRP = bilirubin production and $(1-x)$ expresses the fraction of bilirubin production that is derived from circulating red cells.

Of the bilirubin, 80–90% originates from the breakdown of hemoglobin in the course of red cell destruction and 10–20% from the decomposition of other heme compounds (myoglobin, tissue hemins), as well as from ineffective erythropoiesis (Barrett et al., 1966b; Berk et al., 1970). After the ingestion of labeled glycine there is an early excretion of radioactivity between days 3 and 5, and a second late peak between days 90 and 130, the latter originating from the catabolism of the hemoglobin.

Schwartz (1967), Barrett et al. (1966b) and Berk et al. (1970) demonstrated that the early peak consisted of two components:

(a) a constant fraction of non-red-cell origin, and

(b) an inconstant fraction whose extent is proportional to the rate of red cell production.

The latter seems to be connected with ineffective erythropoiesis. Thus the value of the $(1-x)$ expression increases if red cell production is enhanced.

According to Berk et al. (1970), the value of $(1-x)$ is normally between 0.8 and 0.75. The life-span of the red cells is calculated according to the formula:

$$\frac{\text{total red cell volume (ml)}}{\text{red cell destruction rate (ml/day)}} = \text{average life-span of red cells}$$

The determination of the life-span of red cells based on bilirubin production cannot be used in those pathological processes in which the early peak of radioactivity is disproportionately increased, as in pernicious anemia, erythropoietic porphyrias, "shunt"-hyperbilirubinemic syndromes, thalassemia, sideroblastic anemias, and lead poisoning.

One of the advantages of the ^{14}C-BR method, in addition to its rapidity, is that it also gives information about the capacity of the liver to take up bilirubin from the plasma. Furthermore, the short sampling time makes repeated determinations quite feasible.

Table 15/1 summarizes the information that may be obtained from erythrokinetic studies. Such studies used in combination reveal the great differences that exist in certain pathological conditions between the total and effective erythropoiesis (Giblett et al., 1956). The results are usually expressed in terms of the normal values (Fig. 15/11). The normal erythrokinetic standards referred to 70 kg body weight are shown in Table 15/2.

Table 15/1

Informative value of the erythrokinetic examinations

	Red cell production		Red cell destruction	
	Total	Effective	Total	Effective
Plasma iron transport rate	+			
Red cell iron turnover		+		
E/M quotient	+			
Abs. (corr.) reticulocyte count		+		
Feces ubg. content			+	
Life-span of red cells				+

It has become possible to make an erythrokinetic classification of anemias, as follows:

1. The intact bone marrow adjusts to increased requirements, for example, following hemorrhage, and effective erythropoiesis can rise 6–8 times the normal level if necessary. Beyond this, anemia will necessarily develop.

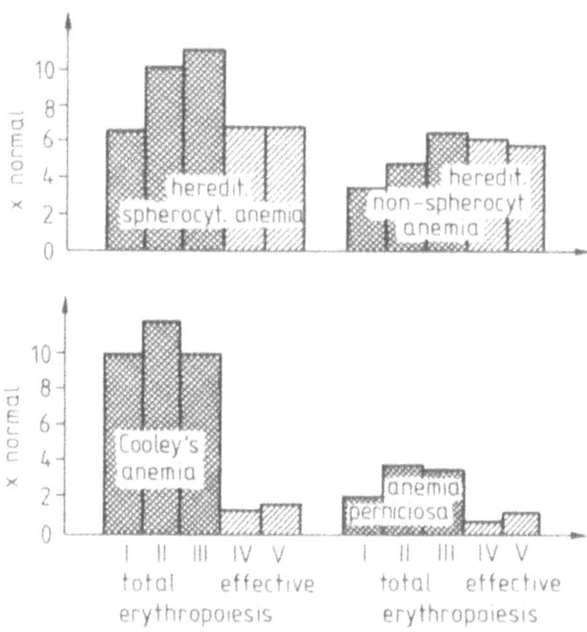

Fig. 15/11. Erythrokinetic characteristics of hemolytic anemias (after Bothwell, T. H. and Finch, C. A.: Iron Metabolism. Little, Brown and Co., Boston 1962). I: E/M quotient; II: plasma iron transport rate; III: urobilinogen excretion; IV: iron utilization in the course of red cell production; V: absolute reticulocyte count. I–III characterize total erythropoiesis, IV–V effective erythropoiesis

2. Effective red cell production cannot keep pace with the demands if:

(a) the bone marrow is damaged, for example by ionizing irradiation or toxic substances;

(b) the available iron is not sufficient for hemoglobin synthesis;

(c) as the result of a variety of disturbances (endocrine, infection, renal disease, thermal injury, etc.) bone marrow function may be depressed, and in such cases even a slightly shortened life-span of the red cells is enough to upset the balance of production and destruction.

3. If there is a disproportionate amount of ineffective erythropoiesis, as in megaloblastic anemia, pyridoxine-responsive anemias, thalassemia, anemia can develop in spite of markedly increased total red cell production.

Table 15/2

Theoretical erythrokinetic data in normal man (from Bothwell and Finch, 1962 – modified)

Initial data	Calculated values	Daily turnover (0.83–1.0% of calculated values)
Blood volume 5000 ml	Total volume of red cells 2000 ml	16.7–20.0 ml red cells
Venous hematocrit 44.5%	Total Hb 660 g	5.6–6.7 g Hb
Normal whole-body hematocrit 40%	Total amount of circulating iron 2224 mg (2311 mg)*	18.5–22.3 mg Fe (19.4–23.2 mg)*
	Total amount of circulating pyrrol pigment 22.5 g	193–234 mg ubg

Changeover constants

Whole-body hematocrit correction constant = 0.92 (× venous hematocrit)
Mean hemoglobin concentration of red cells = 33%
(Mean corpuscular hemoglobin concentration)
Mean red cell volume = 87 μ^3
Iron content of 1 g hemoglobin = 3.38 mg (3.47 mg)*

Daily rate of red cell destruction = 1/120 (0.83%)
Molecular weight of protoporphyrin = 566
Molecular weight of stercobilinogen = 580
Molecular weight of hemoglobin = 66,000
Atomic weight of iron = 56

* modified values

BIBLIOGRAPHY

Ashby, W.: Determination of the length of life of transfused blood corpuscles in man. J. Exp. Med. *29,* 267 (1919).

Barrett, P. V. D., Berk, P. D., Menken, M., Berlin, N. I.: Bilirubin turnover studies in normal and pathologic states using bilirubin-^{14}C. Ann. intern. Med. *68,* 355 (1968).

Barrett, P. V. D., Cline, M. J., Berlin, N. I.: The association of the urobilin "early peak" and erythropoiesis in man. J. clin. Invest. *45,* 1657 (1966).

Barrett, P. V. D., Mullino, F. X., Berlin, N. I.: Studies on the biosynthetic production of bilirubin-C^{14}, an improved method utilizing delta-amino levulinic acid-4-C^{14} in dogs. J. Lab. clin. Med. *68,* 905 (1966).

Berk, P. D., Howe, R. B., Bloomer, J. R., Berlin, N. I.: Studies of bilirubin kinetics in normal adults. J. clin. Invest. *48,* 2176 (1969).

Berk, P. D., Howe, R. B., Bloomer, J. R., Berlin, N. I.: The life span of the red cell as determined with labeled bilirubin. In: Greenwalt, T. J., Jamieson, G. A. (eds.): Formation and Destruction of Blood Cells. Lippincott, Philadelphia–Toronto 1970.

Berlin, N. I.: Life span of the red cell. In: Bishop, C., Surgenor, D. M. (eds.): The Red Blood Cell, pp. 423–450. Academic Press, New York 1964.

Berlin, N. I., Waldmann, T. A., Weissman, S. M.: Life span of red blood cell. Physiol. Rev. *39*, 577 (1959).

Bernát, I.: Az égési anaemia pathogenesise (The Pathogenesis of Anemia after Thermal Injury). Akadémiai Kiadó, Budapest 1971.

Bernát, I., Cornides, I.: Stabil izotópok alkalmazása a biológiai-orvosi kutatásban (The use of stable isotopes in biological-medical research). Honvédorvos *22*, 201 (1970).

Bothwell, T. H., Finch, C. A.: Iron Metabolism. Little, Brown and Co., Boston 1962.

Box, H. C., Schenk, W. C., Wiles, C. E.: Biophysical methods for the assay of life span of red cells. Science *118*, 72 (1953).

Brown, A. K., Zuelzer, W. W., Burnett, H. H.: Studies on the neonatal development of the glucuronide conjugating system. J. clin. Invest. *37*, 332 (1958).

Cline, M. J., Berlin, N. I.: An evaluation of DFP^{32} and Cr^{51} as methods of measuring red-cell life span in man. Blood *22*, 459 (1963).

Cooper, R. A., Jandl, J. H.: Destruction of erythrocytes. In: Williams, W. J., Beutler, E., Erslev, R. W. (eds.): Hematology. McGraw-Hill Book Co., New York 1972.

Dearing, W. H., Needham, G. M., Mason, H. L.: The effect of oral antibiotics on the intestinal bacteria and the formation of urobilinogen. Proc. Mayo Clin. *33*, 646 (1958).

Donaldson, G. U., Johnson, P. F., Tothill, P.: Red cell survival time in man measured by ^{50}Cr and activation analysis. Brit. Med. J. *2*, 585 (1968).

Donohue, D. M., Motulsky, A. G., Giblett, E. R., Pirzio-Biroli, G., Viranuvati, V., Finch, C. A.: The use of chromium as a red-cell tag. Brit. J. Haemat. *1*, 249 (1955).

Dreyfus, J. C., Schapira, G.: Le fer. L'Expansion, Paris 1958.

Finch, C. A.: Some quantitative aspects of erythropoiesis. Ann. N. Y. Acad. Sci. *77*, 410 (1959).

Finch, C. A., Coleman, D. H., Motulsky, A. G., Donohue, D. M., Reiff, R. H.: Erythrokinetics in pernicious anemia. Blood *11*, 807 (1956).

Garby, L.: Analysis of red-cell survival curves in clinical practice and the use of di-iso-propylfluorophosphonate ($DF^{32}P$) as a label for red cells in man. Brit. J. Haemat. *8*, 15 (1962).

Giblett, E. R., Coleman, D. H., Pirzio-Biroli, G., Donohue, D. M., Motulsky, A. G., Finch, C. A.: Erythrokinetics. Quantitative measurements of red-cell production and destruction in normal subjects and patients with anemia. Blood *11*, 291 (1956).

Hallberg, L.: Blood volume, hemolysis and regeneration of blood in pernicious anemia. Studies based on the endogeneous formation of carbon monoxide and determination of the total amount of hemoglobin. Scand. J. clin. Lab. Invest. 7 (Suppl. 6), 1 (1955).

Heilmeyer, L., Oetzel, W.: Blutfarbstoffwechselstudien. II. Dtsch. Arch. klin. Med. *171*, 365 (1931).

Howe, R. B., Berk, P. D., Bloomer, J. R., Berlin, N. I.: Preparation and properties of specifically labeled radiochemically stable ^{3}H-bilirubin. J. Lab. clin. Med. *75*, 499 (1970).

Hurley, T. H., Weisman, R.: The determination of the survival of transfused red cells by a method of differential hemolysis. J. clin. Invest. *33*, 835 (1954).

International Committee for the Standardization in Haematology: Recommended methods for radioisotope red-cell survival studies. Blut *25*, 108 (1972).

Lajtha, L. G.: The Use of Isotopes in Haematology. p. 22. Blackwell, Oxford 1961.

London, I. M., Shemin, D., West, R., Rittenberg, D.: Heme synthesis of red blood cells; dynamics in normal humans and in subjects with polycythemia vera, sickle-cell anemia and pernicious anemia. J. biol. Chem. *179*, 463 (1949).

Nizet, A.: Recherches sur la physiopathologie des hématies; la réticulocytose et la libération de hématies à partir de la moelle osseuse. Acta med. scand. *124*, 590 (1946).

PENNER, J. A.: Investigation of erythrocyte turnover with selenium-75-labeled methionin. J. Lab. Clin. Med. *67*, 427 (1966).
REIFF, R. H., NUTTER, J. Y., DONOHUE, D. M., FINCH, C. A.: The relative number of marrow reticulocytes. Amer. J. clin. Path. *30*, 199 (1958).
RIGAS, J. (1961), cit. BOTHWELL, FINCH, 1962.
SCHWARTZ, S.: Bilirubin Metabolism. Davis, Philadelphia 1967.
SEIP, M.: Reticulocyte studies. The liberation of red blood corpuscles from the bone marrow into the peripheral blood and the production of erythrocytes elucidated by reticulocyte investigations. Acta med. scand. *146* (Suppl. 282), 1 (1953).
SHEMIN, D., RITTENBERG, D.: The biological utilisation of glycine for the synthesis of the protoporphyrin of hemoglobin. J. biol. Chem. *166*, 621 (1946a).
SHEMIN, D., RITTENBERG, D.: The life span of the human red cell. J. biol. Chem. *166*, 627 (1946b).
SJÖSTRAND, T.: A method for determination of carboxyhaemoglobin concentrations by analysis of the alveolar air. Acta physiol. scand. *16*, 201 (1948).
SJÖSTRAND, T.: Endogenous formation of carbon monoxide in man under normal and pathologic conditions. Scand. J. clin. Lab. Invest. *1*, 201 (1949a).
SJÖSTRAND, T.: Endogenous formation of carbon monoxide in man. Nature *164*, 580 (1949b).
SJÖSTRAND, T.: Endogenous formation of carbon monoxide. The CO concentration in the inspired and expired air of hospital patients. Acta physiol. scand. *22*, 137 (1951).
STRUMIA, M. M.: The life span of erythrocytes and their post-transfusional survival. In: TOCANTINS, L. M. (ed.): Progress in Hematology, Vol. 1. Grune and Stratton, New York–London 1956.
WINTROBE, M. M.: Clinical Haematology. 7th ed., p. 198. Lea and Febiger, Philadelphia 1974.
YOUNG, L. E., LAWRENCE, J. S.: Maturation and destruction of transfused human reticulocytes. J. clin. Invest. *24*, 554 (1945).

CHAPTER 16

CYTOCHEMICAL STAINS AND MICROSCOPY

The Prussian blue reaction is the one most widely used for the cytochemical demonstration of iron (Perls, 1867). Other methods use the Turnbull blue reaction (Schmeltzer, 1933; Pearse, 1960) and iron sulfide (Quincke, 1868). In the latter the iron is precipitated as its sulfide with sulfurated hydrogen and treated with silver to make the granules visible by light microscopy. Hausmann (1965) adapted the procedure for hematological purposes, but the disadvantage is that the test is not specific for iron.

With the Prussian blue reaction, iron-containing granules more than 0.2 μm in diameter are visible under the light microscope (Bessis and Breton-Gorius, 1962). With electron microscopy, iron granules of 5 nm diameter (ferritin) can also be demonstrated (Lindner, 1963).

SIDEROCYTES

In 1941 Grüneberg demonstrated fine, iron-containing granules in the red blood cells of mouse and human fetuses and called them siderocytes. Normally 0–3‰ of red cells in the peripheral blood are siderocytes (Crosby, 1957; Brüschke, 1962). In neonates the proportion is 3–17% (Brüschke, 1962), but the majority disappear within 4 days of birth. In premature infants the increased proportion may persist for up to 2 weeks.

The number of siderosomes (iron-containing granules) per red cell varies from 1 to 15, although usually only 1–3 fine 0.2- to 0.5-μm-in-diameter granules can be seen. In pathological conditions the siderosomes are larger and the diameter may reach 2–3 μm.

The ratio of siderocytes is elevated following splenectomy (Doniach et al., 1943; Pappenheimer et al., 1945; McFadzean and Davis, 1947; Douglas and Dacie, 1953). Crosby (1957) showed that the spleen removes the siderosomes from the blood without damaging the red cells themselves. According to Jung (1958), the iron-containing granules are removed by the sinusoidal cells of the spleen. This function is not specific since Howell-Jolly bodies and other inclusions are also removed from the red cells (Crosby, 1957; Merker, 1968).

Excessive numbers of siderocytes are found in the blood in lead poisoning (Crosby, 1957; Brüschke, 1962). This is attributed to disturbed iron incorporation in the protoporphyrin molecule. Increased numbers are also sometimes found in various hemolytic processes (Crosby, 1957; Brüschke, 1962). Siderocytes are not found in the peripheral blood in iron deficiency.

SIDEROBLASTS

Iron granules can also be demonstrated in nucleated red cells in the bone marrow (Dacie and Doniach, 1947). Kaplan et al. (1954) introduced the name sideroblasts. Of the erythroblasts in healthy humans, 20–60% are sideroblasts (Douglas and Dacie, 1953; Kaplan et al., 1954). The average ratio is around 35% (Bernát, 1971). The normal sideroblast contains 1–4, occasionally more, siderosomes, which may occur anywhere in the cytoplasm, and in particular, they are not closely related topographically with the cell nucleus. The granules are sometimes connected by a fine network.

A negative cytochemical stain for iron does not imply absence of siderosomes since only those 0.2 μm or larger are visible by light microscopy.

Pathological sideroblasts that are found in association with some hematological abnormalities contain more and larger iron granules, and these are not randomly situated in the cytoplasm but are grouped characteristically around the nucleus (Hayhoe and Quaglino, 1960; Bowman, 1961), the so-called ringed sideroblasts (Bowman, 1961).

Table 16/1

Changes in the sideroblast ratio in various pathological processes

The ratio of sideroblasts		
Decrease	Moderate increase	Marked increase
Hyposiderosis	Thalassemia	Refractory (normoblastic) sideroblastic anemia
Anemia of infection	Pyridoxine-responsive anemia	Hereditary hypochromic sideroachrestic anemia
Thermal injury	Pyridoxine deficiency	
	Drugs used in therapy of tuberculosis (INH, cycloserine, or pyrazinoic acid)	
	Lead intoxication	
	Hemolytic anemias	
	Aplastic anemias	
	Megaloblastic anemias	
	Myeloproliferative diseases	
	Hemochromatosis	

Sideroblasts are reduced in number or absent in iron deficiency and in the anemia of infection or thermal injury, where iron accumulates in the RE cells and little is incorporated into the erythroblasts. The two types of pathological conditions can be distinguished on the basis of the iron content of the RE cells (Table 8/3).

The proportion of sideroblasts increases if the incorporation of iron into the heme molecule is inhibited and iron accumulates within the cell (Table 16/1).

Iron incorporation may be disturbed by:

(1) decreased heme synthetase activity

(2) a metabolic disturbance of the erythroblasts that renders the intracellular iron incapable of the reduction necessary for incorporation into heme

(3) structural damage of the erythroblasts, preventing the iron from reaching the mitochondria

(4) diminished production of protoporphyrin-III-9, and

(5) decreased, decelerated, or pathological globin synthesis, which gives rise to secondary disturbances in heme synthesis.

SIDEROMACROPHAGES

Reticuloendothelial cells in which iron may be demonstrated by cyto- or histochemical techniques are called sideromacrophages. They can be seen in both bone marrow smears and sections. Under physiological conditions 54–74% of the erythrocytes die in the bone marrow; the remainder are sequestered in the liver and the spleen (Ehrenstein and Lockner, 1959). Normally the iron content of the bone marrow gives a fair indication of the total iron stores (Weinfeld, 1964). However, in certain pathological conditions, e.g., in paroxysmal nocturnal hemoglobinuria (PNH) and in essential pulmonary hemosiderosis, little iron may be demonstrated in the bone marrow, whereas there may be heavy deposits in the kidneys and the lungs, respectively. In the rare condition of atransferrinemia, the cytological methods show little or no iron in the bone marrow, while the liver and other parenchymatous organs show excessive iron accumulation (Heilmeyer et al., 1961, 1965).

In iron deficiency there is no stainable iron in the reticulum cells (see Table 8/3) but, paradoxically, if the patient has received parenteral iron therapy some iron may be demonstrated in the bone marrow from which it has not been fully utilized for hemoglobin synthesis (see also Chapter 8).

BIBLIOGRAPHY

BERNÁT, I.: Az égési anaemia pathogenesise (The Pathogenesis of Anemia after Thermal Injury). Akadémiai Kiadó, Budapest 1971.

BESSIS, M., BRETON-GORIUS, J.: Iron metabolism in the bone marrow as seen by electron microscopy: A critical review. Blood *19*, 635 (1962).

BOTHWELL, T. H., FINCH, C. A.: Iron Metabolism. Little, Brown and Co., Boston 1962.
BOWMAN, W. D.: Abnormal (ringed) sideroblasts in various hematologic and non-hematologic disorders. Blood *18*, 662 (1961).
BRÜSCHKE, G.: Der Siderocyt. Akademie Verlag, Berlin 1962.
CARTWRIGHT, G. E., DEISS, A.: Sideroblasts, siderocytes, and sideroblastic anemias. New Engl. J. Med. *292*, 185 (1975).
CROSBY, W. H.: Siderocytes and the spleen. Blood *12*, 165 (1957).
DACIE, J. V., DONIACH, J.: Basophilic property of iron-containing granules in siderocytes. J. Path. Bact. *59*, 684 (1947).
DONIACH, J., GRÜNEBERG, H., PEARSON, J. E. G.: The occurrence of siderocytes in adult human blood. J. Path. Bact. *55*, 23 (1943).
DOUGLAS, A. S., DACIE, J. V.: The incidence and significance of iron containing granules in human erythrocytes and their precursors. J. clin. Path. *6*, 307 (1953).
EHRENSTEIN, G., LOCKNER, D.: Der physiologische Erythrocytenabbau. Acta haemat. (Basel) *22*, 129 (1959).
GRÜNEBERG, H.: Siderocytes: a new kind of erythrocytes. Nature *148*, 139 (1941).
HAUSMANN, K.: Cytochemische Differenzierung und Lokalisation locker gebundener Schwermetalle in Blut- und Knochenmarkzellen. X. Kongreß der Europäischen Gesellschaft für Hämatologie, Strassburg 1965.
HAYHOE, F. G. J., QUAGLINO, D.: Refractory sideroblastic anemia and erythremic myelosis. Possible relationship and cytochemical observations. Brit. J. Haemat. *6*, 381 (1960).
HEILMEYER, L.: Die sideroachrestischen Anämien. Folia haemat. (Frankfurt) N. F. *6*, 1 (1961a).
HEILMEYER, L.: Durch Eisenstoffwechselstörungen bedingte Anämien. Wien. klin. Wschr. *73*, 181 (1961b).
HEILMEYER, L.: Morphologie, Funktion und Bedeutung der Sideroblasten. Wien. Z. inn. Med. *44*, 1 (1963).
HEILMEYER, L., HEILMEYER, I.: Der Eisenstoffwechsel. In: HEILMEYER, L., HITTMAIR, A. (eds.): Handbuch der gesamten Hämatologie, Vol. I/1. 2nd ed., Urban und Schwarzenberg, München–Berlin–Wien 1959.
HEILMEYER, L., KELLER, W., VIVELL, O., BETKE, K., WÖHLER, F., KEIDERLING, W.: Die kongenitale Atransferrinämie. Schweiz. med. Wschr. *91*, 1203 (1961).
HEILMEYER, L., MERKER, H., WETZEL, H. P., KLEMEN, D., BURMEISTER, P., HAAS, R.: Atransferrinämie bei nephrotischem Syndrom. Dtsch. med. Wschr. *90*, 1649 (1965).
JUNG, F.: Das Schicksal toxisch veränderter roter Blutzellen in der Milz. Klin. Wschr. *36*, 63 (1958).
KAPLAN, E., ZUELZER, W. W., MOURIQUAND, C.: Sideroblasts: a study of stainable non-hemoglobin iron in marrow normoblasts. Blood *9*, 203 (1954).
LINDNER, E.: Elektronenmikroskopischer Nachweis von Schwermetallen, cit. SCHIEBLER, T. H.: Histochemie der Mineralstoffe, Acta histochem. (Jena), Suppl. III, *98* (1963).
MCFADZEAN, A. J. S., DAVIS, L. J.: Iron staining erythrocyte inclusions with special reference to acquired hemolytic anaemia. Glasg. med. J. *28*, 237 (1947).
MERKER, H.: Cytochemie der Blutzellen. In: HEILMEYER, L. (ed.): Blut und Blutkrankheiten. Springer, Berlin–Heidelberg–New York 1968.
PAPPENHEIMER, A. M., THOMPSON, W. P., PARKER, D. D., SMITH, K. E.: Anemia associated with unidentified erythrocytic inclusions after splenectomy. Quart. J. Med. *14*, 75 (1945).
PEARSE, A. G. E.: Histochemistry. Little, Brown and Co., Boston 1960.
PERLS, M.: Nachweis von Eisenoxyd in gewissen Pigmenten. Virchows Arch. path. Anat. *39*, 42 (1867).
QUINCKE, H. J.: Über das Verhalten der Eisensalze im Tierkörper. Arch. Anat. Phys. 757 (1868).
SCHMELTZER, W.: Der mikrochemische Nachweis von Eisen in Gewebselementen mittels Rhodanwasserstoffsäure und die Konservierung der Reaktion in Paraffinöl. Z. wiss. Mikr. *50*, 99 (1933).
WEINFELD, A.: Storage iron in man. Acta med. scand. Suppl. 427 (1964).

CHAPTER 17

ELECTRON MICROSCOPIC INVESTIGATIONS

Electron microscopy has made it possible to visualize molecules of ferritin and hemosiderin, to examine the phases of erythrophagocytosis, and to define the nature of the sideroblasts and siderocytes.

FERRITIN

From electron microscopic studies of a solution of horse spleen ferritin, Farrant (1954) concluded that the ferritin iron is situated at the four angles of a 55 Å-sided square. This feature made it possible to identify the ferritin molecule, partly in dispersed form and partly in aggregates, in various cells such as erythroblasts, reticulum cells, and intestinal epithelium (Bessis and Breton-Gorius, 1956, 1957a, b, c, 1959; Richter, 1957, 1958, 1959b). More recent studies show that the iron in ferritin is not situated at the angles of a square but in the six apices of an octahedron, and the various forms of appearance of the ferritin (4, 5, or 6 black dots) depend on the angle from which it is visualized (Bessis and Breton-Gorius, 1960; Kerr and Muir, 1960; Muir, 1960). In hypersiderosis, in addition to ferritin being found in reticulum cells, erythroblasts, liver cells, and intestinal cells, it may be found in other parenchymatous organs, either dispersed or in smaller or larger clusters that are often surrounded by a one- or two-layered membrane but may occur free in the cytoplasm of cells.

HEMOSIDERIN

Hemosiderin appears in the form of yellow or brown granules under the light microscope and is stained blue by the Prussian blue reaction. Under the electron microscope ferritin is seen as a constant constituent of hemosiderin (Bessis and Breton-Gorius, 1957c, 1959b; Richter, 1957, 1958). Hemosiderin may consist of pure ferritin crystals or of a mixture of ferritin and apoferritin crystals (form 1), or ferritin masses surrounded by a simple or double membrane (form 2). It may have a

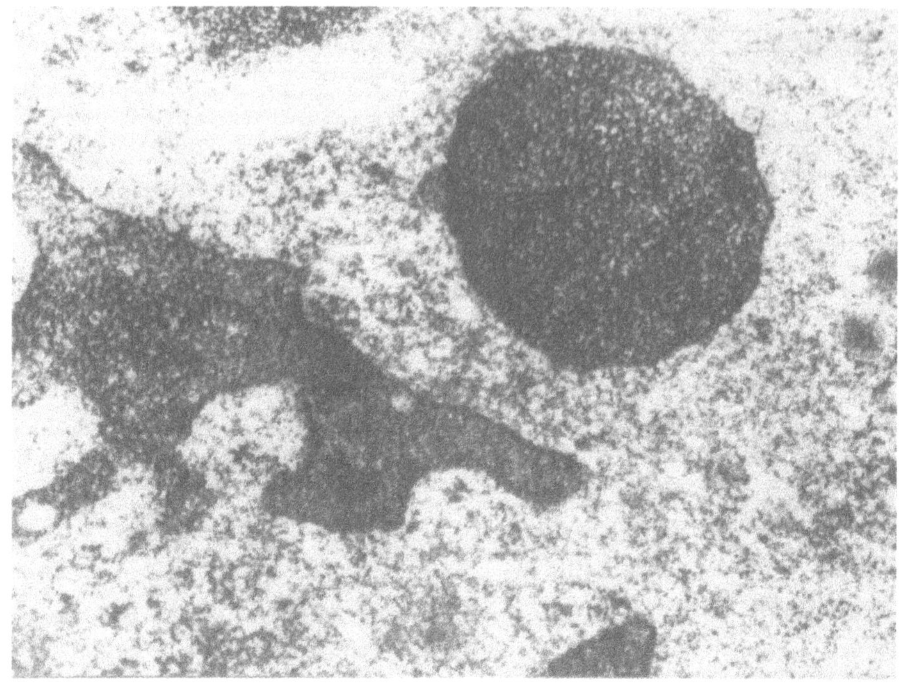

Fig. 17/1. Dispersed ferritin molecules in the cytoplasm of a reticulum cell and randomly distributed ferritin clusters within the cytoplasmic structures. So-called hemosiderin III (after Lelkes, G. and Bernát, I., 1970)

heterogeneous composition and may contain, besides ferritin, lipids and other compounds (form 3) (Fig. 17/1) or other iron compounds (form 4). This latter hemosiderin type is rare.

ERYTHROPHAGOCYTOSIS

Erythrophagocytosis and digestion of red cells can be followed by phase contrast microscopy and microkinetography. The red cells are usually fragmented before phagocytosis and may break up into smaller fragments within the macrophages prior to hemolysis (Policard and Bessis, 1953) (Fig. 17/2).

The remaining stroma is decomposed into small particles, the entire process taking about 10 minutes. Under high magnification, numerous granules can be seen around the fragments of the phagocytosed red cells. With further magnification these prove to be groups of ferritin molecules. Erythrophagocytosis can be seen in

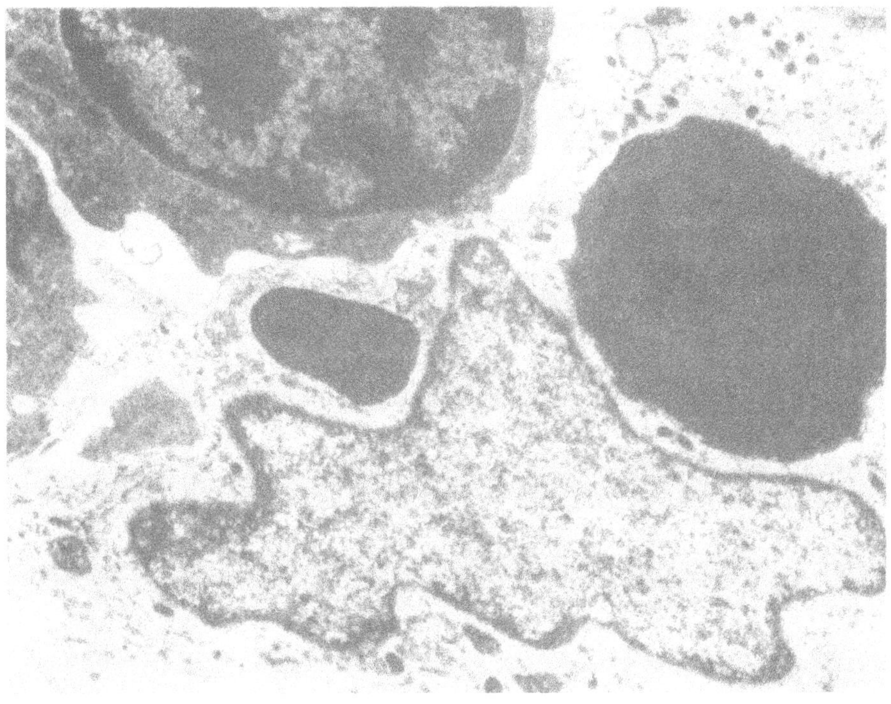

Fig. 17/2. Phagocytosed red cell fragments in the cytoplasm of a bone marrow reticulum cell (after Lelkes, G. and Bernát, I., 1970)

normal bone marrow reticulum cells, where the bulk of red cell destruction occurs normally, and sometimes also in the spleen, but only exceptionally in the Kupffer cells of the liver of normal individuals (Miescher, 1956; Ehrenstein and Lockner, 1958). In pathological conditions, e.g., in hemolytic anemia, erythrophagocytosis takes place mainly in the spleen and in the Kupffer cells of the liver.

IRON TRANSPORT

Iron, when given parenterally, is taken up by reticulum cells. It combines with apoferritin in the cytoplasm, and the characteristic ferritin crystals can be observed on electron microscopy after about 1 or 2 weeks. Iron may be subsequently transferred to transferrin. The iron transport within the organism takes place via three different mechanisms:

1. From the reticulum cells the iron is taken up by transferrin, and the transferrin carries the iron to the erythroblasts or to other reticulum cells for temporary storage as ferritin.

2. Iron can be transported as ferritin through the blood or lymphatic vessels, but this is normally an insignificant mechanism except in conditions like acute liver damage.

3. The iron can be transported within reticulum cells or histiocytes, as in cases of severe iron overload.

The islands of erythroblasts in the bone marrow are not easily recognized in bone marrow smears because of disruption during the preparation of the smears. They can, however, be recognized in marrow sections and are a permanent anatomical feature of normal bone marrow (Bessis, 1958; Bessis and Breton-Gorius, 1959a) (Fig. 17/3). There are usually one or two centrally situated reticulum cells surrounded by erythroblasts, often in two or three rows, the inner ring being the less mature basophilic cells and the middle and outer ring the more mature polychromatophilic and oxyphilic erythroblasts. Marmond and Damasio (1960) suggest that there is a further ring composed of reticulocytes. As they reach this outermost ring the cells become mobile, detach from the island, and reach the sinusoids of the bone marrow by diapedesis (Bessis and Breton-Gorius, 1960).

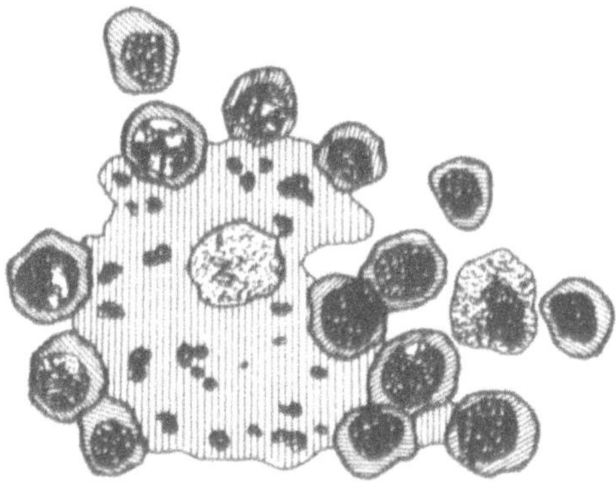

Fig. 17/3. Erythroblast island in bone marrow smear. The centrally located reticulum cell is surrounded by a ring of erythroblasts (after Bernát, I., 1971)

Electron microscopy suggests that the processes of the central reticulum cell embrace partially or totally the surrounding nucleated red cells. Prussian blue staining may indicate the presence of iron granules within the processes of the reticulum cell. If the cytoplasmic fragments lie free among the nucleated red cells, it appears as if the iron is situated extracellularly (Kaplan et al., 1954; Mouriquand, 1958).

In response to hypoxia, the central reticular cell becomes actively phagocytic and contains numerous extruded red cell nuclei. The fine processes of the reticular cell come into close contact with the developing nucleated erythroid cells, and small lymphocytes may appear around the island. In posthypoxic polycythemia a gradual decrease is seen in the number of erythroblasts surrounding the reticular cell. The processes of the cell become retracted, and the cell becomes smaller. It also becomes surrounded by increasing numbers of small lymphocytes (Ben Ishay and Joffey, 1971).

RHOPHEOCYTOSIS

Electron microscopy has further shown that on the surface of erythroblasts adhering to the reticular cells first invaginations and later vacuoles appear within the cytoplasm of the erythroblast. The process resembles pinocytosis as observed under the optical microscope, but the vacuoles are only of the order of 100 Å in diameter. The process is termed rhopheocytosis (the Greek ropheo = I aspirate). As observed, the membrane of the reticulum cell "dissolves" partially or completely,

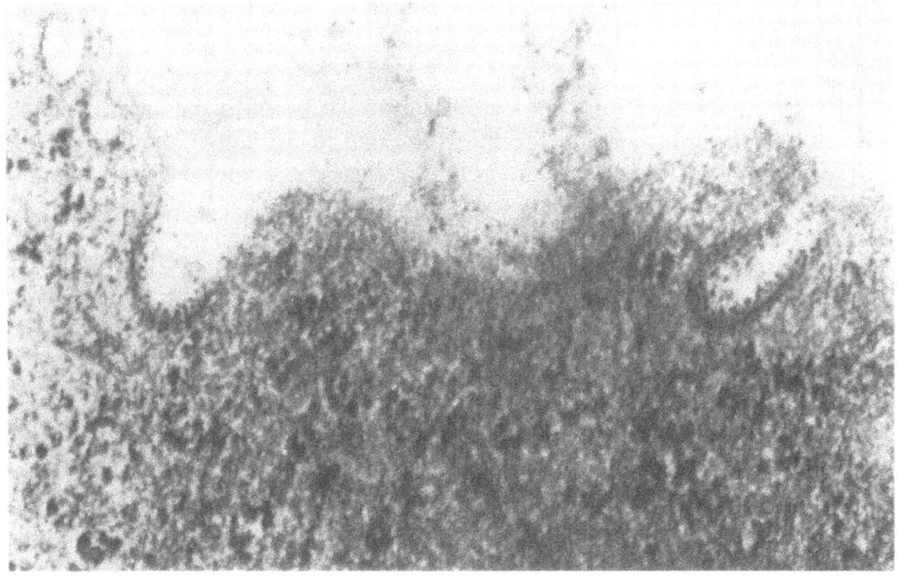

Fig. 17/4. Rhopheocytosis

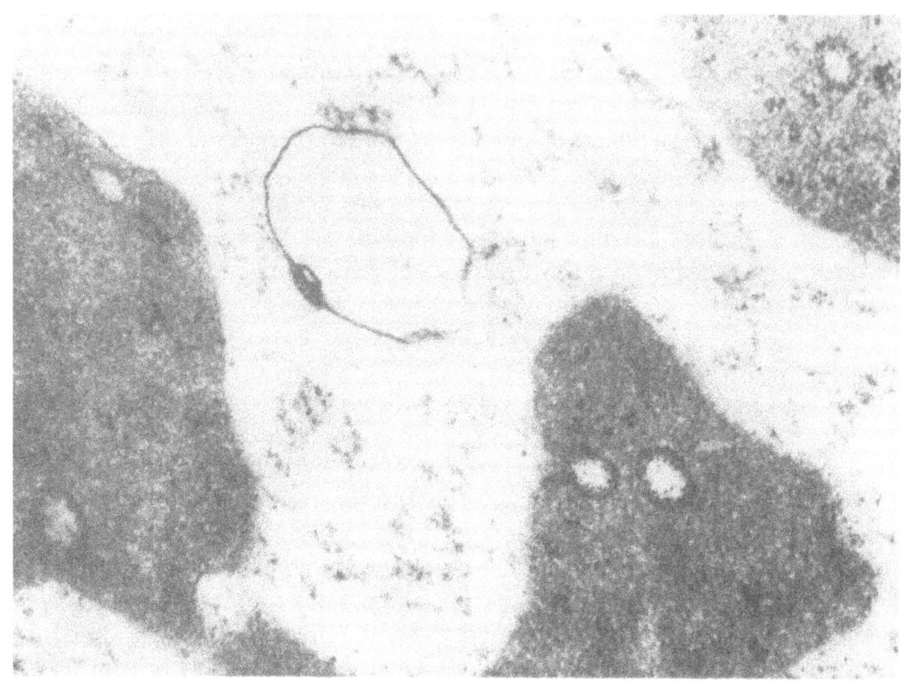

Fig. 17/5. Rhopheocytotic vacuoles. Some of them contain iron granules, while others (left, top) are devoid of iron granules (after Lelkes, G. and Bernát I., 1970)

and ferritin is liberated from the cytoplasm and adheres to the surface of the erythroblast membrane. Following this, the small invaginations develop on the surface of the nucleated red cell, and in the vacuoles that are formed subsequently 30–50 ferritin molecules can be observed (Fig. 17/4 and Fig. 17/5). Vacuoles devoid of ferritin can also be seen.

There has been some criticism of the interpretation of the electron microscopic findings, but Bessis and Breton-Gorius (1962) have argued that a similar process can be observed if the ferritin is introduced and the direction of the process is predetermined, and during the maturation process of erythroblasts the ferritin content increases and the iron progressively accumulates in the cell. It has also been suggested that the reverse process, *exocytosis*, may serve to rid the normoblast of excess iron (Tanaka, 1970).

The present evidence suggests that the greater part of the iron arrives at the normoblast surface bound to transferrin (Katz and Jandl, 1964).

SIDEROBLASTS AND SIDEROCYTES

Electron microscopy indicates that the iron granules seen by light microscopy in sideroblasts and siderocytes are clusters of ferritin lying free in the cytoplasm (Fig. 17/6) or surrounded by membrane. Every erythroblast contains a certain number of ferritin molecules, although only the larger ones are visible by light microscopy. In the younger erythroblasts the ferritin molecules are scattered, but in the course of maturation they tend to aggregate.

In certain pathological conditions such as the sideroblastic anemias the iron accumulates in the mitochondria (Fig. 17/7 and Fig. 17/8), giving rise to the typical ringed sideroblasts (Bowman, 1961).

Ferritin in the erythroblasts may serve as an iron reserve for hemoglobin synthesis (Bessis and Breton-Gorius, 1962). The amount declines in the course of red cell maturation, and in the mature red cell there is little if any ferritin present.

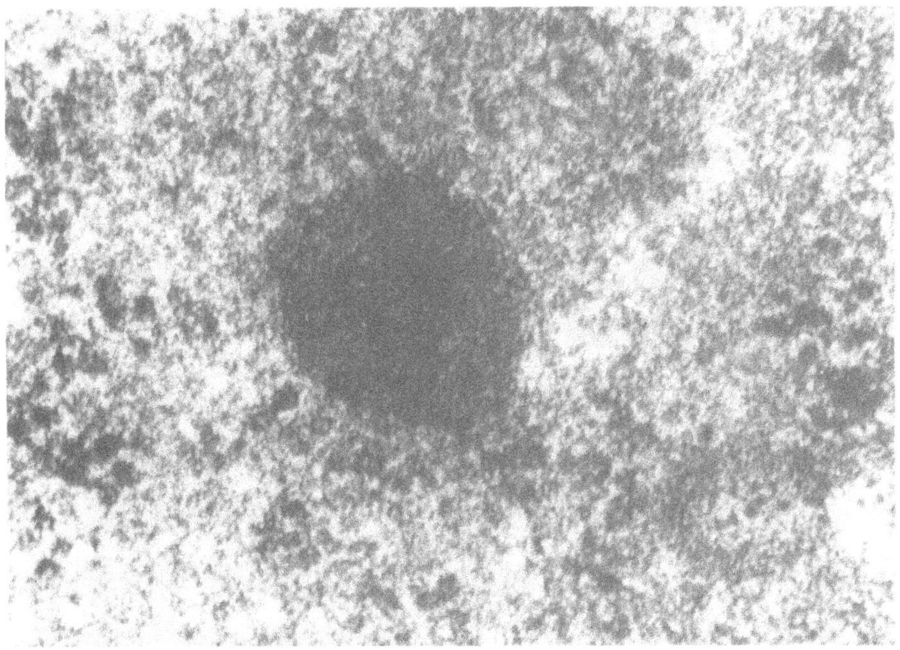

Fig. 17/6. Siderosome (iron aggregate) in a reticulocyte (after Lelkes, G. and Bernát, I., 1970)

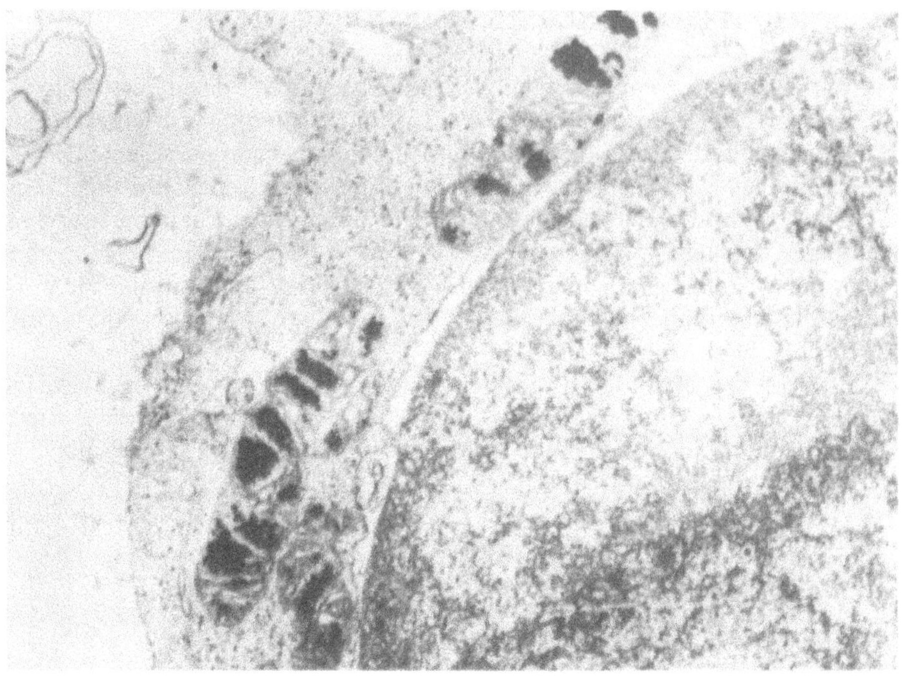

Fig. 17/7. Massive iron accumulation in the mitochondria (acquired sideroblastic anaemia)

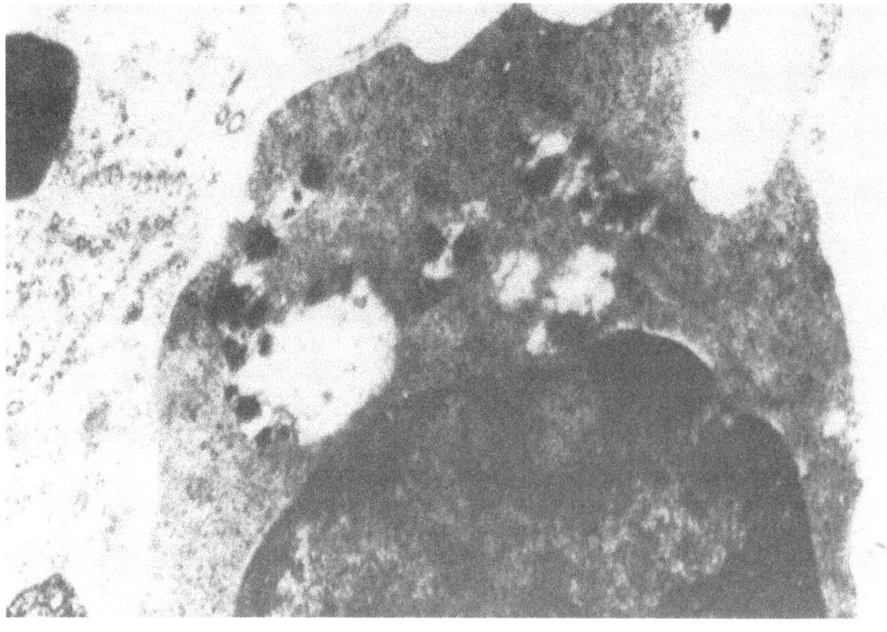

Fig. 17/8. Massive iron deposits in degenerated mitochondria (after Bernát, I., 1971)

BIBLIOGRAPHY

ALPEN, E. L., GRANMORE, D.: Observations on the regulation of erythropoiesis and on cellular dynamics by Fe^{59} autoradiography. In: STOHLMAN, F. (ed.): The Kinetics of Cellular Proliferation, p. 326. Grune and Stratton, New York–London 1959.

BEN ISHAY, Z., JOFFEY, J. M.: Reticular cells of erythroid islands of rat bone marrow in hypoxia and rebound. RES (N. Y.) *10,* 482 (1971).

BERNÁT, I.: Az égési anaemia pathogenesise (The Pathogenesis of Anemia after Thermal Injury), p. 136. Akadémiai Kiadó, Budapest 1971.

BESSIS, M.: L'îlot érythroblastique, unité fonctionnelle de la moelle osseuse. Rev. Hémat. *13,* 8 (1958).

BESSIS, M., BRETON-GORIUS, J.: Granules ferrugineux observés au microscope électronique dans les cellules de la moelle osseuse et dans les sidérocytes. C. R. Acad. Sci. (Paris) *243,* 1235 (1956).

BESSIS, M., BRETON-GORIUS, J.: Étude au microscope électronique des granulations ferrugineuses des érythrocytes normaux et pathologiques. Anémie hémolytique. Hémoglobinopathies. Saturnisme. Rev. Hémat. *12,* 43 (1957a).

BESSIS, M., BRETON-GORIUS, J.: Granules ferrugineux dans les cellules macrophages et les érythrocytes au cours du saturnisme expérimental. Examen au microscope électronique. C. R. Soc. Biol. (Paris) *151,* 275 (1957b).

BESSIS, M., BRETON-GORIUS, J.: Trois aspects du fer dans des coupes d'organes examinées au microscope électronique (Ferritine et dérivés, dans les cellules intestinales, les érythroblastes et les cellules réticulaires). C. R. Acad. Sci. (Paris) *245,* 1271 (1957c).

BESSIS, M., BRETON-GORIUS, J.: Nouvelles observations sur l'îlot érythroblastique et la rhophéocytose de la ferritine. Rev. Hémat. *14,* 165 (1959a).

BESSIS, M., BRETON-GORIUS, J.: Différents aspects du fer dans l'organisme: I. Ferritine et micelles ferrugineuses. II. Différentes formes de l'hémosidérine. J. biophys. Biochem. Cytol. *6,* 231 (1959b).

BESSIS, M., BRETON-GORIUS, J.: Diapédèse des réticulocytes et des érythroblastes. C. R. Acad. Sci. (Paris) *251,* 465 (1960).

BESSIS, M., BRETON-GORIUS, J.: Iron metabolism in the bone marrow as seen by electron microscopy: A critical review. Blood *19,* 635 (1962).

BOWMAN, W. D.: Abnormal (ringed) sideroblasts in various hematologic and non-hematologic disorders. Blood *18,* 662 (1961).

EHRENSTEIN, G., LOCKNER, D.: Sites of the physiological breakdown of the red blood corpuscles. Nature *181,* 911 (1958).

FARRANT, J. L.: An electron microscopic study of ferritin. Biochim. biophys. Acta (Amst.) *13,* 569 (1954).

KAPLAN, E., ZUELZER, W. W., MOURIQUAND, C.: A study of stainable non hemoglobin iron in marrow normoblasts. Blood *9,* 203 (1954).

KATZ, J. H., JANDL, J. H.: The role of transferrin in the transport of iron into the developing red cell. In: GROSS, F. (ed.): Iron Metabolism, p. 103. Springer, Berlin 1964.

KERR, D. N. S., MUIR, A. R.: A demonstration of the structure and disposition of ferritin in the human liver cell. J. Ultrastruct. Res. *3,* 313 (1960).

LAJTHA, L. G., SUIT, H. D. Uptake of radioactive iron (^{59}Fe) by nucleated red cells in vitro. Brit. J. Haemat. *1,* 55 (1955).

LELKES, G., BERNÁT, I.: Electron microscopical study of the bone marrow of burned patients. Haematologia *4,* 295 (1970).

LESSIN, L. S., BESSIS, M.: Morphology of the erythron. In: WILLIAMS, W. J. et al. (eds.): Hematology. McGraw-Hill Inc., New York 1972, 1977.

MARMOND, A., DAMASIO, E.: Fluorescence microscopy in hematology with special regard to the reticulocytopoietic activity of the so-called erythroblastic nests or islands. In: Proc. 8th Congr. Europ. Soc. Haemat., Part 1, p. 66. Karger, Basel 1960.

MIESCHER, P.: Le mécanisme de l'érythroclasie à l'état normal. Rev. Hémat. *11,* 248 (1956).

Mouriquand, C.: Le sidéroblaste: étude morphologique et essai d'interprétation. Rev. Hémat. *13,* 79 (1958).

Muir, A. R.: The molecular structure of isolated and intracellular ferritin. Quart. J. exp. Physiol. *45,* 192 (1960).

Policard, A., Bessis, M.: Fractionnement d'hématies par les leucocytes au cours de leur phagocytose. C. R. Soc. Biol. (Paris) *147,* 982 (1953).

Policard, A., Bessis, M.: Sur un mode d'incorporation des macromolécules par le cellule, visible au microscope électronique: la rhophéocytose. C. R. Acad. Sci. (Paris) *246,* 3194 (1958).

Policard, A., Bessis, M.: La pénétration de substances dans la cellule et ses mécanismes (dialyse, phagocytose, pinocytose, rhophéocytose). Rev. franç. Études clin. biol. *4,* 839 (1959).

Policard, A., Bessis, M., Breton-Gorius, J.: Structures myéliniques observées au microscope électronique sur des coupes de globules rouges en voie de lyse. Exp. Cell. Res. *13,* 184 (1957).

Richter, G. W.: A study of haemosiderosis with aid of electron microscopy with observations on the relationship between hemosiderin and ferritin. J. exp. Med. *106,* 203 (1957).

Richter, G. W.: Electron microscopy of hemosiderin: presence of ferritin and occurrence of crystalline lattices in hemosiderin deposits. J. biophys. biochem. cytol. *4,* 55 (1958).

Richter, G. W.: The cellular transformation of injected colloidal iron complexes into ferritin and hemosiderin in experimental animals. A study with the aid of electron microscopy. J. exp. Med. *109,* 197 (1959).

Stoeckenius, W.: Morphologische Beobachtungen beim intrazellulären Erythrocytenabbau und der Eisenspeicherung in der Milz des Kaninchens. Klin. Wschr. *35,* 760 (1957).

Tanaka, Y.: Bi-directional transport of ferritin in guinea-pig erythroblasts in vitro. Blood *35,* 793 (1970).

CHAPTER 18

IRON DEFICIENCY

Although it had long been suspected that the cause of iron-responsive anemias was iron deficiency of the organism (Dameshek, 1931; Schulten, 1934), proof was only provided in the 1930s. Chlorosis was ascribed to the disturbed endocrine function of the ovaries (Naegeli, 1931; and others) and the therapeutic effect of iron was believed to be through a stimulatory effect on the pathological bone marrow function that developed as a result of the endocrine disorder. Essential hypochromic anemia was ascribed to "gastrogenic intoxication" (Kaznelson et al., 1929), and it was suggested that the beneficial effect of iron was through the influence of the iron upon the microorganisms of the gastrointestinal tract. M. B. Schmidt (1928) was the first to induce experimental iron deficiency in mice, but he did not believe that humans could suffer from this disorder. In 1936 Heilmeyer and Plötner demonstrated that the level of plasma iron was low in all patients with iron-responsive anemia and was normal or high in those with iron-refractory anemia. They also showed that iron given parenterally disappeared rapidly from the blood in iron-responsive anemias and that iron was poorly absorbed in "gastrogenic" chronic hypochromic anemia or "achlorhydrische Anämie." They concluded from their findings that the iron-responsive hypochromic anemias were the consequence of iron deficiency. In the next few years the concept of iron-deficiency disease became generally accepted, and the differentiation of the various clinical forms, e.g., chlorosis and idiopathic or essential hypochromic anemia, lost its significance.

INCIDENCE

Iron deficiency is the most prevalent worldwide deficiency. In some countries it is almost universal, and in different populations its incidence varies from about 20 to 95% (WHO Technical Report, 1959). In India (Ramalingaswami and Patwardhan, 1949) and certain parts of Africa (Gosden and Reid, 1948) more than 50% of the population was found to be anemic, and between 10 and 40% of the maternal mortality during pregnancy and delivery in some regions of India is attributable to anemia (Menon, 1958).

Although most surveys do not make a distinction among the various types of anemia, it is justifiable to presume that the most frequent cause is iron deficiency; e.g., Stott (1960) found that the anemia in Mauritius that affected about half the population was in 90% of cases due to iron deficiency. Very high frequencies of iron deficiency have also been reported from the Philippines, East Pakistan (Bokhari, 1958), and Turkey (Reimann, 1956).

Hyposiderosis is not confined to the developing countries. The incidence in poorer sections of the population studied in the 1930s in Scotland and London was extremely high, almost half the infants and preschool children and adult women being anemic (Davidson et al., 1942; MacKay et al., 1942). Population studies done in the 1960s showed a reduced incidence, but nevertheless throughout Western Europe recent surveys have shown a prevalence of between 15 and 25% of iron-deficiency anemia for women of child-bearing age (Kilpatrick and Hardisty, 1961; Kilpatrick, 1970; Hallberg, 1970b; Seibold, 1970; Vellar, 1970). An estimate has been made that at least 18,000,000 people in the United States are iron-deficient (Fairbanks and Beutler, 1977). In Hungary, Simonovits et al. (1970) have made large-scale surveys for the assessment of the incidence of iron-deficiency anemia; their results are shown in Table 18/1.

Table 18/1

The incidence of iron-deficiency anemia in children (10–14 years), young adults (19–24 years), and pregnant women (from Simonovits et al., 1970)

	Percentile distribution of the examined individuals on the basis of the Hb concentration (g/dl)					Absolute No. of cases
	< 10.6	10.6–12.0	12.1–13.5	13.6–15.0	> 15.0	
Boys (10–14 years)	1.1	20.8	55.2	22.4	0.5	447
Males (19–24 years)	—	1.1	8.0	36.7	54.2	188
Girls (10–14 years)	0.3	21.5	62.0	16.2	—	592
Females (19–24 years)	2.9	14.5	57.3	24.9	0.4	477
Pregnant women	14.4	47.4	34.6	3.6	—	1122

The approximate frequencies of iron deficiency in four high-risk groups are given in Table 18/2. (Percentages are based on data from several sources; they may be considered as being representative of recent studies.)

One method of investigating the incidence of iron deficiency in a population is to follow the response to treatment with iron. This is the method that has been applied by Vellar (1970) and by Garby and his co-workers (Garby, 1973). The results of Vellar's investigations are shown in Tables 18/3, 18/4 and 18/5. Analysis of his data shows that between the ages of 7 and 13 there is no sex difference in the incidence of

Table 18/2

Approximate frequency of iron deficiency with or without anemia (from Fairbanks and Beutler, 1977)

Group	Frequency (per cent)	
	Iron depletion or iron deficiency without anemia	Iron deficiency anemia
Infants	50	25
Children	30*	0–6
Women		
Premenopausal	50	15
Pregnant	90	30**

* Owen et al., 1971.
** In women not receiving iron supplementation during pregnancy.

Table 18/3

Normal blood hemoglobin level, as well as prevalence of iron-deficiency anemia in males between 7 and 20 years of age (from Vellar, 1970)

Age (years)	Normal Hb level (g/dl)	Hb lower than (g/dl)	Prevalence of anemia (%)
7– 9	12.7 ± 1.6	11.0	0.6
10–13	13.2 ± 1.6	11.5	2.1
14–16	15.0 ± 2.0	13.0	2.9
17–20	15.5 ± 2.0	13.5	0.7

Table 18/4

Normal blood hemoglobin level, as well as prevalence of iron-deficiency anemia in females between 7 and 20 years of age (from Vellar, 1970)

Age (years)	Normal Hb level (g/dl)	Hb lower than (g/dl)	Prevalence of anemia (%)
7– 9	12.7 ± 1.6	11.0	0.6
10–13	13.2 ± 1.6	11.5	2.1
14–16	14.2 ± 2.0	12.0	3.4
17–20	14.2 ± 2.0	12.0	1.7

Table 18/5
Prevalence of iron-deficiency anemia between 15 and 60 years of age (from Vellar, 1970)

Age (years)	Males* (%)	Females** (%)
15–19	2.5	2.6
20–29	1.9	3.4
30–39	1.1	12.5
40–49	4.8	3.9
50–59	4.1	6.8
>60	6.1	7.1
Mean	3.1	5.3

* Hb level < 14.0 g/dl
** Hb level < 12.5 g/dl

Normal Hb level: 15.7 g/dl ± 1.8 (males)
14.3 g/dl ± 1.8 (females).

iron-deficiency anemia, but in the 14- to 16-year-old group the incidence becomes higher in girls, and between 17 and 20 years of age there is a definite female predominance. This becomes even more marked between 30 and 39 years of age.

THE CLINICAL PICTURE OF IRON DEFICIENCY

Iron deficiency affects the entire organism, and hypochromic anemia is a late stage of the disease. It is preceded by a period of hyposiderosis without anemia, a state that is more frequent than iron-deficiency anemia itself (Jasinski and Roth, 1954; Seibold, 1970). Severe degrees of iron deficiency undoubtedly reduce the work capacity for an individual, and even milder degrees of anemia may affect physical fitness (Viteri and Torun, 1974). The clinical symptoms of hyposiderosis without anemia depend on the effects of tissue iron deficiency (Vannotti, 1959).

Iron deficiency usually develops slowly and insidiously. Many patients have no specific complaints; others have vague symptoms of tiring easily, headache, irritability, or depression. Tachycardia, shortness of breath, or anginal pain develop later, and vertigo, tinnitus, and a tendency to faint may also be present. Many patients suffer from gastrointestinal discomfort, with loss of appetite or perversion of the appetite ("pica sideropenica") (Glevitsch, 1959; Waldenström, 1964;

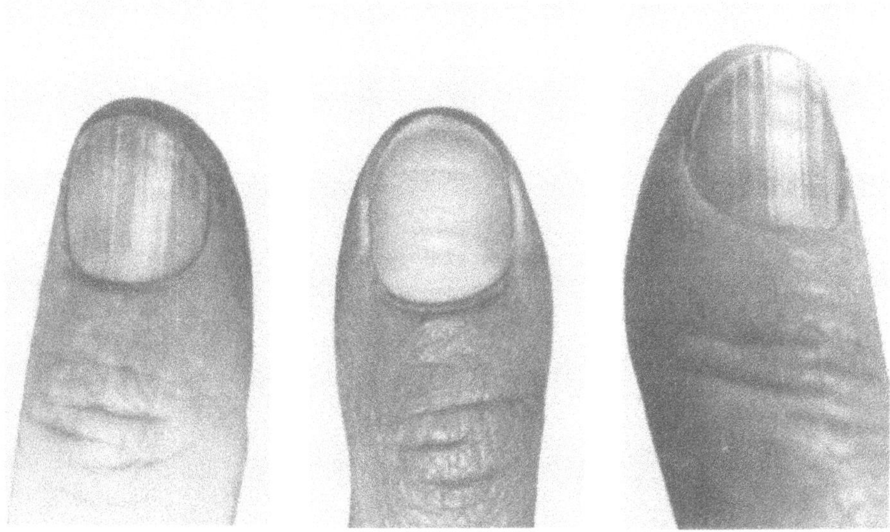

Fig. 18/1. Longitudinal and transverse folds on the nail of the thumb in iron deficiency (after Bernát, I., 1966)

McDonald and Marshall, 1964; Minnich et al., 1968; Reynolds et al., 1968; Brown and Dyment, 1972; Masuya, 1975).

Dryness of the mouth and tongue or a sore tongue may be present, or flatulence, nausea, abdominal cramps, diarrhea, or constipation may be symptoms, although these are likely to be caused by underlying gastrointestinal disease rather than solely iron deficiency.

There may be little to show on physical examination, or there may be marked pallor and sometimes dryness or roughness of the skin, or it may be more transparent and thinner than normal. Occasionally the extremities show mild cyanosis or Raynaud's syndrome (Bernát, 1966). The characteristic chlorotic greenish complexion previously described is seldom seen now.

With long-standing hyposiderosis there may be trophic changes, particularly in the nails (Figs. 18/1 and 18/2), which become brittle, soft, and flattened or spoon-shaped (koilonychia). The incidence of nail changes is variable, but Beveridge et al. (1965) found a 28% incidence in a series of 105 patients, although definite koilonychia was only noted in 18%. Jasinski and Roth (1954) suggest that koilonychia develops only in those iron-deficient women who do hard physical work, and they consider that trauma is required as well as the iron deficiency. The fact that koilonychia develops principally in the index finger and thumb of the right hand would fit in with this hypothesis. Moutier (1951) found the changes more

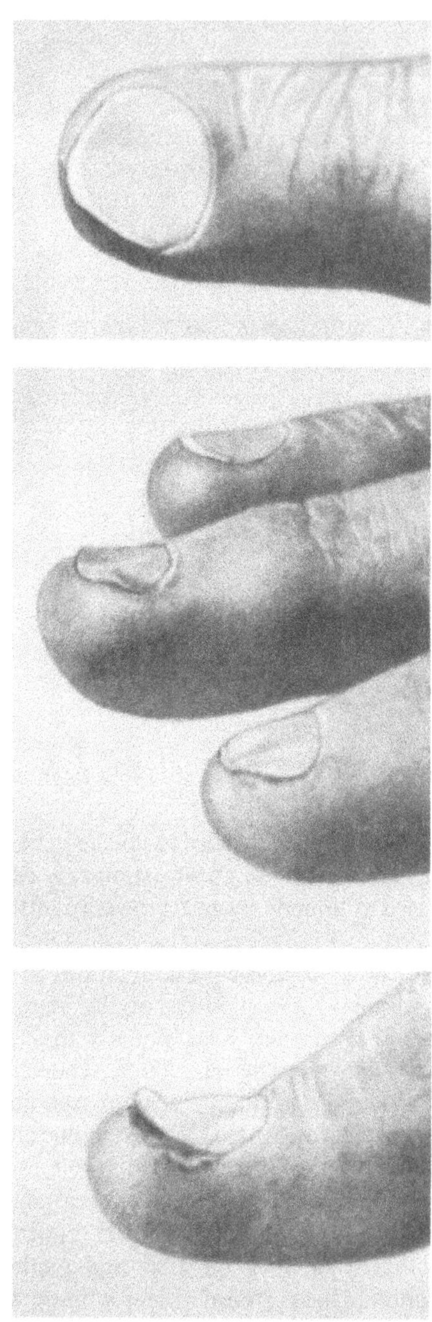

Fig. 18/2. Koilonychia in iron-deficiency disease associated with atrophic rhinitis (after Bernát, I., 1966)

frequent on the index and middle finger. The changes may affect all fingers on both hands. It has been suggested that deficiency in other substances, for example, cystine (Jalili and Al-Kassab, 1959) or zinc (Prasad et al., 1963), might contribute to the changes.

The hair may be brittle, splitting at the ends, and thinning (Fig. 18/3), and there may also be early graying.

The lips are often dry and cracked, and the surface may become uneven (Fig. 18/4). Cheilosis associated with painful, moist cracks at the angles of the mouth, with or without hyperkeratosis, is also characteristic (Fig. 18/5), and is particularly frequent in edentulous patients. It occurs in about 15% of iron-deficient patients (Chisholm, 1973).

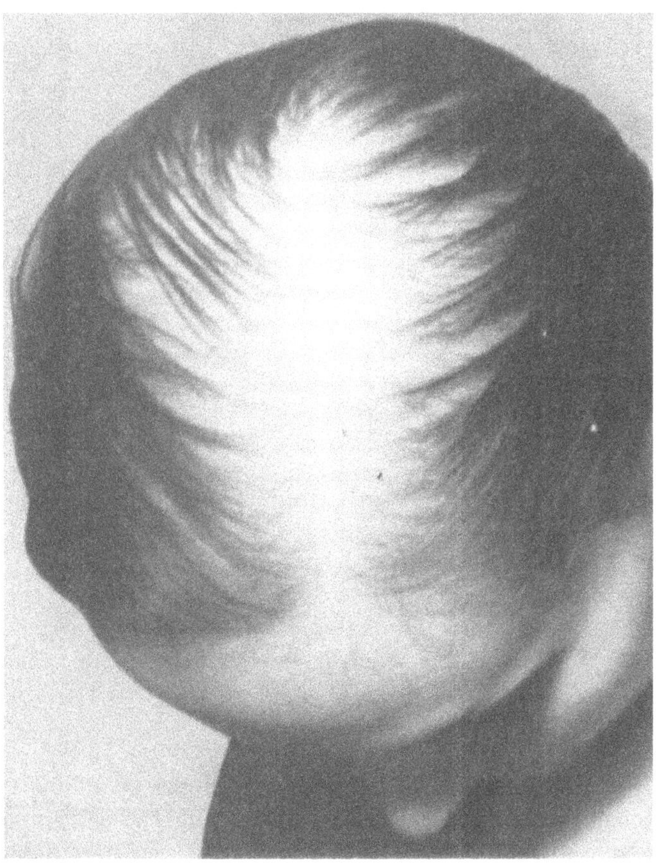

Fig. 18/3. Loss of hair in iron deficiency

Changes in the tongue occur in up to 50% of patients suffering from severe chronic iron deficiency. Biopsy of the tongue shows a loss of filiform papillae, going on to severe atrophy and total absence of papillae in some cases (Taft et al., 1958; Baird et al., 1961). The surface of the tongue may appear smooth, glossy, and

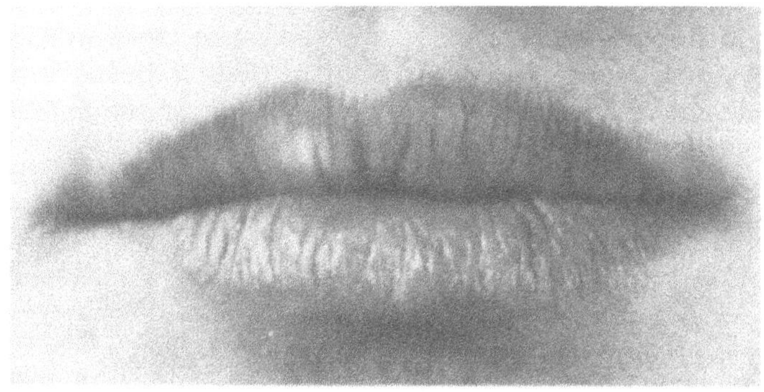

Fig. 18/4. Cheilosis in iron deficiency

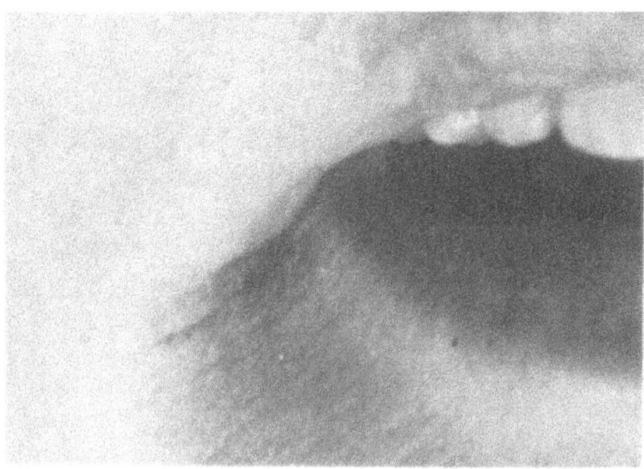

Fig. 18/5. Cracks in the angle of the mouth

mirrorlike (Fig. 18/6), there may be reddening of the tip of the tongue, and sometimes small painful vesicles or erosions develop. The patient complains of soreness and burning of the tongue and occasionally a sensation that the tongue feels swollen. In response to iron treatment the papillae may redevelop and the tongue become entirely normal (Fig. 18/7).

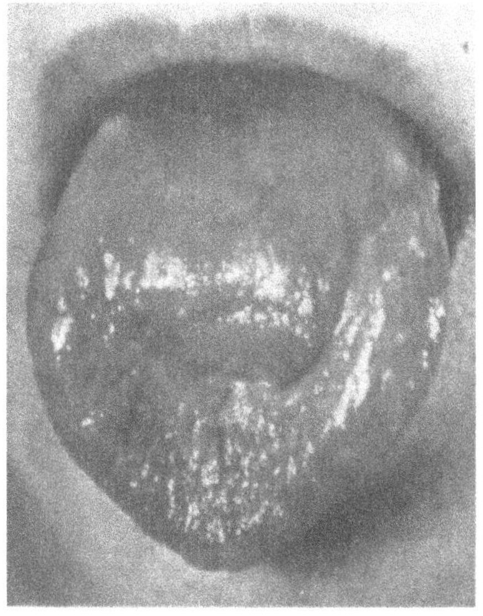

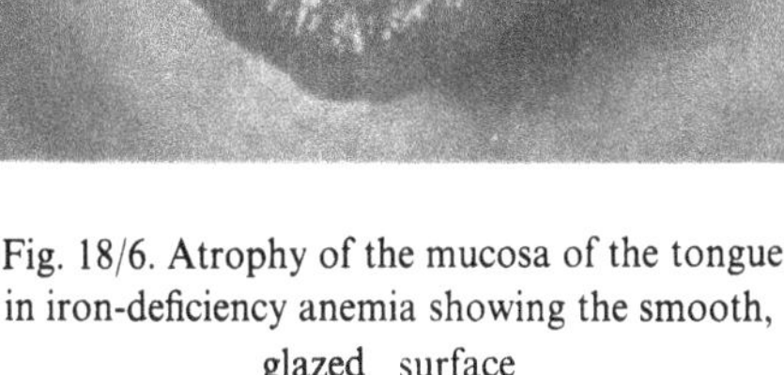

Fig. 18/6. Atrophy of the mucosa of the tongue in iron-deficiency anemia showing the smooth, glazed surface

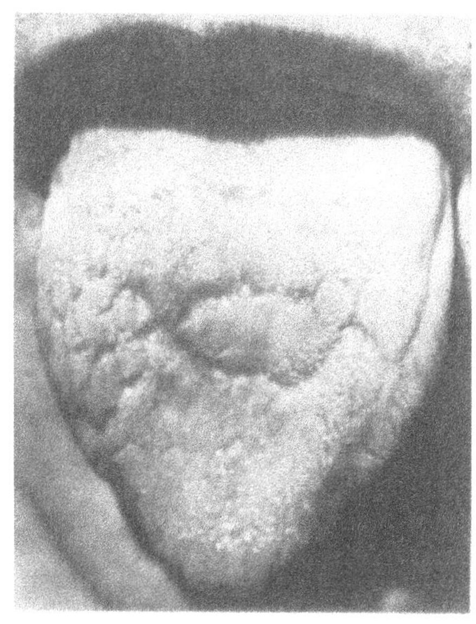

Fig. 18/7. Tongue from the same patient showing regeneration of papillae after iron treatment

Histological studies have shown that the buccal mucosa may be pathologically thin (Jacobs, 1960), and mitoses may become more frequent. There may also be some round cell infiltration in the epithelium, and the diameter of the cells of the buccal mucosa may be less than normal (Boddington, 1959, Dabski, 1960).

Sideropenic dysphagia is a relatively infrequent but rather characteristic complaint (Plummer-Vinson or Paterson-Kelly syndrome), and its incidence has been variously reported as between 5 and 20% (Jones, 1961; Chisholm et al., 1971b). There is a close association with the iron-deficiency state (Chisholm, 1973), and the dysphagia may occur in association with latent iron deficiency as well as in iron-deficiency anemia (Bernát, 1965; Manolidis et al., 1972). The characteristic complaint is a sensation of burning in the postcricoid region, or a sensation that food sticks in this region. The dysphagia is caused by narrowing of the esophagus by a membranous or fibrous web (Suzman, 1933; Waldenström and Kjellberg, 1939). There may be signs of chronic inflammation around the stricture, and the inflammation may extend into the muscular layer (Entwhistle and Jacobs, 1965). The mucosa of the esophagus may be thin, pale, dry, or even hemorrhagic. The radiological appearances have been described by Waldenström and Kjellberg (1939), Danilenko (1950), Jasinski and Roth (1954), and Bernát (1966) (Fig. 18/8a and b).

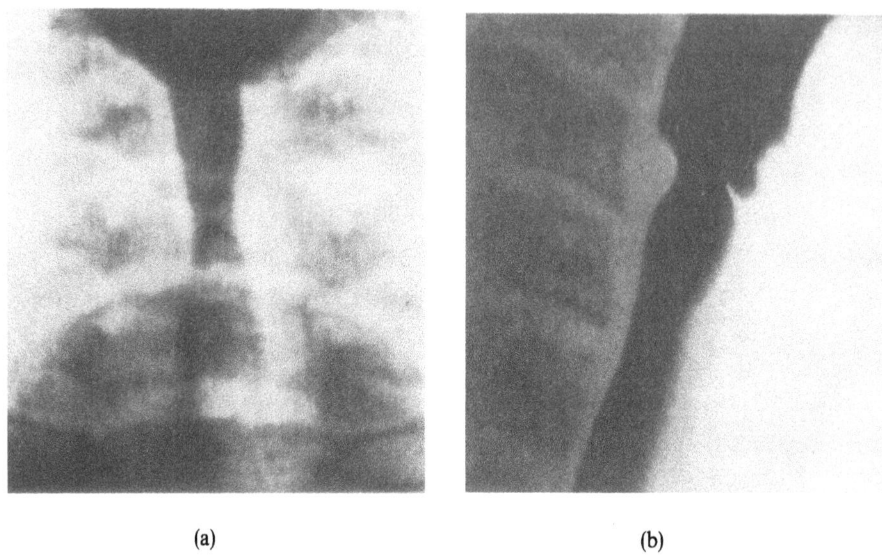

(a) (b)

Fig. 18/8. The radiological appearance of the esophagus in sideropenic dysphagia. (a) Frontal view; (b) sagittal view (after Bernát, I., 1966)

Cineradiography provides a more accurate way of detecting the webs than conventional barium swallow (Chisholm, 1973). The commonest site of a web is at the level of the cricopharyngeal stricture, but they can also occur in the lower lateral food channels or the esophagus. They are usually situated anteriorly or anterolaterally and sometimes encircle the lumen and go on to stricture formation. Chisholm et al. (1971) found multiple webs in about one-third of their patients. The web can usually be distinguished from other postcricoid filling defects by virtue of the fact that the distance the web projects into the pharynx is greater than its thickness.

Iron treatment may result in disappearance of symptoms, but this is not necessarily accompanied by resolution of the web. In some cases esophageal dilatation is required to relieve symptoms.

Sideropenic dysphagia is a premalignant condition, with an incidence of cancer occurring in 4–16% of cases (Chisholm, 1973).

The pathogenesis of atrophic rhinitis has been widely debated but is attributed by Bernát (1965) to iron deficiency. This was based on the observation that 50% of patients are cured following treatment with iron and a further 30% are relieved of their symptoms; furthermore, Bernát found that atrophy of the nasal mucosa could be induced experimentally with an iron-deficient diet (Bernát, 1965). The cilia disappear first, and this is followed by metaplasia of the columnar epithelium to a squamous type, although a few islands of the normal columnar type may remain. With iron supplementation, recovery of the mucosa probably occurs from these foci. If the iron deficiency is prolonged, the squamous epithelium becomes stratified and eventually keratinized. Desquamation may occur as the subepithelial connective tissue undergoes progressive change. Initially there is hyperemia with plasma cells and lymphocytic infiltration. This is followed by fibrosis and hyalinization, and finally, the submucosal layer is reduced in depth and becomes compact and fibrotic, and the mucous glands almost completely disappear (Figs. 18/9–18/12). Similar changes were found in the esophagus and stomach of these animals.

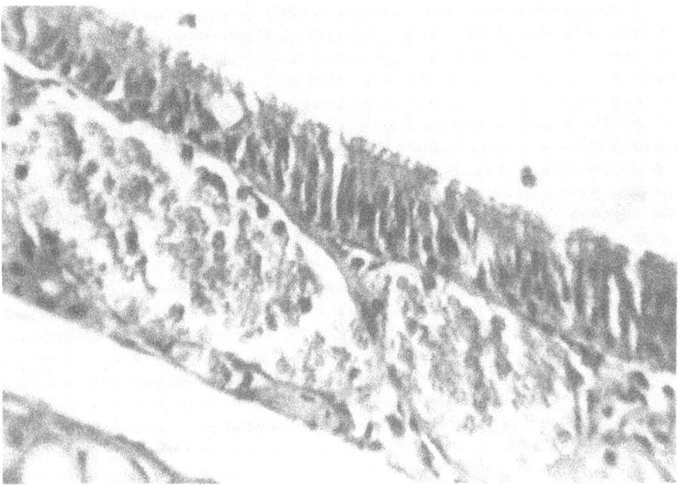

Fig. 18/9. Nasal mucous membrane of a normal mouse. The epithelium is columnar and ciliated (after Bernát, I., 1965)

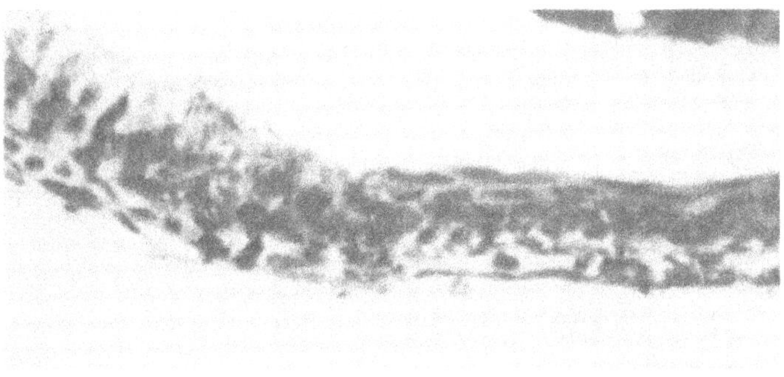

Fig. 18/10. The transition from columnar ciliated epithelium on the left to stratified squamous epithelium on the right in the nasal mucosa of a hyposiderotic mouse (after Bernát, I., 1965)

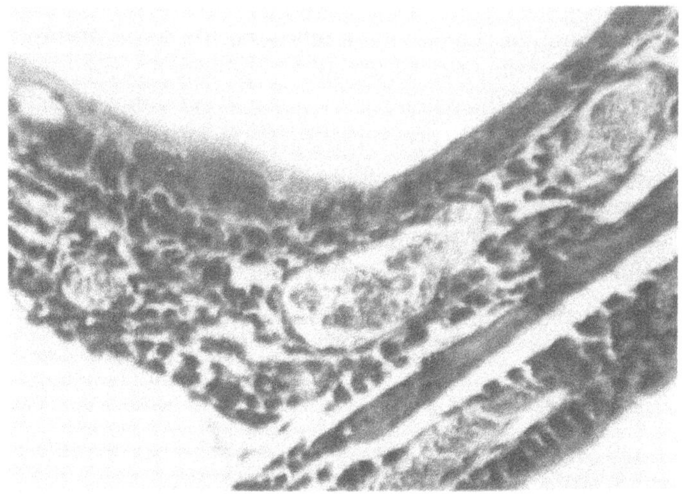

Fig. 18/11. Dystrophic transformation of the nasal mucosa of a sideropenic mouse. There is round-cell infiltration below the epithelium (after Bernát, I., 1965)

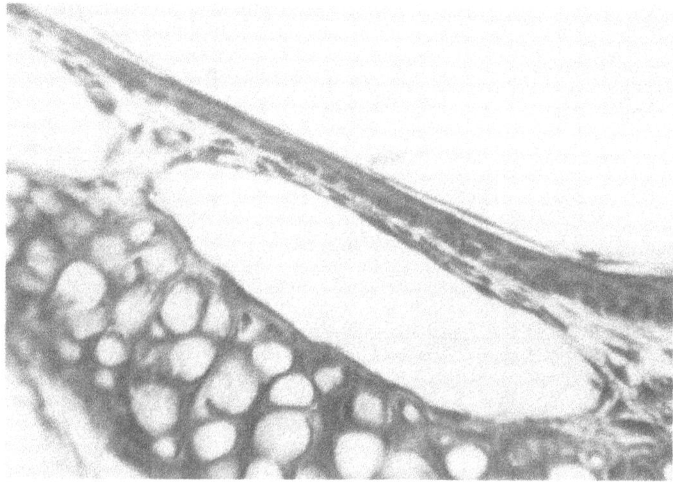

Fig. 18/12. Final stage of mucosal atrophy of the nasal mucous membrane in a hyposiderotic mouse. There was extensive nasal incrustation and a complete absence of subepithelial glandular tissue (after Bernát, I., 1965)

Some writers believe that atrophic rhinitis (ozaena) develops as a result of inherited constitutional characteristics. In this respect they ascribe particular significance to the shape of the face and skull, which "regardless of essential racial differences is of a uniform type" (Fleischmann, 1932a, b). In ozaena the skull is typically brachycephalic, the nose is broad and flattened, the bridge is depressed and shows the saddle deformity. The forehead is domed and the superciliary arches are prominent. The cheek bones protrude. The skull bones are pathologically thinner than normal. In a minority of cases the palate is wide, low, and short, and the sinuses are late in development (Fleischmann, 1932b). In a number of patients the mongoloid facies is very obvious.

Fleischmann (1932b) ascribes the development of this characteristic cranial shape to dysontogenesis caused by ectodermal deficiency. In other words, the development of the skull is arrested at an embryonic stage, and so remains infantile. The "fetal" character of the skull is also stressed by others.

Bernát (1965) has preferred the hypothesis that the shape of the face and skull in ozaena is the direct consequence of the intrauterine iron deficiency on the developing bones, rather than an inherited constitutional characteristic. The changes are, in fact, identical with those described by Reimann (1965) as a result of iron deficiency during pregnancy. The sequence occurs most often when iron deficiency is "endemic" in a population, e.g., when it is due to traditional dietary habits, and so persists from generation to generation. Since iron is an essential factor for growth of many types of cells, iron deficiency *in utero* must have a far-reaching effect on fetal development.

The appearances of the facial and skull bones described by Reimann (1956) as being the result of intrauterine iron deficiency are as follows. It will be noticed how accurately they compare with the typical appearances in ozaena. "The remarkable and characteristic shape of the face and skull occurs in many cases with the result that these hyposiderotic patients have a striking resemblance to each other. Typically, the head is brachycephalic and polyhedral, the face mongoloid, the forehead steep and the cheekbones are prominent. On section, the skull bones show decreased bone formation in the spongiosa layer and thinning out of the cortical layers."

We found among our hyposiderotic patients a number of examples of this typical skull shape; some of them were suffering from ozaena; in others this had not developed. As this abnormality in ossification of the skull is invariably the consequence of intrauterine iron deficiency, it would seem appropriate to describe the characteristic cranial and facial shape as the hyposiderotic skull and the hyposiderotic facies.

It is agreed that ozaena does occur in the absence of typical facial and skull bone changes. However, when the iron deficiency starts *in utero*, then these changes are always present. When there is no lack of iron until some stage in childhood, the body abnormality does not occur.

Turning to our animal experiments, our object was to produce iron deficiency in white mice in order to decide if this would lead to the various changes associated with ozaena in man. We carried out the experiments using the technique described by Schmidt (1928). We paid particular attention to the development of the skull and facial bones, and to the changes in the nasal mucosa (see pp. 224–226).

Perceptible changes were noted in the second hyposiderotic generation. The sideropenic animals showed a marked degree of arrested development. They moved slowly on trembling legs, and sometimes seemed too weak to raise their heads. Their ears, nose, and tail were paler than those of the normal controls, their coats were shaggy and scanty, and there was patchy loss of fur. Furthermore, in a number of animals we found that there were abnormalities of skull growth characteristic of ozaena. The heads of these animals differed significantly from those of normal mice. The skull was abnormally shorter and wider. The nose was short, broad, and stubby with a collapsed "saddle nose" appearance of the bridge (Fig. 18/13).

Our experiments lend considerable support to the hypothesis that the characteristic bone structure of the skull and face in ozaena is the result of intrauterine iron deficiency.

In the human subject the pharyngeal mucosa may look dry and pallid, and the surface may be covered with a greenish yellow crust. Although the laryngeal mucosa

Fig. 18/13. Experimental iron deficiency in the albino mouse. The dorsum of the nose is sunken (congenital saddle nose), the fur is matted, the animal is weak and unable to stand (after Bernát, I., 1966)

is not often affected, atrophy can arise in this area, and the vocal cords become dry and pale and their motility may be disturbed. This may lead to chronic hoarseness, and the changes are sometimes irreversible. The following are two illustrative cases.

R. L., 38-year-old man. Had suffered from headaches and tiredness since the age of 15. Atrophic rhinitis was diagnosed in adolescence. At 18 he developed hoarseness, which became progressive, and at about 32 years of age he noticed deformity of his fingernails. He also complained of burning of his tongue when eating sour food or drinks, and he had occasional painful cracks at the angles of his mouth. He had been impotent for some years.

On examination he was pale. He had hyperkeratosis in the angle of the mouth and around the nostrils. The tongue was smooth and atrophic, and he had koilonychia of all the nails of both hands. The nasal passages were dilated, and the inferior nasal conchae were atrophic. The middle turbinates were covered with mucus and some patches of superficial crusting. The nasal and pharyngeal mucosae were thin and hemorrhagic, and there was a greenish brown crust on the posterior wall of the pharynx. The larynx was moist, the middle third of the vocal cords was thin with a slit between them on phonation, and the voice was weak and hoarse.

Hemoglobin was 11.1 g/dl, MCH 28 pg, the fasting plasma iron was 65 μg/dl, 1 hour after oral iron loading 227, at 3 hours 325, and at 7 hours 215 μg/dl. TIBC was 415 μg/dl with a saturation of 16%.

Cs. N. E., 32-year-old woman. Brought up by her stepmother in poor social circumstances, with a deficient diet based on soup, potatoes, pasta, and vegetables. She had suffered from tiredness and anemia since school age, with a tendency to faint since adolescence. At 15 she became hoarse, the hoarseness progressing to almost complete aphonia. She also suffered from tachycardia and dyspnea on the slightest exertion.

She was a pale, thin woman of small stature. Her lips were dry and cracked and her skin was rough. Her nails were brittle, and the teeth were either absent or irregular in shape and growth. She had marked atrophic rhinitis and an atrophic laryngeal mucosa with yellowish-greenish crusting on the mucosa and lack of mobility of the middle third of the right vocal cord.

Hemoglobin was 11.4 g/dl, fasting plasma iron 60 μg/dl, TIBC 450 μg/dl with a saturation of 13%.

Achlorhydria and gastritis are frequently found in association with iron deficiency, and such patients were thought to form a specific group of "simple achlorhydric anemias" (Witts, 1930; Wintrobe and Beebe, 1933).

The introduction of a simple technique for gastric biopsy has enabled studies to be made in a large series of patients with iron deficiency, and this has shown that all grades of change may be found from superficial gastritis through to atrophic gastritis or gastric atrophy. The mucosa is seldom entirely normal (Joske et al., 1955; Badenoch et al., 1955, 1957; Davidson and Markson, 1955; Lees and Rosenthal, 1958; Rawson and Rosenthal, 1960; Ikkala and Siurala, 1964; Naiman et al., 1964). A follow-up study by Ryss (1971) showed that the functional and morphological changes in the stomach tended to progress in iron deficiency, and in some patients a change from normal appearances to atrophic gastritis was relatively rapid.

Studies by Ghosh et al. (1972) in children with iron deficiency, many of whom had hookworm infestation, suggest that the reduced gastric acidity found in these patients was a reflection of the iron deficiency rather than of any damage caused by the hookworms. Other workers have, in fact, called attention to the rarity of achlorhydria and atrophic gastritis in patients with iron deficiency secondary to hookworm infestation (Cowan et al., 1966). Improvement in gastric secretion and gastric histology following treatment with iron has been reported in some cases of iron-deficiency anemia (Jacobs et al., 1966; Stone, 1968; Gupta et al., 1972). Such improvement, however, appears to be rare in older patients with longer standing iron deficiency in whom the abnormalities are likely to persist or even progress (Lees and Rosenthal, 1958; Beveridge, 1963).

Naiman et al. (1964) found evidence of duodenitis and thought that iron deficiency could cause a reversible enteropathy in children. Similar findings have not been observed in adults (Rawson and Rosenthal, 1960; Halsted et al., 1965).

How far the gastric mucosal lesions and achlorhydria are the cause rather than the effect of iron deficiency is debatable. The fact that occasionally there is regression of the mucosal lesions and a return of acid secretion would favor the changes being secondary to the deficiency. On the other hand, the fact that in many instances the gastric lesions are irreversible or progressive suggests that the gastric changes may be an etiological factor in iron deficiency.

Antibodies to the gastric parietal cells of the stomach are frequently found in patients with histamine-fast achlorhydria and iron deficiency (Dagg et al., 1964). Such patients always have chronic gastritis (Coghill et al., 1965). It is thought that the organ-specific autoantibodies are formed as the result of cell damage, but the tendency to form autoantibodies might also be an etiological factor in producing damage.

It has long been known that iron-deficiency anemia and pernicious anemia may develop in members of the same family (Kaufmann and Thiessen, 1939). Furthermore, iron deficiency may precede the later development of pernicious anemia (Callender and Denborough, 1957). Latent or manifest pernicious anemia can be found in some 5–6% of patients suffering from iron-deficiency anemia (Beveridge et al., 1965; Dagg et al., 1966). Since gastric parietal cell antibodies occur in both diseases, the autoimmune event may be the pathogenetic link between the two conditions (Dagg et al., 1971).

The significance of the involvement of the small intestine in iron deficiency is debatable. Malabsorption of iron as a result of changes in the jejunal mucosa secondary to gluten enteropathy is an etiological factor in some cases of iron deficiency, but apart from such cases changes in the small intestinal mucosa occur rarely if at all. The only exception perhaps is the condition described by Naiman et al. (1964) in children in whom intestinal blood loss and changes in the small gut mucosa were found in iron deficiency, the changes reverting to normal when the blood condition was treated. Some of these cases may have been due to cow's milk intolerance.

An interesting observation made by Watson et al. (1963) and Tunessen et al. (1969) was an increased frequency of excretion of the red pigment betanin following ingestion of beetroots in the presence of iron deficiency. The normal proportion of subjects who excrete the pigment is about 14%, but the incidence in iron deficiency was reported as 80%. Watson et al. ascribe the phenomenon to increased absorption of the pigment as a consequence of pathologically increased permeability of the intestinal mucosa in iron deficiency.

Occasionally iron deficiency may lead to atrophy of the mucous membranes of the vulva, vagina, and uterus (Jasinski and Roth, 1954; Kovács, 1970). Menorrhagia is common, largely because it is the most frequent etiological factor of iron deficiency, and if menstruation is delayed or there is hypo- or amenorrhea, the association of malabsorption should be suspected, although iron deficiency in the absence of malabsorption in adolescents may delay sexual development.

Reimann (1956) has described striking retarded somatic development and sexual infantilism in the severe iron deficiency found in some areas of Turkey. The overall bodily development of the patients was delayed, and some of the male patients showed feminization. The primary and secondary sexual characteristics were poorly developed, and amenorrhea was frequent in girls and the capability of erection weak in the boys. Testicular biopsies showed disturbances of spermatogenesis and dystrophy of the testicular tissue. Although many of these disturbances can be relieved or at least mitigated by iron therapy, it is not entirely clear as to whether the iron deficiency alone is responsible for the changes or whether part of the changes could be due to associated nutritional deficiencies. Reimann (1956) described striking acceleration of growth and development in girls in response to iron treatment, but some of the iron-deficient males remained juvenile in appearance, and occasionally, following severe and prolonged iron deficiency, a eunuchoid picture resulted.

Experimental iron deficiency in mice has been shown by Schmidt (1928) to result in loss of power to procreate.

Some observers have described patients in whom iron deficiency was associated with clinical symptoms suggestive of hyperthyroidism (Moutier, 1951; Kassirski and Alekseev, 1962). Iron therapy alone may lead to the disappearance of such symptoms. We have observed a female patient who presented with loss of weight, nervousness, tremor, tachycardia suggestive of hyperthyroidism. She was found to have moderately elevated protein-bound iodine, but also hypochromic anemia. Her nails were dystrophic and she complained of dysphagia. On iron therapy alone the subjective and objective signs of hyperthyroidism disappeared and the protein-bound iodine returned to normal.

In relation to such cases it is interesting that there is an increased incidence of thyroid antibodies, both thyrocytoplasmic and thyroglobulin antibodies, in patients with iron-deficiency anemia, particularly those with postcricoid webs (Chisholm, 1973), indicating that autoimmune phenomena may play a part in some cases of iron-deficiency anemia.

Disturbed hypothalamic pituitary function has been postulated by some observers (Gasser, 1948; Reimann, 1956). Laub described interstitial fibrosis and atrophy of the hypophysis.

Other findings that have been reported in severe iron deficiency are myocardial changes, which are reversible on treatment (Jasinski and Roth, 1954; Somers, 1959), and also mild fever, which settles on treatment with iron (Reimann, 1949), although this may occur in other types of anemia and is not specific.

Kniseley and Noyes (1972) reported a case of iron-deficiency anemia in which papilledema, thrombocytosis, and transient hemiparesis were noted, all the abnormalities responding to ferrous sulfate treatment.

It is clear that chronic iron deficiency can affect many tissues in the body but particularly those that have a high turnover of cells. How far many of the changes are specific for iron deficiency as opposed to iron deficiency associated with other nutritional defects is difficult to determine. There seems little doubt that the changes are more frequent in patients who live in very poor social circumstances. The finding that many of the symptoms are relieved by treatment with iron (Jasinski and Roth, 1954; Bernát, 1965; Ohira et al., 1979; Basta et al., 1979) does not necessarily exclude other deficiencies, since the iron therapy alone may improve appetite and hence nutrition, or the patients may be brought into the hospital and given a better diet than is customary for them. On the other hand, evidence of reduction of the iron-containing tissue enzymes such as cytochrome C, cytochrome oxidase, aconitase, and succinate dehydrogenase has been found in the tissues associated with iron deficiency (Vannotti, 1949, 1959; Beutler, 1957, 1964; Beutler and Blaisdell, 1958a, b; Kampschmidt et al., 1959; Salmon, 1962; Dagg et al., 1966c; Kalinin, 1970; Fairbanks et al., 1971; Beutler and Fairbanks, 1980), but the understanding of the tissue effects of iron deficiency is still incomplete (Dallman, 1974; Jacobs, A., 1975; Beutler and Fairbanks, 1980).

SPECIAL FEATURES OF IRON-DEFICIENCY ANEMIA OF INFANCY AND CHILDHOOD

As already pointed out, iron deficiency is likely to develop in infants from the age of about 6 months if feeding with milk only is prolonged beyond this time. In order to prevent this iron deficiency, mixed feeding should gradually be introduced. If, however, anemia has already developed, therapeutic iron supplementation may be necessary.

The anemia develops slowly and insidiously, and there are seldom acute symptoms. The infants may appear well nourished although they tend to lose their appetite, and some show retardation in development. Physical examination usually reveals a systolic cardiac murmur and, in about 10% of cases, a palpable spleen (Burman, 1973). In severe cases there may be moderate enlargement of the liver and lymph nodes as well. Trophic disturbances of the skin and mucous membranes are rare, although Naiman et al. (1964) demonstrated chronic duodenitis and atrophy of the mucosa with associated achlorhydria, steatorrhea, and disturbed absorption in some infants, and found the changes to disappear on treatment with iron. Such children were found to have a low resistance to infection.

Impairment in cell-mediated immunity and bactericidal functions of polymorphonuclear neutrophils have been demonstrated in iron-deficient states (Chandra 1973; Joynson et al., 1972; MacDougall et al., 1975; Srikantia et al., 1976). These changes were reversible after iron therapy (Yetgin et al., 1979).

Some workers have noted developmental abnormalities in severely iron-deficient children born to iron-deficient mothers (Reimann, 1956; Shahidi and Diamond, 1960).

LABORATORY FINDINGS

As already discussed, iron deficiency in terms of reduction or absence of the iron stores can occur without anemia, and in such cases the mean corpuscular hemoglobin and mean corpuscular hemoglobin concentration may be within normal limits or only marginally reduced.

Where the iron deficit has progressed beyond this and anemia has developed, the hemoglobin concentration is characteristically more reduced than the red cell count or hematocrit. Thus the mean corpuscular hemoglobin is usually below 30 pg and usually between 20 and 25 pg or even as low as 12 pg in severe cases (Wintrobe, 1974). The mean corpuscular hemoglobin concentration is also reduced to between 25 and 30%.

The blood film shows hypochromia of the red cells (Fig. 18/14). The thickness of the erythrocytes is reduced (1.5–2.0 μ) or occasionally even less, down to 1 μ thick (Wintrobe, 1967). In more advanced cases anisocytosis and poikilocytosis are marked, and a variable degree of microcytosis occurs. The average diameter of the red cells varies between 6.2 and 6.7 μ, and the base of the Price-Jones curve is broadened but the peak may be normal or shifted slightly to the left (Fig. 18/15).

The reticulocytes may be normal or reduced or slightly increased if the patient is actively bleeding. Nucleated red cells are seldom seen in the peripheral blood.

The osmotic resistance is often increased, with some intact cells being detected even in 0.25–0.20% saline (Reimann and Arkun, 1954). Following iron treatment the osmotic resistance returns to normal.

The white blood cell count is normal or reduced, the reduction being in the number of granulocytes. However, in the presence of hemorrhage there may be a transient leucocytosis and possibly also some shift to the left. A sustained leucocytosis should make one suspect the presence of neoplasm as the cause of bleeding.

The platelet count is usually within normal limits, although hemorrhage may produce a thrombocytosis (Schloesser et al., 1965).

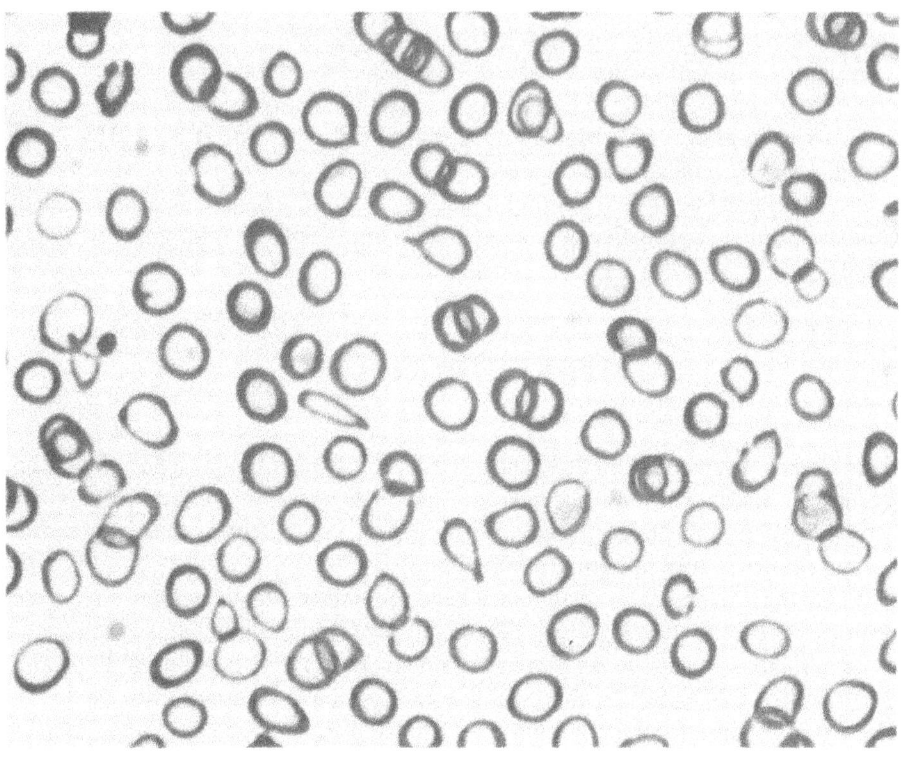

Fig. 18/14. Hypochromic red cells in iron-deficiency anemia

The bone marrow shows mild erythroid hyperplasia sometimes with an increase in the number of basophilic and polychromatophilic nucleated red cells. The normoblasts are small, with scanty, ragged cytoplasm.

Iron stains show that sideroblasts are reduced or absent (Kaplan et al., 1954; Bilger, 1957; Brüschke, 1962), and there is little or no stainable iron in the reticulum cells. Electron microscopic studies show an absence of ferritin in these cells (Bessis and Breton-Gorius, 1962).

Sometimes the ratio of erythroblasts in the bone marrow is reduced before iron treatment ("anémie ferriprive pseudo-aplastique"). Granulopoiesis is usually unaffected, but occasionally some giant metamyelocytes may be found (Wintrobe, 1967).

With regard to biochemical changes, the bilirubin in the plasma is low, and plasma proteins are often also reduced. The most significant change is the reduction

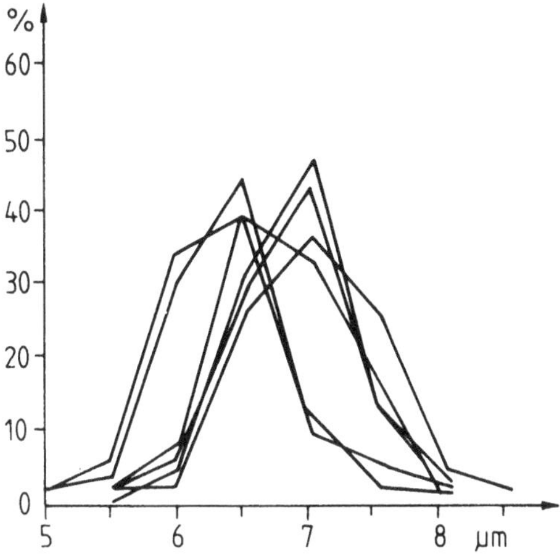

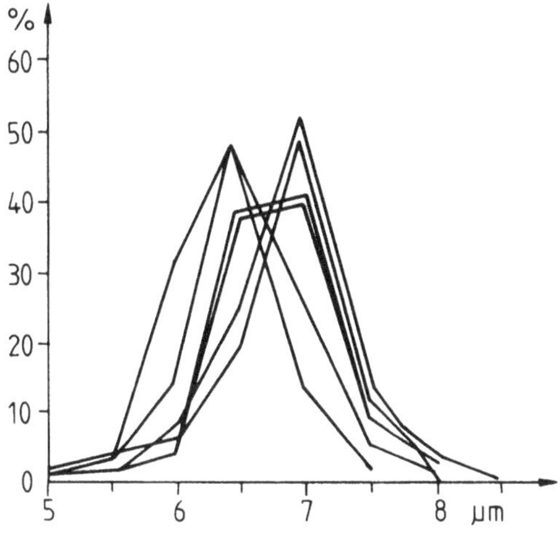

Fig. 18/15. Price-Jones curves in iron deficiency

in plasma iron concentration and increase in the iron-binding capacity of the plasma with a consequent reduction in transferrin saturation (Fig. 18/16).

In response to oral iron loading the plasma iron concentration reaches a high level and decreases at a slower rate than normal (see Fig. 6/26), provided that the patient is not suffering from malabsorption. Plasma copper level is normal or moderately elevated.

Because the cells are deficient in iron, protoporphyrin accumulated and the free protoporphyrin concentration of the red cells may be increased up to 500 μg/dl from the normal value of 20–40 μg/dl. The free erythrocyte protoporphyrin may also be elevated in latent iron deficiency and may remain high in both latent and overt iron

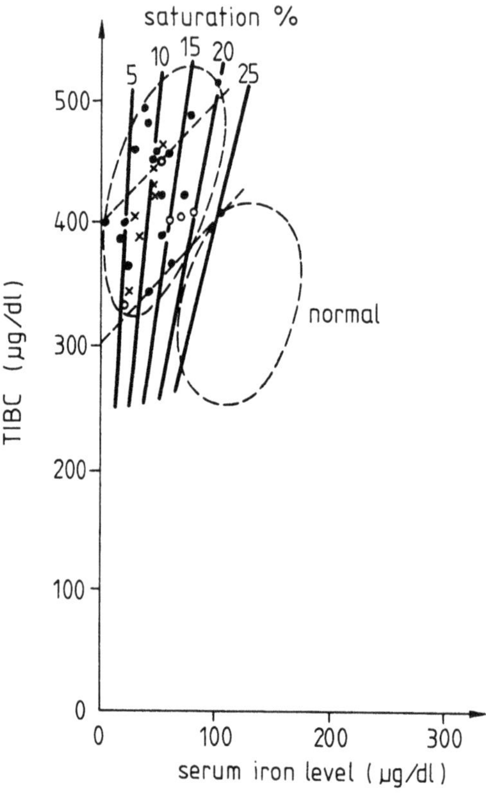

Fig. 18/16. Serum iron and iron-binding capacity in iron-deficiency anemia. The percent saturation of transferrin is low (after Verloop, M. C. et al.: Brit. J. Haemat. *4,* 70, 1958). (×) males (8); (•) females (17); (o) females (normal Hb) (5)

deficiency following treatment until the iron-deficient red cells reach the end of their life-span (Dagg and Goldberg, 1973) (Fig. 18/17). This means that iron deficiency can be established retrospectively for some time after treatment, although it should be kept in mind that the free erythrocyte protoporphyrin increases also in other disturbances of heme synthesis such as sideroblastic anemia and anemia of infection.

The coproporphyrin and uroporphyrin levels of the red cells are also elevated, and the urinary coproporphyrin is increased (100–600 μg/24 hrs). The delta-aminolevulinic acid of the urine is augmented, although the porphobilinogen excretion does not change.

The serum ferritin values are decreased below the normal values of 17–66 μg/l (Addison et al., 1972; Miles et al., 1974; Jacobs and Worwood, 1975; Valberg et al., 1976; Dawkins, 1979) (see Fig. 8/15).

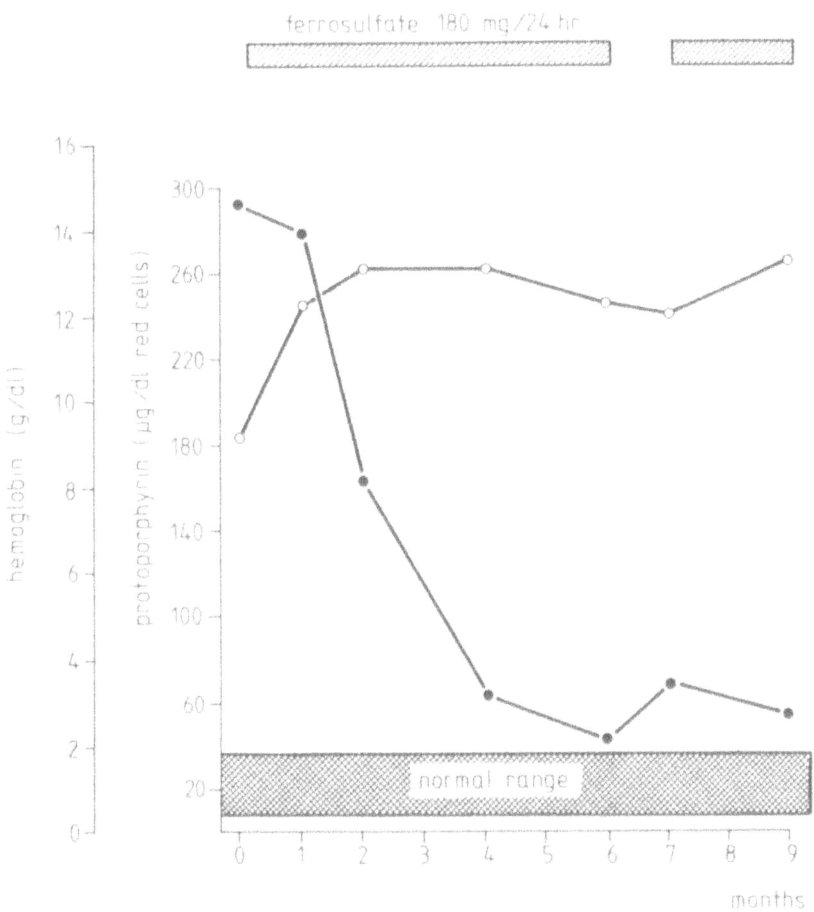

Fig. 18/17. The rise in hemoglobin level (–○–○–) and the decline in free erythrocyte protoporphyrin (–●–●–) in iron deficiency in response to iron therapy. The protoporphyrin level returns to normal only 9 months after the initiation of therapy (after Dagg, J. H. et al.: Brit. J. Haemat. *12*, 326, 1966b)

The life-span of the red cells is either normal (Temperley and Sharp, 1962) or slightly shortened (Rasch et al., 1958; Heimpel, 1965; Huser et al., 1967). Loria et al. (1967) showed that the survival of iron-deficient erythrocytes was reduced in the circulation of healthy individuals, whereas normal erythrocytes survived normally in iron-deficient subjects. This would indicate that any reduction of the red cells is the consequence of an intracorpuscular defect.

LABORATORY FINDINGS OF IRON DEFICIENCY IN INFANCY AND CHILDHOOD

The features of the iron deficiency of infancy are that the red cells are hypochromic and the MCH and MCHC may drop considerably. The mean volume and thickness of the red cells also diminish. The blood smears reveal a moderate anisocytosis and severe hypochromia (see Fig. 18/14). Occasionally target cells and nucleated red cells may also appear. The reticulocyte count, in general, is normal. Erythropoiesis is hyperplastic, the proportion of basophilic and polychromatophilic nucleated red cells is increased, and there is decreased hemoglobin synthesis.

Serum concentrations of myoglobin are significantly lower in iron-deficient than in non-iron-deficient children. No abnormality of serum myoglobin concentration was found in adults with iron-deficiency anemia (Hargreaves et al., 1981).

The iron deficiency has to be distinguished from other hypochromic anemias, notably the thalassemias and hemoglobinopathies (Jacobs and Worwood, 1980), but rare idiopathic or secondary sideroachrestic anemias must also be excluded (Heilmeyer, 1964; Moncrieff et al., 1964; Jacobi, 1967). In such cases the plasma iron does not decrease but is normal or increased, the proportion of sideroblasts in the bone marrow is high, and the cells of the reticuloendothelial system contain much iron, in contrast to iron deficiency where there is little or no demonstrable iron.

FERROKINETICS

When the state of hyposiderosis has resulted in depletion of iron stores and a fall in the plasma iron level, there is a reduction in the half-time of radioiron disappearance from the plasma. In the more severe degrees of iron deficiency the half-time may be as short as 10–15 minutes. The radioiron accumulates rapidly over the bone marrow, but there is no activity detectable over the liver and spleen. Subsequently, close to 100% of the dose of radioiron can be recovered from the circulating erythrocytes (Ventura et al., 1957; Pollycove, 1959). In spite of this, there is evidence that there may be some ineffective erythropoiesis in the presence of iron-deficiency anemia (Pollycove, 1958; Robinson, 1969; Cook et al., 1970). It is assumed that the iron released from cells that have been destroyed prematurely within the bone marrow is re-utilized immediately, thus resulting in the high red cell utilization. Labardini et al. (1973) have shown a lengthening of marrow transit time in iron deficiency that is compatible with this interpretation.

THE DIAGNOSIS OF IRON DEFICIENCY

The subjective complaints of iron deficiency may be very vague, and symptoms such as fatigue may be attributed to overwork or disregarded completely. Many of the symptoms associated with iron deficiency may not be elicited unless deliberately sought for, e.g., a sore tongue, brittle nails, indigestion, or menorrhagia. Excessive menstrual loss is very difficult to assess by individual patients, and they should be questioned about the duration of the menstruation, the passage of clots, how many sanitary towels they use and of what make, and whether the loss is sufficient to be socially inconvenient. Symptoms suggestive of a hiatus hernia, ulcer, or other gastrointestinal lesions should be deliberately enquired about.

During physical examination the condition of the skin, nails, hair, tongue, gingivae, and pharynx should all be noted in addition to general physical examination to look for evidence of gastrointestinal or genitourinary disease, and a rectal examination should always be included, with the addition of a test for occult blood.

If the blood count shows anemia of the hypochromic type, iron deficiency is by far the most likely diagnosis. The alternative diagnosis is anemia of chronic disorders or, much more rarely, thalassemia or sideroblastic anemia.

The question of the definition of anemia can give some difficulty. Many of the figures quoted are based on the examination of apparently healthy individuals and represent only the level that is prevalent in the population under consideration. Other surveys have looked at the results obtained after treatment with iron, and under such circumstances Verloop et al. (1959) found the mean hemoglobin concentration of males to be 16.5 g/dl and that of females 14.5 g/dl. The difficulties in definition of iron deficiency are discussed by Garby (1973).

The diagnosis of iron deficiency in pregnancy presents particular difficulties because of the variable increase in red cell mass and plasma volume. Several studies have shown that iron supplementation during pregnancy results in higher mean hemoglobin values than in control subjects; e.g., Rybo (1973) reported that more than half of the 148 pregnant women given placebo during the second half of pregnancy developed a hemoglobin concentration less than 11.0 g/dl compared with only 10% of 178 patients given 100 mg of iron daily. It has become widely accepted that supplemental iron should be given to all pregnant patients as a prophylactic measure. In the eighth month of pregnancy a hemoglobin of 13.0 g/dl ±10% should probably be regarded as normal (Bothwell and Finch, 1962).

In infants the average value for hemoglobin after iron therapy was found by Sturgeon (1959a) and Hammond and Murphy (1960) to be between 11 and 12 g/dl, and a normal value between 6 and 12 months of age should probably be 12.0 g/dl ±10% (Table 18/6).

Early changes of iron deficiency are not easy to recognize in the blood and, in order to establish an iron-deficient state, it may be necessary to measure the serum iron and total iron-binding capacity and saturation. In addition, the iron status can be assessed by cytochemical examination of the bone marrow. It is seldom necessary to proceed to any other tests, although further abnormalities that can be demonstrated in iron deficiency are a reduction in serum ferritin (Jacobs et al., 1972; Cook et al., 1974; Heinrich, 1979), an increase in red cell protoporphyrin (Dagg and Goldberg, 1973), a lower than normal iron excretion in the urine following an injection of desferrioxamine (Wöhler, 1964), and an increased urinary excretion of an oral dose of ^{57}cobalt (Valberg et al., 1972). The measurement of iron absorption may be useful since an increase in iron absorption is a sensitive index of iron depletion except where the patient suffers from a malabsorption syndrome. A reduced iron absorption may indicate the etiology of the iron deficiency. The various indicators of iron metabolism are summarized in Table 18/7.

Table 18/6
Normal blood hemoglobin level (g/dl)

Age (years)	Males	Females	References
½–1	12.0±1.2	12.0±1.2	Sturgeon, 1959b; Hammond and Murphy, 1960
7–9	12.7±1.6	12.7±1.6	Vellar, 1970
10–13	13.2±1.6	13.2±1.6	Vellar, 1970
14–16	15.0±2.0	14.2±2.0	Vellar, 1970
17–20	15.5±2.0	14.2±2.0	Vellar, 1970
Adults	15.7±1.8	14.3±1.8	Vellar, 1970
Adults	16.5	14.5	Verloop et al., 1959
Pregnant females	—	13.0±1.3	Bothwell and Finch, 1962
Pregnant females	—	14.3	Davis and Jennison, 1954, Fisher and Biggs, 1955

Table 18/7

Some indices of iron metabolism in healthy subjects and in disturbed erythropoiesis (hemoglobin synthesis) (based on personal experience and literary data)

	Normal values	Hyposiderosis	Infectious and tumorous anemia	Disorders of heme and globin synthesis
PlFe (μg/dl)	115 (80–150)	<70 (20–50)	<70 (20–65)	>180 (180–300)
Total iron-binding capacity of plasma (μg/dl)	320 (290–350)	>360 (360–550)	200 (80–280)	250 (200–360)
Saturation coefficient (%)	32 (30–40)	<25 (5–25)	<25 (10–35)	~ 100 (70–100)
Iron content of reticulum cells in bone marrow	+	0	+ +	+ + +
Sideroblasts (%)	30 (20–50)	3 (0–5)	10 (5–15)	80 (50–90)
Course of oral iron loading curve	80–150% rise	steep (>250% rise)	flat (low fasting PlFe; <50% rise)	flat (high fasting PlFe; <50% rise)
Absorption of radioiron from gastrointestinal tract (%)	25–35	30–60	10–25	
Iron content of urine after 100 mg i.v. iron load (mg/24 hr)	4.5–5.5	0.9–4.5	5.0–15.0	
Iron content of urine after i.m. injection of 500 mg DFO (mg/24 hr)	0.5–1.0	0.1–0.5	1.0–8.0	1.0–20.0
Serum ferritin (μg/l)				
Males } mean values	100–120	<10–12	many 100	many 100 to many 1000
Females } mean values	35			
Extreme values	12–300			
^{57}Co excretion (% of test dose)	4–11	>12	<11	

DIFFERENTIAL DIAGNOSIS OF IRON DEFICIENCY

Clinically, patients with iron deficiency may appear to be suffering from primary cardiovascular disease with the complaint of tiredness, dyspnea on exertion, palpitations, and edema of the ankles, or they may be considered as suffering from neurosis. In younger patients particularly, endocrine disease may be suspected because of the frequent retardation of growth. Some of the tissue changes, particularly those of the tongue, may resemble those seen in vitamin B_{12} deficiency, and other epithelial changes, e.g., the cheilosis, may occur in riboflavin deficiency. From the point of view of laboratory findings the most important differential diagnosis is between iron deficiency and the anemia of chronic disorders, where a mild hypochromia and a low serum iron may lead to an erroneous diagnosis. The anemia of chronic disorders, e.g., infections, is distinguished by the fact that the total iron-binding capacity is reduced and iron is demonstrable in the RE cells of the bone marrow.

ETIOLOGY AND PATHOGENESIS OF IRON DEFICIENCY

As already pointed out, the development of iron deficiency implies a state of negative iron balance, which in practical terms in the majority of cases is attributable to blood loss, either physiological or pathological.

A source of blood loss should be looked for, particularly from the gastrointestinal tract and urogenital system. Bleeding gastric and duodenal ulcers, esophageal varices, tumors of the stomach and the small and large intestine, hemorrhoids, ulcerating polyps, ulcerative colitis, regional ileitis, hemorrhagic telangiectasia, and hiatus hernia all may lead to iron-deficiency states. In about 20% of diaphragmatic hernias severe iron deficiency develops that is probably attributable to ischemic or congestive hemorrhage or bleeding from an ulcer in the herniated gastric segment (Kümmerle, 1953). Lesions that are most likely to escape detection as the etiological factor are hiatus hernia, hemorrhagic telangiectasia, and a cecal carcinoma (Felder et al., 1960; Holt et al., 1968).

In parts of the world where parasitic infestations are common, Ankylostoma duodenale and more rarely Trichocephalus dispar infestations may give rise to chronic hemorrhage. Layrisse et al. (1961) found a correlation between the number of eggs passed in the feces and the volume of blood loss. In the presence of 1000 ova of hookworm/1 g of feces they measured 2.4 ml of blood loss, and 5000 or more ova/1 g of feces was always associated with anemia. Anemia in the presence of lighter hookworm loads was presumed by Stott (1960) to be due to nutritional iron deficiency, the incidence of the anemia being markedly reduced by the addition of 6 mg of iron daily to the food.

Some drugs are irritant to the gastrointestinal tract and may cause bleeding, notably aspirin, steroids, and butazolidine (Summerskill and Alvarez, 1958; Bannerman et al., 1964).

Menorrhagia is one of the most frequent causes of iron deficiency and should always be suspected as the cause in women during reproductive life (Beveridge et al., 1965; Göltner, 1975; Rybo, 1970, 1973).

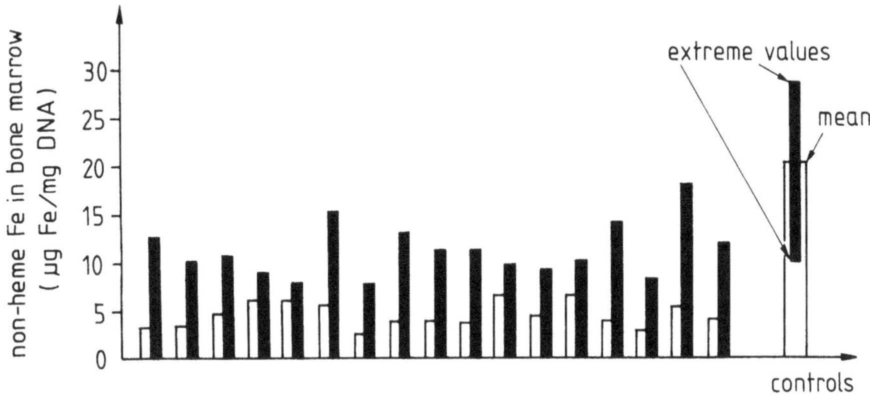

Fig. 18/18. Non-heme iron in the bone marrow of blood donors, (□) before and (■) after 3 months' iron treatment (2 × 50 mg/day) (after Weinfeld, A., in: Hallberg, L. et al.: Iron Deficiency. Academic Press, London–New York 1970)

Bleeding from the renal tract is not a frequent cause of iron deficiency, although prolonged or recurrent hematuria associated with such lesions as neoplasms, hemorrhagic pyelonephritis, or glomerulonephritis may deplete the iron stores.

Rarely, lesions of the respiratory system such as frequent nosebleeds or pulmonary hemosiderosis may produce iron deficiency, although in the latter case there is diversion of the iron to the lung and the total body iron may not be depleted.

It is often forgotten that blood donors may develop latent or manifest hyposiderosis, and some surveys suggest that 30–70% of regular blood donors will do so if they give more than 1 liter of blood per year. Many authors recommend prophylactic iron medication for blood donors (de Vries and de Potter, 1958; Schneider et al., 1958; Heistø and Foss, 1958; Holländer, 1959; Göltner, 1959; and others) (Figs. 18/18 and 18/19).

If excessive iron loss cannot be incriminated as a cause of iron deficiency in any particular case, disturbances in iron absorption should be investigated. Iron absorption may be reduced by hypo- or achlorhydria, by rapid gastrointestinal transit time, by inflammatory or degenerative changes in the mucosa, or by factors in the diet that reduce absorption, e.g., a high phytate content. Although in practice the depression in iron absorption brought about by such factors is usually insufficient to be the sole cause of the anemia, it may be a contributory factor at times when physiological requirements are high.

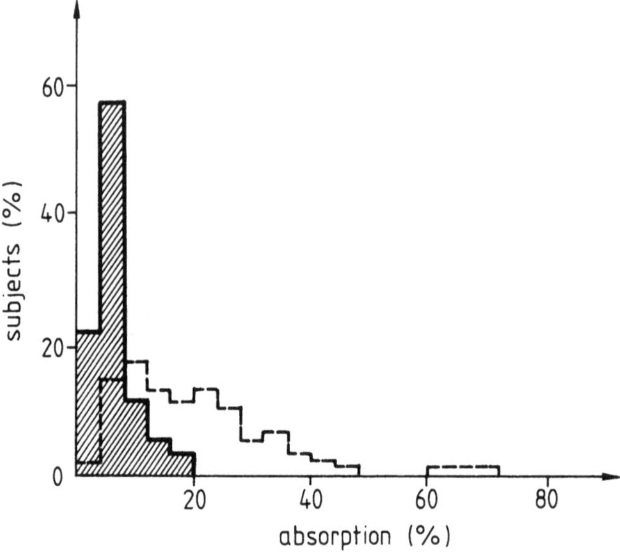

Fig. 18/19. Iron absorption from the gastrointestinal tract of healthy males (▨) and non-anemic blood donors (⌐¬). The increased absorption observed in blood donors is the consequence of latent iron deficiency (after Hallberg, L., in: Hallberg, L. et al.: Iron Deficiency. Academic Press, London–New York 1970a)

Severe disturbance of iron absorption is not common but is seen in the coeliac syndrome (Cooke, 1968; Callender, 1975). In children with coeliac disease anemia is present in the majority of cases and is largely due to iron deficiency. In adults the iron deficiency may be the sole presenting feature, and there may be little or no indication of primary intestinal disease. In sprue the anemia is frequently dimorphic, but the incidence of iron deficiency is reported as up to 17% (Estren, 1957). The disturbance in iron absorption returns to normal in the coeliac syndrome in response to a gluten-free diet, and it may also be relieved by steroids (Badenoch and Callender, 1960), although some patients may continue to show malabsorption of iron even when the manifest steatorrhea is successfully brought under control (McGuigan and Volwiler, 1964). A gluten-free diet alone may well be sufficient in adult coeliac patients, although recovery may be more rapid if other specific deficiencies (e.g., folic acid) are treated at the same time.

Many patients develop iron deficiency following partial gastrectomy, although the incidence in different series has been variously reported between 10 and 80% (Heinrich, 1954; Gouttas et al., 1956; Wallinsten, 1953; Blake and Rechnitzer, 1953; Baird et al., 1959). Most reports indicate the female-to-male ratio for hyposiderosis after partial gastrectomy as about 2:1. The anemia is probably due to a variety of factors. Many of the patients will have been operated on because of hemorrhage, and their iron stores may have been exhausted at the time of operation and not replenished by treatment. Others may have been on a restricted diet for years before, and even after, operation, and there may also be some degree of malabsorption of iron after partial gastrectomy. More recent studies have, however, indicated that those who become iron deficient have continued to lose blood either from gastritis or stomal ulcer.

The influence of nutritional factors has already been discussed in relation to iron absorption, as have the variations in physiological requirements at various ages. The times of high physiological requirement, i.e., infancy, early childhood, puberty, and reproductive life, are those at which there is the greatest susceptibility to develop iron deficiency.

In summary, iron-deficiency disease is usually the result of the combined effect of several factors: first, physiological increased iron requirements; second, a pathological iron loss, i.e., hemorrhage; third, factors in the diet that interfere with iron absorption; and fourth, gastrointestinal disorders causing malabsorption of iron.

It is unnecessary to differentiate among the various forms of clinical manifestation of iron-deficiency disease that in the past have been termed "chlorosis," "essential hypochromic anemia," "chronic hemorrhagic anemia," etc. These have now only historical interest. Iron-deficiency disease should be regarded as a homogeneous pathological process due essentially to an imbalance between the supply and the demand for iron.

THERAPY OF IRON DEFICIENCY

HISTORICAL REVIEW

In antiquity iron was already respected as a curative agent. It was used by the Egyptians, Indians, Greeks, and Romans and was regarded as a present of the gods that fell from the stars to help men. The divine origin is well reflected by the name first used for iron: an-bar, meaning "heaven-fire" (Mircea Eliade: Forgerons et alchimistes. Flammarion, Paris, 1956). The Romans named the iron preparations they used in therapy after Mars, the god of war, and the Roman legionaries regularly drank the cooling water of the armorers because they found that it made them strong and enduring in battle. Galenus and Celsus, among others, mention this favorable effect of the armorers' cooling water, and French physicians still refer to treatment with iron as the therapy of Mars (*thérapie martiale*).

The first authentic recording of iron therapy is 3200 years old, according to which Melampos, a high priest at court, wanted to return the lost masculinity to Iphiclos, king of Argos, by ordering him wine in which the sacrificial axe had been soaked. Hippocrates and Dioscurides ordered for their patients wine in which pieces of red-hot iron or iron grids had been soaked (Kaplan, 1940).

In the time of Paracelsus (1493–1541), iron therapy gained a definitive place in medical practice. Johannes Lange (1520) was the first to treat chlorosis with iron, and in the sixteenth century many iron preparations were available, among them the popular *crocus Martis*, which, according to the descriptions of Monardes (1571), was prepared by soaking iron filings in vinegar for 20 days, following which the iron was ground to a fine dust and mixed with honey or oil. Sydenham (1624–1689) described in detail the various modes of preparation, and application of iron preparations (Hebbert, 1950; Saunders, 1958), from which it appears that in the seventeenth century metallic iron, iron filings, and iron oxide were all used (Lémery, 1683). In the nineteenth century the iron salts, ferrous sulfate, ferrous carbonate, and ferrous ammonium citrate, began to be used (Blaud, 1832; Trousseau, 1882). The waters from iron-containing spas had been popular for centuries and were recommended by Sydenham and Willis (1684).

The scientific basis for iron therapy was established in the eighteenth century, when Lémery and Geoffroy demonstrated the presence of iron in the ashed blood. Menghini (1705–1759) described in "De ferrearum particularum sede in sanguine" that human and animal blood contained more iron than other tissues, and that iron could be found in the red cells. In further work, "De ferrearum particularum progressu in sanguine," he reported that the iron content of blood increased after iron treatment.

In 1832 Blaud made his important communication about the treatment of chlorosis with iron, noting that the amount of hemoglobin increased as the result of iron medication. He emphasized the necessity for prolonged treatment with large doses of iron. Iron treatment became extremely widespread but was discredited because Bunge, Wöhler, and others suggested that inorganic iron salts were not readily absorbed and only organic iron compounds should be used. As a result of this faulty conception iron treatment failed in many cases. Only later, through the work of Heubner (1912), Alder (1920), Naegeli (1920), Meulengracht (1932), and Starkenstein (1926), was the use of inorganic iron compounds reestablished. Extensive biochemical research and clinical investigation in the next decades substantiated the use of such compounds (Reimann and Fritsch, 1931; Schulten, 1934; Fontès and Thivolle, 1936; Heilmeyer and Plötner, 1937; Moore et al., 1939; McCance and Widdowson, 1946).

The first attempts at parenteral iron medication date back to the end of the nineteenth century, but the general and local side effects of the preparations used rendered them unacceptable, and it was not until 1947, when Nissim introduced saccharated oxide of iron, that a useful and safe iron preparation for parenteral use was found. Later, iron dextran and iron sorbitol citrate were introduced.

THE PRINCIPLES OF IRON THERAPY

Hyposiderosis responds to treatment by iron alone, and polypharmacy with the addition of liver extract, vitamin B_{12}, folic acid, etc., may only confuse the diagnosis and produce no therapeutic response. As a preliminary to treatment, however, it is essential to try to determine the cause of the iron imbalance that has led to iron deficiency and to remedy this if possible as well as treating the iron lack.

The majority of patients will respond to simple oral iron medication. The extensive studies of Brise and Hallberg (1962), which compared the absorption from a variety of ferrous and ferric compounds with the absorption from a standard preparation of ferrous sulfate using a double isotope technique, have shown quite clearly that

1. iron from ferrous salts is much better absorbed than from ferric compounds,
2. ferrous sulfate ranks among the best iron compounds from the point of view of absorption,
3. the absorption of iron from ferrous lactate, ferrous fumarate, ferrous glycine sulfate, and ferrous glutamate is similar to that from ferrous sulfate,
4. the absorption of iron from ferrous gluconate, ferrous citrate, ferrous tartarate, and ferrous pyrophosphate is poorer than that from ferrous sulfate, and
5. iron from ferrous succinate is slightly better absorbed than that from ferrous sulfate (Fig. 18/20).

However, when tablets, as opposed to liquid preparations, of ferrous succinate and ferrous sulfate were compared, the absorption of iron did not differ significantly.

Brise and Hallberg considered that the iron absorption was better from ferrous succinate because the succinic acid, liberated, promoted iron transfer through the cells of the mucous membrane. The addition of succinic acid and ascorbic acid were both shown by Brise and Hallberg (1962) to enhance iron absorption, but ascorbic acid has the disadvantage of increasing side effects. Bernát (1968) has shown that sodium dioctylsulfosuccinate also enhances iron absorption, to a greater degree than succinic acid.

Other factors to be considered in the effectiveness of iron preparations are the amount of iron incorporated into the preparation, the type of preparation (e.g., solution, tablets, slow-release preparation), and the time of ingestion, particularly in relation to food intake, the absorption being better when the preparation is taken in the fasting state.

Various forms of slow-release preparation have recently been introduced on the assumption that gastric irritation would be reduced if the iron were not released until it had passed the stomach, and with the suggestion also that the mucosa would not at any one time be exposed to a large dose of iron. Clinical experience has not so

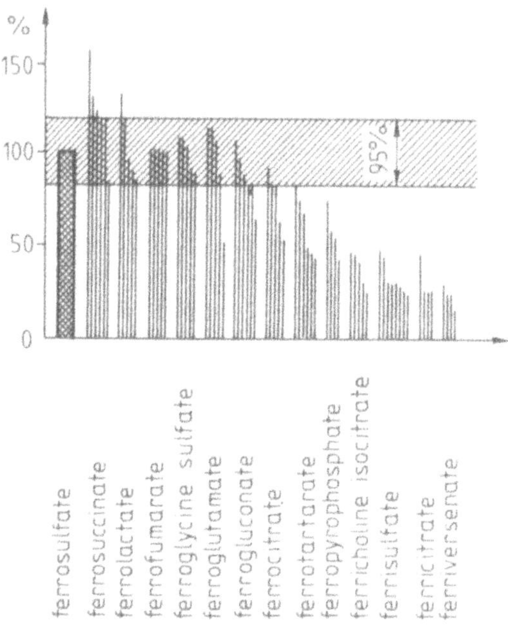

Fig. 18/20. Absorption of various iron preparations related to absorption from ferrous sulfate (after Brise, H. and Hallberg, L.: Acta med. scand. *171* (Suppl. 376), 1 (1962)

far supported claims for the superiority of slow-release preparations, and, indeed, some patients who have been treated with some such preparations have appeared to be refractory to iron but have responded quite promptly to conventional ferrous sulfate (Beutler and Meerkreebs, 1966; Callender, 1969; Hallberg, 1970a; Elwood and Williams, 1970).

DOSAGE OF IRON

For prophylactic use, for example in pregnancy, a dose of 20–50 mg of iron a day is sufficient, but for a therapeutic effect 100–250 mg of elemental iron in the form of a ferrous compound should be given.

For the quantitative evaluation of the therapeutic effect of iron preparations, patients should be selected who have a proven iron deficiency with a hemoglobin deficit of at least 5 g/dl. Patients to be compared should have a similar degree of deficiency, and they should not be actively bleeding during investigation. The etiology of the anemia should preferably be known, as some groups such as posthemorrhagic anemias may respond more easily than, for example, those with achlorhydria and gastric atrophy. The administration of the iron preparation should be for at least 3 weeks, the iron should be taken at a standard time, preferably on an empty stomach, and the compliance of the subject in relation to the iron dosage should be checked.

In clinicopharmacological investigations the dosage may be lower, e.g., 50 mg of iron per day, than the therapeutic doses, as comparative absorption can be better assessed or estimated with the smaller dose.

If the therapeutic dose is adequate, the mean daily hemoglobin increase of iron-deficient patients varies between 0.12 and 0.25 g/dl, although it may go as high as 0.3 g/dl in some straightforward uncomplicated iron deficiencies, particularly in younger subjects (Bothwell and Finch, 1962). Some of the more commonly used iron preparations are as follows:

Conferon (Chinoin)	50 mg Fe^{2+}
Ferrostabil (Schering)	22 mg Fe^{2+}
Ferrlecit 100 (Nattermann)	50 mg Fe^{2+} + 50 mg Fe^{3+}
Ferro-Redoxon (Hoffmann–La Roche)	40 mg Fe^{2+}
Ceferro (Nordmark)	22 mg Fe^{2+}
Ferronicum (Sandoz)	22 mg Fe^{2+}
Ferro 66 (Promonta)	44 mg Fe^{2+}
Resoferrix (Geigy)	37 mg Fe^{2+}
Kendural C (Abbot)	105 mg Fe^{2+}

It is not sufficient simply to continue iron treatment until the hemoglobin reaches normal. If no attempt is made to replenish the iron stores, anemia may recur, particularly if the cause of the anemia, e.g., menorrhagia, is not remediable. Iron treatment should therefore continue for several months after the anemia has been corrected (Bothwell and Finch, 1962; Haskins et al., 1952; Wintrobe, 1974).

The duration of the treatment depends on the extent of the deficiency, which may amount to as much as 2–3 g in some cases (Fig. 18/21). Iron absorption, which is high at the commencement of treatment, falls off gradually as the anemia improves (Norrby and Sölvell, 1974). As a result of this declining absorption, the replenishing of the iron stores may require as much as 5–6 months of additional iron treatment (Beutler et al., 1954; Jasinski and Roth, 1954; Finch, 1961; Dagg and Goldberg, 1973; Wintrobe, 1974).

Side effects are a problem in treatment with iron since about 10–20% of patients complain of nausea, epigastric pain, constipation, or diarrhea. If treatment is started with lower doses and gradually built up to full therapeutic doses, these complaints can sometimes be avoided. Sölvell (1970) studied the side effects of iron in 5000

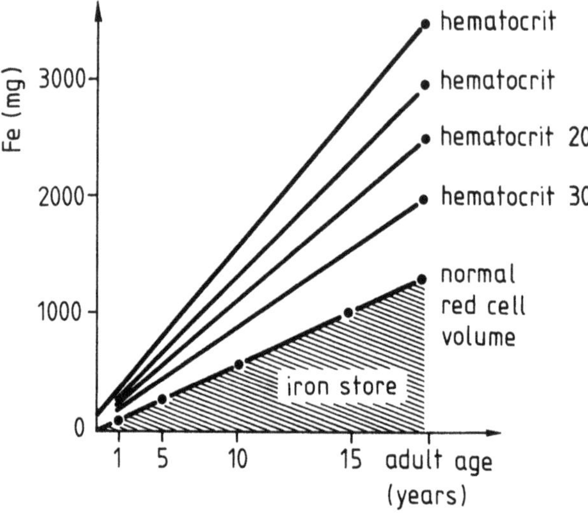

Fig. 18/21. The amount of iron required to make up for the deficiency as a function of age and severity of the anemia. The shaded area indicates the amount of extra iron necessary for iron stores. Total requirement equals the sum of the two factors (after Bothwell, T. H. and Finch, C. A.: Iron Metabolism. Little, Brown and Co., Boston 1962)

individuals. Complaints occurred in 30% taking iron preparations and 15% with placebo, and the frequency of side effects was the same for ferrous sulfate, ferrous fumarate, and ferrous carbonate. The observations of Girdwood in 1952 and Kerr and Davidson (1958) suggest that emotional factors may play a part in provoking side effects in some patients; thus Girdwood found that more patients complained of side effects when given green ferrous sulfate tablets than when given white tablets, and Kerr and Davidson found that the same proportion of nurses taking placebo but believing that they took iron preparations complained as those who received genuine iron preparations, in contrast to a very small proportion who complained if they knew that they had been given a placebo.

PARENTERAL IRON THERAPY

There are few indications for giving parenteral iron, and its use should be strictly limited to situations where

1. gastrointestinal disease renders oral medication impossible,
2. the oral iron preparations give rise to serious complaints, giving poor patient compliance,
3. iron loss is too great or continuous to be compensated for by oral iron treatment, and
4. malabsorption of iron is present, as in coeliac disease.

Even patients with gastrointestinal diseases may tolerate oral iron therapy, but in some patients with ulcerative colitis or Crohn's disease or diverticulitis the tendency for iron to produce diarrhea may contraindicate its use.

Serious side effects with oral iron treatment are rarely encountered, but many patients are unwilling to tolerate even minor abdominal discomfort and may stop treatment or take it only very irregularly.

Although some degree of malabsorption may occur following partial gastrectomy, the majority of patients respond favorably to oral iron unless they are continuing to have occult bleeding.

The most commonly used parenteral iron preparations are saccharated iron oxide, iron dextrin, iron dextran, and iron sorbitol citrate.

Saccharated iron oxide (Ferrivenin® and Proferrin®) was the first safe preparation to be introduced commercially. It is an electronegative colloidal iron hydroxide solution stabilized with sugar and alkali, and each 5 ml contains 100 mg of iron. Because of the sugar content the solution is hypertonic. The preparation should not be diluted with physiological saline or injected with any other drug because the stability of the iron-sugar solution is diminished by such treatment, and side effects will be provoked. The solution should be given very slowly and with extreme care intravenously, making sure that no iron leaks into the surrounding tissue. Following injection the plasma iron concentration rises rapidly, then falls rapidly in the first 1–2 hours, and in the next 3–6 hours declines more slowly. Most of the iron goes to the reticuloendothelial system (Andersson, 1950), and 0.12–7.0% is excreted in the urine (Slack and Wilkinson, 1949; Scott and Govan, 1954; Crowley, 1952). An average of about 5% is excreted. Single doses daily, or every other day, of 60–100 mg iron are given until the estimate deficit of iron has been given.

Iron-dextrin complex (Astrafer®, Dextriferron®) is a complex of ferric hydroxide with hydrolyzed dextrin. The preparation contains 20 mg of iron per milliliter, and it was introduced by Agner et al. (1948). Its clinical use has been investigated by Andersson (1950), Lucas and Hagedorn (1952), and Fielding (1961). It has a similar rate of outflow from the circulation as saccharated iron oxide with a T/2 of 1 hour, and only a small proportion is excreted in the urine.

Iron dextran (Imferon®) is a complex compound of ferric hydroxide with dextran. The preparation contains 50 mg of iron per milliliter. Its principal use is by intramuscular injection from which it is absorbed via the lymphatics (Goldberg, 1958). Upon injection 50–90% is absorbed within 4–5 days, while the rest of the iron complex is mobilized only slowly from the muscles, and 20–40% may remain as long as 40–50 days following injection. A small proportion may even remain as long as 100 days (Martin et al., 1955; Garby and Sjölin, 1957; Grimes and Hutt, 1957; Evans and Ramsay, 1957; Stevens, 1958; Karlefors and Norden, 1958; Will, 1968). The serum iron level starts to rise within 2–3 hours of injection, and, depending upon the amount given, it reaches a value of some 700–1500 μg/dl. About 30–40% of the dose appears in the red cells some 4 weeks after injection, but incorporation may continue up to 50–60 days. The plasma iron level remains elevated for at least 4 weeks and then starts to decline gradually (McCurdy, 1970).

Following intravenous injection (Wood et al., 1968; Will and Groden, 1968; McCurdy, 1970) there is a massive elevation of the plasma iron level, occasionally up to several thousands of μg/dl. The T/2 time of iron dextran from the plasma is 2–3 days. Most of the iron dextran leaving the circulation goes to the liver and the spleen, where the iron is split off from the dextran and, bound to transferrin, returned to the blood. In the circulating blood 1–2% of the iron binds directly to transferrin (Hillman and Henderson, 1969b). In the first 3 weeks 30–35% of the iron is utilized in the course of erythropoiesis, but later the rate of utilization declines, and in 50 days about 50–60% of the dose has been utilized. A small proportion is stored for some time in the reticuloendothelial system.

Although earlier studies had suggested that the intravenous use of iron dextran might give anaphylactic type of reactions, within the past few years the technique of total dose infusion of iron dextran has been introduced as a method of replacing the total iron deficit with one intravenous infusion. Extreme care should be taken to test for sensitivity with a small initial dose before proceeding to the total dose infusion. With suitable care, however, the technique may be useful in selected cases, e.g., in pregnancy where the patient is intolerant of oral iron or where there is continued chronic bleeding, e.g., in hemorrhagic telangiectasia (Miller et al., 1966) (Fig. 18/22).

The iron sorbitol–citric acid complex (Jectofer®) contains 50 mg of iron per milliliter. The preparation is stabilized by dextran. Following intramuscular injection the iron is quickly absorbed via the blood vessels and lymphatics. From the site of injection absorption takes place in two phases, a rapid first phase and a slower subsequent one (McCurdy, 1964). The serum iron level rises to 300–500 μg/dl within 2–3.5 hours (Andersson, 1961; Lindvall and Andersson, 1961), and the transferrin is usually completely saturated (McCurdy, 1964). As much as 70% of the dose has disappeared from the site of absorption in 3 hours, and after 10 hours there is only a small amount left at the site of injection (Pringle et al., 1962; Wetherley-Mein et al., 1962). The iron level returns to normal in 24 hours. A great part of the iron present in the preparation reaches the bone marrow directly and is then utilized quickly in the course of hemoglobin synthesis (Andersson, 1962; Pringle et al., 1962).

Of the dose, 25–30% is excreted in the urine within the first 24 hours, and this must be taken into account when calculating the required dosage.

The total amount of iron required to make up for the deficit must take into account not only the level of hemoglobin but also the additional amount of iron needed to replenish iron stores. Approximately 180–200 mg of iron is required to raise the concentration of hemoglobin by 1 g/dl; thus, if the hemoglobin deficit

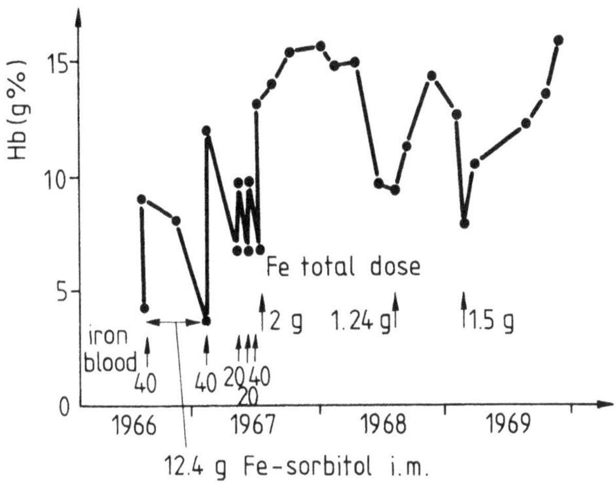

Fig. 18/22. Total dose infusion of iron dextran for treatment of recurrent iron-deficiency anemia in idiopathic pulmonary hemosiderosis. Iron sorbitol i.m. and blood transfusions gave only partial remission. TDI (total dose infusion) had a lasting result

before treatment was 6 g, then 6 × 0.2 will be the approximate amount that will be required to bring the hemoglobin up to normal, i.e., 1.2 g, and to this we should add 800 mg–1 g for the iron stores.

As already mentioned, with the iron sorbitol preparation in particular, an addition has to be made for the proportion lost in the urine, i.e., some 25–30% more than the calculated amount must be given.

The response to treatment with parenteral iron therapy is hardly greater than that achieved by effective oral iron treatment, the mean response being 0.15–0.30 g/dl daily (Andersson, 1950, 1961; Baird and Podmore, 1954; Brown and Moore, 1956). In children the rate of hemoglobin synthesis may be somewhat higher. In comparative studies Cope and his collaborators (1956) found a mean daily rise of 0.14 g/dl with oral treatment, 0.15 g/dl with intravenous, and 0.13 g/dl with intramuscular treatment (McCurdy, 1965). In a comparison of the response in monozygotic twins with iron deficiency, one of whom was treated with intravenous iron and the other with oral iron (Fig. 18/23), no difference in the two forms of therapy was observed (Bothwell and Finch, 1962). On the other hand, Pritchard (1961) found that parenteral iron therapy was somewhat more effective in treatment,

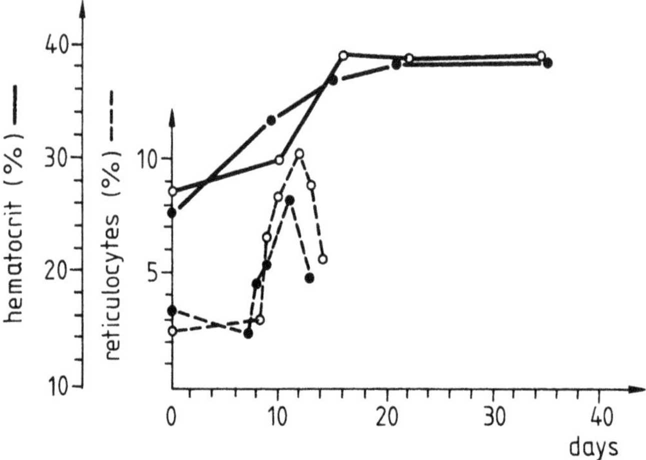

Fig. 18/23. Comparison of the effect of oral (●) and parenteral (○) iron therapy in monozygotic twins (after Bothwell, T. H. and Finch, C. A.: Iron Metabolism. Little, Brown and Co., Boston 1962)

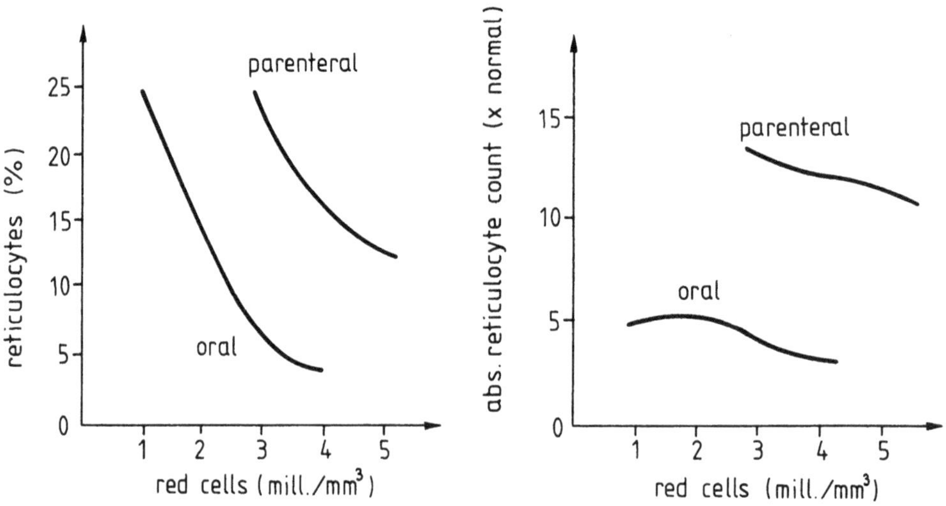

Fig. 18/24. Reticulocytosis developing in the course of oral and parenteral iron therapy (after Heath, C. W.: Arch. intern. Med. *51*, 459, 1933; Goetsch, A. T. et al.: Blood *1*, 129, 1946)

producing a mean daily hemoglobin increase of 0.26 g/dl compared with 0.20 g/dl with orally administered ferrous gluconate. Some workers have found a slightly higher increase in reticulocyte count following parenteral iron therapy as compared to oral medication (Heath, 1933; Goetsch et al., 1946) (Fig. 18/24). Figueroa (1964) found a difference only if intravenous preparations were employed.

THE TOXICITY OF PARENTERAL IRON PREPARATIONS

The side effects of parenteral iron preparations are shown in Table 18/8. Headache and flushing may be observed if an intravenous injection is given too quickly (Fielding, 1961; Newcombe, 1967). Fever, arthralgia, and lymphadenopathy may follow iron dextran injections (McCurdy, 1965). Pains in the limbs and sometimes nausea, vomiting, and diarrhea result mainly from cases in which iron sorbitol citrate has been injected in massive doses (McCurdy, 1964). These symptoms are probably related to the amount of ionized iron liberated from the preparation.

Severe anaphylactic reactions have been reported in connection with iron dextran and dextriferron injections (Becker et al., 1966; Blazar and del Riego, 1962; Conrad, cited by McCurdy, 1970). Seftel (1965) observed phlebitis not only at the site of injection but also more distant from it. Forristal and Witt (1968) described pleocytosis in the CSF after an intravenous injection of iron dextran.

Table 18/8

Toxicity of parenteral iron preparations

	Iron dextran	Dextriferron	Iron sorbitex
Systemic reactions			
Flushing, headache, hypotension	+	+	
Joint pain	+		+
Nausea, vomiting, diarrhea	+	+	+
Urticaria, bronchospasm, anaphylactic reaction	+	+	+
Late arthralgia, lymphadenopathy	+		
Fever	+	+	
Local reactions			
Pain	+	+	+
Skin staining	+		+

Early on in the use of parenteral iron preparations it was observed that sarcomata could be induced with iron dextran in experimental animals, and fears were expressed that tumors might be induced also in man (Lundin, 1961; Fielding, 1962; Haddow et al., 1964; Bonser, 1967; Carter et al., 1968). However, these fears have not been substantiated.

TREATMENT OF IRON DEFICIENCY IN CHILDREN

Iron deficiency of infancy responds readily to appropriate treatment. There is no need to give any medication other than an adequate dose of a ferrous iron preparation, 5 mg of ferrous iron per kg body weight per day. This results in reticulocytosis in 3–4 days, reaching a peak between 5 and 10 days (Erlandson, 1962). This is followed by a gradual rise in hemoglobin level. A suitable prescription for infants is:

Ferrous sulfate		100 mg
Ascorbic acid		5 mg
Orange syrup		1 ml
Chloroform water		0.1 ml
Purified water	to	5.0 ml

(de Gruchy, 1978). The dose should be given three times a day. Response is in general more rapid than in adults; the hemoglobin level may increase by as much as 0.3–0.4 g/dl/day. The treatment should be continued for 1–2 months after the hemoglobin has reached a normal level.

Treatment with parenteral iron should be restricted to cases in which oral iron treatment is contraindicated because of gastrointestinal disease or lack of cooperation of the mother. For intramuscular injection the relatively lower molecular weight iron sorbitol compounds, e.g., Jectofer, are recommended; for intravenous treatment Ferrlecit has been found the most suitable. Injections can be given every other day, and in most cases 10–20 injections in all are needed. Betke (1961) recommends the following formula for the total amount of iron required:

$$\underset{(mg)}{\text{Fe}} = \underset{(kg)}{\text{body weight}} \times (16 - \underset{(g/dl)}{\text{Hb level of the patient}}) \times 3.5$$

SOME EFFECTS OF IRON THERAPY

With an adequate therapeutic response, the iron-deficient patient usually begins to feel general improvement within 2–3 weeks, and the symptoms associated with tissue iron deficiency also improve. The papillae of the tongue may regenerate (see Figs. 18/6 and 18/7) and other dermal and mucosal changes may improve (Fig. 18/25). Dysphagia caused by a postcricoid web often disappears in spite of the fact that the web may still be demonstrable by cine-X-ray.

Failure to respond to treatment is most commonly due to the fact that the patient is not taking the iron prescribed or that it is an ineffective preparation, e.g., some of the slow-release preparations. Alternatively there may be persistent bleeding that cannot be adequately replaced by iron, or there may be occult malabsorption. Finally, the diagnosis may be erroneous; e.g., the patient may suffer from sideroblastic anemia or thalassemia.

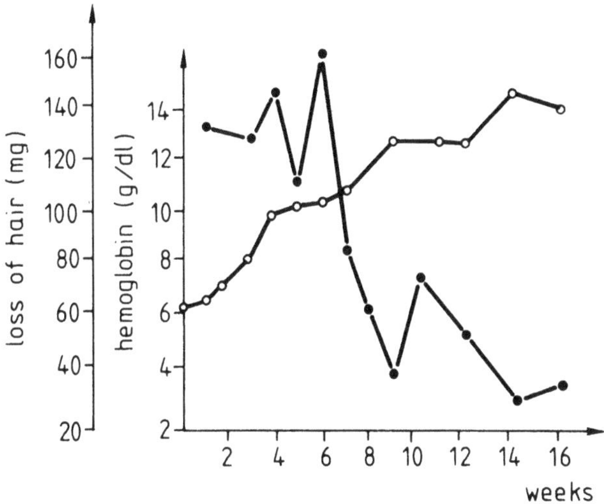

Fig. 18/25. Massive loss of hair in iron-deficiency anemia. In response to iron treatment the hemoglobin rise is rapid, but the regrowth of hair is delayed up to about 6 weeks and is therefore not due to anemia *per se* (after Fielding, J., in: Hallberg, L. et al.: Iron Deficiency. Academic Press, London–New York 1970). (–●–●–) loss of hair (mg/week); (–○–○–) Hb level (g/dl)

ACUTE IRON INTOXICATION

Iron poisoning occurs most commonly in children who mistake the sugar-coated pills for sweets. The iron has a necrotizing effect on the gastrointestinal mucosa and results in vomiting, abdominal pain and melena, shock, and, in severe cases, even death. The lethal dose is estimated at 3–16 grams (Aldrich, 1958). The child usually starts to vomit 1–1.5 hours after taking the pills (Craig, 1955). If death does not occur during the initial phase, there is a latent period that may last for as much as 24 hours before acidosis, evidence of hepatocellular damage, and spasms may develop that, in untreated cases, may lead to fatal coma (Whitten et al., 1965). Without specific treatment the mortality is estimated at 45–50% (Aldrich, 1958). The mechanism of the toxic effect is not fully known. It has been postulated that ferritin release causes irreversible shock (Mazur and Shorr, 1948), but Bronson and Sisson (1960) could not confirm this. However, free ionic iron in high concentration inhibits the *in vitro* activity of various intracellular enzyme systems (Racker and Krimsky, 1948; Morrow et al., 1969), and the toxic effect may be explained by this (Dagg et al., 1971).

TREATMENT OF IRON OVERDOSE

An immediate injection of 2 g desferrioxamine should be given intramuscularly. If the patient is seen soon enough after ingestion of tablets, it may be useful to induce vomiting mechanically and to perform gastric lavage with 1% sodium bicarbonate to recover as many tablets as possible. Desferrioxamine has proved very efficacious and, indeed, lifesaving in treatment (Henderson et al., 1963; McEnery and Mack, 1964; Dugdale and Powell, 1964; Shapiro and Barbezat, 1964; Whitten et al., 1965; Jacobs et al., 1965; Leiken et al., 1967; Barr and Fraser, 1968). After gastric lavage 5–12 grams of desferrioxamine should be given through the tube in an attempt to chelate the unabsorbed iron within the gut, and at the same time an intravenous infusion of 15 mg/kg/hour is set up and continued for at least 24 hours. The total amount given intravenously will depend on the clinical response. If an intravenous infusion is impossible, the desferrioxamine can be given intramuscularly in four equally divided doses. Parenteral administration of desferrioxamine should not exceed 6 g in any 24-hour period.

BIBLIOGRAPHY

Addison, G. M., Beamish, M. R., Hales, C. N., Hodgkins, M., Jacobs, A., Llewellin, P.: An immunoradiometric assay for ferritin in the serum of normal subjects and patients with iron deficiency and iron overload. J. clin. Path. *25*, 326 (1972).

Agner, K., Andersson, N. S. E., Nordenson, N. G.: Intravenöse Eisentherapie. Acta haemat. (Basel) *1*, 193 (1948).

Aksoy, M., Cambi, N., Erden, S.: Roentgenographic bone changes in chronic iron deficiency anemia. A study in 12 patients. Blood *27*, 677 (1966).

Alder, A.: Zur Dosierung des Eisens. Schweiz. med. Wschr. *1*, 663 (1920).

Aldrich, R. A.: Acute iron toxicity. In: Wallerstein, R. O., Mettier, S. R. (eds.): Iron in clinical Medicine. University of California Press, Berkeley–Los Angeles 1958.

Anderson, J. R., Buchanan, W. W., Gondie, R. B.: In: Autoimmunity. Thomas, Springfield, Illinois 1967.

Andersson, N. S. E.: Experimental and clinical investigations into the effect of parenterally administered iron. Acta med. scand. *138* (Suppl. 241), 1 (1950).

Andersson, N. S. E.: Clinical investigation on a new intramuscular haematinic. Brit. med. J. *2*, 275 (1961).

Andersson, N. S. E.: Side-effect of iron sorbitol citrate. Brit. med. J. *2*, 1260 (1962).

Arcasoy, A. et al.: Decreased iron and zinc absorption in Turkish children with iron deficiency and geophagia. Acta haemat. (Basel) *60*, 76 (1978).

Badenoch, J., Callender, S. T.: Iron metabolism in steatorrhoea. The use of radioactive iron in studies of absorption and utilization. Blood *9,* 123 (1954).

Badenoch, J., Callender, S. T.: Effect of corticosteroids and gluten-free diet on absorption of iron in idiopathic steatorrhoea and coeliac disease. Lancet *1,* 192 (1960).

Badenoch, J., Callender, S. T., Evans, J. R., Spray, G. H., Richards, W. C. D., Turnbull, A., Wakisaka, G., Witts, L. J.: Gastric biopsy and radioactive vitamin B_{12} in the study of the stomach in anaemia. Rev. hémat. *10,* 194 (1955).

Badenoch, J., Evans, J. R., Richards, W. C. D.: The stomach in hypochromic anaemia. Brit. J. Haemat. *3,* 175 (1957).

Bainton, D. F., Finch, C. A.: The diagnosis of iron deficiency anemia. Amer. J. Med. *37,* 62 (1964).

Baird, I. M., Blackburn, E. K., Wilson, G. M.: The pathogenesis of anaemia after partial gastrectomy. I. Development of anaemia in relation to time after operation, blood loss and diet. Quart. J. Med. *28,* 21 (1959).

Baird, I. M., Dodge, O. G., Palmer, F. J., Wawman, R. J.: The tongue and oesophagus in iron-deficiency anaemia and the effect of iron therapy. J. clin. Path. *14,* 603 (1961).

Baird, I. M., Podmore, D. A.: Intramuscular iron therapy in iron-deficiency anemia. Lancet *2,* 942 (1954).

Baker, S. J., Mathan, V. I.: Prevalence, pathogenesis and prophylaxis of iron deficiency in the tropics. In: Kief, H. (ed.): Iron Metabolism and its Disorders. Excerpta Medica, Amsterdam–Oxford; American Elsevier, New York 1975.

Bannerman, R. M., Beveridge, B. R., Witts, L. J.: Anaemia associated with unexplained occult blood loss. Brit. med. J. *1,* 1417 (1964).

Barr, G. D., Fraser, D. K. B.: Acute iron poisoning in children. Role of chelating agents. Brit. med. J. *1,* 737 (1968).

Basta, S., Soekirman, K. D., Scrimshaw, N. S.: Iron deficiency anaemia and the productivity of adult males in Indonesia. Amer. J. clin. Nutr. *32,* 916 (1979).

Basu, S. K.: Rapid administration of iron-dextran in late pregnancy. Lancet *1,* 1430 (1963).

Basu, S. K.: Administration of iron-dextran complex by continuous intravenous infusion. J. Obstet. Gynaec. Brit. Cwlth *72,* 253 (1965).

Beard, M. E. J., Weintraub, L. R.: Hypersegmented neutrophilic granulocytes in iron deficiency anaemia. Brit. J. Haemat. *16,* 161 (1969).

(Bechterev, V. M.) Бехтерев, В. М.: О психических расстройствах глотания. Обзор психиатрии 5 (1901).

Becker, C. E., MacGregor, R. R., Walker, K. S., Jandl, J. H.: Fatal anaphylaxis after intramuscular iron-dextran. Ann. intern. Med. *65,* 745 (1966).

Bekier, A., Holdener, E.: Der ^{57}Co-Exkretions- und Resorptionstest in der Diagnose der Eisenmangelanämie. Nucl.-Med. (Stuttg.) *15,* 126 (1976).

Ben-Ishay, D.: Toxic reactions to intramuscular administration of iron-dextran. Lancet *1,* 476 (1961).

Bernát, I.: Über die Bedeutung der Hyposiderose in der Pathogenese der Ozaena. Z. ges. inn. Med. *15,* 1058 (1960).

Bernát, I.: Ozaena és hyposiderosis (Ozaena and hyposiderosis). Congress of Hematology, Budapest 1961.

Bernát, I.: Ozaena — A Manifestation of Iron Deficiency. Pergamon Press, Oxford–London–Edinburgh–New York–Paris–Frankfurt 1965.

Bernát, I.: Die Ozaena — eine Manifestation der Eisenmangelkrankheit. Akadémiai Kiadó, Budapest; VEB Verlag Volk und Gesundheit, Berlin 1966.

Bernát, I.: Unpublished results (1968).

Bernát, I.: Az égési anaemia pathogenesise (The Pathogenesis of Anemia after Thermal Injury). Akadémiai Kiadó, Budapest 1971.

Bernát, I., Fáber, V.: Pathology and histopathology of ozaena. Acta med. Acad. Sci. hung. *18,* 139 (1962).

BERNÁT, I., KOVÁCS, E.: Vizsgálatok a normális és kóros vasanyagcsere köréből. II. és III. (Investigations on the normal and pathological iron metabolism. II and III). Katonaorvosi Szle. *8*, 725 und 882 (1956).
BERNÁT, I., VALLÓ, J.: Ozaena: the causes of its familial occurrence. Acta med. Acad. Sci. hung. *20*, 89 (1964).
BESSIS, M., BRETON-GORIUS, J.: Iron metabolism in the bone marrow as seen by electron microscopy: A critical review. Blood *19*, 635 (1962).
BESSIS, M., BRETON-GORIUS, J., BARAT-SAVARD, N.: L'îlot érythroblastique et la rhophéocytose dans l'anémie ferriprive. Rev. Hémat. *15*, 233 (1960).
BETKE, K.: Iron deficiency in children. In: HALLBERG, L., HARWERTH, H. G., VANNOTTI, A. (eds.): Iron Deficiency. Academic Press, London–New York 1970.
BEUTLER, E.: Iron enzymes in iron deficiency. I. Cytochrome C. J. clin. Invest. *36*, 874 (1957).
BEUTLER, E.: Tissue effects of iron deficiency. In: GROSS, F. (ed.): Iron Metabolism. Springer, Berlin–Göttingen–Heidelberg 1964.
BEUTLER, E., BLAISDELL, R. K.: Iron enzymes in iron deficiency. III. Catalase in rat red cells and liver with further observations on cytochrome. J. Lab. clin. Med. *52*, 694 (1958a).
BEUTLER, E., BLAISDELL, R. K.: Iron enzymes in iron deficiency. II. Catalase in human erythrocytes. J. clin. Invest. *37*, 833 (1958b).
BEUTLER, E., FAIRBANKS, V. F.: The effects of iron deficiency. In: JACOBS, A., WORWOOD, M. (eds.): Iron in Biochemistry and Medicine. Vol. 2. Academic Press, London–New York 1980.
BEUTLER, E., MEERKREEBS, G.: Doses and dosing. New Engl. J. Med. *274*, 1152 (1966).
BEUTLER, E., DRENNAN, W., BLOCK, M.: The bone marrow and liver in iron deficiency anemia. J. Lab. clin. Med. *43*, 427 (1954).
BEVERIDGE, B. R.: The natural history of hypochromic anaemia with special reference to some epithelial changes. Ph. D. Thesis, University of Oxford 1963.
BEVERIDGE, B. R., BANNERMAN, R. M., EVANSON, J. M., WITTS, L. J.: Hypochromic anaemia. Quart. J. Med. N. S. *134*, 145 (1965).
BILGER, R.: Die Siderocyten. In: HEILMEYER, L., HITTMAIR, A.: Handbuch der gesamten Hämatologie, Vol. 1/1, p. 235. Urban und Schwarzenberg, München–Berlin–Wien 1957.
BLAKE, J., RECHNITZER, P. A.: The haematological and nutritional effects of gastric operations. Quart. J. Med. *22*, 419 (1953).
BLAUD, P. (1832), cit. DREYFUS, SCHAPIRA, 1958.
BLAZAR, A. S., DEL RIEGO, M. G.: Severe idiosyncratic reactions to an intravenous iron preparation. Obstet. and Gynec. *20*, 156 (1962).
BODDINGTON, M. M.: Changes in buccal cells in the anaemias. J. clin. Path. *12*, 222 (1959).
BOKHARI, S. M. H.: Iron deficiency in recruits, soldiers and their families. Rev. int. Serv. Santé Armées *31*, 221 (1958).
BONNAR, J., GOLDBERG, A., SMITH, J. A.: Do pregnant women take their iron? Lancet *1*, 457 (1969).
BONSER, G. M.: Cancer hazards of the pharmacy. Brit. med. J. *4*, 129 (1967).
BORKA, I.: Terhes-szűrés anaemiára (Studies on anemia during pregnancy). Transfusio *4* (1970).
BOTHWELL, T. H., FINCH, C. A.: Iron Metabolism. Little, Brown and Co., Boston 1962.
BOTHWELL, T. H., GLYN THOMAS, R.: Sideropenic dysphagia. S. Afr. med. J. *32*, 614 (1958).
BOTHWELL, T. H., CALLENDER, S., MALLETT, B., WITTS, L. J.: The study of erythropoiesis using tracer quantities of radioactive iron. Brit. J. Haemat. *2*, 1 (1956).
BRAT, L.: Loosersche Umbauzonen und essentielle hypochrome Anämie. Fortschr. Röntgenstr. *77*, 204 (1952).
BRISE, H., HALLBERG, L.: Iron absorption studies. II. Acta med. scand. *171* (Suppl. 376), 1 (1962).
BRONSON, W. R., SISSON, T. R. C.: Studies on acute iron poisoning. Amer. J. Dis. Child. *99*, 18. (1960).
BROWN, E. B., MOORE, C. W.: Parenterally administered iron in the treatment of hypochromic anemia. In: TOCANTINS, L. M. (ed.): Progress in Hematology, Vol. 1. Grune and Stratton, New York 1956.
BROWN, E. B., DUBACH, R., MOORE, C. W.: Studies on iron transportation and metabolism. XI. Critical analysis of mucosal block by large doses of inorganic iron in human subjects. J. Lab. clin. Med. *52*, 335 (1958).

BROWN, K., LUBIN, B., SMITH, R., OSKI, F.: Prevalence of anemia among preadolescent and young adolescent urban black Americans. J. Pediat. *81*, 714 (1972).
BROWN, W. D., DYMENT, P. G.: Pagophagia and iron deficiency anemia among adolescent girls. Pediatrics *49*, 766 (1972).
BRUMFITT, W.: Primary iron-deficiency anaemia in young men. Quart. J. Med. N. S. *29*, 1 (1960).
BRÜSCHKE, G.: Der Siderocyt. Akademie Verlag, Berlin 1962.
BURKO, J., MELLINS, H. Z., WATSON, J.: Skull changes in iron deficiency anemia simulating congenital hemolytic anemia. Amer. J. Roentgenol. *86*, 447 (1961).
BURMAN, D.: Iron deficiency in infants and children. In: CALLENDER, S. T. (ed.): Iron Deficiency and Iron Overload (Clinics in Haematology). Saunders, London–Philadelphia–Toronto 1973.
CALLENDER, S. T.: Quick and slow-release iron; a double-blind trial with a single daily dose regimen. Brit. med. J. *4*, 531 (1969).
CALLENDER, S. T.: Oral iron therapy. Brit. J. Haemat. *18*, 123 (1970).
CALLENDER, S. T.: Iron deficiency due to malabsorption of food iron. In: KIEF, H. (ed.): Iron Metabolism and its Disorders. Excerpta Medica, Amsterdam–Oxford; American Elsevier, New York 1975.
CALLENDER, S. T., DENBOROUGH, M. A.: A family study of pernicious anaemia. Brit. J. Haemat. *3*, 88 (1957).
CARTER, R. L., PERCIVAL, W. H., ROE, F. J. C.: The effects of iron dextran on squirrel monkeys (Saimirai sciurea). Brit. J. Cancer *22*, 116 (1968).
CASPARIS, C.: Über eine familiäre Skelettveränderung mit Eisenmangelanämie. Diss. Zürich 1957.
CELADA, A. et al.: Inorganic iron absorption in subjects with iron deficiency anaemia, achylia gastrica and alcoholic cirrhosis using a whole-body counter. Acta haemat. (Basel) *60*, 182 (1978).
CHANDRA, R. K.: Arch. Dis. Child. *48*, 864 (1973).
CHARLTON, R. W., BOTHWELL, T. H., SEFTEL, H. C.: Dietary iron overload. In: CALLENDER, S. T. (ed.): Clinics in Haematology, Vol. 2/2. Saunders, London–Philadelphia–Toronto 1973.
CHEVALLIER, P., MOUTIER, F., STEWART, W., SEVAUX, A., ELY, Z.: L'atrophie gastrique et la thérapeutique antianémique dans les affections sans anémie. Conception d'une maladie à manifestations diverses dont les anémies essentielles ne sont que des formes cliniques graves. Bull. Soc. franç. Derm. Syph. *8*, 719 (1934).
CHISHOLM, M.: J. Obstet. Gynaec. Brit. Cwlth *73*, 191 (1966).
CHISHOLM, M.: Tissue changes associated with iron deficiency. In: CALLENDER, S. T. (ed.): Iron Deficiency and Iron Overload (Clinics in Haematology), Vol. 2/2. Saunders, London–Philadelphia–Toronto 1973.
CHISHOLM, M., ARDRAN, G. M., CALLENDER, S. T., WRIGHT, R.: Iron deficiency and autoimmunity in postcricoid webs. Quart. J. Med. N. S. *40*, 421 (1971).
CHISHOLM, M., WRIGHT, R., CALLENDER, S. T., ARDRAN, G.: Iron deficiency and autoimmunity in relation to postcricoid webs. (Abstr.) XIIth Congr. Int. Haemat., New York 1968.
CHOWN, B.: The fetus can bleed. Three clinico-pathologic pictures. Amer. J. Obstet. Gynec. *70*, 1298 (1955).
COGHILL, N. F., DONIACH, D., ROITT, I. M., MOLLIN, D., WYNN-WILLIAMS, A.: Autoantibodies in simple atrophic gastritis. Gut *6*, 48 (1965).
COLEMAN, D. H., STEVENS, A. R., FINCH, C. A.: The treatment of iron deficiency anemia. Blood *10*, 567 (1955).
COOK, J. D., FINCH, C. A., SMITH, N. J.: Evaluation of the iron status of a population. Blood *48*, 449 (1976).
COOK, J. D., LIPSCHITZ, D. A., MILES, L. E. M., FINCH, C. A.: Serum ferritin as a measure of iron stores in normal subjects. Amer. J. clin. Nutr. *27*, 681 (1974).
COOK, J. D., MARSAGLIA, G., ESCHBACH, J. W., FUNK, D. D., FINCH, C. A.: Ferrokinetics: A biologic model for plasma iron exchange in man. J. clin. Invest. *49*, 197 (1970).
COOKE, W. T.: In: Progress in Gastroenterology. Grune and Stratton, New York–London 1968.
COPE, W., GILLHESPY, R. O., RICHARDSON, R. W.: Treatment of iron deficiency anemia. Comparison of methods. Brit. med. J. *2*, 638 (1956).

COPPO, L.: Klinische und histologische Beobachtungen am Ösophagus von Ozaena-Kranken. Valsalva *34*, 66 (1958).
COTES, P. M., MOSS, G. F., MUIR, A. R., SCHEUER, P. J.: Distribution of iron in maternal and foetal tissues from pregnant rhesus monkeys treated with a simple intravenous infusion of (^{59}Fe) iron dextran. Brit. J. Pharmacol. *26*, 633 (1966).
COURMOULIS, M., GISINGER, E.: Eisenmangel und Fettresorption. Wien. Z. inn. Med. *36*, 81 (1955).
COWAN, B., JOSEPH, S., SATIJA, V. K.: The gastric mucosa in anaemia in Punjabis. Gut *7*, 234 (1966).
CRAIG, J. O.: Poisons children swallow. Brit. med. J. *2*, 1496 (1955).
CROSBY, W. H.: Pica. J. Amer. med. Ass. *235*, 2765 (1976).
CROWLEY, J.: Iron absorption tests in anaemia: use of intravenous iron preparations. Edinb. med. J. *59*, 478 (1952).
DABSKI, H.: Morphological changes in the epithelium of the oral cavity in pernicious anaemia and in hypochromic anaemia. Pol. Tyg. lek. *15*, 942 (1960).
DAGG, J. H.: Iron deficiency, clinical and experimental studies. Diss. University of Glasgow 1968.
DAGG, J. H., GOLDBERG, A.: Detection and treatment of iron deficiency. In: CALLENDER, S. T. (ed.): Clinics in Haematology, Vol, 2/2, p. 367. Saunders, London–Philadelphia–Toronto 1973.
DAGG, J. H., CUMMING, R. L. C., GOLDBERG, A.: Disorders of iron metabolism. In: GOLDBERG, A., BRAIN, M. C. (eds.): Recent Advances in Haematology. Churchill–Livingstone, Edinburgh–London 1971.
DAGG, J. H., GOLDBERG, A., ANDERSON, J. R., BECK, J. S., GRAY, K. G.: Autoimmunity in iron deficiency anaemia. Brit. med. J. *1*, 1349 (1964).
DAGG, J. H., GOLDBERG, A., GIBBS, W. N., ANDERSON, J. R.: Detection of latent pernicious anaemia in iron deficiency anaemia. Brit. med. J. *2*, 619 (1966a).
DAGG, J. H., GOLDBERG, A., LOCKHEAD, A.: Value of erythrocyte protoporphyrin in diagnosis of latent iron deficiency. Brit. J. Haemat. *12*, 326 (1966b).
DAGG, J. H., JACKSON, J. M., CURRY, B., GOLDBERG, A.: Cytochrome oxidase in latent iron deficiency (sideropenia). Brit. J. Haemat. *12*, 331 (1966c).
DAGG, J. H., MORROW, J. J., MCFARLANE, D. B., GOLDBERG, A.: Sideropenia (latent iron deficiency). Abstr. Quart. J. Med. N. S. *36*, 600 (1967).
DALLMAN, P. R.: Tissue effects of iron deficiency. In: JACOBS, A., WORWOOD, M. (eds.): Iron in Biochemistry and Medicine, pp. 437–475. Academic Press, New York 1974.
DALLMAN, P. R., BEUTLER, E., FINCH, C. A.: Effects of iron deficiency exclusive of anaemia. Brit. J. Haemat. *40*, 179 (1978).
DAMESHEK, W.: Primary hypochromic anemia. A new type of idiopathic anemia. Amer. J. med. Sci. *182*, 520 (1931).
(DANILENKO, S. S.) Даниленко, С. С.: Расстройство глотания при малокровии. Diss. Moscow 1950.
DARBY, W. J.: The oral manifestations of iron deficiency. J. Amer. med. Ass. *130*, 830 (1946).
DAVIDSON, L. S. P., FULLERTON, H. W.: Chronic nutritional hypochromic anaemia. Edinb. med. J. *45*, 1 (1938).
DAVIDSON, L. S. P., DONALDSON, G. M. M., DYAR, M. J., LINDSAY, S. T., MCSORLEY, J. G.: Nutritional iron deficiency anaemia in wartime; haemoglobin levels of 831 infants and children. Brit. med. J. *2*, 505 (1942).
DAVIDSON, W. M. B., MARKSON, J. L.: The gastric mucosa in iron-deficiency anaemia. Lancet *2*, 639 (1955).
DAVIS, L. R., JENNISON, R. F.: Response of the "physiological anaemia" of pregnancy to iron therapy. J. Obstet. Gynec. Brit. Emp. *61*, 103 (1954).
DAVIS, L. R., MARTEN, R. H., SARKANY, I.: Iron-deficiency anaemia in European and West Indian infants in London. Brit. med. J. *2*, 1426 (1960).
DAWKINS, S.: Serum ferritin concentration in normal subjects. Clin. Lab. Haemat. *1*, 41 (1979).
DE GRUCHY, G. L.: Clinical Haematology in Medical Practice. Blackwell, Oxford–Edinburgh 1978.
DELAMORE, I. W., SHEARMAN, D. J.C.: Chronic iron deficiency anaemia and atrophic gastritis. Lancet *1*, 889 (1965).
DE LEEUW, N. H. W., LOWENSTEIN, L., HSIEH, Y.: Iron deficiency and hydremia in normal pregnancy. Medicine (Baltimore) *45*, 291 (1966).

DE VRIES, S. I., DE POTTER, E.: Studies on iron metabolism in blood donors. Vox Sang. (Basel) N. S. *3*, 392 (1958).
DE WIJN, J. F., DE JONGSTE, J. L., MOSTERD, W., WILLBRAND, D.: Hemoglobin, packed cell volume, serum iron and iron-binding capacity of selected athletes during training. Nutr. Metabol. (Basel) *13*, 129 (1971).
DIEZ-EWALD, M., LAYRISSE, M.: Mechanisms of hemolysis in iron deficiency anemia. Blood *32*, 884 (1968).
DRESCH, C.: Prevalence of iron deficiency in France. In: HALLBERG, L., HARWERTH, H. G., VANNOTTI, A. (eds.): Iron Deficiency. Academic Press, London–New York 1970.
DREYFUS, J. C., SCHAPIRA, G.: Le fer. L'Expansion, Paris 1958.
DUGDALE, A. E., POWELL, L. W.: Acute iron poisoning: its effect and treatment. Med. J. Aust. *2*, 990 (1964).
Editorial: Carcinogenic risk of iron-dextran. Brit. med. J. *1*, 788 (1960).
ELWOOD, P. C., WILLIAMS, G.: A comparative trial of slow-release and conventional iron preparations. Practitioner *204*, 812 (1970).
ENTWHISTLE, C. C., JACOBS, A.: Histological findings in the Paterson-Kelly syndrome. J. clin. Path. *18*, 408 (1965).
ERLANDSON, M. E.: Iron metabolism and iron deficiency anemia. Paediat. clin. N. Amer. *9*, 673 (1962).
ESTREN, S.: The blood and bone marrow in idiopathic sprue. J. Mt Sinai Hosp. *24*, 304 (1957).
EVANS, L. A. J., RAMSAY, N. W.: Absorption studies on radioactive iron dextran in pregnancy. Lancet *2*, 1192 (1957).
FAIRBANKS, V. F., BEUTLER, E.: Iron Deficiency. In: WILLIAMS, W. J. et al. (eds.): Hematology. McGraw-Hill Inc., New York 1977.
FAIRBANKS, V. F., FAHEY, J. L., BEUTLER, E.: Clinical Disorders of Iron Metabolism. Grune and Stratton, New York 1971.
FAY, J., CARTWRIGHT, G. E., WINTROBE, M. M.: Studies on free erythrocyte protoporphyrin, serum iron, serum iron-binding capacity and plasma copper during normal pregnancy. J. clin. Invest. *28*, 487 (1949).
FELDER, S. L., MASLEY, P. M., WOLFF, W. I.: Anemia as a presenting symptom of esophageal hiatal hernia of the diaphragm. Arch. Intern. Med. *105*, 873 (1960).
FIELDING, J.: Intravenous iron-dextran in iron-deficiency anaemia. Brit. med. J. *2*, 279 (1961).
FIELDING, J.: Sarcoma induction by iron-carbohydrate complexes. Brit. med. J. *1*, 1800 (1962).
FIELDING, J.: In: HALLBERG, L., HARWERTH, H. G., VANNOTTI, A. (eds.): Iron Deficiency. Academic Press, London–New York 1970.
FIELDING, J. F., BADENOCH, J.: The hematological response to a gluten-free diet in adult celiac disease with triple deficiency anemia. J. Irish med. Ass. *66*, 213 (1973).
FIGUEROA, W. G.: Parenteral treatment of iron deficiency. In: DREYFUS, J. C., SCHAPIRA, G. (eds.): Iron Metabolism, p. 436. Springer, Berlin–Göttingen–Heidelberg 1964.
FINCH, C. A. (1958; 1961) cit. BOTHWELL, FINCH, 1962.
FINCH, C. A.: Diagnostic value of different methods to detect iron deficiency. In: HALLBERG, L., HARWERTH, H. G., VANNOTTI, A. (eds.): Iron Deficiency. Academic Press, London–New York 1970.
FISCHER, R., THEDERING, F.: Präoperativer Einsatz von Kobalt-Eisen zur Behandlung des larvierten Eisenmangels. Chirurg *26*, 371 (1955).
FISHER, M., BIGGS, R.: Iron deficiency in pregnancy. Brit. med. J. *1*, 385 (1955).
FLEISCHMANN, O.: Inwieweit kommt ein Ektodermaldefekt als Voraussetzung für das Auftreten einer Ozaena in Betracht. Mschr. Ohrenheilk. *66*, 1060 (1932a).
FLEISCHMANN, O.: Betrachtungen über die Ozaenagenese. Arch. Ohr.-, Nas.- u. Kehlk.-Heilk. *133*, 199 (1932b).
FLETCHER, F., LONDON, E.: Intravenous iron. Brit. med. J. *1*, 984 (1954).
FONARGE, L.: Sideropenic dysphagia. Scalpel (Brux.) III/8, 174 (1958).
FONTÈS, G., THIVOLLE, L.: Recherches expérimentales sur la thérapeutique de l'anémie grave par carence martiale et notamment par hémorragie. Sang *10*, 1056 (1936).
FORRISTAL, T., WITT, M.: Pleocytosis after iron dextran. Lancet *1*, 1428 (1968).

FORSHAW, J. W.: Idiopathic hypochromic anaemia in males. Brit. Med. J. *4893*, 908 (1954).

FOWLER, W. M., BARER, A. P.: The etiology and treatment of idiopathic hypochromic anemia. Amer. J. med. Sci. *194*, 625 (1937).

FOWLER, W. M., BARER, A. P., SPIELHAGEN, C. F.: Retention and utilization of small amounts of orally administrated iron. Arch. Intern. Med. *59*, 1024 (1937).

FOY, H., KONDI, A.: Anaemias of the tropics. Relation to iron intake, absorption and losses during growth, pregnancy and lactation. J. trop. Med. Hyg. *60*, 105 (1957).

FOY, H., KONDI, A., HARGREAVES, A.: Anaemias of Africans. Trans. roy. Soc. trop. Med. Hyg. *46*, 327 (1952).

GANZONI, A.: Neue Aspekte des Eisenmangels. Schweiz. med. Wschr. *100*, 691 (1970).

GANZONI, A.: Die durch Störungen des Eisenstoffwechsels bedingten Anämien. In: QUEISSER, W. (ed.): Das Knochenmark. Thieme, Stuttgart 1978.

GARBY, L.: Iron deficiency: Definition and prevalence. In: CALLENDER, S. T. (ed.): Iron Deficiency and Iron Overload (Clinics in Haematology). Vol. 2/2, Saunders, London–Philadelphia–Toronto 1973.

GARBY, L., SJÖLIN, S.: Some observations on the distribution kinetics or radioactive colloidal iron. Acta med. scand. *157*, 319 (1957).

GASSER, C.: Achylische Chloranämie im Kindesalter. Helv. paediat. Acta *3*, 167 (1948).

GERNEZ, A.: L'intérêt du syndrome de Plummer-Vinson en cancérologie. Paris méd. *18*, 202 (1949a).

GERNEZ, A.: Dysphagie sidéropénique et membranes oesophagiennes. Presse méd. *57*, 362 (1949b).

GHOSH, S., DAGA, S., KASTHURI, D., MISRA, R. C., CHUTTANI, H. K.: Gastrointestinal function in iron deficiency states in children. Amer. J. Dis. Child. *123*, 14 (1972).

GIRDWOOD, R. H.: Treatment of anaemia. Brit. Med. J. *1*, 599 (1952).

GLANZMANN, E.: Zur Behandlung der Kinderanämien mit ascorbinsaurem Eisen. Schweiz. med. Wschr. 436 (1937).

GLEVITSCH, E.: Pica in iron deficiency anemia. T. norske Laegeforen. *79*, 398 (1959).

GLOVER, J., JACOBS, A.: Activity pattern of iron deficiency rats. Brit. med. J. *2*, 5814, 627 (1972).

GOETSCH, A. T., MOORE, C. V., MINNICH, V.: Observations on the effect of massive doses of iron given intravenously to patients with hypochromic anemia. Blood *1*, 129 (1946).

GOLDBERG, A.: Porphyrin metabolism in relation to iron deficiency. In: HALLBERG, L., HARWERTH, H. G., VANNOTTI, A. (eds.): Iron Deficiency. Academic Press, London–New York 1970.

GOLDBERG, A., LOCHHEAD, A., DAGG, J. H.: Histamin-fast achlorhydria and iron absorption. Lancet *1*, 848 (1963).

GOLDECK, H., REMY, D., LABHARD, H.: Eisenmangel und Schwangerschaft. Dtsch. med. Wschr. *79*, 211 (1954).

GOLDWATER, L. J.: Short history of iron therapy. Ann. med. Hist. *7*, 261 (1935).

GÖLTNER, E.: Das Serumeisen bei Dauerblutspendern. Med. Klin. *54*, 351 (1959).

GÖLTNER, E.: Iron requirement and deficiency in menstruating and pregnant women. In: KIEF, H. (ed.): Iron Metabolism and its Disorders. Excerpta Medica, Amsterdam–Oxford; American Elsevier, New York 1975.

GOSDEN, M., REID, J. D.: Account of blood count results in Sierra Leone. Trans. roy. Soc. trop. Med. Hyg. *41*, 637 (1948).

GOUTTAS, A., DASCALAKIS, T., TSEVRENIS, H., COSTEAS, F.. VAKRINOS, E., YADZIDIS, H., ANTIPAS, E.: Le comportement du fer plasmatique au cours des anémies des gastrectomiées avant et après traitement. Verhandlungen des 5. Kongresses der Europäischen Gesellschaft für Hämatologie, p. 125, 1956.

GRIMES, A. J., HUTT, M. S. R.: Metabolism of ^{59}Fe-dextran complex in human subjects. Brit. Med. J. *2*, 1074 (1957).

GUEST, G. M., BROWN, E. W.: Erythrocytes and hemoglobin of the blood in infancy and childhood. III. Factors in variability, statistical studies. J. Dis. Child. *93*, 486 (1957).

GUPTA, S. P., CHUGH, T. D., DHAWAN, R. K.: The stomach in chronic iron deficiency anemia. Amer. J. Gastroent. *57*, 41 (1972).

HADDOW, A., HORNING, E. S.: On the carcinogenicity of an iron-dextran complex. J. nat. Cancer Inst. *24*, 109 (1960).

HADDOW, A., ROE, F. J. C., MITCHLEY, B. C. V.: Induction of sarcomata in rabbits by intramuscular injection of iron dextran ("Imferon"). Brit. med. J. *1*, 1593 (1964).
HAHN, P. H. et al.: Relative absorption and utilization of ferrous and ferric iron in anemia as determined with radioactive isotope. Amer. J. Physiol. *143*, 191 (1945).
HALLBERG, L.: Die Eisenresorption—einige neuere physiologische und therapeutische Erkenntnisse. In: HEILMEYER, L., KEIDERLING, W., HOFFMANN, G. (eds.): Radio-Isotope in der Hämatologie, p. 47. Schattauer, Stuttgart 1963.
HALLBERG, L.: Oral iron therapy. Factors affecting the absorption. In: HALLBERG, L., HARWERTH, H. G., VANNOTTI, A. (eds.): Iron Deficiency. Academic Press, London–New York 1970a.
HALLBERG, L.: Prevalence of iron deficiency in Sweden. In: HALLBERG, L., HARWERTH, H. G., VANNOTTI, A. (eds.): Iron Deficiency. Academic Press, London–New York 1970b.
HALLBERG, L., BRISE, H., SOLVELL, L.: A new method for studies on iron absorption in man. Proc. VII. Congr. Int. Soc. Hemat., Grune and Stratton, New York–London 1960.
HALLBERG, L. et al.: Oral iron with succinic acid in the treatment of iron-deficiency anaemia. Scand. J. Haemat. *8*, 104 (1971).
HALSTED, J. A., PRASAD, A. S., NADIMI, N.: Gastrointestinal ferritin in iron deficiency anemia. Arch. intern. Med. *116*, 253 (1965).
HAMMOND, D., MURPHY, A.: The influence of exogenous iron on formation of hemoglobin in the premature infant. Pediatrics *25*, 362 (1960).
HARGREAVES, R. M., STREET, M., HOY, T., JACOBS, A., MCLAREN, C.: Myoglobin depletion in childhood iron deficiency. Brit. J. Haemat. *47*, 399 (1981).
HASCHEN, R. J.: Exopeptidasen und Glutamat-Oxalacetat-Transamidase der menschlichen Erythrocyten bei Anämien. Acta biol. med. germ. *9*, 15 (1962).
HASKINS, D., STEVENS, A. R., FINCH, S., FINCH, C. A.: Iron metabolism. Iron stores in man as measured by phlebotomy. J. clin. Invest. *31*, 543 (1952).
HAUSSMANN, K. et al.: Diagnostische Kriterien des prälatenten und manifesten Eisenmangels. Klin. Wschr. *49*, 1164 (1971).
HAWKINS, C. F., PEENEY, A. L. P., COOKE, W. T.: Refractory hypochromic anaemia and steatorrhoea; treatment with intravenous iron. Lancet *2*, 387 (1950).
HAYEM, P.: Les maladies du sang. Paris 1880.
HEATH, C. W.: Oral administration of iron in hypochromic anemia. Arch. intern. Med. *51*, 459 (1933).
HEBBERT, F. J.: Some historical aspects of iron therapy. Glasg. med. J. *1*, 385 (1950).
HEILMEYER, L.: Die Störungen der Bluthämsynthese, Thieme, Stuttgart 1964.
HEILMEYER, L.: Disturbances in Heme Synthesis. Thomas, Springfield, Illinois 1966.
HEILMEYER, L.: Die Hypochromanämien. In: HEILMEYER, L., BEGEMANN, H. (eds.): Blut und Blutkrankheiten. Springer, Berlin–Heidelberg–New York 1970.
HEILMEYER, L., BEGEMANN, H. (eds.): Blut und Blutkrankheiten. Springer, Berlin–Heidelberg–New York 1970.
HEILMEYER, L., HARWERTH, H. G.: Clinical manifestations of iron deficiency. In: HALLBERG, L., HARWERTH, H. G., VANNOTTI, A. (eds.): Iron Deficiency. Academic Press, London–New York 1970.
HEILMEYER, L., KOCH, H.: Eisenstoffwechseluntersuchungen. Dtsch. Arch. klin. Med. *185*, 89 (1939).
HEILMEYER, L., PLÖTNER, K.: Eisenmangelzustände und ihre Behandlung. Klin. Wschr. 1669 (1936).
HEILMEYER, L., PLÖTNER, K.: Das Serumeisen und die Eisenmangelkrankheit. Fischer, Jena 1937.
HEILMEYER, L., WEISSBECKER, L.: Funktion und Stoffwechsel der Schwermetalle. In: BÜCHNER, F., LETTERER, E., ROULET, E.: Handbuch der allgemeinen Pathologie, Vol. IV/2. Springer, Berlin–Göttingen–Heidelberg 1957.
HEIMPEL, J. (1965), cit. HEILMEYER, 1970.
HEINRICH, G.: Die enterale Eisenresorptionsstörung nach Magenresection. Chirurg *25*, 490 (1954).
HEINRICH, H. C.: Prevalence of iron deficiency in Germany (Discussion). In: HALLBERG, L., HARWERTH, H. G., VANNOTTI, A. (eds.): Iron Deficiency, p. 439. Academic Press, London–New York 1970.
HEINRICH, H. C.: Diagnostic relevance of radioiron absorption and serum ferritin in iron deficiency and iron overload. Med. Welt *30*, 89 (1979).

HEINRICH, H. C., BARTELS, H., HEINISCH, B.: Klin. Wschr. *46*, 199 (1968).
HEISTØ, H., FOSS, O.: Iron deficiency and treatment in blood donors studied by the iron absorption test. Scand. J. clin. Lab. Invest. *10*, 102 (1958).
HENDERSON, F., VIETTI, T. J., BROWN, E. B.: Desferrioxamine in the treatment of acute toxic reaction to ferrous gluconate. J. Amer. med. Ass. *186*, 1139 (1963).
HENDERSON, I. D.: Sideropenic dysphagia. Lancet *1*, 493 (1954).
HERVEY, G. W., MCINTIRE, R. T., WATSON, V.: Low haemoglobin levels in women as revealed by blood donor record. J. Amer. med. Ass. *149*, 1127 (1952).
HEUBNER, W. (1912), cit. SCHULTEN, PRIBILLA: Eisentherapie. In: KEIDERLING, W. (ed.): Eisenstoffwechsel. Thieme, Stuttgart 1959.
HILLMAN, R. S., HENDERSON, P. A.: Control of marrow production by the level of iron supply. J. clin. Invest. *48*, 454 (1969a).
HILLMAN, R. S., HENDERSON, P. A.: Characteristics of iron dextran utilisation in man. Blood *34*, 357 (1969b).
HITTMAIR, A.: Eisenmangel. Med. Klin. *55*, 677 (1960).
HOGAN, G. R., JONES, B.: The relationship of koilonychia and iron deficiency in infants. J. Pediat. *77*, 1054 (1970).
HOGBERG, K. G., LINDVALL, S.: The distribution of parenteral iron haematinics in non-pregnant, pregnant, and lactating rats. Brit. J. Pharmacol. *22*, 275 (1964).
HOGLUND, S., EHN, L., LIEDEN, G.: Studies in iron absorption. VII. Iron deficiency in young men. Acta haemat. (Basel) *44*, 193 (1970).
HOLLÄNDER, L.: Zeitgemäße Sicherungen bei Spendern und Empfängern von Blutkonserven. Z. Präv.-Med. *4*, 201 (1959).
HOLT, J. M., MAYET, F. G. H., WARNER, G. T., CALLENDER, S. T., GUNNING, A. J.: Iron absorption and blood loss in patients with hiatus hernia. Brit. med. J. *3*, 22 (1968).
HOSKING, C. S.: Radiology in the management of acute iron poisoning. Med. J. Aust. *1*, 576 (1969).
HUSER, H. J., RIEBER, E. E., BERMAN, A. R.: Experimental evidence of excess hemolysis in the course of chronic iron deficiency anemia. J. Lab. clin. Med. *69*, 405 (1967).
HUSSAIN, R., WALKER, R. B., LAYRISSE, M., CLARK, P., FINCH, C. A.: Nutritial value of food iron. Amer. J. clin. Nutr. *16*, 464 (1965).
IKKALA, E., SIURALA, M.: Gastric lesion in iron deficiency anaemia. Acta haemat. (Basel) *31*, 313 (1964).
JACOBI, H.: Gestörte Hämoglobinsynthese. In: Handbuch der Kinderheilkunde, Vol. 6, p. 883. Springer, Berlin–Heidelberg–New York 1967.
JACOBS, A.: The buccal mucosa in anaemia. J. clin. Path. *13*, 463 (1960).
JACOBS, A.: Iron deficiency: effects on tissues and enzymes. In: KIEF, H. et al. (eds.): Iron Metabolism and its Disorders. Excerpta Medica, Amsterdam–Oxford, American Elsevier, New York 1975.
JACOBS, A., WORWOOD, M.: Ferritin in serum. New Engl. J. Med. *292*, 951 (1975).
JACOBS, A., WORWOOD, M.: Iron in Biochemistry and Medicine, Vol. 2, p. 219. Academic Press, London–New York 1980.
JACOBS, A., LAWRIE, J. H., ENTWHISTLE, C. C., CAMPBELL, H.: Gastric acid secretion in chronic iron-deficiency anaemia. Lancet *II*, 190 (1966).
JACOBS, A., MILLER, F., WORWOOD, M., BEAMISH, M. R., WARDROP, C. A.: Ferritin in the serum of normal subjects and patients with iron deficiency and iron overload. Brit. Med. J. *4*, 206 (1972).
JACOBS, J., GREENE, H., GENDEL, B. R.: Acute iron intoxication. New Engl. J. Med. *273*, 1124 (1965).
JALILI, M. A., AL-KASSAB, S.: Koilonychia and cystine contents of nails. Lancet *2*, 108 (1959).
JASINSKI, B.: Die Bedeutung der Eisenresorptionsversuche für die Diagnose und Differentialdiagnose der Eisenmangelanämien, insbesondere für die Erkennung der Eisenmangelzustände ohne Anämie. Schweiz. med. Wschr. *79*, 291 (1949).
JASINSKI, B., DIENER, E.: Zur Frage der Häufigkeit des larvierten Eisenmangels bei Frauen, insbesondere bei Graviden und bei Wöchnerinnen. Gynaecologia (Basel) *133*, 293 (1952).
JASINSKI, B., ROTH, O.: Die larvierte Eisenmangelkrankheit. Schwabe, Basel 1954.

JONES, R. F.: The Paterson–Brown–Kelly syndrome, its relationship to iron deficiency and post-cricoid carcinoma. J. Laryng. *75,* Part I, 529; Part II, 544 (1961).
JOSKE, R. A., FINCKH, E. S., WOOD, I. J.: Gastric biopsy; study of 1000 consecutive successful gastric biopsies. Quart. J. Med. *24,* 269 (1955).
JOYNSON, D. H. M., JACOBS, A., WALKER, D. M., DOLBY, A. E.: Lancet *2,* 1058 (1972).
KALININ, V. I.: Tissues of the oral cavity in iron deficiency anaemia. Stomatologica *49,* 20 (1970).
KAMPSCHMIDT, R. F., ADAMS, M. E., GOODWIN, W. L.: Cytochrome C concentration in the tissues of normal and tumour-bearing rats. Arch. Biochem. Biophys. (USA) *82,* 42 (1959).
KAPLAN, E., ZUELZER, W. W., MOURIQUAND, C.: Sideroblasts: A study of stainable nonhemoglobin iron in marrow normoblasts. Blood *9,* 203 (1954).
KAPLAN, O.: Histoire du traitement des anémies par le fer. Thèse, Médicine, Paris 1940.
KARLEFORS, T., NORDEN, A.: Studies on iron-dextran complex. Acta med. scand. Suppl. 342 (1958).
(KASSIRSKI, I. A., ALEKSEEV, G. A.) Кассирский, И. А., Алексеев, Г. А.: Клиническая гематология. Медгиз, Moscow 1962; Медицина, Moscow 1970.
KAUFMANN, O., THIESSEN, R.: Zur Erbbiologie der perniziösen Anämie. Z. klin. Med. *136,* 474 (1939).
KAZNELSON, P., REIMANN, F., WEINER, W.: Achylische Chloranaemie. Klin. Wschr. *1929,* 1071.
KEIDERLING, W., SCHMIDT, H. A., LEE, M.: Radioisotope in Klinik und Forschung, Vol. 2. Urban und Schwarzenberg, München–Berlin 1956.
KERR, D. N. S., DAVIDSON, S.: Gastrointestinal intolerance to oral iron preparations. Lancet *2,* 489 (1958).
KILPATRICK, G. S.: Prevalence of anaemia in the United Kingdom. In: HALLBERG, L., HARWERTH, H. G., VANNOTTI, A. (eds.): Iron Deficiency. Academic Press, London–New York 1970.
KILPATRICK, G. S., HARDISTY, R. M.: The prevalence of anaemia in the community. A survey of a random sample of the population. Brit. med. J. *2,* 778 (1961).
KIRKMAN, H. N., RILEY, H. D. Jr.: Posthemorrhagic anemia and shock in the newborn. Pediatrics *24,* 92 (1959); *24,* 97 (1959).
KNIGHT, R.: The use of Ferastral in iron-deficient, non pregnant Gambian adults. Scand. J. Haemat. *19,* Suppl. 32, 348 (1977).
KNISELEY, H. JR., NOYES, W. D.: Iron deficiency papilledema, thrombocytosis, and transient hemiparesis. Arch. intern. Med. *129,* 483 (1972).
KOHN, R., HEILMEYER, L., CLOTTEN, R.: Reversible pyridoxin-sensible symptomatische sideroachrestische Anämie unter Isoniazidbehandlung bei einer käsigen Lymphknotentuberkulose mit Pleuritis exsudativa. Dtsch. med. Wschr. *87,* 1765 (1962).
KOVÁCS, L.: Die Rolle des Eisenstoffwechsels in der Pathogenese des unspezifischen Fluor im Pubertätsalter. Diss., Budapest 1970.
KRAUSE, U.: Iron deficiency after gastrectomy. In: HALLBERG, L., HARWERTH, H. G., VANNOTTI, A. (eds.): Iron Deficiency. Academic Press, London–New York 1970.
KÜMMERLE, F.: Zur Klärung der Anämie bei Zwerchfellhernien. Dtsch. med. Wschr. *78,* 487 (1953).
LABARDINI, J., PAPAYANNOPOULOU, T., COOK, J. D.: Marrow radioiron kinetics. Haematologia (Budapest) *7,* 301 (1973).
LANGE, H. F., SKJAEGGESTAD, Ö.: Iron deficiency anemia in old age. Acta med. scand. *162,* 321 (1958).
LANZKOWSKY, P.: Radiological features of iron deficiency anemia. Amer. J. Dis. Child. *116,* 16 (1968).
LANZKOWSKY, P., MCKENZIE, D.: Iron deficiency anaemia in Cape colored and African children in Cape Town. S. Afr. med. J. *33,* 21 (1959).
LAYRISSE, M., PAZ, A., BLUMENFELD, N., ROCHE, M.: Hookworm anemia; Iron metabolism and erythrokinetics. Blood *18,* 61 (1961).
LEES, F., ROSENTHAL, F. D.: Gastric mucosal lesions before and after treatment in iron deficiency anaemia. Quart. J. Med. N. S. *27,* 19 (1958).
LEIBEL, R., GREENFIELD, D., POLLITT, E.: Biochemical and behavioural aspects of sideropenia. Brit. J. Haemat. *41,* 145 (1979).
LEIKEN, S., WOSSOUGH, P., MOCHIR-FATEMI, F.: Chelation therapy in acute iron poisoning. J. Pediat. *71,* 425 (1967).
LÉMERY (1683), cit. HEBBERT, 1950.

LENNARTSSON, J. et al.: Characteristics of anaemic women. The population study of women in Göteborg 1968–1969. Scand. J. Haemat. *22,* 17 (1979).
LEONARD, B. J.: Hypochromic anaemia in R. A. F. recruits. Lancet *1,* 899 (1954).
LIE-INJO LUAN ENG: Chronic iron deficiency anaemia with bone changes resembling Cooley's anaemia. Acta haemat. (Basel) *19,* 263 (1958).
LINDVALL, S., ANDERSSON, N. S. E.: Studies on a new intramuscular haematinic iron-sorbitol. Brit. J. Pharmacol. *17,* 358 (1961).
LIPSCHITZ, D. A., COOK, J. D., FINCH, C. A.: A clinical evaluation of serum ferritin as an index of iron stores. New Engl. J. Med. *290,* 1213 (1974).
LORIA, A., SANCHEZ-MEDAL, L., LIOKER, R., DE RODRIGUEZ, E., LABORDINI, J.: Red-cell life span in iron deficiency anaemia. Brit. J. Haemat. *13,* 294 (1967).
LUCAS, J. E., HAGEDORN, A. B.: Intravenous use of high molecular ferric carbohydrate compound in treatment of hypochromic anemia. Blood *7,* 358 (1952).
LUHBY, A.: Megaloblastic anemia of infancy. III. Clinical considerations and analysis. J. Pediat. *54,* 617 (1959).
LUND, C. J.: Studies on the iron deficiency anemia of pregnancy, including plasma volume, total hemoglobin, erythrocyte protoporphyrin, treated and untreated normal and anemic patients. Amer. J. Obstet. Gynec. *62,* 947 (1951).
LUNDIN, P. M.: The carcinogenic action of complex iron preparations. Brit. J. Cancer *15,* 838 (1961).
MACDOUGALL, L. G., ANDERSON, R., MCNAB, G. M., KATZ, J.: J. Pediat. *86,* 833 (1975).
MACKAY, H. M. M., DOBBS, R. H., WILLS, L., BINGHAM, K.: Anaemia in women and children on war-time diets. Lancet *2,* 32 (1942).
MANOLIDIS, L., DANILIDIS, J., KEKES, G.: Dysphagia in sideropenia. Z. Laryng. Rhinol. *51,* 437 (1972).
MARTIN, L. E., BATES, C. M., BERESFORD, C. R., DONALDSON, J. D., MCDONALD, F. F., DUNLOP, D., SHEARD, P., LONDON, E., TWIGG, G. D.: The pharmacology of an iron-dextran intramuscular haematinic. Brit. J. Pharmacol. *10,* 375 (1955).
MASUYA, T.: Iron metabolism in tropical medicine. Jap. J. Trop. Med. Hyg. *3,* 205 (1975).
MAZUR, A., SHORR, E.: Hepato-renal factors in circulatory homeostasis. IX. The identification of the hepatic vasodepressor substance, VDM, with ferritin. J. biol. Chem. *176,* 771 (1948).
MCCANCE, R. A., WIDDOWSON, E. M.: Variations du fer suivant l'âge et le sexe. Journées Thérapeutiques de Paris, p. 93, Doin 1946.
MCCURDY, P. R.: Parenteral iron therapy. II. A new iron-sorbitol citric acid complex for intramuscular injection. Ann. intern. Med. *61,* 1053 (1964).
MCCURDY, P. R.: Oral and parenteral iron therapy. A comparison. J. Amer. med. Ass. *191,* 859 (1965).
MCCURDY, P. R.: Parenteral iron therapy. In: HALLBERG, L., HARWERTH, H. G., VANNOTTI, A. (eds.): Iron Deficiency. Academic Press, London–New York 1970.
MCDONALD, R., MARSHALL, S. R.: The value of iron therapy in pica. Pediatrics *34,* 558 (1964).
MCENERY, J. T., MACK, R. B.: Use of desferrioxamine in a case of acute ferrous sulfate poisoning. Illinois med. J. *126,* 550 (1964).
MCFARLANE, D. B., PINKERTON, P. H., DAGG, J. H., GOLDBERG, A.: Incidence of iron deficiency, with and without anaemia, in women in general practice. Brit. J. Haemat. *13,* 790 (1966).
MCGUIGAN, J. E., VOLWILER, W.: Celiac sprue, malabsorption of iron in the absence of steatorrhoea. Gastroenterology *47,* 636 (1964).
MENON, K. (1958), cit. BOTHWELL, FINCH, 1962.
MENZEL: Einseitige Choanalatresie mit Ozaena der nicht verschlossenen Seite. Mschr. Ohrenheilk. *66,* 1403 (1932).
MEULENGRACHT, E.: Simple achylic anaemia. Acta med. scand. *78,* 387 (1932).
MEYERS, S. G., PRICE, A. H., MACK, H. C., FOSTER, L. J., SHARP, E. A.: Chronic hypochromic anemia in women. Ann. intern. Med. *11,* 1590 (1938).
MIDDLETON, E. J., NAGY, E., MORRISON, A. B.: Studies on the absorption of orally administered iron from sustained-release preparations. New Engl. J. Med. *274,* 136 (1966).
MIEHLKE, A., DIEPEN, R.: Ozaena vergesellschaftet mit hypophysär-hypothalamischen Störungen. Arch. Ohr.- Nas.- u. Kehlk.-Heilk. *160,* 178 (1951).

MILES, L. E. M., LIPSCHITZ, D. A., BIEBER, C. P., COOK, J. D.: Measurement of serum ferritin by a 2-site immunoradiometric assay. Ann. Biochem. *61*, 209 (1964).

MILES, L. E. M., LIPSCHITZ, D. A., BIEBER, C. P., COOK, J. D.: Ann. Biochem. *61*, 209 (1974).

MILLER, V., RUNCIE, J., THOMSON, T. J.: Recurrent iron deficiency anaemia treated by iron dextran. J. Therap. *3*, 11 (1966).

MINNICH, V., OKÇUOGLU, A., TARCON, Y., ARKASOY, A., CIU, S., YÖRÜKOGLU, Y., RENDA, F., DEMIRAG, B.: Pica in Turkey. II. Effect of clay on iron absorption. Amer. J. clin. Nutr. *21*, 78 (1968).

MONCRIEFF, A., KOUMIDES, O., CLAYTON, B., PATRICK, A., REINWICK, A., ROBERTS, G.: Lead poisoning in children. Arch. Dis. Childh. *39*, 1 (1964).

MOORE, C. V.: Importance of nutritional factors in the pathogenesis of iron deficiency anemias. Amer. J. clin. Nutr. *3*, 3 (1955).

MOORE, C. V.: The importance of nutritional factors in the pathogenesis of iron deficiency anemia. Scand. J. clin. Lab. Invest. *9*, 292 (1957).

MOORE, C. V., ARROWSMITH, W. R., QUILLIGAN, J. J., READ, J. T.: Studies on iron transportation and metabolism. I. Chemical methods and normal values for plasma iron and "easily split-off" blood iron. J. clin. Invest. *16*, 613 (1937).

MOORE, C. V., ARROWSMITH, W. R., WELCH, J., MINNICH, V.: Studies on iron transportation and metabolism. IV. Observations on the absorption of iron from the gastrointestinal tract. J. clin. Invest. *18*, 553 (1939).

MOORE, C. V., DUBACH, R., MINNICH, V., ROBERTS, H. K.: Absorption of ferrous and ferric radioactive iron by human subjects and by dogs. J. clin. Invest. *23*, 755 (1944).

MORROW, J. J., URATA, G., GOLDBERG, A.: The effect of lead and ferrous and ferric iron on delta-aminolaevulinic acid synthetase. Clin. Sci. *37*, 533 (1969).

MOSELEY, J. E.: Skull changes in chronic iron deficiency anemia. Amer. J. Roentgenol. *85*, 649 (1961).

MOUTIER, F.: La dysphagie sidéropénique: du syndrome de Kelly-Paterson (ex-syndrome du Plummer-Vinson) au cancer hypopharyngé. Arch. Mal. Appar. dig. Suppl. *5*, 53 (1951).

NAEGELI, O.: Zur Frage der Eisenwirkung bei Anämie, speziell bei Chlorose. Schweiz. med. Wschr. *1*, 661 (1920).

NAEGELI, O.: Blutkrankheiten und Blutdiagnostik, 5th ed. Springer, Berlin 1931.

NAIMAN, J. L., OSKI, F. A., DIAMOND, L. K., VAWTER, G. F., SCHWACHMAN, H.: The gastrointestinal effects of iron-deficiency anemia. Paediatrics *33*, 83 (1964).

NEWCOMBE, R.: Precautions in the intravenous use of iron dextran. Postgrad. med. J. *43*, 372 (1967).

NISSIM, J. A.: Intravenous administration of iron. Lancet *2*, 49 (1947).

NISSIM, J. A., ROBSON, J. M.: Preparation and standardization of saccharated iron oxide for intravenous administration. Lancet *1*, 686 (1949).

NORRBY, A., SÖLVELL, L.: Iron absorption studies in iron deficiency. Scand. J. Haemat. Suppl. *5*, 125 (1974).

OERTEL, J., GERHARTZ, H.: Die Ferritinkonzentration im Serum bei verschiedenen Typen der Eisenmangelanämie. Dtsch. med. Wschr. *102*, 1147 (1977).

OERTEL, J., BOMBIK, B. M., STEPHAN, M., GERHARTZ, H.: Ferritin in bone marrow and serum in iron deficiency and iron overload. Blut *37*, 113 (1978).

OHIRA, Y., EDGERTON, V. R., GARDNER, G. W., SENEWIRATE, B., BERNARD, R. J., SIMPSON, D. R.: Work capacity, heart rate and blood lactate responses to iron treatment. Brit. J. Haemat. *41*, 365 (1979).

OTT, W.: Die Shockgefährdung durch larvierten Eisenmangel. Helv. chim. Acta *22*, 183 (1955).

OWEN, G. M. et al.: Preschool children in the United States. Who has iron deficiency? J. Pediatr. *79*, 563 (1971).

PARKES-WEBER, F.: Case of achlorhydric anaemia in a male followed up for 20 years. Brit. med. J. *2*, 1529 (1954).

PLÖTNER, K.: Renale Eisenausscheidung bei Eisenmangelanämien nach intravenöser Eisenbelastung. 5. Kongreß der Europäischen Gesellschaft für Hämatologie, Freiburg/Br. 1955, Springer, Berlin–Göttingen–Heidelberg 1956.

POLLACK, S., GEORGE, J. N., REBA, R. C., KAUFMANN, R. M., CROSBY, W. H.: The absorption of nonferrous metals in iron deficiency. J. clin. Invest. *4*, 1470 (1965).

POLLYCOVE, M.: Iron kinetics. In: WALLERSTEIN, R. O., METTIER, S. R. (eds.): Iron in Clinical Medicine, pp. 43–57. University of California Press, Berkeley 1958.
POLLYCOVE, M.: Ferrokinetics: Techniques. In: KEIDERLING, W. (eds.): Eisenstoffwechsel, p. 20. Thieme, Stuttgart 1959.
PRASAD, A. S., MIALE, A., FARID, Z., SANDSTEAD, H. H., SCHULER, A. R., DARBY, W. J.: Biochemical studies on dwarfism, hypogonadism and anemia. Arch. int. Med. *111*, 407 (1963).
PRINGLE, A., GOLDBERG, A., MACDONALD, E., JOHNSTON, S.: ^{59}Fe iron sorbitol citric-acid complex in iron-deficiency anaemia. Lancet *2*, 749 (1962).
PRITCHARD, J. A.: The response to iron in iron deficiency. Anemia due to gynecologic disease. J. Amer. med. Ass. *175*, 478 (1961).
RACKER, F., KRIMSKY, I.: Inhibition of coupled phosphorylation in brain homogenates by ferrous sulphate. J. biol. Chem. *173*, 519 (1948).
RAMALINGASWAMI, V., PATWARDHAN, V. N.: Diet and health of South Indian plantation labour. Indian J. med. Res. *37*, 51 (1949).
RASCH, C. A., COTTON, R. R., GRIGGS, R. C., HARRIS, J. W.: The survival of autotransfused Cr^{51} labeled erythrocytes in infants with severe iron deficiency anemia. J. clin. Med. *52*, 938 (1958).
RATH, C. R., FINCH, C. A.: Sternal marrow hemosiderin; method for determination of available iron stores in man. J. Lab. clin. Med. *33*, 81 (1948).
RATH, C. R., FINCH, C. A.: Measurement of iron binding capacity of serum in man. J. clin. Invest. *28*, 79 (1949).
RAWSON, A., ROSENTHAL, F. D.: The mucosa of the stomach and small intestine in iron deficiency. Lancet *1*, 730 (1960).
REIMANN, F.: Das Eisenmangelfieber. Acta haemat. (Basel) *2*, 247 (1949).
REIMANN, F.: Wachstumsanomalien und Mißbildungen bei Eisenmangelzuständen (Asiderosen). 5. Kongreß der Europäischen Gesellschaft für Hämatologie, Freiburg/Br. 1955, p. 546. Springer, Berlin–Göttingen–Heidelberg 1956.
REIMANN, F., ARKUN, N. S.: Die osmotische Resistenz der Erythrozyten bei den Asiderosen. Z. klin. Med. *151*, 559 (1954).
REIMANN, F., ARKUN, N. S.: Die Einwirkung der Eisenbehandlung auf das Verhalten der osmotischen Resistenz der Erythrocyten bei den eisenempfindlichen chronischen Chloranaemien (Asiderosen). Z. klin. Med. *153*, 589 (1956).
REIMANN, F., FRITSCH, F.: Vergleichende Untersuchungen zur therapeutischen Wirksamkeit der Eisenverbindungen bei den secundären Anämien. Z. klin. Med. *115*, 13 (1930).
REIMANN, F., FRITSCH, F.: Experimentelle und klinische Untersuchungen über die Wirkung des Ferrum reductum. Z. klin. Med. *117*, 304 (1931).
REIMANN, F., FRITSCH, F., SCHICK, K.: Eisenbilanzversuche bei Gesunden und bei Anämischen. II. Untersuchungen über das Wesen eisenempfindlicher Anämien ("Asiderosen") und der therapeutischen Wirkung des Eisens bei diesen Anämien. Z. klin. Med. *131*, 1 (1937).
REINUS, A.: Über den Tonus des vegetativen Nervensystems und über das Sexualhormon bei Ozaena. Mschr. Ohrenheilk. *63*, 523 (1929).
REIZENSTEIN, P.: Metabolic studies of iron-poly (sorbitol-gluconic acid) complex, Ferastral. Scand. J. Haemat. *19*, Suppl. 32, 163 (1977).
REYNOLDS, R. D., BINDER, H. J., MILLER, M. B., CHANG, W. W. Y., HORAN, S.: Pagophagia and iron deficiency anemia. Ann. intern. Med. *69*, 435 (1968).
RICHMOND, H. G.: Induction of sarcoma in the rat by iron-dextran complex. Brit. med. J. *1*, 947 (1959).
ROBINSON, C. E. G., BELL, D. N., STURDY, J. H.: Hazards of iron-dextran. Brit. med. J. *1*, 744 (1961).
ROBINSON, S. H.: Increased formation of early-labeled bilirubin in rats with iron deficiency anemia; evidence for ineffective erythropoiesis. Blood *33*, 909 (1969).
(ROSSOLIMO, G. I.) Россолимо, Г. И. Амиотактическая дисфагия как особый вид расстройства глоатния. Клин. Мед. *1900*, 3, 1.
ROTHLIN, E., UNDRITZ, E.: Experimenteller Beitrag zur Kenntnis der larvierten ferripriven Anämie (Sideropenie ohne Anämie). Schweiz. med. Wschr. *77*, 58 (1947).

RUSTUNG, E.: Studies on serum iron. Acta derm.-venerol. (Stockh.) *29,* 1 (1946).
RYBO, G.: Menstrual loss of iron. In: HALLBERG, L., HARWERTH, H. G., VANNOTTI, A. (eds.): Iron Deficiency. Academic Press, London–New York 1970.
RYBO, G.: Physiological causes of iron deficiency in women: menstruation and pregnancy. In: CALLENDER, S. T. (ed.): Iron Deficiency and Iron Overload (Clinics in Haematology). Vol. 2/2. Saunders, London–Philadelphia–Toronto 1973.
RYSS, E. S.: Dynamic studies of functional and morphological changes of the gastric mucous membrane in hyposiderotic anemia. Klin. Med. (Mosk.) *49,* 33 (1971).
SACREZ, R., LAUSECKER, CH., WARTER, P.: A propos d'un cas de dysphagie grave avec sidéropénie évoluant depuis la naissance. Arch. franç. Pédiat. *11,* 954 (1954).
SADDI, R., SCHAPIRA, G.: Iron requirements during growth. In: HALLBERG, L., HARWERTH, H. G., VANNOTTI, A. (eds.): Iron Deficiency. Academic Press, London–New York 1970.
SALMON, H. A.: The cytochrome C content of the heart, kidney, liver and skeletal muscle of iron-deficient rats. J. Physiol. (Lond.) *164,* 17 (1962).
SASS, M. D., SPEAR, P. W.: Red cell transaminase levels in anemia. III. Acute and chronic blood loss. J. Lab. clin. Med. *58,* 586 (1961).
SAUNDERS, J. B.: Iron and the development of medicine. In: WALLERSTEIN, R. O., METTIER, S. R. (eds.): Iron in Clinical Medicine. University of California Press, Berkeley–Los Angeles 1958.
SCHÄFER, K.: Der Eisenstoffwechsel des wachsenden Organismus. Ergebn. Inn. Med. Kinderheilk. N. F. *4,* 706 (1953).
SCHLOESSER, L. L., KIPP, M. A., WENZEL, F. J.: Thrombocytosis in iron deficiency anemia. J. Lab. clin. Med. *66,* 107 (1965).
SCHMIDT, M. B.: Der Einfluß eisenarmer und eisenreicher Nahrung auf Blut und Körper. Fischer, Jena 1928.
SCHNEIDER, W., BÖWING, G., BENNHOLD, I.: Untersuchungen des Eisenhaushaltes von Blutspendern mittels peroraler Eisenbelastungen. Medizinische 1305 (1958).
SCHULMAN, J.: Iron requirements in infancy. J. Amer. med. Ass. *175,* 118 (1961).
SCHULTEN, H.: Über die essentielle hypochrome Anämie und verwandte Krankheitsbilder. Ergebn. Inn. Med. Kinderheilk. *46,* 236 (1934).
SCHULZ, J., SMITH, N.: A quantitative study of the absorption of food iron in infants and children. Amer. J. Dis. Child. *95,* 109 (1958).
SCOTT, D. E., PRITCHARD, J. A.: Iron deficiency in healthy young college women. J. Amer. med. Ass. *199,* 897 (1967).
SCOTT, D. E., PRITCHARD, J. A., SALTIN, A. E., HUMPHREYS, J. M.: Iron deficiency during pregnancy. In: HALLBERG, L., HARWERTH, H. G., VANNOTTI, A. (eds.): Iron Deficiency, p. 491. Academic Press, London–New York 1970.
SCOTT, E. M., WRIGHT, R. C., HANARI, B. T.: Anemia in Alaskan Eskimos. Brit. med. J. *2,* 1083 (1949).
SCOTT, J. M., GOVAN, A. D. T.: Anaemia of pregnancy treated with intramuscular iron. Brit. med. J. *2,* 1257 (1954).
SEFTEL, H. C.: Reactions to intravenous iron dextran. Brit. med. J. *1,* 657 (1965).
SEIBOLD, M.: Prevalence of iron deficiency in Germany. In: HALLBERG, L., HARWERTH, H. G., VANNOTTI, A. (eds.): Iron Deficiency. Academic Press, London–New York 1970.
SEIBOLD, M. M., HANNAPPEL, A., FLORIAN, H. J., SCHMID, E.: Die Verbreitung von Eisenmangelzuständen bei der weiblichen Bevölkerung. Münch. med. Wschr. *107,* 816 (1965).
SHAHIDI, N. T., DIAMOND, L. K.: Skull changes in infants with chronic iron deficiency anaemia. New Engl. J. Med. *262,* 137 (1960).
SHAPIRO, N., BARBEZAT, G. O.: A case of acute iron poisoning treated with desferrioxamine B. S. Afr. med. J. *38,* 461 (1964).
SIMONOVITS, I. (1968–1970): Personal communications.
SIMONOVITS, I.: Az anaemia epidemiológiája (The epidemiology of anemia). Orvosképzés *45,* 212 (1970).
SIMONOVITS, I., LÉPES, P., SIMON, T., BUDAI, B., BODNÁR, L., SZÁSZ, D.: Epidemiológiai vizsgálatok anaemiára (Studies on the epidemiology of anemia). Transfusio *4,* 4 (1970).

SIMONOVITS, I. et al.: Szűrővizsgálatok anaemiára és vashiányra Magyarországon (Studies on anemia and iron deficiency in Hungary). Transfusio *10,* 39 (1977).
SIMPSON, R., BLUNT, A.: Acute ferrous sulphate poisoning treated with edathamil calcium-sodium. Lancet *2,* 1120 (1960).
SISSON, T., LUND, C.: Influence of maternal iron deficiency on the newborn. Amer. J. clin. Nutr. *6,* 376 (1958).
SKOUGE, E.: Klinische und experimentelle Studien über das Serumeisen. Dybwad, Oslo 1939.
SLACK, H. G. B., WILKINSON, J. F.: Intravenous treatment of anaemia with iron-sucrose preparation. Lancet *1,* 11, 163 (1949).
SMITH, N. J., HUNTER, R. E.: Iron requirements during growth. In: HALLBERG, L., HARWERTH, H. G., VANNOTTI, A. (eds.): Iron Deficiency. Academic Press, London–New York 1970.
SMITH, N. J., ROSELLO, S., SAY, M. B., YEYS, K.: Iron storage in the first five years of life. Pediatrics *16,* 166 (1955).
SÖLVELL, L.: Oral iron therapy — side effects. In: HALLBERG, L., HARWERTH, H. G., VANNOTTI, A. (eds.): Iron Deficiency. Academic Press, London–New York 1970.
SOMERS, K.: Acute reversible heart failure in severe iron-deficiency anemia associated with hookworm infestation in Uganda Africans. Circulation *19,* 672 (1959).
SORBIE, J., OLATUNBOSUN, D., CORBETT, W. E. N., VALBERG, L. S.: Cobalt excretion test for the assessment of body iron stores. Canad. med. Ass. J. *104,* 777 (1971).
SORBIE, J., VALBERG, L. S., CORBETT, W. E. N., LUDWIG, J.: Serum ferritin, cobalt excretion and body iron status. Canad. med. Ass. J. *112,* 1173 (1975).
SRIKANTIA, S. G., PRASAD, J. S., BHASKARAM, C., KRISNAMACHARI, K. A. V. R.: Lancet *1,* 1307 (1976).
STARKENSTEIN, E.: Beiträge zur Pharmakologie des Eisens. Arch. exp. Path. Pharmak. *118,* 131 (1926).
STEVENS, A. R.: In: WALLERSTEIN, R. O., METTIER, S. R. (eds.): Iron in Clinical Medicine, p. 144. University of California Press, Berkeley–Los Angeles 1958.
STONE, W. D.: Gastric secretory response to iron therapy. Gut *9,* 99 (1968).
STOTT, G.: Anaemia in Mauritius. Bull. Wld Hlth Org. *23,* 781 (1960).
STURGEON, P.: Studies of iron requirements in infants and children. I. Normal values for serum iron copper and free erythrocyte protoporphyrin. Pediatrics *13,* 107 (1954).
STURGEON, P.: Iron metabolism. A review with special consideration of iron requirements during normal infancy. Pediatrics *18,* 267 (1956).
STURGEON, P.: Studies of iron requirements in infants. III. Influence of supplemental iron during normal pregnancy on mother and infant. A. The mother. Brit. J. Haemat. *5,* 31 (1959a).
STURGEON, P.: Studies of iron requirements in infants. III. Influence of supplemental iron during normal pregnancy on mother and infant. B. The infant. Brit. J. Haemat. *5,* 45 (1959b).
SUMMERSKILL, W. H. J., ALVAREZ, A. S.: Salicylate anaemia. Lancet *2,* 925 (1958).
SURJÁN, L.: Az ozaena aetiológiája és kezelése (Etiology and therapy of ozaena). Diss., Budapest 1956.
SUZMAN, M. M.: Syndrome of anemia, glossitis and dysphagia. Arch. intern. Med. *51,* 1 (1933).
SZENES, H.: Zum Plummer-Vinsonschen Syndrom. Mschr. Ohrenheilk. *83,* 206 (1949).
TAFT, L. I., HUGHES, A., WOOD, I. J.: Tongue biopsy. A technique using a rigid suction tube. Lancet *II,* 69 (1958).
TAYMOR, M. L., STURGIS, S. H., YAHIA, C.: The etiologic role of chronic iron deficiency in production of menorrhagia. J. Amer. med. Ass. *187,* 323 (1964).
TEMPERLEY, I. J., SHARP, A. A.: The life span of erythrocytes in iron deficiency anaemia. J. clin. Path. *15,* 346 (1962).
TERRIER, B.: L'atrophie caractéristique de la muqueuse buccale au début de l'anémie ferriprive. Praxis (Bern) *44,* 668 (1955).
TIBBLIN, E. et al.: Haemoglobin concentration and peripheral blood cell counts in women. The population study of women in Göteborg 1968–1969. Scand. J. Haemat. *22,* 5 (1979).
TROUSSEAU, A.: De la chlorose vraie et des fausses chloroses. Clinique Médicale de l'Hôtel-Dieu *3,* 533 (1882).
TUNESSEN, W. W., SMITH, C., OSKI, F. A.: Betaninuria in iron deficiency anemia. Amer. J. Dis. Child. *117,* 424 (1969).

VALBERG, L. S., LUDWIG, F., OLATONBOSUN, D.: Alteration in cobalt absorption in patients with disorders of iron metabolism. Gastroenterology *56,* 241 (1969).

VALBERG, L. S., SORBIE, J., CORBETT, W. E. N., LUDWIG, F.: Cobalt test for the detection of iron deficiency anemia. Ann intern. Med. *77,* 181 (1972).

VALBERG, L. S., SORBIE, J., LUDWIG, F., PELLETIER, D.: Can. Med. Ass. J. *114,* 417 (1976).

VAN GEERTRUYDEN, J.: Anémies après gastrectomie subtotale. Acta clin. belg. *13,* 171 (1958).

VANNOTTI, A.: Funktionelle Beziehungen zwischen Hämoglobin- und Cytochrom-C-Stoffwechsel. Schweiz. med. Wschr. *79,* 261 (1949).

VANNOTTI, A.: Die Eisenfunktion der Zellhämine. In: KEIDERLING, W. (ed.): Eisenstoffwechsel. Thieme, Stuttgart 1959.

VELLAR, O. D.: Prevalence of iron deficiency in Norway. In: HALLBERG, L., HARWERTH, H. G., VANNOTTI, A. (eds.): Iron Deficiency. Academic Press, London–New York 1970.

VENTURA, S., MARMONI, U., PURICELLI, G., SUARDI, L., MATIOLI, G.: Studio analitico del ricambio del ferro in condizioni normali e patologice. III. Haematologica (Pavia) *42,* 563 (1957).

VERLOOP, M. C. et al.: Comparison of the "iron absorption test" with the determination of the iron binding capacity of serum in the diagnosis of iron deficiency. Brit. J. Haemat. *4,* 70 (1958).

VERLOOP, M. C., BLOKZUIS, E. W. M., BOS, C. C.: Causes of the "physiological" anemia of pregnancy. Acta haemat. (Basel) *22,* 158 (1959).

VERLOOP, M. C., LIEM, K. S., DE WIJN, J. F.: Iron depletion and anaemia due to iron deficiency. In: HALLBERG, L., HARWERTH, H. G., VANNOTTI, A. (eds.): Iron Deficiency. Academic Press, London–New York 1970.

VINSON, P. P.: Hysterical dysphagia. Minn. Med. *5,* 107 (1922).

VINSON, P. P. et al.: Hysterical dysphagia. Virginia med. Mth. *64,* 142 (1937).

VITERI, F. E., TORUN, B.: Clin. Haemat. *3,* 609 (1974).

WALDENSTRÖM, J.: Iron and epithelium. Some clinical observations. Acta med. scand. Suppl. *90,* 380 (1938).

WALDENSTRÖM, J.: Incidence of "iron deficiency" (sideropenia) in some rural and urban populations. Acta med. scand. Suppl. *170,* 252 (1946).

WALDENSTRÖM, J.: Eisenmangelkrankheit. In: KEIDERLING, W. (ed.): Eisenstoffwechsel. Thieme, Stuttgart 1959.

WALDENSTRÖM, J.: Die Eisenmangelzustände und ihre Behandlung. 70. Tagung der Deutschen Gesellschaft für Innere Medizin. Wiesbaden 1964.

WALDENSTRÖM, J., KJELLBERG, S. R.: Roentgenological diagnosis of sideropenic dysphagia (Plummer-Vinson's syndrome). Acta radiol. (Stockh.) *20,* 615, 618 (1939).

WALKER, A. R. P., ARVIDSSON, U. B.: Iron "overload" in the South African Bantu. Trans. roy. Soc. trop. Med. Hyg. *47,* 536 (1953).

WALLINSTEN, S.: Anemi och sideropeni efter ventrikelresektioner. Nord. med. *50,* 1601 (1953).

WATERS, W. E., WITHEY, J. L., KILPATRICK, G. S., WOOD, P. H. N., ABERNATHY, M.: Ten year haematological follow-up; mortality and haematological changes. Brit. med. J. *4,* 761 (1969).

WATSON, W. C., LUKE, R. G., INALL, J. A.: Beeturia; its incidence and a clue to its mechanism. Brit. med. J. *2,* 971 (1963).

WEINFELD, A.: Iron stores. In: HALLBERG, L., HARWERTH, H. G., VANNOTTI, A. (eds.): Iron Deficiency, p. 359. Academic Press, London–New York 1970.

WETHERLEY-MEIN, G., BUCHANAN, J. G., GLASS, U. H., PEARCE, L. C.: Metabolism of ^{59}Fe-sorbitol complex in man. Brit. med. J. *1,* 1796 (1962).

WHITTEN, C. F., GIBSON, G. W., GOOD, M. H., GOODWIN, J. F., BROUGH, A. J.: Studies in acute iron poisoning. 1. Desferrioxamine in the treatment of acute iron poisoning; clinical observations, experimental studies and theoretical considerations. Pediatrics *36,* 322 (1965).

WILL, G.: The absorption, distribution and utilization of intramuscularly administered iron-dextran: a radioisotope study. Brit. J. Haemat. *14,* 395 (1968).

WILL, G., GRODEN, B. M.: The treatment or iron deficiency anaemia by iron-dextran infusion: a radioisotope study. Brit. J. Haemat. *14,* 61 (1968).

WILLIAMS, J.: The effect of ascorbic acid on iron absorption in postgastrectomy anaemia and achlorhydria. Clin. Sci. *18*, 521 (1959).
WINTROBE, M. M.: Clinical Hematology. Kimpton, London 1967.
WINTROBE, M. M.: Clinical Hematology. Lea and Febiger, Philadelphia 1967, 1974.
WINTROBE, M. M., BEEBE, R. T.: Idiopathic hypochromic anemia. Medicine *12*, 187 (1933).
WITTS, L. J.: Simple achlorhydric anaemia. Guy's Hosp. Rep. *80*, 253 (1930).
WÖHLER, F.: Diagnosis of iron storage diseases with desferrioxamine (Desferal test). Acta haemat. (Basel) *32*, 321 (1964).
WOOD, J. K., MILNER, P. F. A., PATHAK, V. N.: The metabolism of iron-dextran given as a total dose infusion to iron deficient Jamaican subjects. Brit. J. Haemat. *14*, 119 (1968).
WOODRUFF, C., BRIDGEFORTH, F.: Relationship between the hemogram of the infant to that of the mother during pregnancy. Pediatrics *12*, 681 (1953).
World Health Organization Technical Report Series No. 182. Iron Deficiency Anemia. Report of a study group. Geneva 1959.
YETGIN, S., ALTAY, C., CILIV, G., LALELI, Y.: Acta Haemat. *61*, 10 (1979).

CHAPTER 19

ANEMIA OF INFECTION

Infections, particularly chronic infections, are frequently associated with anemia. Bacterial infections or conditions associated with severe sepsis may produce a fairly severe anemia of rapid onset, but the anemia of chronic disorders is usually not severe or progressive (Cartwright, 1966; Cartwright and Lee, 1971). In general, it is normochromic or slightly hypochromic, the PCV is seldom less than 30%, and the mean hemoglobin concentration of the red cells usually ranges between 28 and 32%. Hypochromia of the degree seen in iron-deficiency disease is unusual. Anisocytosis and poikilocytosis are not conspicuous, and signs of bone marrow regeneration such as polychromatophilia or the presence of nucleated red cells in the blood are absent. The number of reticulocytes is diminished or normal (Cartwright et al., 1946, 1951, 1954). An increased reticulocyte count is exceptional.

In the bone marrow there may be some indication of inhibition of red cell production (Cartwright and Wintrobe, 1952; Jeffrey, 1953), and the sideroblast ratio is lower than normal (5–20%), while there is an increased deposition of iron in the reticuloendothelial cells (Cartwright, 1966) (see Fig. 8/12). Electron microscopic studies show that the erythroblasts and reticulocytes may take up a relatively large amount of iron by rhopheocytosis, but the ferritin is finely dispersed and does not occur in clusters (Bessis and Breton-Gorius, 1962, 1964).

In response to infection there is a reduction in both the plasma iron concentration (Heilmeyer and Plötner, 1937; Thoenes, 1941; Hirvonen, 1941; Heilmeyer et al., 1941; Vahlquist, 1941; Brøchner-Mortensen and Stein, 1942; Cartwright et al., 1946; Cartwright and Lee, 1971) and the plasma iron-binding capacity (Laurell, 1947, 1952; Rath and Finch, 1949; Hagberg, 1953; Cartwright and Lee, 1971) (Fig. 19/1), and the saturation of transferrin is usually lower than normal (Bainton and Finch, 1964; Cartwright, 1966; Cartwright and Lee, 1971). A similar picture may be seen in association with malignancy, following major operations, and in severe allergic processes (Schäfer, 1948; Heilmeyer and Plötner, 1937; Heilmeyer et al., 1941; Keiderling and Scharpf, 1953), and it can also be produced experimentally by vaccines, toxins, and protein breakdown products (Heilmeyer and Plötner, 1937; Heilmeyer et al., 1941; Keiderling et al., 1955, 1956; Schäfer, 1942; Elin et al., 1977; Torrance et al., 1978). Stressor agents such as ACTH and glucocorticoids may also produce similar changes (Cartwright and Wintrobe, 1949, 1952).

The changes are not influenced by the administration of either oral or parenteral iron (Heilmeyer and Plötner, 1937; Skouge, 1939) or by the intravenous infusion of $beta_1$-globulin (Cartwright and Wintrobe, 1949, 1952). The values return to normal

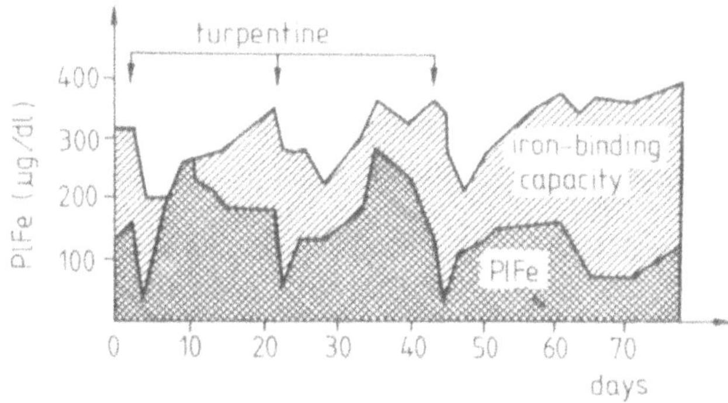

Fig. 19/1. The effect of turpentine abscess upon the plasma iron concentration (lower curve) and total iron-binding capacity (upper curve). After the turpentine injections first the iron level and then the transferrin concentration decline (after Cartwright, G. E. and Wintrobe, M. M.: J. clin. Invest. *28*, 86, 1949)

spontaneously, however, once the underlying disorder is remedied (Keiderling and Schmidt, 1959). In acute infections of short duration the iron metabolism regains its balance before anemia can become manifest.

Ferrokinetic investigations have shown an increased rate of outflow of iron from the plasma (Keiderling et al., 1955, 1956, 1959; Cartwright et al., 1951; Freireich et al., 1957; Haurani et al., 1963, 1965; Roberts et al., 1963), and the T/2 is reduced to 10–60 minutes (normal 70–130 minutes). Keiderling and Schmidt (1959) found the plasma iron transport rate to be increased two to three times; others have found it to be normal or only slightly increased (Wintrobe, 1967). The red cell iron turnover is unchanged or increased one- or twofold, but this is not sufficient to prevent the occurrence of anemia. Cartwright and Lee (1971) have emphasized the three factors that are concerned in the pathogenesis of the anemia. First, an impairment of red cell survival, second, a reduced marrow response, and third, the sequestration of iron in the RE cells and impaired flow of iron to the bone marrow.

The red cell life-span is only moderately diminished, as shown by the Ashby and ^{51}Cr methods and by the use of the cohort labeling method with ^{59}Fe (Cartwright,

1966; Hollingsworth and Hollingsworth, 1955; Keiderling et al., 1955, 1956; Dinant and de Maat, 1978). The reduction in red cell survival appears to be due to extracorpuscular factors, as indicated by observations in patients with rheumatoid arthritis in whom the survival of normal cells was reduced, whereas red cells taken from the patients with rheumatoid arthritis survived normally in a healthy human subject (Roy et al., 1955; Freireich et al., 1951). The nature of the extracorpuscular factors has not been determined, but presumably the red cells are damaged either directly by pathogens or indirectly by toxic substances produced in the course of the disease (Marti, 1968).

The anemia of infection fails to produce the normal erythropoietin response, as has been shown by Ward et al. (1971) in both man and animals. However, cobalt and anoxia, which stimulate erythropoietin production, will cause a hemopoietic response in the presence of infection, as will the injection of purified erythropoietin. Thus the marrow is capable of a response but the stimulus to production of erythropoietin appears to be lacking (Wintrobe et al., 1947, 1948, 1958; Gutnisky and van Dyke, 1963; Cartwright, 1966).

Central to the type of anemia developed is the impairment of flow of iron from the reticuloendothelial cells, the so-called RE iron block (Freireich et al., 1957; Weinstein, 1959; Kampschmidt and Schultz, 1961, 1965; Haurani et al., 1963, 1965; Umeda, 1965; Konijn and Hershko, 1977).

Transferrin, ferritin, and hemosiderin iron are presumed to exert a nonspecific antibacterial and antitoxic protective effect at the site of inflammation and in the cells of the reticuloendothelial system (Heilmeyer et al., 1958; Umeda, 1965). Hemosiderin detoxicates bacterial toxins, e.g., tetanus, diphtheria, and botulinum, and the products of tissue decomposition (Heilmeyer and Wöhler, 1961; Janoff, 1964). This nonspecific detoxicating system exerts its effect until specific antibodies are formed. In the histiocytes the iron binds to the lysosomes and assumes a role in the lysis of bacteria. In the course of phagocytosis the iron activates the hydrolases of the lysosomes and neutralizes the toxins liberated during the decomposition of the phagocytosed bacteria (Janoff, 1964).

Kurnick et al. (1972) have suggested that a relative impairment of protein synthesis affects the albumin, transferrin, and erythropoietin levels, all of which are lowered in patients with chronic disorders, and that this impairment is causally related to the anemia.

Other changes associated with the anemia of infection are in the porphyrin metabolism. There is an increase in the free protoporphyrin and coproporphyrin concentration of the red cells, as well as in the porphobilinogen, delta-aminolevulinic acid, and 2-COOH-porphyrin content of the urine, while the uroporphyrin and coproporphyrin content of the urine remain within normal limits (Heilmeyer, 1964) (Fig. 19/2). On the other hand, Campbell et al. (1978), in rheumatoid disease, found that the delta-aminolevulinic synthetase activity did not increase in response to anemia, and the free protoporphyrin and the activity of ferrochelatase were both normal.

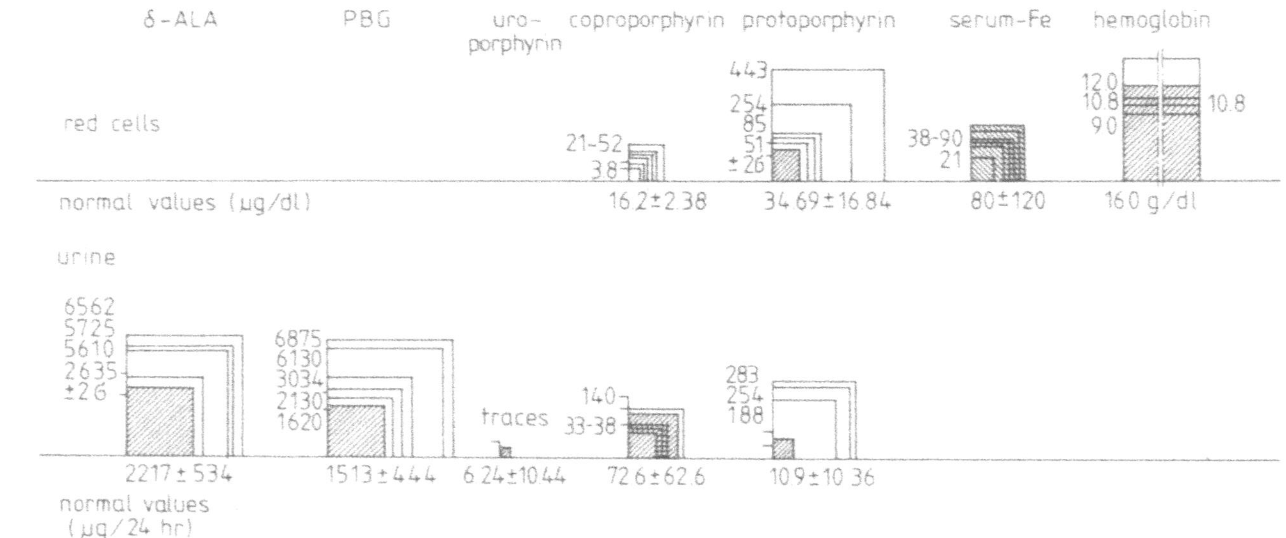

Fig. 19/2. Quantitative changes of heme precursors in the red cells and urine in anemia of infection (after Heilmeyer, L.: Die Störungen der Bluthämsynthese, Thieme, Stuttgart 1964)

The serum copper level is extremely high in most cases of anemia of infection, ranging between 150 and 250 $\mu g/dl$. This finding, among others, may be useful in differentiating iron-deficiency anemia from the anemia of infection (Fig. 19/3).

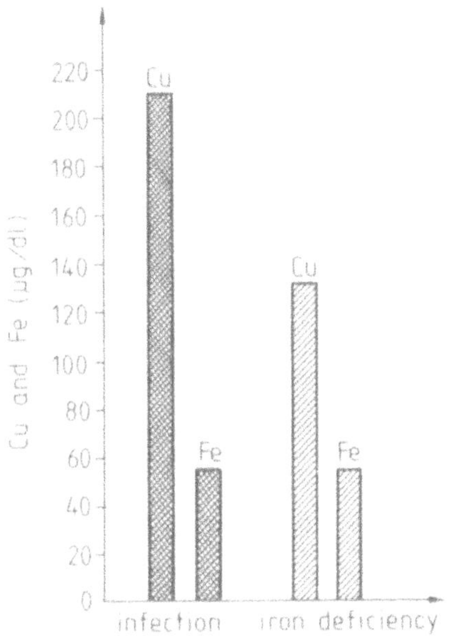

Fig. 19/3. The copper and iron level of plasma in anemia of infection and iron-deficiency anemia (after Heilmeyer, L.: Blut und Blutkrankheiten. Springer, Berlin–Heidelberg–New York 1970)

DIFFERENTIAL DIAGNOSIS

Anemia of infection is often confused with iron-deficiency anemia. In both, the plasma iron level is low, the free protoporphyrin concentration of the red cells is high, and the anemia may be hypochromic. They differ in that transferrin is reduced in the presence of infection and enhanced in iron deficiency, the oral iron loading curve is flat in contrast to the steep or rising curve in hyposiderosis, and the plasma copper concentration is elevated, whereas in iron deficiency it is normal or only slightly increased (see Fig. 19/3). The increase of stainable iron in the bone marrow reticulum cells (Fig. 19/4) and the increase of serum ferritin are important distinguishing features.

True iron deficiency may, of course, complicate the anemia of chronic disorders, as, for example, in rheumatoid arthritis, where gastrointestinal hemorrhage may result from aspirin medication.

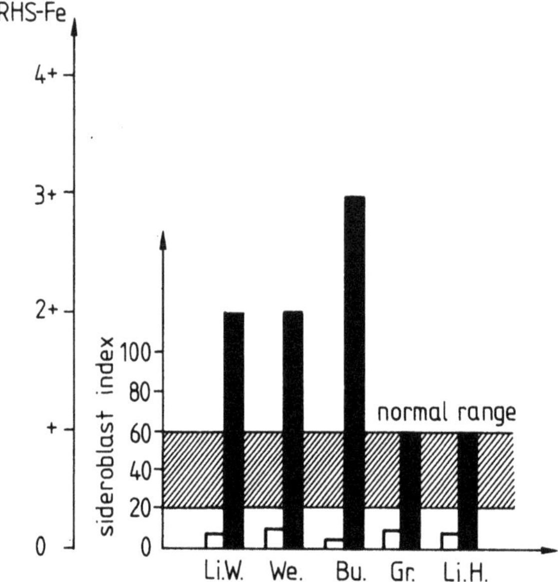

Fig. 19/4. Abnormalities of iron in the bone marrow in anemia associated with infection and malignancy. The sideroblast index (□) is lower than normal but the RE iron (■) is increased (after Merker, H., in: Heilmeyer, L.: Blut und Blutkrankheiten. Springer, Berlin–Heidelberg–New York 1970)

TREATMENT

Cobalt was at one time recommended for treatment, and was thought to exert its effect via the stimulation of erythropoietin production (Weissbecker, 1950). The toxicity of cobalt, however, makes its use undesirable, and treatment should be directed toward dealing with the underlying disease process. If it can be controlled, then the anemia will recover spontaneously. Iron therapy is of use only if there is a complicating iron deficiency.

BIBLIOGRAPHY

Andersson, N. S. E.: Acta med. scand. Suppl. 241 (1950).
Bainton, D. F., Finch, C. A.: The diagnosis of iron deficiency anemia. Amer. J. Med. *37,* 62 (1964).
Beresford, C. H. et al.: Iron absorption and pyrexia. Lancet *2,* 568 (1971).

Bessis, M. C., Breton-Gorius, J.: Iron metabolism in the bone marrow as seen by electron microscopy: A critical review. Blood *19*, 635 (1962).
Bessis, M., Breton-Gorius, J. (1962), cit. Heilmeyer, 1964.
Besta, B., Valenti, S.: L'anemia tubercolare: patogenesi, clinica e terapia. Minerva Med. 1037 (1955).
Brøchner-Mortensen, K., Stein, K. S.: Serumeisen bei akuten und chronischen Infektionen. Nord. med. 235 (1942).
Brown, E. B., Moore, C. V.: Parenterally administered iron in the treatment of hypochromic anemia. In: Tocantins, L. M. (ed.): Progress in Hematology, Vol. 1. Grune and Stratton, New York 1956.
Bullen, J. J. et al.: Iron-binding proteins and infection. Brit. J. Haemat. *23*, 389 (1972).
Campbell, B. C. et al.: Haem biosynthesis in rheumatoid disease. Brit. J. Haemat. *40*, 563 (1978).
Cartwright, G. E.: The anemia of chronic disorders. Semin. hemat. *3*, 351 (1966).
Cartwright, G. E., Lee, G. R.: The anemia of chronic disorders. Brit. J. Haemat. *21*, 147 (1971).
Cartwright, G. E., Wintrobe, M. M.: Chemical, clinical and immunological studies on the products of human plasma fractionation. XXXIX. The anemia of infection. Studies on the iron-binding capacity of serum. J. clin. Invest. *28*, 86 (1949).
Cartwright, G. E., Wintrobe, M. M.: The anemia of infection. XVII. A review. Advances in Internal Medicine. Vol. 5, p. 165. Year Book Publishers, Chicago 1952.
Cartwright, G. E., Gubler, C. J., Wintrobe, M. M.: Studies on copper metabolism. XI. J. clin. Invest. *33*, 685 (1954).
Cartwright, G. E., Hamilton, L. D., Gubler, C. J., Fellows, V. M., Ashenbrucker, H., Wintrobe, M. M.: The anemia of infection. XIII. Studies on experimentally produced acute hypoferremia in dogs and the relationship of the adrenal cortex to hypoferremia. J. clin. Invest. *30*, 161 (1951).
Cartwright, G. E., Lauritzen, M. A., Humphreys, S. R., Jones, P. J., Merrill, J. M., Wintrobe, M. M.: The anemia associated with chronic infection. Science *103*, 72 (1946).
Coleman, D. H., Stevens, A. R., Finch, C. A.: The treatment of iron deficiency anemia. Blood *10*, 567 (1955).
Crowley, J.: Iron absorption tests in anaemia: use of intravenous iron preparations. Edinb. med. J. *59*, 478 (1952).
Dinant, H. J., de Maat, C. E. M.: Erythropoiesis and mean red-cell lifespan in normal subjects and in patients with the anaemia of acute rheumatoid arthritis. Brit. J. Haemat. *39*, 437 (1978).
Douglas, S. W., Adamson, J. W.: The anemia of chronic disorders. Studies of marrow regulation and iron metabolism. Blood *45*, 55 (1975).
Elin, R. J., Wolff, S. M., Finch, C. A.: Effect of induced fever on serum iron and ferritin concentrations in man. Blood *49*, 147 (1977).
Engel, M.: Klinische Erfahrungen mit dem intravenös injizierten Ferronascin "Roche". Schweiz. med. Wschr. *76*, 1079 (1946).
Fillet, G. et al.: Storage iron kinetics. VII. A biological model for reticuloendothelial iron transport. J. clin. Invest. *53*, 1527 (1974).
Freireich, E. J., Ross, J. F., Bayles, T. B., Emerson, D. P., Finch, S. C.: Radioactive iron metabolism and erythrocyte survival studies of the mechanisms of the anaemia associated with rheumatoid arthritis. J. clin. Invest. *36*, 1043 (1951).
Freireich, E. J. et al.: The effect of inflammation on the utilization of erythrocyte and transferrin-bound radio-iron for red cell production. Blood *12*, 972 (1957).
Greif, S.: Hochdosierte parenterale Eisentherapie. Wien. klin. Wschr. *62*, 145 (1950).
Gutnisky, A., van Dyke, D.: Normal response to erythropoietin or hypoxia in rats made anemic with turpentine abscess. Proc. Soc. exp. Biol. Med. *112*, 75 (1963).
Hagberg, B.: Studies on the plasma transport of iron. Uppsala 1953.
Handjani, A. M. et al.: Serum iron in acute myocardial infarction. Blut *23*, 363 (1971).
Haurani, F. J., Young, K., Tocantins, L. M.: Reutilization of iron in anemia complicating malignant neoplasms. Blood *22*, 73 (1963); J. Lab. clin. Med. *65*, 560 (1965); Amer. J. med. Sci. *249*, 537 (1965).
Heilmeyer, L.: Die Störungen der Bluthämsynthese. Thieme, Stuttgart 1964.

Heilmeyer, L.: Die Hypochromanämien. In: Heilmeyer, L., Begemann, H. (eds.): Blut und Blutkrankheiten. Springer, Berlin–Heidelberg–New York 1970.
Heilmeyer, L., Plötner, K.: Das Serumeisen und die Eisenmangelkrankheit. Fischer, Jena 1937.
Heilmeyer, L., Wöhler, F.: Über die Entgiftungsfunktion des Speichereisens. I. Klin. Wschr. *39,* 563 (1961).
Heilmeyer, L., Keiderling, W., Stüwe, G.: Kupfer und Eisen als körpereigene Wirkstoffe und ihre Bedeutung beim Krankheitsgeschehen. Fischer, Jena 1941.
Heilmeyer, L., Keiderling, W., Wöhler, F.: Der Eisenstoffwechsel beim Infekt und die Entgiftungsfunktion des Speichereisens. Dtsch. med. Wschr. *83,* 1965 (1958).
Hershko, C. et al.: Storage iron kinetics. VI. The effect of inflammation on iron exchange in the rat. Brit. J. Haemat. *28,* 67 (1974).
Hirvonen, M.: Untersuchungen über den Serumeisengehalt bei einigen gewöhnlichen Infektionskrankheiten. Acta med. scand. *106,* 495 (1941).
Hollingsworth, J. W., Hollingsworth, D. R.: Study of total red cell volume and erythrocyte survival using radioactive chromium in patients with advanced pulmonary tuberculosis. Ann. intern. Med. *42,* 810 (1955).
Holly, R. G.: Intravenous iron: Evaluation of use of saccharated iron oxide in iron deficiency states in obstetrics and gynecology. Blood *6,* 1159 (1951).
Hutchinson, H. G., Lowther, C. P., Alexander, W. D.: On the appearance in the marrow of iron administered intravenously. J. clin. Path. *7,* 281 (1954).
Jacobs, A., Worwood, M.: Ferritin: clinical aspects. In: Kief, H. (ed.): Iron Metabolism and its Disorders, p. 90. Excerpta Medica, Amsterdam–Oxford; American Elsevier, New York 1975.
Janoff, A.: The role of iron in macrophages. J. theor. Biol. *7,* 168 (1964).
Jeffrey, M. R.: Some observations on anemia in rheumatoid arthritis. Blood *8,* 502 (1953).
Kampschmidt, R. F., Schultz, G. A.: Hypoferremia in rats following injection of bacterial endotoxin. Proc. Soc. exp. Biol. Med. *106,* 870 (1961); *110,* 191 (1962); *113,* 142 (1963); *116,* 420 (1964); Amer. J. Physiol. *208,* 68 (1965).
Keiderling, W., Scharpf, H.: Über die klinische Bedeutung der Serumkupfer- und Serumeisenbestimmung bei neoplastischen Krankheitszuständen. Münch. med. Wschr. *95,* 437 (1953).
Keiderling, W., Schmidt, H. A. E.: Die infektiöse und neoplastische Eisenstoffwechselstörung. In: Keiderling, W. (ed.): Eisenstoffwechsel. Thieme, Stuttgart 1959.
Keiderling, W., Lee, M., Schmidt, H. A. E.: Über die Dynamik des Eisenstoffwechels und seine Beeinflussung durch entzündliche Reize. Verh. dtsch. Ges. inn. Med. *62,* 356 (1956).
Keiderling, W., Schmidt, H. A. E., Lee, M.: Untersuchungen über die Infektanämie mit Radioeisen und Radiochrom. 5. Europäischer Hämatologen-Kongreß, 1955.
Keiderling, W., Schmidt, H. A. E., Lee, M.: Untersuchungen über die Dynamik des Erythrozytenumsatzes mit Radioeisen (Fe^{59}) und Radiochrom (Cr^{51}). In: Radioaktive Isotope in Klinik und Forschung Vol. 2, p. 24. Urban und Schwarzenberg, München–Berlin 1956.
Konijn, A. M., Hershko, C.: Ferritin synthesis in inflammation. I. Pathogenesis of impaired iron release. Brit. J. Haemat. *37,* 7 (1977).
Kubanek, B.: Konservative Behandlung der Infekt- und Tumoranämie. Dtsch. med. Wschr. *104,* 238 (1979).
Kurnick, J. E., Ward, H. P., Pickett, J. C.: Mechanism of the anemia of chronic disorders. Correlation of hematocrit value with albumin, vitamin B_{12}, transferrin, and iron stores. Arch. intern. Med. *130,* 323 (1972).
Laurell, C. B.: Studies on the transportation and metabolism of iron in the body. Acta physiol. scand. *14,* Suppl. 46, 1 (1947).
Laurell, C. B.: Plasma iron and the transport of iron in the organism. Pharmacol. Rev. *4,* 371 (1952).
Locke, A., Main, E. R., Rosbash, D. O.: The copper and non-hemoglobinous iron contents of the blood serum in disease. J. clin. Invest. *11,* 527 (1932).
Lukens, J. N.: Control of erythropoiesis in rats with adjuvant-induced chronic inflammation. Blood *41,* 37 (1973).

Marti, H. R.: Infektanämien. Schweiz. med. Wschr. *98*, 377 (1968).
Merker, H.: Cytochemie der Blutzellen. In: Heilmeyer, L., Begemann, H. (eds.): Blut und Blutkrankheiten. Vol. 1. Springer, Berlin–Heidelberg–New York 1970.
Rath, C. R., Finch, C. A.: Measurement of iron binding capacity of serum in man. J. clin. Invest. *28*, 79 (1949).
Roberts, F. D., Hagedorn, A. B., Slocumb, C. H., Owen, C. A.: Evaluation of the anemia of rheumatoid arthritis. Blood *21*, 470 (1963).
Roy, L. M. H., Alexander, W. R. M., Duthie, J. J. R.: Nature of anemia in rheumatoid arthritis. I. Metabolism of iron. Ann. rheum. Dis. *14*, 63 (1955).
Schade, S. G.: Normal incorporation of iron into intestinal ferritin in inflammation. Proc. Soc. exp. Biol. (Med.) *139*, 620 (1972).
Schäfer, K. H.: Untersuchungen über den exogenen Eisenstoffwechsel bei fieberhaften Infekten im Kindesalter. Klin. Wschr. *1940*, 979.
Schäfer, K. H.: Über den Einfluß von Infektionen und ähnlichen Vorgängen auf den Eisenstoffwechsel. I. Z. exper. Med. *110*, 678 (1942).
Schäfer, K. H.: Einfluß allergischer Vorgänge auf das Serumeisen. Mschr. Kinderheilk. *96*, 18 (1948).
Schäfer, K. H.: Der Eisenstoffwechsel des wachsenden Organismus. Ergebn. inn. Med. Kinderheilk. N. F. *4*, 706 (1953).
Scott, J. M., Govan, A. D. T.: Anaemia of pregnancy treated with intramuscular iron. Brit. med. J. *2*, 1257 (1954).
Skouge, E.: Klinische und experimentelle Untersuchungen über das Serumeisen. Oslo 1939.
Slack, H. G. B., Wilkinson, J. F.: Intravenous treatment of anaemia with iron-sucrose preparation. Lancet *1*, 11, 163 (1949).
Snick, J., Masson, P. L., Heremans, J. F.: The affinity of lactoferrin for the RES as the molecular basis for the hyposideraemia of inflammation. In: Chrichton, R. R. (ed.): Proteins of Iron Storage and Transport in Biochemistry and Medicine. North-Holland, Amsterdam 1975.
Thedering, F., Gross, R.: Fortschritte mit der intravenösen Eisentherapie. Z. ges. inn. Med. *4*, 634 (1949).
Thoenes, F.: Physiologie und Pathologie des Eisenstoffwechsels im Wachstumsalter. Dtsch. Z. Verdau.- u. Stoffwechselkr. *4*, 209 (1941).
Thomas, W. J. et al.: Free erythrocyte porphyrin: hemoglobin ratios, serum ferritin, and transferrin saturation levels during treatment of infants with iron-deficiency anemia. Blood *49*, 455 (1977).
Torrance, J. D., Charlton, R. W., Simon, M. O., Lynch, S. R., Bothwell, T. H.: The mechanism of endotoxin-induced hypoferraemia. Scand. J. Haemat. *21*, 403 (1978).
Umeda, T.: Transferrin metabolism in infection. J. Kyushu Hematological Society *15*, 153 (1965).
Vahlquist, B. C.: Das Serumeisen. Acta paediat. *28*, 1 (1941).
Ward, H. P. et al.: Serum level of erythropoietin in anemias associated with chronic infection, malignancy and primary hematopoietic disease. J. clin. Invest. *50*, 332 (1971).
Weinberg, E. D.: Roles of iron in host-parasite interactions. J. infect. Dis. *124*, 401 (1971).
Weinstein, I. M.: A correlative study of the erythrokinetics and disturbances in iron metabolism associated with the anemia of rheumatoid arthritis. Blood *14*, 950 (1959).
Weissbecker, L.: Die Kobalttherapie. Dtsch. med. Wschr. *75*, 116 (1950).
Wintrobe, M. M.: Clinical Hematology. Lea and Febiger, Philadelphia 1967, 1975.
Wintrobe, M. M. et al.: The anemia of infection; uptake of radioiron in iron-deficient and in pyridoxine-deficient pigs before and after acute inflammation. J. clin. Invest *26*, 103 (1947).
Wintrobe, M. M. et al.: The anemia of infection. J. clin. Invest. *26*, 114, 121 (1947); Blood *2*, 323 (1947); *27*, 245 (1958); J. Lab. clin. Med. *33*, 532 (1948).
Wintrobe, M. M., Greenberg, G. R., Cartwright, G. E.: Pathogenesis of anemia of infection. Trans. Ass. Amer. Phycns *59*, 110 (1947).
Zucker, S., Lysik, R. M.: Anemia in cancer. In: Sutmick, A. I., Engström, P. F. (eds.): Oncologic Medicine, p. 261. University Park Press, Baltimore 1976.

CHAPTER 20

ANEMIA OF THERMAL INJURY

Accidental burns, either in industry or as a result of household accidents, are frequent, and anemia is a common complication (Moore et al., 1946; Ludwig et al., 1965). The anemia is thought to delay the healing of the injury, to hinder the adhesion of transplants, and in general to have an unfavorable influence on the final outcome (Monsaingeon, 1963).

DEVELOPMENT, TYPE, AND COURSE OF THE ANEMIA OF THERMAL INJURY

Small, superficial burns do not usually give rise to anemia, but more extensive and deep burns are followed by a rapid onset of anemia after a transitory hemoconcentration. The hemoglobin tends to fall faster than the reduction in red cell count, and the normochromic anemia gradually becomes hypochromic (Bernát, 1971). The initial picture results from hemolysis, which is the direct effect of the heat injury. Brooks et al. (1950) demonstrated agglutination of the red cells not only in the vicinity of the thermal injury but also in vessels more remote from the burn; they also observed considerable erythrophagocytosis in the spleen, liver, and bone marrow of experimental animals who had been exposed to massive thermal injury. Microspherocytosis and fragmentation of the red cells can be seen some hours after the trauma (Brown, 1946). The fragmented cells disappear rapidly from the circulation but the microspherocytes may remain up to 2–4 days. Three phases can be distinguished in the course of the anemia. In the first, the hemolysis is predominant. This is followed by a time at which erythropoiesis and red cell destruction are largely in equilibrium and, finally, red cell production is predominant (Fig. 20/1). The moderately hypochromic cells, produced following the injury, can be distinguished from the normochromic red cells originating from before the injury (Bernát et al., 1968). The free protoporphyrin in the red cells rises gradually, the increase in free protoporphyrin being correlated with the decrease in the hemoglobin content of the red cells (Fig. 20/2).

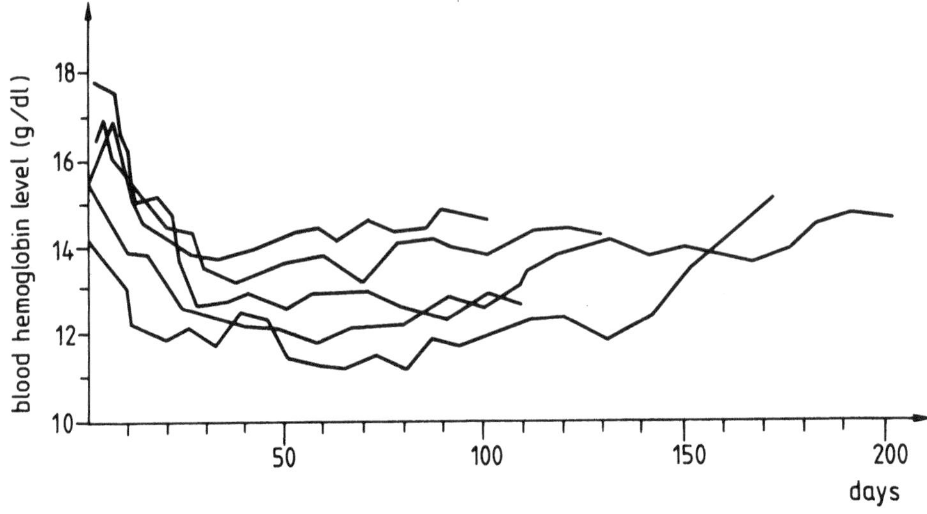

Fig. 20/1. Changes in the blood hemoglobin level in thermal injury (after Bernát, I., 1971)

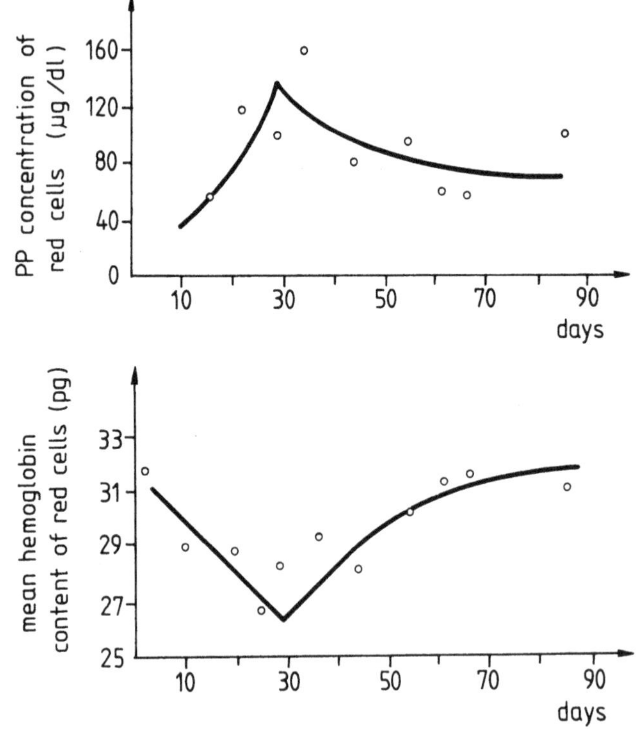

Fig. 20/2. Changes in the free protoporphyrin concentration and mean hemoglobin content of the red cells in thermal injury (after Bernát, I., 1971)

There is a slight increase in the absolute reticulocyte count soon after the injury, but later the values tend to be lower than normal, indicating that there is a reduction in erythropoietic response. It may increase to about twice normal, but far less than would be expected with any other type of hemolytic anemia.

The hemolytic phase is accompanied by reduced osmotic and mechanical resistance of the red cells (Shen et al., 1943; Moore et al., 1946; Ham et al., 1948;

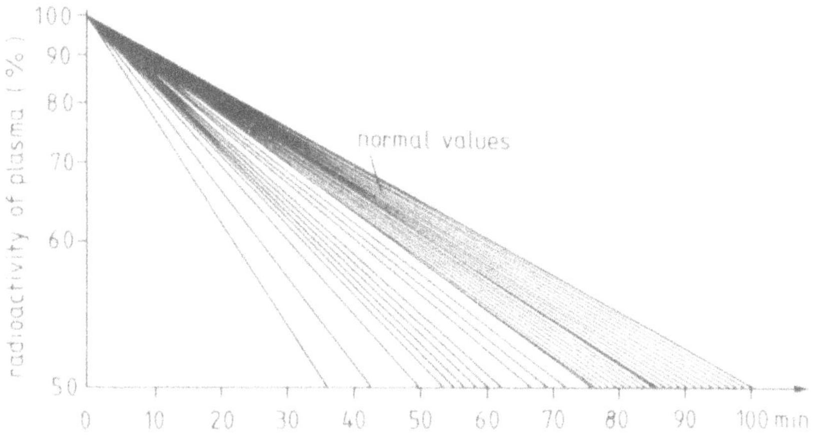

Fig. 20/3. Shortening of the half-time (T/2) of plasma ^{59}Fe clearance in rabbits with thermal injury (after Bernát, I. and Fehérvári, T., 1968)

Alpen et al., 1954; and others), a rise in serum hemoglobin and bilirubin levels (Muir, 1961), hemoglobinuria (Shen et al., 1943; Sevitt, 1956), the enhancement of endogenous carbon monoxide production and alveolar carbon monoxide concentration (Sjöstrand, 1948, 1949a, b, 1951; Troell et al., 1955), an increase in fecal urobilinogen content (Moore et al., 1946), and a reduction in the life-span of the red cells (Davis et al., 1955).

The hemoglobinuria seen in severe injuries is frequently biphasic or intermittent (Sevitt, 1957). The first phase corresponds to the immediate breakdown of damaged red cells and occurs usually within 6–12 hours, while the second phase, due to delayed destruction of more moderately affected erythrocytes, occurs from 12 to 48 hours (Davies and Topley, 1956; Topley et al., 1962).

The pathogenesis of the anemia is clearly complex. Although the initial phase is due to hemolysis (Lederer, 1960; Schubothe and Kähler, 1960; Wintrobe, 1967; and others), the more prolonged anemia has been attributed to a multiplicity of factors

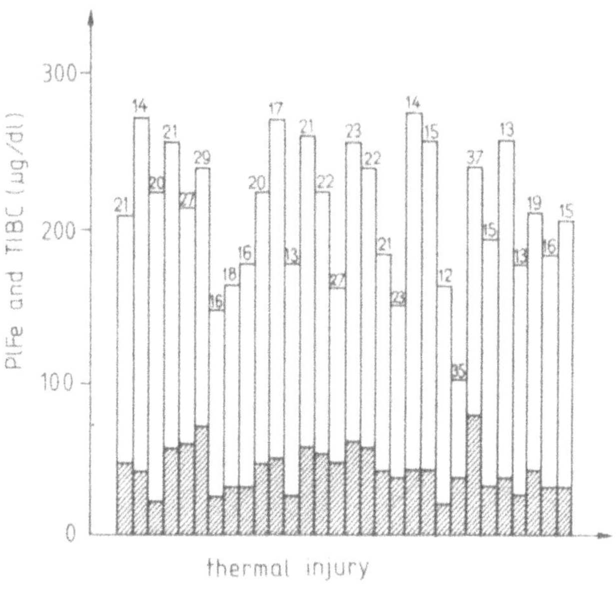

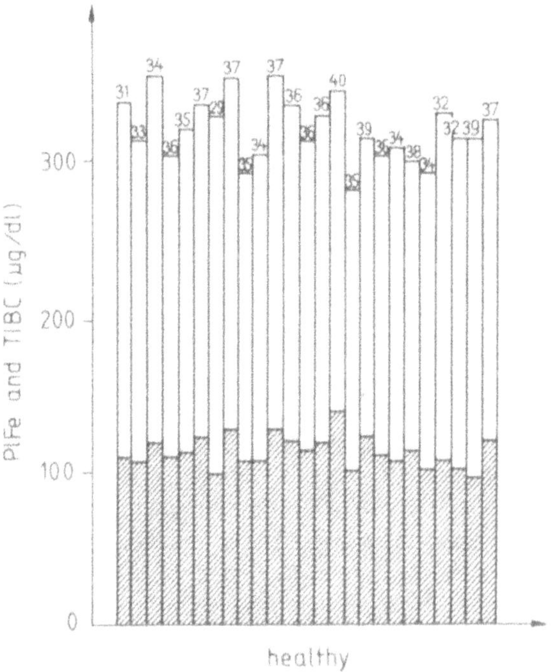

Fig. 20/4. Plasma iron concentration (shaded areas in columns) and iron-binding capacity (full height of columns) in patients with thermal injury and in healthy man (after Bernát, I., 1971)

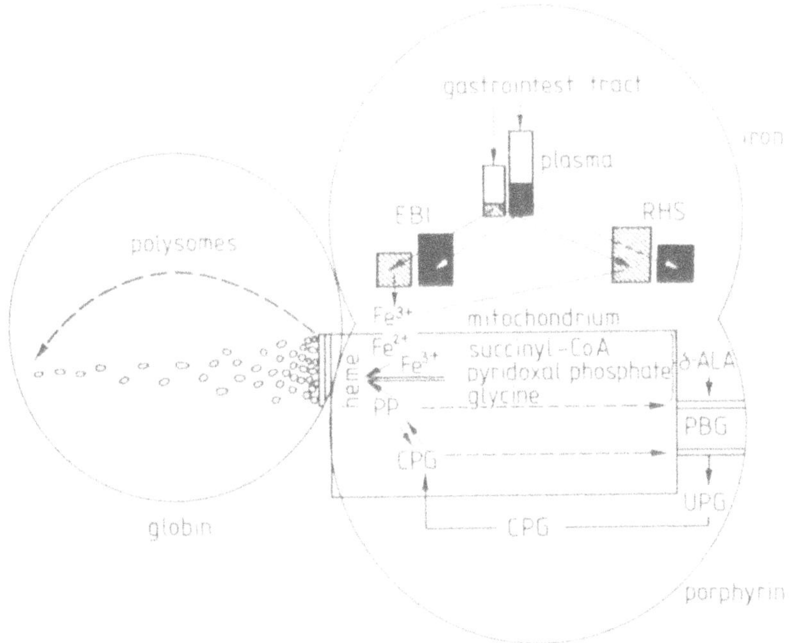

Fig. 20/5. Schematic representation of disturbed hemoglobin synthesis. Healthy subjects (■); patients with thermal injury (▨). EBl = erythroblast; RHS = reticulohistiocytic system; - - → feedback

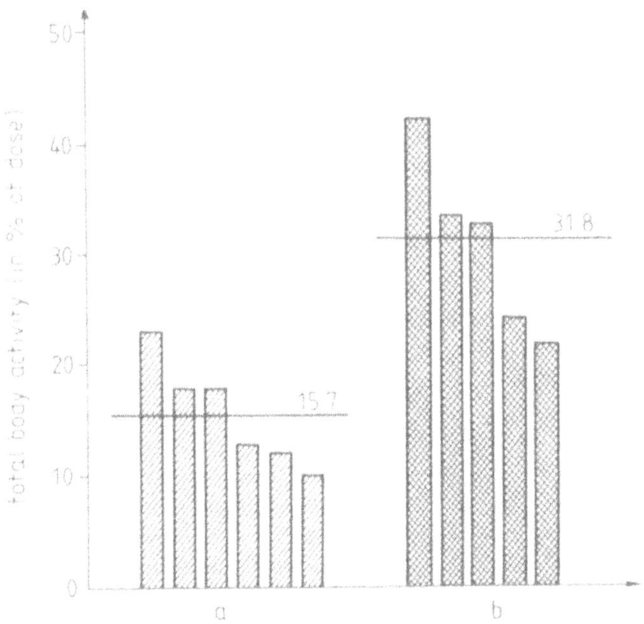

Fig. 20/6. Absorption of radioiron from the intestinal tract of (a) patients with thermal injury, and (b) healthy man (after Bernát, I., 1971)

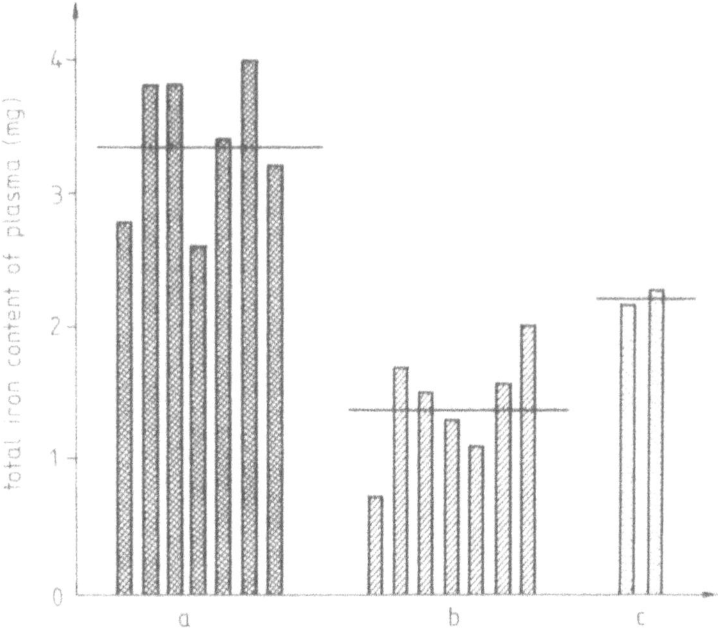

Fig. 20/7. Total plasma iron content in (a) healthy man, (b) patients with thermal injury, and (c) two convalescent patients. The horizontal lines indicate mean values (after Bernát, I. et al., 1968)

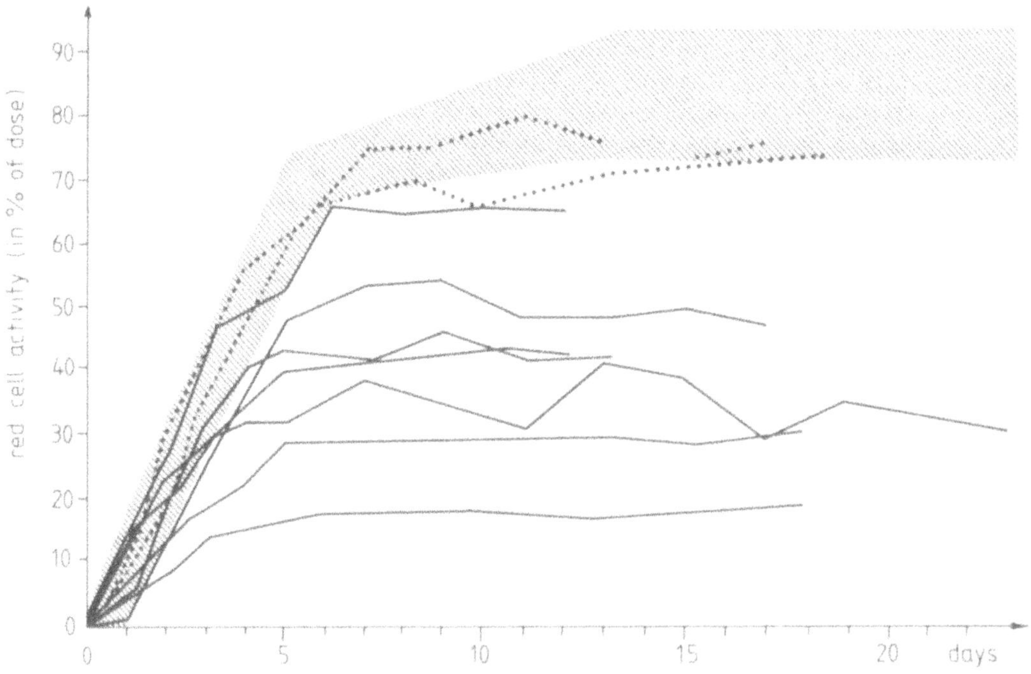

Fig. 20/8. Red cell radioactivity of patients with thermal injury after the i.v. injection of ^{59}Fe as a function of time. The shaded area represents the normal values (after Bernát, I. et al., 1968)

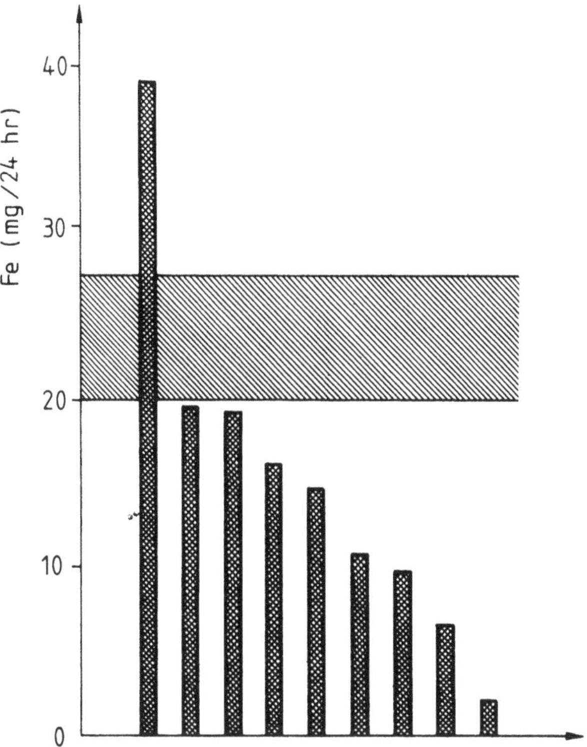

Fig. 20/9. The absolute amount of iron utilized daily in the course of erythropoiesis of patients with thermal injury. The horizontal shaded area represents the normal values (after Bernát I. et al., 1968)

such as infection, toxic effects, protein deficiency and disturbed protein metabolism, iron deficiency, and blood loss (Altemeier and Carter, 1942; Moore et al., 1946; Cope, 1947; Robsheit-Robbins et al., 1947; Braithwaite and Moore, 1948; Sevitt, 1957).

The studies of Bernát (1965) have shown that although the features of the anemia of thermal injury have something in common with those due to chronic infection, the metabolic disturbance cannot be attributed to infection since the disorder develops before the toxic and septic phase following the injury. A rapid outflow of iron from the circulation, a drop in plasma iron level, and reduction in the iron-binding capacity of the plasma can be demonstrated as early as 2–6 hours after the initial injury (Bernát, 1966) (Fig. 20/3 and Fig. 20/4). The effect of heat as a stressor appears to activate the reticuloendothelial system, producing a change in kinetics of iron turnover and distribution of iron in the organism (Fig. 20/5). The iron is directed to the reticuloendothelial system (see Table 8/3), and the increased outflow

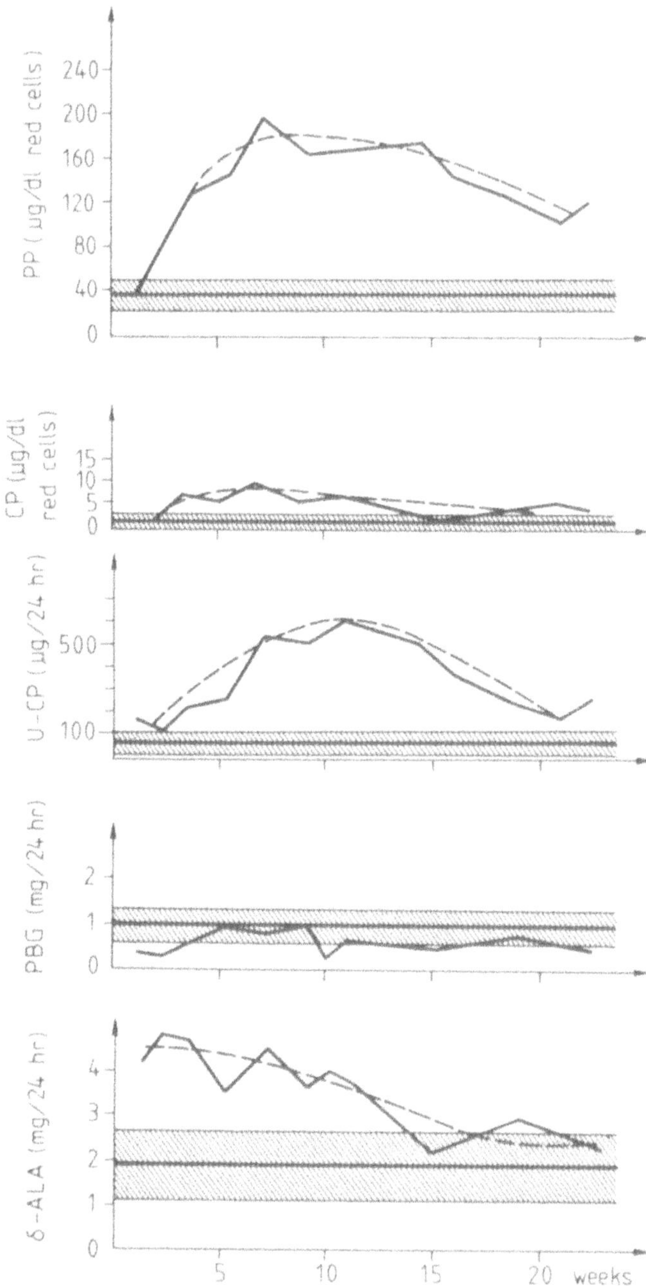

Fig. 20/10. Quantitative changes in the heme precursors in the red cells and urine during the course of thermal injury (after Bernát, I., 1971)

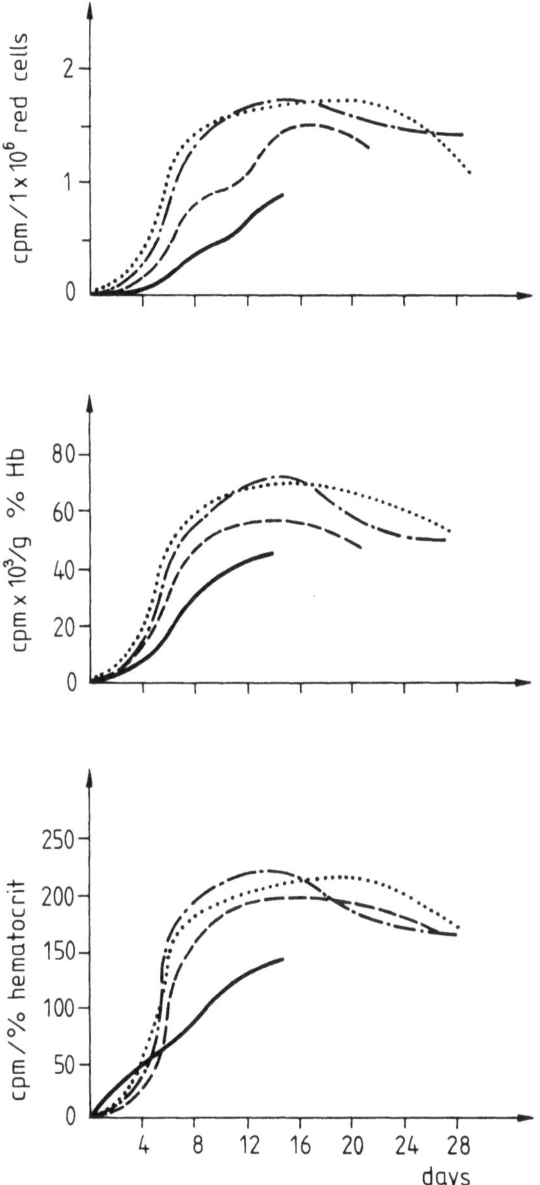

Fig. 20/11. Incorporation of 1-^{14}C-glycine into globin in healthy rabbits and in rabbits with thermal injury 1 to 3 weeks after thermal injury (after Bernát, I., 1971)

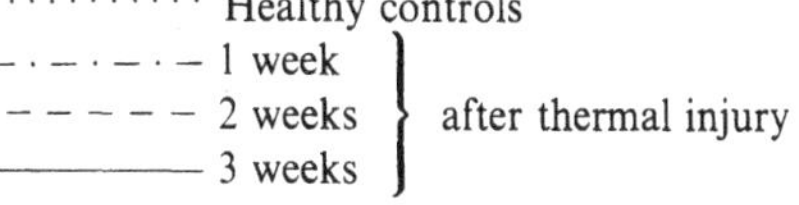

of iron from the circulation and diminished absorption of iron from the gastrointestinal tract (Fig. 20/6) result in a decreased plasma iron pool (Fig. 20/7). The incorporation of iron into the erythroid precursors is reduced (Fig. 20/8), and effective erythropoiesis is diminished (Fig. 20/9). Hemoglobin synthesis is reduced in spite of considerable iron reserves in the form of ferritin or hemosiderin in the iron depots.

There is a decrease in the amount of intracellular iron in the nucleated red cells, and probably a reduction of heme synthetase, leading to the increase in free erythrocyte protoporphyrin (Fig. 20/10).

The delta-ALA-dehydrogenase and PBG-desaminase activity decreases, resulting in an increase in delta-aminolevulinic acid in the red cells and increased excretion through the kidneys (Fig. 20/10). The accumulation of free protoporphyrin inhibits transformation of coproporphyrinogen, and there is a rise in intracellular coproporphyrin as well as an increased excretion in the urine. There is also a decrease in globin formation (Fig. 20/11) (Bernát, 1971).

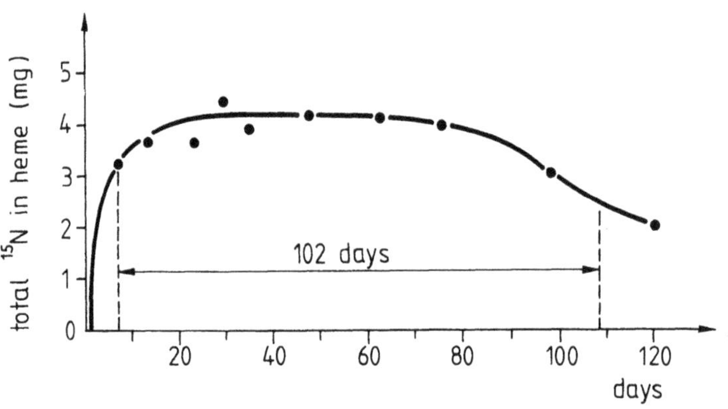

Fig. 20/12. ^{15}N incorporated in the heme after the intake of glycine labeled with heavy nitrogen isotope in a patient with thermal injury (after Bernát, I., 1971)

The disturbed erythropoiesis results in the production of a proportion of cells with a reduced life-span (see Fig. 15/8 and Fig. 15/9), but the average life-span of the red cells is between 102 and 111 days (Fig. 20/12), which means that the decrease does not play a significant role in the pathogenesis of the anemia.

THERAPY

The anemia of thermal injury cannot be influenced by the administration of iron, vitamin B_{12}, or folic acid, and blood transfusion has only a temporary effect (Fig. 20/13). The difficulties encountered in treatment have been stressed by Moore et al. (1946), Siegrist et al. (1957), and Ludwig et al. (1965). Anemia may persist for some time even after recovery from the immediate injury.

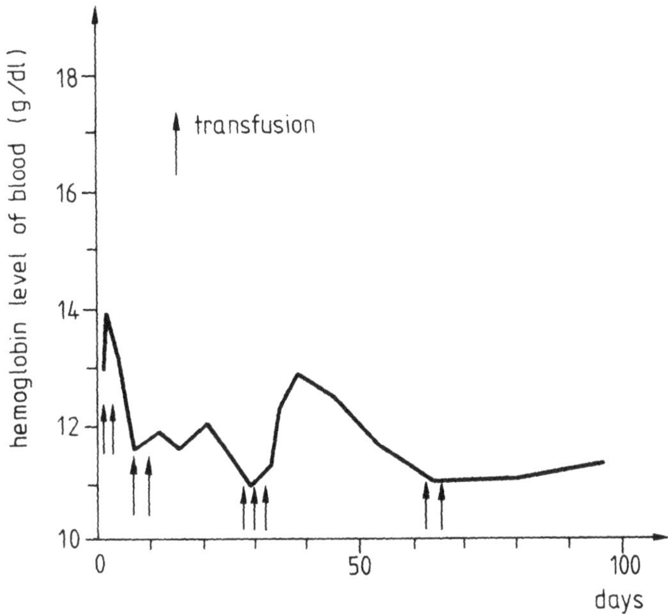

Fig. 20/13. Changes in the hemoglobin level in a patient with 6% 2nd-degree burns and 20% 3rd-degree burns, treated with repeated blood transfusions (after Bernát, I., 1971)

BIBLIOGRAPHY

ALPEN, E. L., ALEXANDER, J. A., DAVIS, A. K.: Combined effects of total body X-irradiation and radiant energy thermal burns on the osmotic and mechanical fragility of the erythrocyte. Amer. J. Physiol. *179,* 541 (1954).

ALTEMEIER, W. A., CARTER, B. N.: Infected burns with haemorrhage. Ann. Surg. *115,* 1118 (1942).

BERNÁT, I.: Az égési anaemia elkülönítése a toxiko-infectiós, illetve fehérjehiányos vérszegénységtől (Differentiation of anemia after thermal injury from toxico-infectious and protein-deficient anemia). Honvédorvos *21,* 156 (1969).

BERNÁT, I.: Az égési anaemia pathogenesise (The Pathogenesis of Anemia after Thermal Injury). Akadémiai Kiadó, Budapest 1971.

Bernát, I., Fehérvári, T.: Anaemia after thermal injury. III. Iron kinetics. Haematologia *2*, 147 (1968).
Bernát, I., Dózsán, G., Magyari, J., Novák, J.: Anaemia after thermal injury. IV. Iron kinetics in burned patients. Haematologia *2*, 279 (1968).
Bernát, I., Dózsán, G., Novák, J., Elek, S.: Anaemia after thermal injury. II. Acta med. Acad. Sci. hung. *22*, 253 (1966).
Bernát, I., Novák, J., Elek, S., Dózsán, G.: Anaemia after thermal injury. I. Acta med. Acad. Sci. hung. *21*, 121 (1965).
Bernát, I., Novák, J., Fáber, V., Dózsán, G., Elek, S.: Neue Beiträge zur Pathogenese der Verbrennungsanämie. Folia haemat. (Lpz.) *86*, 85 (1966).
Braithwaite, F., Moore, F. D.: Some observations on anemia in patients with burns. Brit. J. plast. Surg. *1*, 81 (1948).
Brooks, F., Dragstedt, L. R., Warner, L., Knisely, M. H.: Sludged blood following severe thermal burns. Arch. Surg. *61*, 387 (1950).
Brown, A.: In: Studies of Burns and Scalds. M. R. C. Spec. Report Series, Nr. 249, 1945.
Brown, A.: Morphological changes in red cells in relation to severe burns. J. Path. Bact. *58*, 367 (1946).
Cope, O.: Anemia in burns. Surg. Gynec. Obstet. *84*, 999 (1947).
Davies, J. W. L., Topley, E.: The disappearance of red cells in patients with burns. Clin. Sci. *15*, 135 (1956).
Davis, W. M., Alpen, E. L., Davis, A. K.: Studies of radioiron utilization and erythrocyte life span in rats following thermal injury. J. clin. Invest. *34*, 67 (1955).
Ham, T. H., Shen, S. C., Fleming, E. M., Castle, W. B.: Studies on the destruction of red blood cells. IV. Thermal injury; action of heat in causing increased spheroidicity; osmotic and mechanical fragility, and hemolysis of erythrocytes; observations on the mechanisms of destruction of such erythrocytes in dogs and in patients with fatal thermal burns. Blood *3*, 373 (1948).
Heilmeyer, L.: In: Keiderling, W., Hoffmann, G. (eds.): Radio-Isotope in der Hämatologie, p. 96. Schattauer, Stuttgart 1963.
Heilmeyer, L., Keiderling, W., Wöhler, F.: Der Eisenstoffwechsel beim Infekt und die Entgiftungsfunktion des Speichereisens. Dtsch. med. Wschr. *83*, 1965 (1958).
James III., G. W., Abbott, L. D., Brooks, J. W., Evans, E. I.: The anemia of thermal injury. III. Erythropoiesis and hemoglobin metabolism studied with N^{15}-glycine in dog and man. J. clin. Invest. *33*, 150 (1954).
James III, G. W., Purnell, O. J., Evans, E. I.: The anemia of thermal injury. I. Studies of pigment excretion. J. clin. Invest. *30*, 181 (1951).
Kimber, R. J., Lander, H.: The effect of heat on human red-cell morphology, fragility and subsequent survival in vivo. J. Lab. Med. *64*, 922 (1964).
Lederer, J.: Les anémies hémolytiques toxo-infectieuses. In: Heilmeyer, L., Hittmair, A. (eds.): Handbuch der gesamten Hämatologie, Vol. III/1, p. 606. Urban und Schwarzenberg, München–Berlin 1960.
Ludwig, K., Máday, P., Mészáros, J.: Perifériás vérkép és csontvelőkép változás az égésbetegségben (Changes of the bone marrow and peripheral blood count in anemia after thermal injury). 3rd Congress of Hematology, Budapest 1965.
Monsaingeon, A.: Les brûlés. Masson, Paris 1963.
Moore, F. D., Peacock, W. C., Blakely, E., Cope, O.: The anemia of thermal burns. Ann. Surg. *124*, 811 (1946).
Muir, I. F. K.: Red cell destruction in burns with particular reference to the shock period. Brit. J. plast. Surg. *14*, 273 (1961).
Nissen, R., Staub, H.: Foreword to Allgöwer, M., Siegrist, J.: Verbrennungen. Pathophysiologie, Pathologie, Klinik, Therapie. Springer, Berlin–Göttingen–Heidelberg 1957.
Pawelski, S., Konopka, L., Nasilowski, W., Rechowicz, K.: Kinetyka zelaza radioaktywnego Fe^{59} i Cr^{51} krwinek czerwonych w niedokrwistosciach u osob oparzonych (Kinetics of ^{59}Fe and ^{51}Cr in anemia after thermal injury). Pol. Tyg. lek. *21*, 901 (1966).

ROBSHEIT-ROBBINS, F. S., MILLER, L. L., WHIPPLE, G. H.: Plasma protein and hemoglobin synthesis. J. exp. Med. *85*, 243 (1947).

SCHUBOTHE, H., KÄHLER, H. J.: Die Hämoglobinurien. In: HEILMEYER, L., HITTMAIR, A. (eds.): Handbuch der gesamten Hämatologie, Vol. III/1, p. 654. Urban und Schwarzenberg, München–Berlin 1960.

SEVITT, S.: Distal tubular necrosis with little or no oliguria. J. clin. Path. *9*, 12 (1956).

SEVITT, S.: Burns. Pathology and therapeutic applications. Butterworth, London 1957.

SHEN, S. C., HAM, T. H., FLEMING, E. M.: Studies on the destruction of red blood cells. III. Mechanism and complications of hemoglobinuria in patients with thermal burns; spherocytosis and increased osmotic fragility of red blood cells. New Engl. J. Med. *229*, 701 (1943).

SIEGRIST, J., MIESCHER, P., ALLGÖWER, M. (1957), cit. ALLGÖWER, M., SIEGRIST, J.: Verbrennungen. Pathophysiologie, Pathologie, Klinik, Therapie, p. 89. Springer, Berlin–Göttingen–Heidelberg 1957.

SJÖSTRAND, T.: A method for determination of carboxyhaemoglobin concentration by analysis of the alveolar air. Acta physiol. scand. *16*, 201 (1948).

SJÖSTRAND, T.: Endogenous formation of carbon monoxide in man. Nature *164*, 580 (1949a).

SJÖSTRAND, T.: Endogenous formation of carbon monoxide in man under normal and pathological conditions. Scand. J. clin. Lab. Invest. *1*, 201 (1949b).

SJÖSTRAND, T.: Endogenous formation of carbon monoxide. The CO-concentration in the inspired and expired air of hospital patients. Acta physiol. scand. *22*, 137 (1951).

TOPLEY, E., JACKSON, D. McG., CASON, J. S., DAVIES, J. W. L.: Assessment of red-cell loss in the first two days after severe burns. Ann. Surg. *155*, 581 (1962).

TROELL, L., NORLANDER, O., JOHANSON, B.: Red cell destruction in burns. With special reference to changes in the endogenous formation of carbon monoxide. Acta chir. scand. *109*, 1 (1955).

WAGNER, H. N., RAZZAK, M. A., GAERTNER, R. A., CAINE, W. P., FEAGIN, O. T.: The removal of erythrocytes from the circulation. Arch. intern. Med. *110*, 90 (1962).

WINTROBE, M. M.: Clinical Hematology. Lea and Febiger, Philadelphia 1967, 1975.

WINTROBE, M. M., GREENBERG, G. R., HUMPHREYS, S. R., JONES, P. J., MERRILL, J. M.: The anemia of infection. J. clin. Invest. *26*, 103 (1947).

CHAPTER 21

PROTEIN-DEFICIENCY ANEMIA

Protein malnutrition results in a moderate anemia. Studies in children with kwashiorkor (Coward and Whitehead, 1972) have shown that the hemoglobin concentration does not fall until there is actual body wasting. The role of pure protein deficiency is difficult to determine in the human subject because it is likely to be associated with other nutritional deficiencies, e.g., folic acid, iron, vitamin B_{12}. A study in Vietnam (Le Xuan Chat et al., 1971), involving pregnant women, nonpregnant women, and men, showed a deficiency of iron, B_{12}, and folic acid in 40%, a dual deficiency in 20%, and an isolated deficiency of one of the substances in 24%, the major cause of the deficiencies being an inadequate diet and/or chronic parasitic infection rather than pregnancy. Because of these complications, pure protein-deficiency anemia is probably better studied in the experimental animal.

Human studies in which attempts have been made to exclude other factors have suggested that the red cells are normochromic (Adams et al., 1967; Coward and Whitehead, 1972), the reticulocyte count is near normal (Adams et al., 1967), and the erythroid marrow either normal or cytopenic (Ghitis et al., 1963).

In protein-deficient animals, if the body weight is maintained from other sources of calories, there is a reduction in the bone marrow erythroid cells and in the reticulocyte count (Ito and Reissman, 1966; Reissman, 1964; Aschkenasy, 1963). There is also a reduction in plasma iron turnover (Donati et al., 1964). The incorporation of iron into red cells is reduced (Aschkenasy, 1963), and there is a gradual reduction in the total amount of red cells (Ito and Reissman, 1966).

Robsheit-Robbins et al. (1940) showed that in anemic protein-deficient dogs the rate of hemoglobin production could be increased 3- to 4-fold by stimulation of the bone marrow, and iron-deficient rats maintained on a protein-free diet will show a hemoglobin response following administration of iron (Pearson et al., 1937). Heath and Taylor (1936) have shown a similar response in human subjects. The bone marrow will also respond to erythropoietin stimulation in the presence of protein deficiency (Orten and Orten, 1945; Aschkenasy, 1963; Ito and Reissman, 1966).

VITAMIN E DEFICIENCY

Vitamin E deficiency can induce changes characteristic of iron deficiency in rabbits, including a low plasma iron concentration, an increase in the total iron-binding capacity, and a decreased saturation of transferrin. There is also an increase in erythrocyte protoporphyrin concentration, rapid clearance of transferrin-bound ^{59}Fe from the plasma, an increase in plasma iron turnover, and increased red cell utilization. The iron-deficient erythropoiesis in vitamin E-deficient rabbits is not the result of depletion of total body iron stores, but there is a sequestration of iron in the skeletal muscle (Chou et al., 1978).

BIBLIOGRAPHY

Adams, E. B., Scragg, J. N., Naidoo, B. T., Liljestrand, S. K., Kockram, V. I.: Observations on the aetiology and treatment of anaemia in kwashiorkor. Brit. med. J. *3,* 451 (1967).

Aschkenasy, A.: Études sur la production d'érythropoietine chez le rat carence en protéines. Rev. franç. Étud. clin. biol. *8,* 985 (1963).

Chou, A. C. et al.: Abnormalities of iron metabolism and erythropoiesis in vitamin E-deficient rabbits. Blood *52,* 187 (1978).

Coward, W., Whitehead, R. G.: Changes in haemoglobin concentrations during the development of kwashiorkor. Brit. J. Nutr. *28,* 463 (1972).

Donati, R. M., Chapman, C. V., Warnecke, M. A., Galagher, N. I.: Iron metabolism in acute starvation. Proc. Soc. exp. Biol. Med. *117,* 50 (1964).

Finch, C. A.: Protein deficiency and anemia. Plenary Session Papers. XIIth Congr. Internat. Soc. Hematol., New York 1968.

Ghitis, J., Piazuelo, E., Vitale, J. J.: Cali-Harvard nutrition project. III. The erythroid atrophy of protein deficiency in monkeys. Amer. J. clin. Nutr. *12,* 452 (1963).

Ghitis, J., Velez, H., Linares, F., Sinisterra, L., Vitale, J. J.: Cali-Harvard nutrition project. II. The erythroid atrophy of kwashiorkor and marasmus. Amer. J. clin. Nutr. *12,* 445 (1963).

Heath, C. V., Taylor, F. H. L.: The nitrogen metabolism in anemia during the regeneration of blood. J. clin. Invest. *15,* 411 (1936).

Ito, K., Reissman, K. R.: Quantitative and qualitative aspect of steady state erythropoiesis induced in protein-starved rats by long-term erythropoietin injection. Blood *27,* 343 (1966).

Le Xuan Chat, Le Xuan Ahn, Bun Mat et al.: Nutritional anemias in Vietnam. Rev. franç. Hémat. *11,* 807 (1971).

Orten, J. M., Orten, A. U.: The production of polycythemia by cobalt in rats made anemic by a diet low in protein. Amer. J. Physiol. *144,* 464 (1945).

Pearson, P. B., Elvehjem, C. A., Hart, E. B.: The relation of protein to hemoglobin building. J. biol. Chem. *119,* 749 (1937).

Reissman, K. R.: Protein metabolism and erythropoiesis. II. Blood *23,* 146 (1964).

Robsheit-Robbins, F. S., Madden, S. C., Rowe, A. P., Turner, A. P., Whipple, G. W.: Hemoglobin and plasma protein. Simultaneous production during continued bleeding as influenced by diet low in protein and other factors. J. exp. Med. *72,* 479 (1940).

CHAPTER 22

PERNICIOUS ANEMIA

Iron metabolism is disturbed in the presence of megaloblastic marrow associated with untreated pernicious anemia. Red cell maturation and differentiation is reduced (Nathan and Gardner, 1962), the absolute and percentage reticulocyte count is lower than normal (Murphy and Howard, 1957; Heilmeyer, 1964), and many of the reticulocytes are immature (Heilmeyer, 1970).

The plasma iron transport rate, reflecting total erythropoiesis is increased 3- to 5-fold (Finch et al., 1956; Myhre, 1964) or occasionally even 5- to 10-fold (Huff et al., 1950; Wasserman et al., 1952), but the radioiron incorporation into red cells is reduced to 10–50% (Finch et al., 1956), indicating that a considerable proportion of the erythropoiesis is ineffective (see Fig. 14/11-IV). The iron is accumulated in the reticulum cells of the bone marrow, the liver, and other tissues. The plasma iron concentration is usually higher than normal unless there is a complicating iron-deficiency state (Heilmeyer and Plötner, 1937; Moore et al., 1937; Vahlquist, 1941; Vannotti and Delachaux, 1942; Waldenström, 1945; Büchmann, 1959). A dual deficiency is more common in women but it always deserves thorough investigation, particularly because of the association of pernicious anemia with carcinoma of the stomach (Büchmann, 1959; Waldenström, 1945; and others).

In the untreated patient, in response to oral iron loading there is little rise in plasma iron level (Goldeck and Remy, 1951), but following therapy, when the plasma iron is reduced, iron loading may result in a considerable rise.

Free protoporphyrin concentration of the red cells is usually normal or reduced, while the coproporphyrin is reduced or absent (Watson, 1950; Schmid et al., 1950; Kehl and Günther, 1954; Saito et al., 1958). Heilmeyer (1964), however, found the coproporphyrin in the red cells to be somewhat higher than normal. This discrepancy has not been clarified.

In association with the ineffective erythropoiesis, the porphyrin concentration in the bone marrow is increased (Gewitz et al., 1962), and in both the bone marrow and the blood about 25–30% of the porphyrins belong to the type I isomer, which is unusual in other types of anemia. Specific treatment alleviates these disturbances in porphyrin synthesis.

The red cells in the untreated patient are abnormal in that they show increased saponin, thiourea, and glycerol permeability and have a shortened life-span (Bang and Orskov, 1937; Ponder and Rhoads, 1938; London et al., 1949; Hamilton et al., 1954; Schlegel and Böttner, 1955; Schlegel and Kappest, 1956; Schmid et al., 1964). This shortened life-span is reflected in hyperbilirubinemia and increased urobilin in the feces and urine (Heilmeyer, 1964). The urobilin excretion may rise 2–20 times the normal.

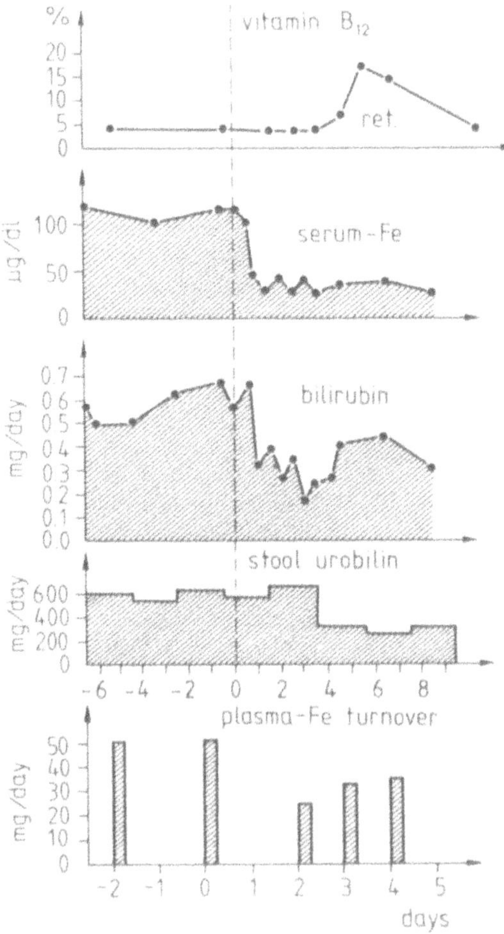

Fig. 22/1. The effect of vitamin B_{12} upon erythrokinetics in pernicious anemia. In the peripheral blood the ratio of reticulocytes increases with the parallel decrease in the other indices. The change is the consequence of the transformation of ineffective to effective erythropoiesis (after Finch, C. A. et al: Blood *11*, 807, 1956)

Following treatment there is a rapid fall in plasma iron concentration, this being an early and characteristic indicator of response (Hawkins, 1955; Büchmann, 1959) (Fig. 22/1) that precedes the development of a reticulocyte response (Moore, 1937; Vannotti, 1942; Waldenström, 1945; Büchmann, 1959; Heilmeyer, 1964). If the plasma iron drops below the normal value and remains there, iron supplementation is indicated. In addition to the fall in plasma iron, the iron-binding capacity returns to normal, and the red cell utilization of iron is markedly enhanced (Finch et al., 1956). This reflects the change from ineffective to effective erythropoiesis (see Fig. 14/11-IV and III).

BIBLIOGRAPHY

BANG, O., ORSKOV, S. I.: Variations in the permeability of the red cells in man, with particular reference to the conditions obtained in pernicious anemia. J. clin. Invest. *16*, 279 (1937).

BEAVEN, G. H., ELLIS, M. J., WHITE, J. C.: Studies on human foetal haemoglobin. II. Foetal haemoglobin levels in healthy children and adults and in certain haematological disorders. Brit. J. Haemat. *6*, 201 (1960).

BÜCHMANN, P.: Eisenstoffwechsel bei haemolytischer Anämie. In: KEIDERLING, W.: Eisenstoffwechsel, pp. 204–218. Thieme, Stuttgart 1959.

CHANARIN, I.: The Megaloblastic Anemias. Davis, Philadelphia 1969.

FINCH, C. A., COLEMAN, D. H., MOTULSKY, A. G., DONOHUE, D. M., REIFF, R. H.: Erythrokinetics in pernicious anemia. Blood *11*, 807 (1956).

FINCH, C. A. et al.: Ferrokinetics in man. Medicine *49*, 17 (1970).

FISCHER, H., HELBIG, W.: Diagnostik und Therapie megaloblastischer Anämien. Z. Ges. inn. Med. *32*, 425 (1977).

GEWITZ, R., COLLICA, C., MIESCHER, P. A.: Erythrokinetische Studien in der Diagnostik anämischer Zustände. Med. Klin. *57*, 1515 (1962).

GOLDECK, H., REMY, D.: Eisenstoffwechsel und Eisenmangel bei der regenerierenden Perniciosa. Dtsch. Arch. klin. Med. *198*, 422 (1951).

HAMILTON, H. E., DE GOWIN, E. L., SHETTS, R. F., JANNEY, C. D., ELLIS, J. A.: Studies with inagglutinable erythrocyte counts. VI. Accelerated destruction of normal adult erythrocytes in pernicious anemia, contribution of hemolysis to the oligocythemia. J. clin. Invest. *33*, 191 (1954).

HAWKINS, C. F.: Value of serum iron levels in assessing effect of haematinics in the macrocytic anemias. Brit. med. J. *1*, 383 (1955).

HEILMEYER, L.: Die Störungen der Bluthämsynthese. Thieme, Stuttgart 1964.

HEILMEYER, L.: Klinik des erythrocytären Systems. In: HEILMEYER, L., BEGEMANN, H. (eds.): Blut und Blutkrankheiten, Vol. 2/2. Springer, Berlin–Heidelberg–New York 1970.

HEILMEYER, L., PLÖTNER, K.: Das Serumeisen und die Eisenmangelkrankheit. Fischer, Jena 1937.

HUFF, R. L., HENNESSY, T. G., AUSTIN, R. E., GARCIA, J. F., ROBERTS, B. M., LAWRENCE, J. H.: Plasma and red cell iron turnover in normal subjects and in patients having various hematopoietic disorders. J. clin. Invest. *29*, 1041 (1950).

JOSEPHSON, A. M., MASRI, M. S., SINGER, L., DWORKIN, D., SINGER, K.: Starch block electrophoretic studies of human hemoglobin solutions. II. Results in cord blood, thalassemia and other hematologic disorders: Comparison with Tiselius electrophoresis. Blood *13*, 543 (1958).

KEHL, R., GÜNTHER, B.: Zur papierchromatografischen Isomeren-Analyse der Porphyrine. Hoppe-Seylers Z. physiol. Chem. *297*, 254 (1954).

LONDON, I. M., SHEMIN, D., WEST, R., RITTENBERG, D.: Heme synthesis and red blood cell dynamics in normal humans and in subjects with polycythemia vera, sickle-cell anemia and pernicious anemia. J. biol. Chem. *179*, 463 (1949).

MOORE, C. V., ARROWSMITH, W. R., QUILLIGAN, J. J., READ, J. T.: Studies in iron transportation and metabolism. I. J. clin. Invest. *16*, 613 (1937).

MOORE, C. V., DOAN, C. A., ARROWSMITH, W. R.: Studies in iron transportation and metabolism. II. The mechanism of iron transportation. Its significance in iron utilization in anemic states of varied etiology. J. clin. Invest. *16*, 627 (1937).

MURPHY, W. P., HOWARD, J.: A comparison of the effect of the vitamin B_{12} with that of liver extracts in the treatment of pernicious anemia during relapse and for maintenance. New Engl. J. Med. *247*, 838 (1957).

MYHRE, E.: Studies on the erythrokinetics in pernicious anemia. Scand. J. clin. Lab. Invest. *16*, 391 (1964).

NATHAN, D. G., GARDNER, F. H.: Erythroid cell maturation and hemoglobin synthesis in megaloblastic anemia. J. clin. Invest. *41*, 1086 (1962).

PONDER, E., RHOADS, C. P.: Red cell resistance to lysins in pernicious anemia. Proc. Soc. exp. Biol. Med. *38*, 540 (1938).

RECHENBERGER, J.: Über die Eisenbindungskapazität des Blutserums. IV. Das Verhalten der Eisenbindungskapazität und des Serumeisens bei der unbehandelten und regenerierenden perniciösen Anämie. Dtsch. Z. Verdau.- u. Stoffwechselkr. *16*, 223 (1956).

RICHTER, K.: Der Hämoglobingehalt des Blutes bei Anämien, Polycythaemia vera und Polyglobulien. Klin. Wschr. *36*, 911 (1958).

SAITO, G., MOREO, L., PERINI, A.: Contributo alla studio delle porfirine libere eritrocitarie nell'anemia de Biermer. Haematologica *43*, 51 (1958).

SCHLEGEL, B., BÖTTNER, H.: Zur Dynamik der intravitalen Erythrocytolyse. Med. Klin. *36*, 1518 (1955).

SCHLEGEL, B., KAPPEST, P.: Untersuchungen zur intravitalen Erythrolyse. Klin. Wschr. *34*, 805 (1956).

SCHMID, J. R., MOESCHLIN, S., HAEGI, V.: Pernicious anemia: an erythrokinetic and autoradiographic study, using H^3-thymidine, H^3-uridine and H^3-cytidine. Acta haemat. (Basel) *32*, 65 (1964).

SCHMID, R., SCHWARTZ, S., WATSON, C. J.: Porphyrins in the bone marrow and circulating erythrocytes in experimental anemias. Proc. Soc. exp. Biol. Med. *75*, 705 (1950).

VAHLQUIST, B.: Das Serumeisen. Acta paediat. Uppsala *28*, Suppl. 5 (1941).

VANNOTTI, A., DELACHAUX, A.: Der Eisenstoffwechsel und seine klinische Bedeutung. Basel 1942.

WALDENSTRÖM, J.: Perniciös anämi och cancer ventriculi. Nordisk Medicine *25*, 7 (1945).

WASSERMAN, L. R., RASHKOFF, I. A., LEAVITT, D., MAYER, J., PORT, S.: The rate of removal of radioactive iron from the plasma—an index of erythropoiesis. J. clin. Invest. *31*, 32 (1952).

WATSON, C. J.: The erythrocyte coproporphyrin. Variation in respect to erythrocyte protoporphyrin and reticulocytes in certain types of the anemias. Arch. intern. Med. *86*, 797 (1950).

WILMS, K.: Die megaloblastären Anämien. In: QUEISSER, W. (ed.): Das Knochenmark, pp. 414–428. Thieme, Stuttgart 1978.

WINTROBE, M. M.: Clinical Hematology. Lea and Febiger, Philadelphia 1967, 1975.

CHAPTER 23

HEMOLYTIC ANEMIAS

Disturbances in iron metabolism occur in the hemolytic anemias, both congenital and acquired. There is usually an elevated plasma iron level, and a normal or low plasma iron concentration is rare (Heilmeyer, 1964) (see Fig. 7/5). A rise in plasma iron concentration in a given phase of the disease indicates predominance of hemolysis over red cell production. Excessive erythropoiesis results in a change of plasma iron level in the opposite direction (Bothwell and Finch, 1962). The rise in temperature that is frequently associated with hemolytic crises does not influence the iron concentration of the plasma.

The total iron-binding capacity of the plasma is usually lower than normal (Laurell, 1947). The transferrin saturation is increased, and more iron is transferred to the storage depots (Finch et al., 1949; Huff et al., 1950). Most of the iron that accumulates in the reticuloendothelial system is stored in the form of hemosiderin. Iron absorption may increase in spite of the saturated iron stores, particularly in the presence of a large proportion of ineffective erythropoiesis, as in the thalassemias, with resulting severe hypersiderosis.

Plasma iron clearance is accelerated (Bothwell et al., 1956) (see Fig. 14/4B), and the plasma iron transport rate may increase to 6- to 8-fold that of healthy individuals. Instead of the normal 70–90% of red cell iron incorporation, it may be reduced to 25–40%. Dilution of the radioiron by a large plasma iron pool may be partly responsible for these lowered values so that they are not necessarily a true indication of the ineffective erythropoiesis.

Hemolysis is a feature of malaria, but in this condition the iron metabolism reflects the effects of infection rather than hemolysis (Fuhrmann and Knüttgen, 1950). Although at the onset of the febrile attack, at the time of sporulation, the plasma iron may be higher than normal, the level drops rapidly and tends to remain at below the normal level. Morczek (1952) found the plasma iron concentration to be low even in the febrile period in his patients. Presumably in malaria the reticuloendothelial activity is so stimulated in response to the infection that it prevents the rise in serum iron level that normally occurs with hemolysis. This reticuloendothelial effect is complemented by the increased red cell production in response to hemolysis, which will also tend to decrease the plasma iron level.

The patterns of survival of isotope-labeled erythrocytes were examined in patients suffering from two variants of nonspherocytic hemolytic anemia with decreased erythrocyte pyruvate kinase (PK) activity (Wazewska-Czyzewska and Gumińska, 1979).

Table 23/1
The ferrokinetic parameters in hemolytic anemia

Half-time of plasma iron clearance	Shortened
Plasma iron concentration	Increased (rarely normal or diminished)
Plasma iron transport rate	2–8 times normal
Incorporation rate into the red cells (%)	25–40; steeply rising incorporation curve, but after a short peak a rapidly declining course
Distribution of radioiron in the organism (on the basis of body surface measurements)	Initially: bone marrow > liver > spleen; later: spleen > liver > bone marrow

In one variant, with primary PK deficit, random destruction of erythrocytes was predominant in the process of hemolysis. In the second variant with primary magnesium-activated adenosine triphosphatase activity, erythrocytes were destroyed by senescence. Two subpopulations of labeled erythrocytes with different destruction rates were observed in all patients examined, except one with the second variant, with very mild hemolysis.

The ferrokinetic findings in hemolytic anemias are summarized in Table 23/1.

BIBLIOGRAPHY

Bothwell, T. H., Finch, C. A.: Iron Metabolism. Little, Brown and Co., Boston 1962.

Bothwell, T. H., Callender, S. T., Mallett, B., Witts, L. J.: The study of erythropoiesis using tracer quantities of radioactive iron. Brit. J. Haemat. *2*, 1 (1956).

Büchmann, P.: Eisenstoffwechsel bei haemolytischer Anämie. In: Keiderling, W. (ed.): Eisenstoffwechsel, pp. 204–218. Thieme, Stuttgart 1959.

Finch, C. A., Gibson, J. G., Peacock, W. C., Fluharty, R. G.: Iron metabolism. Utilization of intravenous radioactive iron. Blood *4*, 905 (1949).

Finch, C. A. et al.: Ferrokinetics in man. Medicine *49*, 17 (1970).

Fuhrmann, G., Knüttgen, H.: Serumeisenbefunde bei Malaria tertiana. Z. Tropenmed. Parasit. *1*, 515 (1950).

Heilmeyer, L.: Die Störungen der Bluthämsynthese. Thieme, Stuttgart 1964.

Huff, R. L., Hennessy, T. G., Austin, R. E., Garcia, J. F., Roberts, B. M., Lawrence, J. H.: Plasma and red cell iron turnover in normal subjects and in patients having various hematopoietic disorders. J. clin. Invest. *29*, 1041 (1950).

Laurell, C. B.: Studies on the transportation and metabolism of iron in the body with special reference to the iron-binding component in human plasma. Acta physiol. scand. *14* (Suppl. 46) (1947).

Moore, C. V., Arrowsmith, W. R., Quilligan, J. J., Read, J. T.: Studies in iron transportation and metabolism. J. clin. Invest. *16,* 613 (1937).

Morczek, A.: Untersuchungen über den Eisenstoffwechsel. II. Die Eisenausscheidung bei verschiedenen Krankheiten. Dtsch. Z. Verdau.- u. Stoffwechselkr. *12,* 14 (1952).

Vahlquist, B.: Das Serumeisen. Acta paediat. Uppsala *28,* Suppl. 5 (1941).

Vannotti, A., Delachaux, A.: Der Eisenstoffwechsel und seine klinische Bedeutung. Basel 1942.

Wazewska-Czyzewska, M., Guminska, S.: Congenital non-spherocytic haemolytic anaemia variants with primary and secondary pyruvate kinase deficiency. Brit. J. Haemat. *41,* 115 (1979).

Wöhler, F.: The treatment of haemochromatosis with desferrioxamine. In: Gross, F. (ed.): Iron Metabolism. Springer, Berlin–Göttingen–Heidelberg 1964.

CHAPTER 24

REFRACTORY HYPOCHROMIC ANEMIAS

THE SIDEROBLASTIC ANEMIAS

Thirty to forty years ago a group of diseases was described in which a hypochromic anemia occurred that did not respond to iron therapy. The anemia was attributed to disturbed iron utilization. It has later become evident that the common feature of this group of anemias is a disturbance of heme synthesis. Iron accumulates in the mitochondria of the nucleated red cells and characteristically forms a ring of iron-staining granules surrounding the nucleus. There is hyperplasia of the nucleated red cells in the bone marrow, and siderosis is found in the liver and other organs. The picture resembles somewhat that produced by pyridoxine deficiency in experimental animals. This group of anemias was originally termed sideroachrestic anemia, but it is now more usually termed sideroblastic anemia. Both genetic and acquired forms are found.

The acquired form was first described by Björkman (1956), and the inherited form was defined by Heilmeyer et al. (1957, 1958) and Garby and his collaborators (1957). Symptomatic sideroblastic anemias may occur with pyridoxine deficiency, lead poisoning, in association with neoplasms, particularly as a preleukemic feature; certain drugs such as antituberculous drugs and chloramphenicol may produce a sideroblastic anemia (Verwilghen et al., 1965). McGibbon and Mollin (1965) have reported a sideroblastic picture in association with a malabsorption syndrome.

HEREDITARY HYPOCHROMIC SIDEROBLASTIC ANEMIA

This condition can occur as a sex-linked anemia with a recessive inheritance, the female carriers being completely healthy or having some abnormal features of the red cells but not being anemic (Losowski and Hall, 1965). A similar disease occurring in women probably belongs to another subgroup (Heilmeyer, 1970). The anemia may have an early onset but often is relatively asymptomatic and can escape recognition until later in life. The features are hypochromic anemia, usually of only a moderate degree (9–13 g/dl), and the MCH is between 20 and 27 pg. The picture tends to be dimorphic, a proportion of the cells appearing normal although the majority are hypochromic. Anisocytosis and poikilocytosis occur, and there may be target cells present. The reticulocyte count is normal or slightly elevated. Serum bilirubin is normal. There is usually an increased osmotic resistance, as in iron-deficiency anemia and thalassemia; the white cell count and platelets are normal.

The bone marrow is always hyperplastic, and on staining for iron the typical ring sideroblasts make the diagnosis apparent. Electron microscopic studies show that the iron is in the mitochondria, and these are damaged and undergo degenerative alteration (see Fig. 17/7).

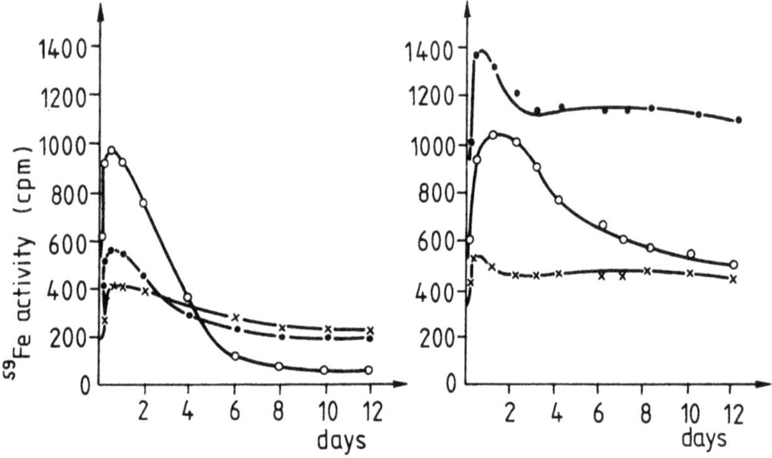

Fig. 24/1. Distribution of radioactivity over the bone marrow (○ – ○ – ○), spleen (× – × – ×) and liver (● – ● – ●) in healthy man (left) and in hereditary sideroachrestic anemia (right) (after Heilmeyer, L., in: Handbuch der inneren Medizin. Springer, Berlin–Heidelberg–New York 1970)

The serum iron is usually high, and there may be reduction of the iron-binding capacity resulting in a high saturation. The ferrokinetic investigations show that a high proportion of the iron is deposited in the liver, and incorporation into the bone marrow erythroblasts is diminished (Figs. 24/1 and 24/2) (Heilmeyer et al., 1958).

Coproporphyrin in the red cells and urine increases, while the protoporphyrin level of the red cells remains normal or is decreased (Garby et al., 1957; Heilmeyer, 1964a, 1970) (Fig. 24/3). The disturbance in conversion of coproporphyrinogen to protoporphyrinogen is presumably due to a congenital defect in coproporphyrinogen oxidase. Occasionally there is a rise in red cell protoporphyrin concentration, but this has usually been attributed to other causes such as hemorrhage, iron deficiency, or infection (Heilmeyer, 1970). In 1964 Heilmeyer described an unusual case of a 30-year-old female patient with a pronounced decrease in the concentration of all porphyrins in the red cells, and he postulated a disturbance in the first phase of heme synthesis. Pyridoxine deficiency, which might lead to a similar disturbance, was excluded in this case.

The erythropoiesis in the sideroblastic anemias is largely ineffective, and as a consequence iron absorption is often increased and severe siderosis may develop. The iron overload is particularly marked in the liver but may affect other organs as well, with the eventual development of secondary hemochromatosis.

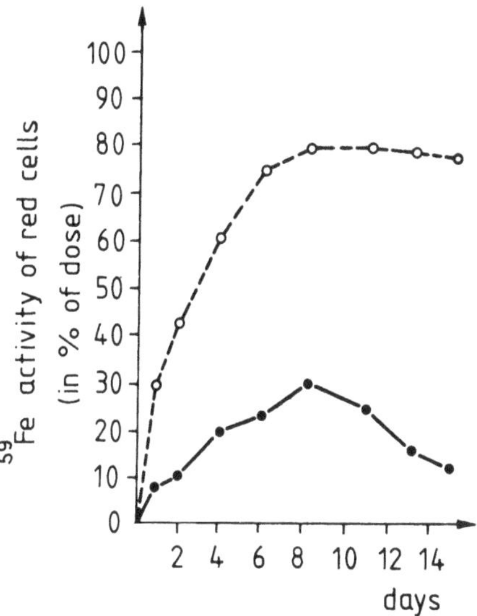

Fig. 24/2. Radioactivity of the red cells after the i.v. injection of ^{59}Fe in healthy man (o– –o– –o) and in hereditary sideroachrestic anemia (•–•–•) (after Heilmeyer, L., in: Handbuch der inneren Medizin. Springer, Berlin–Heidelberg–New York 1970)

In a young patient, provided that the various causes of a symptomatic form of sideroblastic anemia can be excluded, the differential diagnosis has to be from thalassemia, which also presents as a refractory hypochromic anemia. The typical appearance of the sideroblasts, however, will make the distinction easy.

Some patients have been found to be pyridoxine-responsive, and it is always worthwhile to attempt treatment with large doses of this vitamin. Transfusions may have to be resorted to, but because of the danger of iron overload they should be avoided if possible. Crosby and Sheely (1960) have treated some patients with venesection with some improvement, and again, particularly in the younger subjects, this approach may be tried. If transfusions prove necessary, treatment with desferrioxamine may be helpful to reduce the iron load.

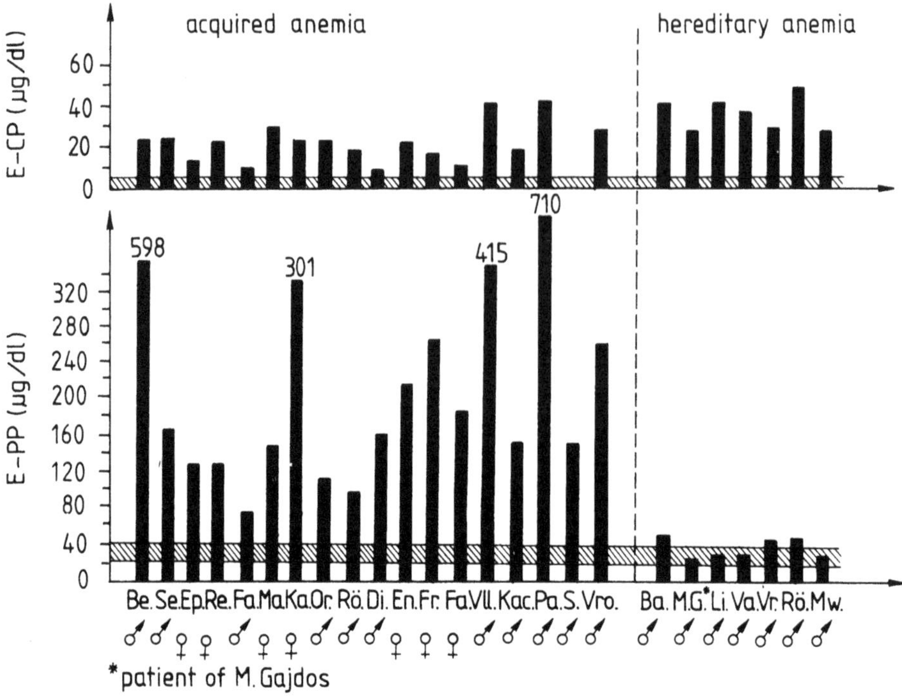

Fig. 24/3. Free coproporphyrin and protoporphyrin level of red cells in 18 patients with acquired sideroblastic anemia and 7 patients with hereditary sideroachrestic anemia (after Hellmeyer, L., in: Handbuch der inneren Medizin. Springer, Berlin–Heidelberg–New York 1970)

ACQUIRED SIDEROBLASTIC ANEMIA

This condition has been given a variety of names: chronic refractory anemia with sideroblastic bone marrow (Björkman, 1956), anemia refractoria sideroblastica (Heilmeyer et al., 1958), and refractory normoblastic anemia (Dacie et al., 1959). The disease develops usually between 50 and 70 years of age and affects both males and females. In most cases the disease is more symptomatic than in the hereditary form and may be quite severe, the hemoglobin ranging between 3 and 12 g/dl. The anemia is usually strikingly dimorphic, so the MCH may be around normal. The combination of normochromic cells and pronounced hypochromic cells suggests that the disturbance of heme synthesis affects only one of two red cell populations. This is supported by electron microscopic studies according to which, although most of the erythroblasts show structural damage of the mitochondria with extensive iron accumulation, other cells are morphologically normal (Bessis and Breton-Gorius, 1964). White cell and platelet counts are usually normal or slightly reduced, but occasionally a pronounced leucopenia and thrombocytopenia may occur (Mollin, 1965).

The bone marrow is very hyperplastic, and a high proportion of the erythroblasts are ring sideroblasts, the sideroblast index usually being between 130 and 150 (Merker and Krauss, 1964; Bousser et al., 1965). Heavy deposits of iron in the reticuloendothelial system are often observed (see Fig. 8/12 and Table 8/3).

The disease may remain in a chronic phase for some years but a relatively high proportion of cases transform into an acute leukemia, which is usually myeloblastic leukemia or erythroleukemia. Björkman (1959) suggests that sideroblastic anemia can be regarded as a preleukemic condition. On the other hand, Hayhoe and Quaglino (1960) have drawn attention to differences in the results of cytochemical examinations in typical acquired sideroblastic anemia and erythroleukemia. In the former, the sideroblast index is usually high and relatively few erythroblasts are PAS-positive. In the neoplastic condition, the sideroblast index is lower and the number of PAS-positive nucleated cells is much higher. Merker (1968) and Bousser et al. (1965) have made similar observations.

Both congenital and primary acquired sideroblastic anemias are a heterogeneous group of disorders with different defects in heme synthesis, reduced activity of delta-aminolevulinic acid synthetase being common but not invariable in both types. Those that are pyridoxine-responsive may be due to the presence of an abnormal ALA-synthetase apoenzyme, which requires excessive amounts of the coenzyme, pyridoxal phosphate, to achieve normal activity. Other abnormalities of heme synthesis may also be present (Konopka and Hoffbrand, 1979).

With regard to treatment in the older patients, pyridoxine should be tried, as some cases will respond, while others may respond to administration of vitamins B_1, B_2, and B_6 (Frerichs and Beck, 1964). In other cases large doses of folic acid or vitamin C have proved effective (Verloop et al., 1962; McGibbon and Mollin, 1965). The results, however, tend to last, and the anemia readily recurs on stopping treatment. As in the younger patients, desferrioxamine may be used to treat or delay the onset of hemosiderosis.

PYRIDOXINE-RESPONSIVE ANEMIAS

As already mentioned, some of the sideroblastic anemias may show a response to treatment with pyridoxine. True vitamin B_6-deficiency anemias, which can be cured with relatively small doses of vitamin B_6, have been described (Snyderman et al., 1950; Spencer et al., 1961; Verloop and Rademaker, 1960; Gehrmann, 1958, 1962; and others). In every respect these anemias correspond to experimental pyridoxine-deficiency anemia (Wintrobe et al., 1943; Cartwright and Wintrobe, 1948; Poppen et al., 1952), and can be described as symptomatic sideroblastic anemia.

Vitamin B_6 in the form of pyridoxal-5-phosphate plays an important role in the first step of heme synthesis, in the union of glycine and succinic acid. Because of this, in vitamin B_6 deficiency the entire heme synthesis is affected, with the eventual development of hypochromic anemia accompanied by hypersideremia and hyperplastic bone marrow erythropoiesis. In the red cells the concentration of

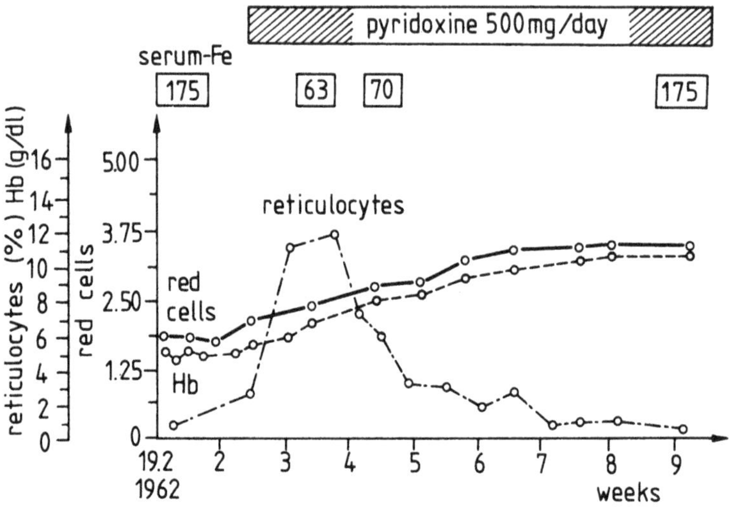

Fig. 24/4. Reticulocytosis and moderate response in red cell count and hemoglobin in sideroblastic anemia treated with pyridoxine (after Heilmeyer, L., in: Handbuch der inneren Medizin. Springer, Berlin–Heidelberg–New York 1970)

protoporphyrin is reduced, as are the other precursors of heme synthesis (Cartwright and Wintrobe, 1948). In response to treatment with vitamin B_6 the pathologically low values rapidly return to normal and the anemia is corrected (Spencer et al., 1961).

In contrast to the true pyridoxine deficiency, other pyridoxine-responsive sideroblastic anemias are not accompanied by other signs of vitamin B_6 deficiency, and in spite of prolonged administration of large doses of pyridoxine remission is only transitory (Fig. 24/4). The pyridoxine metabolic disorder is probably not an etiological factor in the development of the disease (Maier, 1957; Bishop and Bethell, 1959; Bickers et al., 1962; Vuylsteke et al., 1961; Heilmeyer, 1964a; Bourne et al., 1965; and others).

THE DISTURBANCE OF HEME SYNTHESIS IN THALASSEMIA

The thalassemic syndromes are common in some parts of the world, particularly in the Mediterranean area and Southeast Asia. They are genetically determined diseases that in the homozygous form give rise to severe hemolytic anemia. The heterozygous state is asymptomatic apart from a mild anemia.

The underlying defect is one of a disturbance of the globin chain synthesis. In the more common beta thalassemia, there is reduced synthesis of beta chains, with the resultant excess of alpha chain and an enhanced synthesis of gamma and delta chains with the accumulation of Hb-F and Hb-A_2. In alpha thalassemia the production of the alpha chains is reduced.

The excess of globin chain precipitates in the red cells, causing their premature removal from the circulation. The mean cell volume and mean hemoglobin concentration of the red cells are reduced, and target cells, poikilocytes, and schizocytes are frequent. The bone marrow erythropoiesis is hyperplastic, the ratio of sideroblasts increases, and the iron accumulates in the reticulum cells of the liver, spleen, and bone marrow. Severe iron overload is a major complication of the disease (Bannerman, 1961; Witzleben and Wyatt, 1961) (Fig. 24/5). The damage from the iron overload may begin very early in life, since Iancu et al. (1977) have

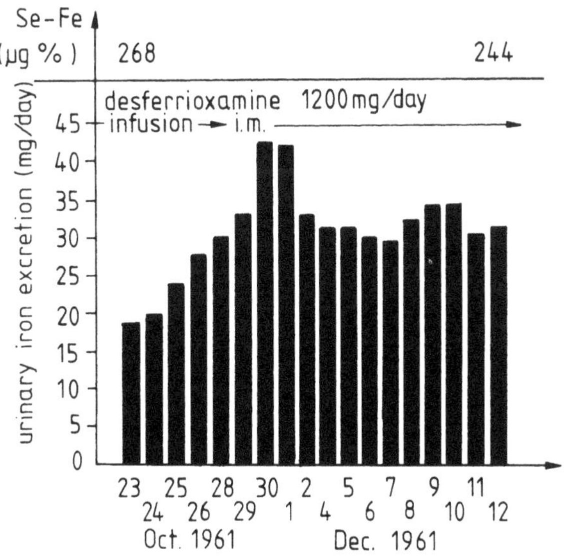

Fig. 24/5. Urinary iron excretion in thalassemia after the i.v. and i.m. injection of DFO. The increased iron excretion reflects the hypersiderosis (after Wöhler, F., in: Gross, F.: Iron Metabolism. Springer, Berlin–Göttingen–Heidelberg 1964)

shown that abnormal collagen formation may be found in liver biopsies with early iron deposition before there is evidence of parenchymal cell damage.

Electron microscopic studies of the bone marrow have shown that ferritin not only occurs randomly in the cytoplasm of the nucleated red cells or in the form of siderosomes but is deposited in large amounts in the mitochondria. Rhopheocytosis is active (Bessis and Breton-Gorius, 1962).

The life-span of the red cells is considerably reduced in the homozygous form of the disease. The serum bilirubin level is usually elevated, the serum iron is high, and the total iron-binding capacity is usually saturated. In the heterozygous state the serum iron level is usually normal or somewhat increased.

Bannerman et al. (1959) were the first to study the disorder of heme synthesis using radioiron and ^{14}C-glycine. They found a more profound disturbance of heme synthesis in thalassemia than in other types of hemolytic anemia. Heilmeyer (1964a) had measured the concentration of heme precursors in the peripheral red cells together with their excretion in the 24-hour urine of 15 patients whom they classified as thalassemia minor and thalassemia minima. The free protoporphyrin concentration of the erythrocytes, with the exception of one case, was considerably augmented. The values ranged between 45 and 438 μg/dl of red cells. The free coproporphyrin level of the red cells was also elevated (1.8–94.0 μg/dl of red cells), and the coproporphyrin was excreted in large amounts in the urine in the majority of cases (57–308 μg/24 hours). The excretion of delta-aminolevulinic acid was increased in all patients (3.140–5.143 μg/24 hours). The porphobilinogen excretion was variable.

The rise in the protoporphyrin level was not the result of an increased proportion of reticulocytes, but combined with the high iron content of the erythrocytes and the nucleated red cells it indicated a disturbance in the combination of protoporphyrin with iron (Bannerman et al., 1959; Heilmeyer, 1964a).

In normal physiological conditions the components of hemoglobin, protoporphyrin, iron, and globin can be found in slight excess but in appropriate proportions in the nucleated red cells. If the proportion of any of the three substances is reduced or production decelerated, there will be intracellular accumulation of the other two. In thalassemia, where there is reduction of globin production, protoporphyrin and iron accumulate in the nucleated red cells. Heilmeyer (1964a) concluded that the disordered heme synthesis could be directly ascribed to the reduction in globin production and was not an independent process.

In alpha thalassemia Heilmeyer (1964a) found the uroporphyrin, coproporphyrin, and protoporphyrin concentrations of the red cells to be equally high (uroporphyrin 75.5, coproporphyrin 24.0, protoporphyrin 444.0 μg/dl of red cells).

PATHOLOGICAL HEME SYNTHESIS ASSOCIATED WITH LEAD AND OTHER TOXIC SUBSTANCES

A number of toxic substances cause symptomatic sideroblastic anemia by the inhibition of heme synthesis. Gajdos et al. (1958), for example, have shown that sodium cyanide, novarsenobenzol, malonic acid, alpha-ketoglutaric acid, and other compounds inhibit the conversion in part of delta-aminolevulinic acid to porphobilinogen, and porphobilinogen to porphyrin. Sodium fluoride and monoiodoacetic acid impede only the course of the former reaction. Heilmeyer and Kohn (1964) reported the reversible disorder of heme synthesis arising in the course of INH treatment of a tuberculous patient.

The use of alkylating agents (chlorambucil, nitrogen mustard, and busulphan) was reported by Bowman (1962) to give rise to pathological sideroblasts in the bone marrow and hypoplastic erythropoiesis in some of the cases. Such changes might be associated with disturbed heme synthesis.

From the clinical point of view, by far the most important of the toxic substances leading to disturbed heme synthesis is lead. In acute lead poisoning anemia develops within some weeks. It is associated with a reticulocytosis and the appearance of basophilic stippling in the red cells and normoblasts. The anemia may be associated with a marked leucocytosis.

In chronic lead poisoning the anemia may be either normocytic or microcytic. The mean hemoglobin concentration of the red cells is usually lower than normal (28–30%), but the hypochromia is not as marked as in iron-deficiency anemia. The hemoglobin level seldom drops below 9–10 g/dl, and the hematocrit value is not less than 30–35%. The reticulocytes may be slightly increased, and basophilic stippling is very characteristic. These granules in the red cells contain RNA and are ribosome aggregates (Sano, 1958). Anisocytosis, poikilocytosis, and polychromasia all occur, and occasional nucleated red cells may be seen in the peripheral blood. The red cell life-span is reduced (Westerman et al., 1965).

Charache and Weatherall (1966) found a fast hemoglobin similar to Hb-A_3 in the blood of about 4% of children with lead poisoning.

The bone marrow shows erythroid hyperplasia and an increase in siderosomes in the nucleated red cells (Bessis and Breton-Gorius, 1956).

The lead damages both globin and heme synthesis (Eriksen, 1955; Goldberg et al., 1956; Kassenaar et al., 1957; Waxman and Rabinovitz, 1966). It inhibits various stages of heme synthesis. It reduces the activity of delta-aminolevulinic acid dehydrase (Lichtman and Feldman, 1963), as a result of which the amount of delta-ALA in the bone marrow erythroblasts, in the peripheral blood, and in the plasma is increased, and its excretion in the urine is higher than in any other disease (Dresel and Falk, 1953; Goldberg et al., 1956; Gibson et al., 1958; Haeger-Aronsen, 1960). In some other sideroblastic anemias the daily excretion of delta-aminolevulinic acid may amount to 10–12 mg and is even higher in porphyria cutanea tarda, but in lead poisoning a daily excretion of 100–150 mg is not infrequent. This may be of great diagnostic significance, and Heilmeyer (1964a) has found a close correlation between the amount of lead in the body and the delta-ALA excreted.

In rats the conversion of porphobilinogen to uroporphyrin has been shown to be inhibited, giving rise to an increased excretion of PBG. In human studies, however, the PBG excretion has been found to be variable. Watson et al. (1953) and Rubino (1961) found a moderate increase in excretion, but Waldenström and Vahlquist (1939) and Haeger-Aronsen (1960) found normal values.

Schmid and Schwartz (1952) and Chisolm (1964) found increased excretion of uroporphyrin I, indicating a block prior to coproporphyrin production in lead poisoning. However, one of the most characteristic signs of lead poisoning is the rise in coproporphyrin excretion (Grotepass, 1932), which is the consequence of disturbed CP–PP conversion and of the resulting high coproporphyrin content of the red cells (Kreimer-Birnbaum and Grinstein, 1965). The coproporphyrin concentration of the red cells is usually between 50 and 100 μg/dl and the daily coproporphyrin content of the urine 500–3600 μg (Heilmeyer, 1964a). The red cells and the urine both contain almost exclusively coproporphyrin III, and the isomer CP–I is excreted only in small amounts. The intermediary substance that is excreted can be demonstrated partly as colorless coproporphyrinogen (Watson et al., 1951) and partly as coproporphyrinogen–Zn complex.

Iron incorporation into the protoporphyrin molecule is greatly inhibited (Goldberg et al., 1956). This is explained by the decreased activity of heme synthetase (Boyett and Butterworth, 1962). This step in heme synthesis is the most vulnerable point of the entire process, the "Achilles' heel" of the biosynthesis (Heilmeyer, 1964a). The free protoporphyrin concentration of the red cells is elevated as the result of the disturbed iron incorporation, and values may reach 100–900 μg/dl (Waldenström, 1937; Heilmeyer, 1964a).

In summary, lead exerts its influence upon hemoglobin synthesis primarily by reducing the activity of delta-ALA dehydrase and heme synthetase, and presumably it also has some inhibiting effect on the activity of uroporphyrinogen decarboxylase, as illustrated by the following scheme:

$$\text{delta-ALA} \nRightarrow \text{PBG} \rightarrow \text{UP} \nrightarrow \text{CP} \nrightarrow \text{PP} + \text{Fe} \nRightarrow \text{heme}.$$

According to Kreimer-Birnbaum and Grinstein (1965), lead has at least four proven points of attack, and the presumption of a further three is justified. It seems that the lead exerts its effect upon the SH groups of delta-ALA dehydrase, UPG carboxylase, and heme synthetase. The effect is reversible, at least in part, by cysteine (Heilmeyer, 1964a). The electron microscopic investigations of Bessis and Breton-Gorius (1957) have revealed the severe damage of the mitochondria, where a significant part of the heme biosynthesis takes place. The mitochondria are swollen, the cristae are broken off, and iron agglomerates adhere to their membranes.

Other heavy metals can bind SH groups without eliciting any disturbance of heme synthesis, and it seems possible that this is because they are less able to enter the cytoplasm of the cells and damage the mitochondria (Heilmeyer, 1964a).

SIDEROBLASTIC ANEMIAS ARISING IN CONNECTION WITH ANTITUBERCULOUS DRUGS

Heilmeyer and Kohn (1964) reported a case of a 40-year-old woman with tuberculosis. She had a mild hypochromic anemia (Hb 8–9 g/dl) with a hypoplastic bone marrow and normal serum iron concentration. The anemia was refractory to treatment. After the start of treatment with INH, 50 mg/day, the anemia became more severe. The hemoglobin dropped to 5 g/dl, and the serum iron concentration rose from 90 μg/dl to 273 μg/dl. There was no evidence of hemolysis, but the bone marrow sideroblasts increased from 32% to 78% with a preponderance of pathological forms. The free erythrocyte protoporphyrin was 153 μg, the coproporphyrin level 45 μg/dl red cells. After withdrawal of the INH the anemia improved rapidly following a marked reticulocytosis (117‰). The serum iron dropped to 66 μg/dl.

Pyridoxine deficiency was presumed to play a role in the development of the anemia provoked by the INH, and the patient was given 200 mg of pyridoxine daily. This resulted in a further reticulocytosis; the abnormal sideroblasts disappeared from the bone marrow almost completely, and the protoporphyrin and coproporphyrin level of the peripheral blood dropped to 63 and 6.3 μg/dl, respectively.

INH is an antagonist to pyridoxal-5-phosphate, forming a compound with it that is then excreted through the kidneys. INH is therefore liable to give rise to vitamin B_6 deficiency. However, the theory that the anemia could be due solely to pyridoxine deficiency is untenable because in this case the free protoporphyrin concentration of the red cells should decrease instead of increase. It would appear, therefore, that in addition to the deficiency of B_6 there has to be a decrease in the activity of heme synthetase and coproporphyrinogen oxidase, which leads to a rise in protoporphyrin and coproporphyrin concentration.

The reason for the relatively low incidence of anemia during INH treatment is attributable to the fact that probably many milder cases go undetected.

SHAHIDI–NATHAN–DIAMOND ANEMIA

Shahidi et al. (1964) described a hypochromic anemia in two siblings, a boy and a girl, aged $12^1/_2$ and 5 years, respectively. The anemia in many respects resembled iron deficiency, but it was associated with a rise in serum iron and proved to be resistant to iron medication.

There was a considerable reduction in hemoglobin (5.8–8.4 g/dl), a normal or moderately increased red cell count (4.8–5.9 million), and a marked reduction in red cell volume (48–62 μ^3). The relative reticulocyte count was 9–44‰. The plasma iron level was 170–250 μg/dl and the TIBC 300–450 μg/dl. The transferrin was normal both quantitatively and qualitatively. The free protoporphyrin concentration of the red cells was raised, and apart from the increased plasma iron level all the findings were in keeping with iron-deficiency anemia. The diagnosis of thalassemia was excluded on the basis of normal Hb-F and Hb-A_2 levels.

Heavy iron deposition was found in the cells of the liver parenchyma but there was no iron in the Kupffer cells. Ferrokinetic studies showed that only 40–65% of the radioiron was incorporated into the peripheral red cells, and a large proportion of the radioiron was deposited in the liver.

They suggested that the disorder was due to an inability of macrophages to take up iron, so that there was a reduction in the iron transport between the reticulum cells and the erythroblasts (Bessis and Breton-Gorius, 1962).

FANCONI'S ANEMIA

Skikne et al. (1978) have studied the erythrokinetic features in patients with Fanconi's anemia over periods of 6 months to 11 years. All their patients were pancytopenic with a depression of the granulocytic and megakaryocytic elements of the bone marrow. In two studies the erythroid activity was depressed, but in a further three a significant erythroid marrow response was present. Later studies showed that erythropoiesis was increased but was markedly ineffective. ^{51}Cr studies indicated a reduced red cell life-span in all the patients studied.

GENETICALLY DETERMINED MICROCYTIC HYPOCHROMIC ANEMIAS

Hereditary microcytic and hypochromic anemia of the mouse was described by Pinkerton and Bannerman (1967, 1968), Pinkerton et al. (1970), and Bannerman et al. (1972). The features were hyposideremia, increased iron-binding capacity of the plasma, high erythrocyte protoporphyrin level, and depleted iron stores. Simple iron deficiency was not the cause since there was neither rapid clearance of iron from the plasma nor a high utilization of radioiron, and the response to parenteral iron treatment was incomplete. The defect is attributed to an impairment of the cellular uptake of iron involving the placental iron transfer and the transfer of iron from the intestinal lumen to the mucosa and from the plasma to the erythroblasts (Kingston et al., 1978).

The hereditary hypochromic microcytic anemia of Belgrade laboratory (b/b) rats was studied by Sladić-Simić et al. (1972). In these rats the iron absorption from the intestine is enhanced in spite of a high serum iron concentration. The plasma iron turnover is increased. The b/b rats treated with iron are viable, with a nearly normal reproductive capacity, but their circulating erythrocytes remain hypochromic and microcytic with a shortened life-span, indicating that the genetic lesion is at the level of the erythroid cell.

BIBLIOGRAPHY

Albahary, C.: Lead and hemopoiesis. Amer. J. Med. *52*, 369 (1972).

Bannerman, R. M.: Thalassemia. A survey of some aspects. Grune and Stratton, New York–London 1961.

Bannerman, R. M., Edwards, J. A., Kreimer-Birnbaum, M. et al.: Hereditary microcytic anaemia in the mouse; studies in iron distribution and metabolism. Brit. J. Haemat. *23*, 235 (1972).

Bannerman, R. M., Grinstein, M., Moore, C. V.: Haemoglobin synthesis in thalassemia. Brit. J. Haemat. *5*, 102 (1959).

Berk, P. D. et al.: Hematologic and biochemical studies in a case of lead poisoning. Amer. J. med. *48*, 137 (1970).

Bernard, J., Lortholang, P., Lévy, J. P., Boiron, M., Najean, Y., Tanzer, J.: Les anémies normochromes sidéroblastiques. Nouv. Rev. franç. Hémat. *3*, 723 (1963).

Bessis, M., Breton-Gorius, J.: Étude au microscope électronique du sang et des organes hémopoétiques dans le saturnisme expérimental. Sem. Hôp. Paris *4*, 441 (1956).

Bessis, M., Breton-Gorius, J.: Étude au microscope électronique du sang et des organes hémopoétiques dans le saturnisme expérimental. Rev. hémat. *12*, 43 (1957); *14*, 165 (1959).

Bessis, M., Breton-Gorius, J.: Iron metabolism in the bone marrow as seen by electron microscopy. Blood *19*, 635 (1962).

Bessis, M., Breton-Gorius, J. (1964), cit. Heilmeyer, 1964a.

Bessis, M., Jensen, W. N.: Sideroblastic anaemia, mitochondria and erythroblastic iron. Brit. J. Haemat. *11*, 49 (1965).

Bickers, J. N., Brown, L., Sprague, C. C.: Pyridoxine responsive anemia. Blood *19*, 304 (1962).

Bishop, R. C., Bethell, F. H.: Hereditary hypochromic anemia with transfusion hemosiderosis treated with pyridoxine. A case report. New Engl. J. Med. *261*, 486 (1959).

Björkman, S. E.: Chronic refractory anemia with sideroblastic bone marrow, a study of 4 cases. Blood *11*, 250 (1956).

Björkman, S. E.: Anaemia refractoria sideroblastica. In: Keiderling, W. (ed.): Eisenstoffwechsel. Thieme, Stuttgart 1959.

Bourne, M. S., Elves, M. W., Israels, M. C. G.: Familial pyridoxine responsive anaemia. Brit. J. Haemat. *11*, 1 (1965).

Bousser, I., Zittoun, R., Guillerm, M.: X. Congr. Soc. Europ. Hémat., Résumés, p. 236. Strasbourg 1965.

Bowman, W. D.: Abnormal (ringed) sideroblasts in various hematologic and non-hematologic disorders. Blood *18*, 662 (1962).

Boyett, J. D., Butterworth, C. E.: Lead poisoning and hemoglobin synthesis. Amer. J. med. *32*, 884 (1962).

Brain, M. C., Herdan, A.: Tissue iron stores in sideroblastic anaemia. Brit. J. Haemat. *11*, 107 (1965).

Cartwright, G. E., Deiss, A.: Sideroblasts, siderocytes and sideroblastic anemia. New Engl. J. Med. *292*, 185 (1975).

CARTWRIGHT, G. E., WINTROBE, M. M.: Studies on free erythrocyte protoporphyrin, plasma copper and plasma iron in normal and in pyridoxine-deficient swine. J. biol. Chem. *172,* 557 (1948); *176,* 571 (1948).
CATOVSKY, D. et al.: Sideroblastic anaemia and its association with leukaemia and myelomatosis. Brit. J. Haemat. *20,* 385 (1971).
CHARACHE, S., WEATHERALL, D. J.: Fast hemoglobin in lead poisoning. Blood *28,* 377 (1966).
CHISOLM, J. J.: Disturbances in the biosynthesis of heme in lead intoxication. J. Pediat. *64,* 174 (1964).
CROSBY, W. H., SHEELY, T. W.: Hypochromic iron loading anemia. Brit. J. Haemat. *6,* 56 (1960).
DACIE, J. V., SMITH, M. D., WHITE, J. G., MOLLIN, D. L.: Refractory normoblastic anemia. A clinical and hematological study of seven cases. Brit. J. Haemat. *5,* 56 (1959).
DRESEL, E. I. B., FALK, I. E.: Conversion of beta-amino-laevulinic acid to porphobilinogen in a tissue system. Nature *172,* 1185 (1953).
EDWARDS, J. A., BANNERMAN, R. M.: Hereditary defect of intestinal iron transport in mice with sex-linked anemia. J. clin. Invest. *49,* 1869 (1970).
EDWARDS, J. A., HOKE, J. E.: Inherited defect in intestinal mucosal iron uptake in mice with hereditary microcytic anemia. Proc. Soc. exp. Biol. *141,* 81 (1972).
EDWARDS, J. A., HOKE, J. E.: Mucosal iron binding proteins in sex-linked anemia and microcytic anemia of the mouse. J. Med. *9,* 353 (1978).
ERIKSEN, L.: The effect of lead on the in vitro biosynthesis of heme and free erythrocyte porphyrins. Scand. J. clin. Lab. Invest. *7,* 80 (1955).
FRERICHS, H., BECK, K.: Ein Beitrag zur Kasuistik der Vitamin B_6-empfindlichen sideroachrestischen Anämien. Med. Welt *1964,* 777.
GAJDOS, I. A., GAJDOS-TÖRÖK, M., BÉNARD, J. H.: Porphyries. Étude chimique et biologique. Paris 1958.
GARBY, L., SJÖLIN, S., VAHLQUIST, B.: Chronic refractory hypochromic anaemia with disturbed haem metabolism. Brit. J. Haemat. *3,* 55 (1957).
GARDNER, F. H., NATHAN, D. G.: Hypochromic anemia and hemochromatosis. Amer. J. med. Sci. *243,* 81 (1962).
GEHRMANN, G.: Pyridoxin-Mangelanämie beim Menschen. Folia haemat. (Frankfurt) N. F. *2,* 225 (1958).
GEHRMANN, G.: Das Pyridoxin-Mangelsyndrom beim Menschen. Ergebn. inn. Med. Kinderheilk. *19,* 274 (1962).
GIBSON, K. D., LAVER, W. G., NEUBERGER, A.: Initial stages in the biosynthesis of porphyrins. Biochem. J. *70,* 71 (1958).
GOLDBERG, A., ASHENBRUCKER, H. E., CARTWRIGHT, G. E., WINTROBE, M. M.: Studies on the biosynthesis of heme in vitro by avian erythrocytes. Blood *11,* 821 (1956).
GOULD, S. E., KULLMAN, H. J., SHECKET, H. A.: Effect of lead therapy on blood cells of cancer patients. Amer. J. med. Sci. *194,* 304 (1937).
GROTEPASS, W.: Zur Kenntnis des im Harn auftretenden Porphyrins bei Bleivergiftung. Hoppe-Seylers Z. Physiol. Chem. *205,* 193 (1932).
HADEN, H. T.: Pyridoxine-responsive sideroblastic anemia due to antituberculous drugs. Ann. intern. Med. *120,* 602 (1967).
HAEGER-ARONSEN, B.: Studies on urinary excretion of delta-amino-laevulic acid and other haem precursors in lead workers and lead intoxicated rabbits. Scand. J. clin. Lab. Invest. *12,* Suppl. 47 (1960).
HAYHOE, F. G., QUAGLINO, D.: Refractory sideroblastic anaemia and erythraemic myelosis: possible relationship and cytochemical observations. Brit. J. Haemat. *6,* 381 (1960).
HEILMEYER, L.: Blut und Blutkrankheiten. In: Handbuch der inneren Medizin, Vol. 2. Springer, Berlin 1941.
HEILMEYER, L.: Die Störungen der Bluthämsynthese. Thieme, Stuttgart 1964a.
HEILMEYER, L.: Human hyposideraemia. In: GROSS, F. (ed.): Iron Metabolism, pp. 201–213. Springer, Berlin–Göttingen–Heidelberg 1964b.

HEILMEYER, L.: Blut und Blutkrankheiten. In: Handbuch der inneren Medizin, Vol. 2/2. Springer, Berlin–Heidelberg–New York 1970.
HEILMEYER, L., KOHN, R.: In: HEILMEYER, L.: Die Störungen der Bluthämsynthese. Thieme, Stuttgart 1964.
HEILMEYER, L., WÖHLER, F.: Drawing of iron in haemochromatosis by means of desferrioxamine. Freiburg med. Res. *1*, 61 (1962).
HEILMEYER, L., EMMRICH, J., HENNEMANN, H. H., LEE, M. H., BILGER, R.: Über eine neuartige hypochrome Anämie bei zwei Geschwistern auf der Grundlage einer Eisenverwertungsstörung. Vortrag auf der 12. Jahresversammlung der Schweizer Hämatologischen Gesellschaft, Schaffhausen. Ref.: Schweiz. med. Wschr. *87*, 1237 (1957).
HEILMEYER, L., KEIDERLING, W., BILGER, R., BERNAUER, H.: Über chronische refraktäre Anämien mit sideroblastischem Knochenmark (anaemia refractoria sideroblastica). Folia haemat. (Frankfurt) N. F. *2*, 29 (1958).
HEILMEYER, L., KEIDERLING, W., MERKER, H., CLOTTEN, R., SCHUBOTHE, H.: Die Anaemia refractoria sideroblastica und ihre Beziehung zur Lebersiderose und Hämochromatose. Acta haemat. (Basel) *23*, 1 (1960).
HERSHKO, C., RACHMILEWITZ, E. A.: Mechanism of desferrioxamine-induced iron excretion in thalassaemia. Brit. J. Haemat. *42*, 125 (1979).
HINES, J. D., GRASSO, J. A.: The sideroblastic anemias. Semin. Hemat. *7*, 86 (1970).
IANCU, T. C., NEUSTEIN, H. B., LANDING, B. H.: The liver in thalassaemia major: ultrastructural observation. In: Iron Metabolism. Ciba Symposium 51. Elsevier, Amsterdam 1977.
JENSEN, W. N., MORENO, G. D., BESSIS, M.: An electron-microscopic description of basophilic stippling in red cells. Blood *25*, 933 (1965).
KASSENAAR, A., MORELL, H., LONDON, I. M.: Incorporation of glycine into globin and the synthesis of heme in vitro in duck erythrocytes. J. biol. Chem. *229*, 423 (1957).
KHALEELI, M. et al.: Sideroblastic anemia in multiple myeloma: a preleukemic change. Blood *41*, 17 (1973).
KINGSTON, P. J., BANNERMAN, C. E. M., BANNERMAN, R. M.: Iron deficiency anaemia in newborn *Ala* mice: a genetic defect of placental iron transport. Brit. J. Haemat. *40*, 265 (1978).
KONOPKA, L., HOFFBRAND, A. V.: Haem synthesis in sideroblastic anaemia. Brit. J. Haemat. *42*, 73 (1979).
KREIMER-BIRNBAUM, M., GRINSTEIN, M.: Porphyrin metabolism in experimental lead poisoning. Biochim. biophys. Acta (Amst.) *111*, 110 (1965).
LARIZZA, P.: Contributo alla conscenza dell'anemia sideroachrestica idiopatica non ereditaria (anaemia refractoria sideroblastica). Haematologica *47*, Suppl. 249 (1962).
LICHTMAN, H. C., FELDMAN, F.: In vitro pyrrole and porphyrin synthesis in lead poisoning and iron deficiency. J. clin. Invest. *42*, 830 (1963).
LOSOWSKI, M. S., HALL, R.: Hereditary sideroblastic anaemia. Brit. J. Haemat. *11*, 70 (1965).
MAIER, C.: Megaloblastäre Vitamin-B_6-Mangelanämie bei Hämochromatose. Schweiz. med. Wschr. *87*, 1284 (1957).
MANIS, J. G.: Intestinal iron transport defect in the mouse with sex-linked anemia. Amer. J. Physiol. *220*, 135 (1971).
MCCURDY, P. R., DONOHOE, R. F.: Pyridoxine-responsive anemia conditioned by isonicotinic acid hydrazide. Blood *27*, 352 (1966).
MCCURDY, P. R. et al.: Reversible sideroblastic anemia caused by pyrazinoic acid. Ann. intern. Med. *64*, 1280 (1966).
MCGIBBON, B. H., MOLLIN, D. L.: Sideroblastic anaemia in man: Observations on seventy cases. Brit. J. Haemat. *11*, 59 (1965).
MERKER, H.: Cytochemie der Blutzellen. In: HEILMEYER, L. (ed.): Blut und Blutkrankheiten, Vol. 2/1. Springer, Berlin–Heidelberg–New York 1968.
MERKER, H., KRAUSS, H. J.: Über die diagnostische Bedeutung von Sideroblasten, Perjodsäure-Schiff-positiven Erythroblasten und eisenspeichernden Reticulumzellen bei verschiedenen Anämien (1964).

Cit. HEILMEYER, L.: Die Hypochromanämien. In: HEILMEYER, L. (ed.): Blut und Blutkrankheiten, Vol. 2/2. Springer, Berlin–Heidelberg–New York 1970.

MILLS, H., LUCIA, S. P.: Familial hypochromic anemia associated with post-splenectomy erythrocytic inclusion bodies. Blood *4,* 891 (1949).

MOLLIN, D. L.: Sideroblasts and sideroblastic anemia. Brit. J. Haemat. *11,* 41 (1965).

PINKERTON, P. H., BANNERMAN, R. M.: Hereditary defect in iron absorption in mice. Nature *216,* 482 (1967).

PINKERTON, P. H., BANNERMAN, R. M.: Histological evidence of disordered iron transport in the X-linked hypochromic anaemia of mice. J. Path. Bact. *95,* 155 (1968).

PINKERTON, P. H., BANNERMAN, R. M., DOEBLIN, T. D., BENISCH, B. M., EDWARDS, J. A.: Iron metabolism and absorption studies in the X-linked anaemia of mice. Brit. J. Haemat. *18,* 211 (1970).

POPPEN, K. J., GREENBERG, C. D., RINEHART, J. F.: The blood picture of pyridoxine deficiency in the monkey. Blood *7,* 436 (1952).

PRATO, V., MAZZA, U.: Some aspects of porphyrin metabolism in thalassaemia. Symposion on the Normal and Pathological Porphyrin Metabolism, p. 44. Panminerva Medica, Torino 1961.

RIEDLER, G. F., STRAUB, P. W.: Abnormal iron incorporation, survival, protoporphyrin content and fluorescence of one red cell population in preleukemic sideroblastic anemia. Blood *40,* 345 (1972).

RUBINO, J. F.: The role of lead in porphyrin. Panminerva med. *1961,* 40.

RUNDLESS, R. W., FALLS, J. F.: Hereditary (sex-linked) anemia. Amer. J. med. Sci. *211,* 641 (1946).

SANO, S.: Studies on the nature of the basophilic stippled cells in lead poisoning. Acta Sch. med. Univ. Kioto *35,* 149, 158 (1958).

SCHMID, R., SCHWARTZ, S.: Experimental porphyria. III. Proc. Soc. exp. Biol. Med. *81,* 685 (1952).

SCHMID, R., SCHWARTZ, S., WATSON, C. J.: Porphyrin content of bone marrow and liver in the various forms of porphyria. Arch. intern. Med. *93,* 167 (1954).

SHAHIDI, N. T., NATHAN, D. G., DIAMOND, L. K.: Iron deficiency anemia associated with an error of iron metabolism in two siblings. J. clin. Invest. *43,* 510 (1964).

SHAW, M. T., FOADI, M.: The disappearance of ring sideroblasts in myeloblastic leukemia in remission. Scand. J. Haemat. *8,* 177 (1971).

SKIKNE, B. S., LYNCH, S. R., BEZWODA, W. R., BOTHWELL, T. H., BERNSTEIN, R., KATZ, J., KRAMER, S., ZUCKER, M.: Fanconi's anaemia, with special reference to erythrokinetic features. S. Afr. med. J. *53,* 43 (1978).

SLADIĆ-SIMIĆ, D., ZIVKOVIĆ, N., PAVIĆ, D.: Iron metabolism in Belgrade laboratory (b/b) rats. Rev. Europ. Etud. Clin. Biol. *17,* 197 (1972).

SNYDERMAN, E., CARRETER, R., HOLT, L. M.: Pyridoxine deficiency in the human being. Fed. Proc. *9,* 37 (1950).

SPENCER, S., RAAB, O., HAUT, A., CARTWRIGHT, G. E., WINTROBE, M. M.: Pyridoxine responsive anemia. Blood *18,* 185 (1961).

STEINER, M., BALDINI, M., DAMESHEK, W.: Abstracts X. Congr. Internat. Soc. Haematol., Vol. 12. Stockholm 1964.

SULLIVAN, A. A., WEINTRAUB, L. R.: Sideroblastic anemias. An approach to diagnosis and management. Med. clin. N. Amer. *57,* 335 (1973).

VERLOOP, M. C., RADEMAKER, W.: Anemia due to pyridoxine deficiency in man. Brit. J. Haemat. *6,* 66 (1960).

VERLOOP, M. C., BIERENGA, M., DIEZERAAD-NJOO, A.: Primary or essential sideroachrestic anaemias; pathogenesis and therapy. Acta haemat. (Basel) *27,* 129 (1962).

VERLOOP, M. C., PLOEM, W., LEUNIS, J.: Hereditary hypochromic hypersideraemic anaemia. In: Iron Metabolism. CIBA Symposion, p. 376. Springer, Berlin–Göttingen–Heidelberg 1964.

VERWILGHEN, R., REYBROUK, G., CALLENS, L., COSEMANS, J.: Antituberculous drugs and sideroblastic anaemia. Brit. J. Haemat. *11,* 92 (1965).

VEYRAT, R., MAURICE, D. A.: Anémie hypochrome mégaloblastique grave avec hypersidérémie et hémochromatose, corrigée par la pyridoxine. Schweiz. med. Wschr. *91,* 1215 (1961).

VUYLSTEKE, J., VERLOOP, M. C., DROGENDIJK, A. C.: Favourable effect of pyridoxine and ascorbic acid in a patient with sideroblastic anemia refractory and haemochromatosis. Acta med. scand. *169,* 113 (1961).

WALDENSTRÖM, J.: Studien über Porphyrie. Acta med. scand. Suppl. *82,* 1 (1937).

WALDENSTRÖM, J., VAHLQUIST, B.: Studien über die Entstehung der roten Hämpigmente (Uroporphyrin und Porphobilin) bei der akuten Porphyrie aus ihrer farblosen Vorstufe (Porphobilinogen). Hoppe-Seylers Z. physiol. Chem. *260,* 189 (1939).

WATSON, C. J., HAWKINSON, V., BOSSENMEIER, I.: Some studies of the nature and clinical significance of porphobilinogen. Amer. Phycns *66,* 144 (1953).

WATSON, C. J., PIMENTA DE MELLO, R., SCHWARTZ, S., HAWKINSON, V., BOSSENMEIER, I.: Porphyrin-chromogens or precursors in urine, blood, bile and feces. J. Lab. clin. Med. *37,* 831 (1951).

WAXMAN, H. S., RABINOVITZ, M.: Control of reticulocyte polyribosome content and hemoglobin synthesis by heme. Biochim. biophys. Acta (Amst.) *129,* 369 (1966).

WESTERMAN, M. P. et al.: Concentration of lead in bone in plumbism. New Engl. J. Med. *273,* 1246 (1965).

WINTROBE, M. M.: Clinical Hematology. Lea and Febiger, Philadelphia 1967, 1975.

WINTROBE, M. M., FOLLIS, R. H., MILLER, M. H., STEIN, H. J., ALCAYAGA, R., HUMPHREYS, S., SUKSTA, A., CARTWRIGHT, G. E.: Pyridoxine deficiency in swine with particular reference to anemia, epileptiform convulsions and fatty liver. Bull. Johns Hopkins Hosp. *72,* 1 (1943).

WITZLEBEN, C. L., WYATT, J. P.: The effect of long survival on the pathology of thalassaemia major. J. Path. Bac. *82,* 1 (1961).

WÖHLER, F.: The treatment of haemochromatosis with desferrioxamine. In: GROSS, F. (ed.): Iron Metabolism. Springer, Berlin–Göttingen–Heidelberg 1964.

CHAPTER 25

DISTURBED IRON METABOLISM IN ACUTE RADIATION INJURY

Ionizing irradiation evokes a progressive pancytopenia (Table 25/1). There is an early reduction in both red cell count and hemoglobin level at 400 R, producing a drop in total red cell volume in dogs to 55% of the initial value within 10 days (Fig. 25/1). The reticulocytes disappear completely from the circulation within 3 days and the ratio of the youngest erythroid and myeloid cells in the bone marrow is

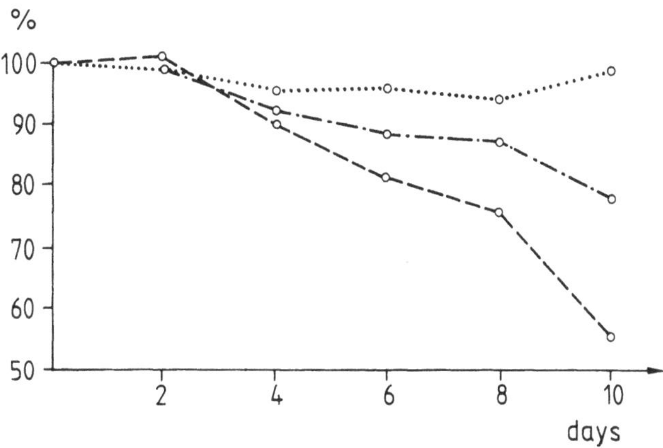

Fig. 25/1. Changes in the total red cell volume (o– –o– –o), plasma volume (o···o···o) and blood volume (o– · –o) after irradiation as a function of time, expressed as a percent of the initial value (after Louwagie, A. C. et al.: Haematologia *5*, 67, 1971a)

markedly reduced. About 85–90% of bone marrow cells are destroyed in 24 hours, and the absolute number remains low until the death of the experimental animal. A rise in the number of plasma cells, lymphocytes, and reticulum cells precedes regeneration in the bone marrow of the surviving animals (Louwagie et al., 1971a).

Table 25/1

Peripheral blood values before and after irradiation (dogs) (from Louwagie et al., 1971a)

Time of examination	No. of animals (n)	Hb (g/dl)		Hematocrit (%)		RBC (× 10³/μl)		Reticulocytes (in μl)		WBC (in μl)		Thrombocytes (× 10³/μl)	
		M	SD	M	SD	M	SD	M	SD	M	SD	M	SD
Before irradiation	33	18.32	1.79	49.18	4.19	7,277.9	711.7	27,691.6	21,672	12,860	4,365	205.7	80.7
After irradiation (days)													
1	33	16.28	1.88	43.72	5.28	6,365.2	992.5	13,343	19,004	7,903	3,791	190.2	78
3	33	15.74	1.83	41.37	4.18	5,921.5	818.9	1,482	4,529	4,015	1,794	174*	84.15
6	33	14.07	1.82	37.48	5.44	5,334.9	687.3	793	2,524	993	590	99.4	60.4
9	31	12.52	1.81	33.71	5.04	4,870.0	678.3	1,991	8,405	693	665	13.3**	17.7
13	12	10.86	1.88	29.54	5.86	4,305.8	891.1	4,124	11,370	370	275	9.25	13.65

* n = 32
** n = 30

Table 25/2

Bone marrow cellularity per mg aspirated bone marrow before and during the first 24 hrs after irradiation (from Louwagie et al., 1971a)

	Before irradiation		After irradiation							
			2 hr		4 hr		18 hr		24 hr	
	M	SD	M	SD	M	SD	M	SD	M	SD
Nucleated cells	158,429	180,142	69,063	54,824	37,940	18,750	12,836	5,197	16,433	8,012
Normoblasts	45,034	25,489	22,505	19,024	10,870	7,370	1,564	884	2,567	1,976
Pro- and basophilic normoblasts	3,378	1,296	1,079	994	464	385	121	162	37	42
Polychromatophilic normoblasts	38,798	21,054	19,765	17,554	9,484	6,822	1,050	768	1,810	1,348
Pycnotic normoblasts	2,858	3,571	1,661	192	922	1,107	393	589	717	722

Nucleated red cells are more radiosensitive than the other bone marrow cells, as demonstrated in mice and rats (Bloom, 1948; Bloom and Jacobson, 1948; Fliedner, 1958), rabbits (Rosenthal et al., 1951), and dogs (Louwagie et al., 1971a) (Table 25/2).

The serum iron level begins to rise within 24 hours of irradiation with 400 R, and the peak is reached about the third day, after which it begins to fall and drops below

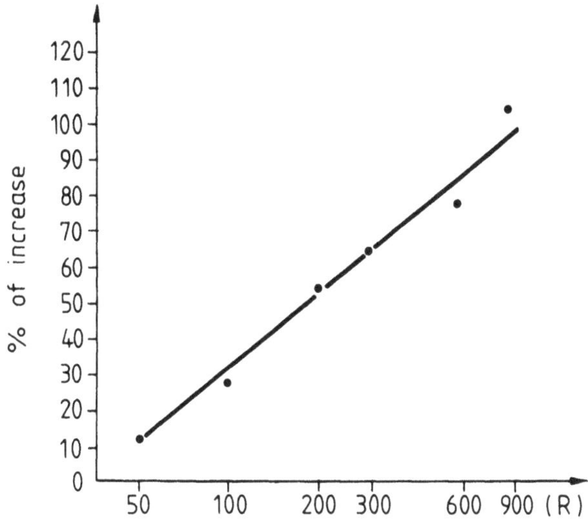

Fig. 25/2. Relation of plasma iron to the dose of irradiation (after Sztanyik, L. and Mándi, E., 1962)

the initial value during the second week. Initial hypersideremia is ascribed to reduced iron utilization and to the iron released in the course of the destruction of the decaying normoblasts (Louwagie et al., 1971b). The early rise in serum iron level is proportional to the radiation load (Sztanyik and Mándi, 1962) (Fig. 25/2).

The T/2 of plasma iron disappearance is prolonged, and the iron turnover drops to about 55% of the preirradiation value (Table 25/3), although the ratio of the nucleated red cells drops to 5.7% of the starting value (see Table 25/2). This is explained by the fact that

1. the surviving normoblasts and reticulocytes incorporate iron to the same degree in spite of irradiation injury, and

2. a large part of the transport iron flows into the reticuloendothelial system and to other tissues (Louwagie et al., 1971b).

Sztanyik and his team (1962) followed the iron transport after irradiation in rabbits and found that the rate of outflow of iron decreased within 24 hours (Fig. 25/3). The prolongation of the T/2 depends not only on the dose but also on the time

Table 25/3

Iron kinetics in normal and irradiated dogs (from Louwagie et al., 1971b)

	Normal dogs			Irradiated dogs											
				Time of ^{59}Fe administration											
				before irradiation						after irradiation					
				6 hr			1 hr			1 hr			24 hr		
	n	M	SD	n	M	SD	n	M	SD	n	M	SD	n	M	SD
Plasma iron level	9	142	43.48	8	147	52.28	8	163	54.52	8	170	79.74	9	247	71.26
^{59}Fe T/2	9	58	9.62	8	59	22.74	8	66	15.60	8	95	26.86	9	193	46.85
Plasma iron turnover	9	1.37	0.42	8	1.47	0.69	8	1.37	0.73	8	0.96	0.57	9	0.75	0.75
^{59}Fe utilization (on the 9th day)	9	87.11	7.88	8	44.91	18.04	8	26.08	9.22	7	20.02	9.95	6	8.24	3.36

Table 25/4

Iron incorporation in normal and irradiated dogs (from Louwagie et al., 1971b)

Days after irradiation	Normal dogs			Irradiated dogs											
				Time of ^{59}Fe administration											
				before irradiation						after irradiation					
				6 hr			1 hr			1 hr			24 hr		
	n	M	SD	n	M	SD	n	M	SD	n	M	SD	n	M	SD
1	9	22.30	13.79	8	18.94	8.67	8	12.44	6.56	8	7.72	3.85			
2	9	57.08	19.42	8	26.89	14.43	4	11.13	4.99	5	11.21	3.69	9	6.39	4.01
3	9	74.13	17.57	8	29.60	10.49	8	19.37	11.63	8	13.04	5.77	8	3.98	2.96
4	4	84.02	12.42	8	34.98	14.37	8	21.54	10.19	8	16.30	6.16	8	4.85	4.64
6	9	87.65	7.66	8	41.83	16.56	8	27.86	10.44	8	20.78	9.95	8	7.35	3.26
9	9	87.11	7.88	8	44.91	18.04	8	26.08	9.22	7	20.02	9.95	6	8.24	3.36

since the irradiation, the prolongation being more marked at 48 hours than at 24 hours. With doses between 100 and 300 R, the T/2 begins to return toward normal at 72 hours, but with larger doses it continues to rise (Fig. 25/4). Blackwell et al. (1962) have reported similar findings.

If the radioiron is injected 24 hours after irradiation, iron utilization 9 days later is about 8%. If it is given 48 hours after irradiation, only about 1–5% of the ^{59}Fe

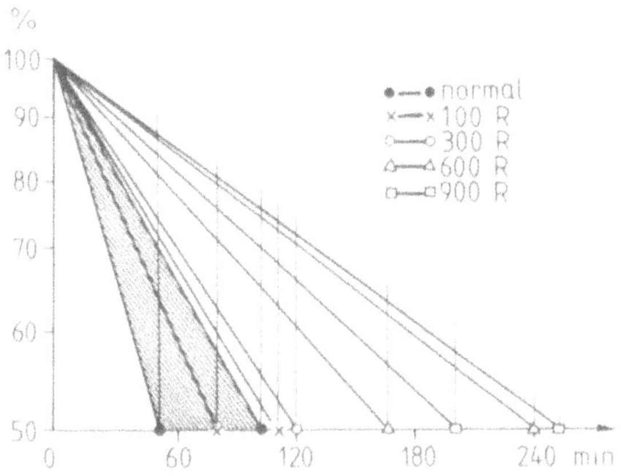

Fig. 25/3. Lengthening of plasma iron clearance with increasing radiation dose (after Sztanyik, L. et al., 1962)

appears subsequently in the red cells (Geszti et al., 1962). If the iron is injected before or shortly after irradiation, there is less reduction in iron utilization (Lajtha and Suit, 1955; Suit et al., 1957; Louwagie et al., 1971a) (see Table 25/4).

A definite dose dependence can be noted in the depression of the incorporation curves (Geszti et al., 1962; Sztanyik and Mándi, 1966). In mice 50 R will produce a reduction in iron incorporation to about 50–60% as opposed to the normal 65–85%, and 100 R reduces the iron incorporation to 30–35%. Also, 300 R reduces it still further to 8–15%, and after 600 R there is virtually no iron incorporation in the red cells. A similar dose dependence can be demonstrated in rabbits (Geszti et al., 1962) (Fig. 25/5).

It was originally thought that the circulating red cells were resistant to ionizing irradiation, but Cook (1965) and Árky et al. (1969) have shown that permeability of the red cell membrane increases both *in vivo* and *in vitro* as a result of massive irradiation, while its osmotic, thermal, and mechanical resistance decreases. More recent investigations have shown that the changes can be produced even by small exposure to irradiation. The radiosensitivity of mature red cells is supported by the

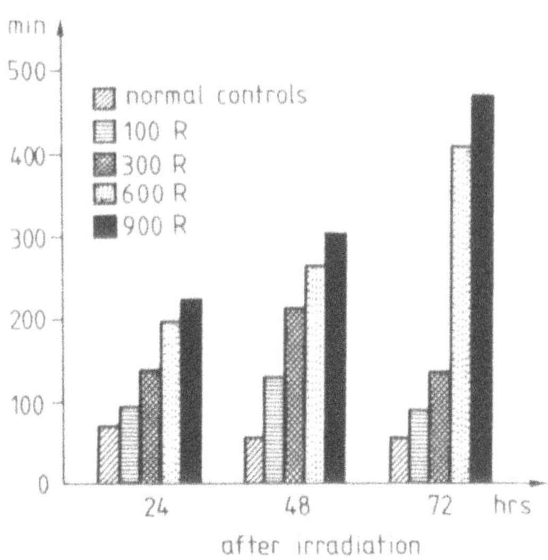

Fig. 25/4. Changes in the T/2 as a function of the dose and the time elapsed since irradiation (after Sztanyik, L. et al., 1962)

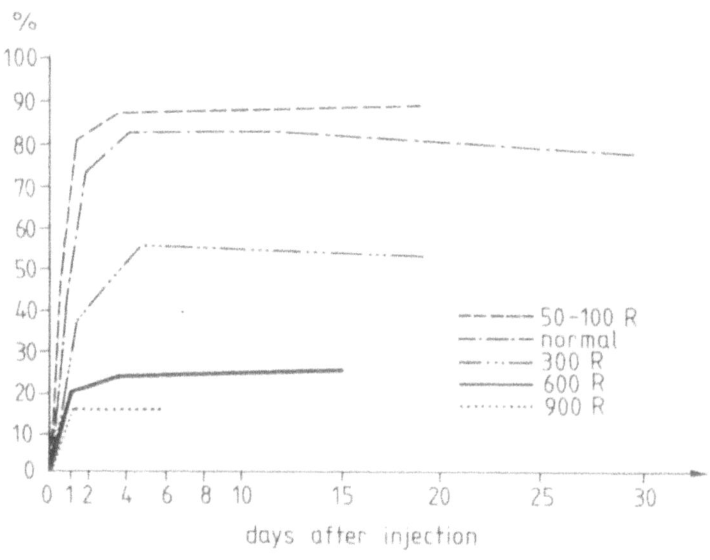

Fig. 25/5. Reduction in the radioiron incorporation with increasing doses of irradiation in rabbits (after Geszti, O. et al., 1962)

fact that a single X-ray examination may evoke a significant rise in plasma hemoglobin concentration and a decrease in the resistance of the red cells to ultrasound (Geszti et al., 1971).

BIBLIOGRAPHY

ÁRKY, I., SZÁSZ, I., GÁRDOS, GY., SZELÉNYI, J. G., BREUER, J. H., VÁRTERÉSZ, V., HOLLÁN, S. R.: Biochemical changes in blood by in vitro X-irradiation. Haematologia *3*, 51 (1969).

BERNÁT, I. et al.: Az egésztest-besugárzás hatása a plazma vaskoncentrációjára (Effect of whole body irradiation on the plasma iron concentration). VIIIth Congr. Hung. Soc. Haemat., Budapest 1978.

BLACKWELL, L. H., SINCLAIR, W. K., HUMPHREY, R. M.: Effect of whole body X-radiation on plasma iron disappearance in the rat. Amer. J. Physiol. *203*, 87 (1962).

BLOOM, W.: Histopathology of irradiation from external and internal sources. National Nuclear Energy Series Div. IV, 221, p. 808. McGraw-Hill, Inc., New York 1948.

BLOOM, W., JACOBSON, L. O.: Some hematologic effects in irradiation. Blood *3*, 586 (1948).

COOK, J. S.: The quantitative interrelationships between iron fluxes, cell swelling, and radiation dose in ultraviolet hemolysis. J. gen. Physiol. *48*, 719 (1965).

FLIEDNER, T. M.: Markzellsuspension bei strahlenbedingter Knochenmarkschädigung. Strahlentherapie *106*, 212 (1958).

GESZTI, O., ELŐD, I., BOJTOR, I., PREDMERSZKY, T., LOVÁNYI, I.: Auswirkung der medizinischen Strahlenbelastung auf die Membranpermeabilität der Erythrozyten im Blutkreislauf. Strahlentherapie (1971).

GESZTI, O., SZTANYIK, L., MÁNDI, E.: Adatok a röntgen-besugárzás erythropoesisre gyakorolt hatásához (Contributions to the effect of X-ray irradiation on the erythropoiesis). Haemat. hung. *2*, 41 (1962).

LAJTHA, L. G., SUIT, H. D.: Uptake of radioactive iron (^{59}Fe) by nucleated red cells in vitro. Brit. J. Haemat. *1*, 55 (1955).

LOUWAGIE, A. C., WAES, J. A., VAN VUCHELEN, J., VERWILGHEN, R. L.: Bone marrow syndrome after lethal whole body irradiation in the dog. I. Haematologia *5*, 67 (1971a).

LOUWAGIE, A. C., WAES, J. A., VAN VUCHELEN, J., VERWILGHEN, R. L.: Bone marrow syndrome after lethal whole body irradiation in the dog. II. Iron kinetics. Haematologia *5*, 79 (1971b).

ROSENTHAL, R. L., PICKERING, B. G., GOLDSCHMIDT, L.: A semiquantitative study of bone marrow in rats following total body X-irradiation. Blood *6*, 600 (1951).

SUIT, H. D., LAJTHA, L. G., OLIVER, R., ELLIS, F.: Studies on the Fe uptake by normoblasts and the failure of X-irradiation to affect uptake. Brit. J. Haemat. *3*, 165 (1957).

SZTANYIK, L., MÁNDI, E.: Változások a transzport-vas koncentrációjában besugárzott állatokon (Changes of the transport iron concentration in irradiated animals). Honvédorvos *14*, 228 (1962).

SZTANYIK, L., MÁNDI, E.: Az erythropoesis sugárkárosodásának vizsgálata egérkísérletekben radioaktív vas-izotóppal. I. Fe^{59} beépülése normál és röntgenbesugárzott egerek vörösvérsejtjeibe és raktárszerveibe (Investigations on radiation damage to the erythropoiesis in mouse experiments with radioiron. I. ^{59}Fe incorporation into the red cells and storage organs of normal and X-irradiated mice). Honvédorvos *18*, 117 (1966).

SZTANYIK, L., MÁNDI, E., GESZTI, O.: A vastranszport korai változása besugárzott állatokon (Early changes of iron transport in irradiated animals). Haemat. hung. *2*, 27 (1962).

CHAPTER 26

IRON METABOLISM IN POLYCYTHEMIA VERA AND SECONDARY POLYCYTHEMIAS

In untreated polycythemia vera the plasma iron is normal or somewhat reduced. It may, however, be profoundly reduced as a result of spontaneous bleeding or regular venesection. Oral iron loading produces a steep rise of the plasma iron level (Pribilla and Wolfers, 1955).

Ferrokinetic studies show that the plasma iron clearance rate is enhanced, the T/2 being on an average 35 min (Huff et al., 1950) (Fig. 26/1), and the plasma iron transport rate may increase. Various authors have reported it to be from 0.7 to 5.1 times the normal value (Huff et al., 1950; Wassermann et al., 1952; Bothwell et al., 1956; Kiely et al., 1961). The incorporation of iron is rapid and nearly complete, the surface measurements indicating that in the early stages erythropoiesis is confined to the bone marrow (Lawrence, 1955; Bothwell et al., 1956; Lajtha, 1961; Bothwell and Finch, 1962; Keiderling et al., 1963; Horst et al., 1963; Telfer and Schiffmann, 1963; Varela et al., 1963; Brunner, 1965; Tubiana et al., 1965; Henry, 1966; Pollycove et al., 1966). The difference between the amount of iron flowing daily through the plasma and the amount actually required for hemoglobin synthesis of red cells with a normal life-span is significant and has been attributed by most authors to ineffective erythropoiesis. Some studies suggested that in polycythemia vera there were two populations of cells, one with a very short average life-span, which would contribute to the increased plasma iron transport rate (Huff et al., 1950; Berlin, 1951). However, other studies have failed to show the presence of a dual population (London et al., 1949; Nathan and Berlin, 1959). Sharney et al. (1954) suggest that the high values of plasma iron transport are due to the fact that the curve of plasma radioiron outflow is not exponential in this condition. The changes in the plasma iron turnover correlate with fluctuations in the rate of total erythropoiesis, which in the later stages of the disease is usually diminished (Varela et al., 1963). The erythrokinetics in the various phases of the disease have been studied by a number of workers (Tubiana et al., 1965; Henry, 1966; Pollycove et al., 1966; Burger and Schmelzer, 1976). In the initial stages of polycythemia vera the following features are found:

1. red cell production is significantly enhanced,
2. erythropoiesis is restricted to the bone marrow, and
3. the life-span of the red cells is normal.

In later stages of the disease the life-span of the red cells may be reduced, with destruction of red cells in the spleen. Extramedullary erythropoiesis appears primarily in the spleen but also to some extent in the liver.

The transition of polycythemia to myelofibrosis is indicated by increasing uptake of iron by the liver and spleen and decreasing uptake by the bone marrow (Fig. 26/2).

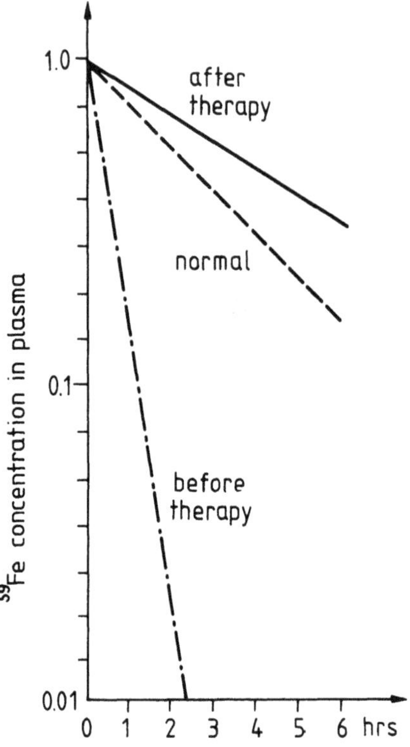

Fig. 26/1. Plasma iron clearance in healthy man (— — —), in polycythemia (— · — · —) before treatment and (———) after treatment (after Dreyfus, J. C. and Schapira, G.: Le fer. L'Expansion. Paris 1958)

The total erythropoiesis, as indicated by the degree of iron turnover, is increased but effective erythropoiesis is diminished (Szur and Smith, 1961; Doering and Lorenz, 1963; Najean et al., 1978).

Secondary polycythemia, as seen, for example, in association with emphysema, shows an increase in plasma iron transport rate to about twice normal (Hammersten et al., 1958), while in the polycythemia secondary to renal disease it

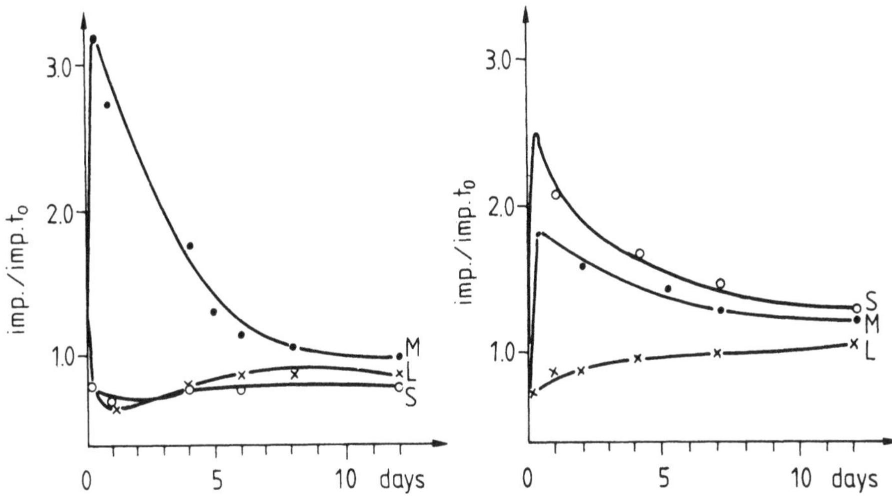

Fig. 26/2. Changes of radioactivity over the bone marrow (M), liver (L), and spleen (S) in healthy man (left) and in a patient whose spleen is the site of extramedullary hematopoiesis (right)

may be 2–4 times normal, and following successful treatment by nephrectomy it returns to normal (Finch et al., 1949; Jones et al., 1960).

Ferrokinetic investigations may be useful in the differentiation of secondary from primary polycythemia. In the secondary polycythemias the T/2 is usually not as rapid as in polycythemia vera, and the plasma iron transport rate is only moderately increased (Huff et al., 1950). In the erythrocytosis produced by hypoxia the plasma iron clearance rate is returned to normal by oxygen inhalation (Lawrence et al., 1952).

BIBLIOGRAPHY

Bateman, S. et al.: Splenic red cell pooling: a diagnostic feature in polycythaemia. Brit. J. Haemat. *40,* 389 (1978).

Berlin, N. I., Lawrence, J. H., Lee, H. C.: The life span of the red blood cell in chronic leukemia and polycythemia. Science *114,* 385 (1951).

Bothwell, T. H., Finch, C. A.: Iron Metabolism. Little, Brown and Co., Boston 1962.

Bothwell, T. H., Callender, S., Mallett, B., Witts, L. J.: The study of erythropoiesis using tracer quantities of radioactive iron. Brit. J. Haemat. *2,* 1 (1956).

Brunner, H. E.: Die Differentialdiagnose der Erythrocythämien durch Untersuchung der Ferro- und Erythrocytenkinetik mit radioaktivem Eisen und Chrom. Klin. Wschr. *43,* 429 (1965).

Burger, T., Schmelzer, M.: Changes in erythropoiesis during the course of polycythaemia vera. Folia haemat. (Lpz.) *103,* 726 (1976).

Doering, P., Lorenz, B.: Diagnostik der Osteomyeloreticulose bei Polycythämie mit Radioeisen. Folia haemat. (Frankfurt) N. F. *8,* 346 (1963).

DREYFUS, J. C., SCHAPIRA, G.: Le fer. L'Expansion, Paris 1958.

FINCH, C. A., GIBSON, J. G., PEACOCK, W. C., FLUHARTY, R. G.: Iron metabolism. Utilization of intravenous radioactive iron. Blood *4*, 905 (1949).

HAMMERSTEN, J. F., WHITECOMB, W. H., JOHNSON, P. C., LOWELL, J. R.: The hematologic adaptation of patients with hypoxia due to pulmonary emphysema. Amer. Rev. Tuberc. *78*, 391 (1958).

HENRY, J. A.: Études des modifications de l'érythropoïese dans la maladie de Vaquez traitée par le phosphore radioactif. Acta clin. belg. Suppl. *2* (1966).

HORST, W., RÖSLER, H., VILLANUEVA-MEYER, H.: 201 Fälle von Polycythämie: P^{32}-Behandlungsergebnisse, Untersuchungen zur Ferrokinetik (Cr^{51} und Fe^{59}) und über erythropoetische Plasmafaktoren vor, unter und nach Therapie. In: KEIDERLING, W., HOFFMANN, G. (eds.): Radio-Isotope in der Hämatologie, p. 361. Schattauer, Stuttgart 1963.

HUFF, R. L., HENNESSY, T. G., AUSTIN, R. E., GARCIA, J. F., ROBERTS, B. M., LAWRENCE, J. H.: Plasma and red cell iron turnover in normal subjects and in patients having various hematopoietic disorders. J. clin. Invest. *29*, 1041 (1950).

JONES, N. F., PAYNE, R. W., HYDE, R. D., PRICE, T. M.: Renal polycythaemia. Lancet *1*, 299 (1960).

KEIDERLING, W., REISSNER, I., DISCHLER, W., HOFFMANN, G.: Klinische Studien über die Kinetik des Eisens. In: KEIDERLING, W., HOFFMANN, G. (eds.): Radio-Isotope in der Hämatologie. Schattauer, Stuttgart 1963.

KIELY, J. M., STROEBEL, C. F., HAULON, D. G., OWEN, C. A.: Clinical value of plasma-iron turnover rate in diagnosis and management of polycythemia. J. nucl. Med. *2*, 1 (1961).

LAJTHA, L. G.: The Use of Isotopes in Haematology. Blackwell, Oxford 1961.

LAWRENCE, J. H.: Polycythemia. Physiology, diagnosis and treatment based on 303 cases. In: Modern Medical Monographs. Grune and Stratton, New York–London 1955.

LAWRENCE, J. H., ELMLINGER, P. J., FULTON, G.: Oxygen and the control of red cell production in primary and secondary polycythemia. Effects on the iron turnover patterns with Fe^{59} as tracer. Cardiologica *21*, 337 (1952).

LONDON, I. M., SHEMIN, D., WEST, R., RITTENBERG, D.: Heme synthesis and red blood cell dynamics in normal humans and in subjects with polycythemia vera, sickle cell anemia and pernicious anemia. J. biol. Chem. *179*, 463 (1949).

NAJEAN, Y., CACCHIONE, R., CASTRO-MALASPINA, H., DRESCH, C.: Erythrokinetic studies in myelofibrosis: their significance for prognosis. Brit. J. Haemat. *40*, 205 (1978).

NATHAN, D. G., BERLIN, N. I.: Studies of the production and life span of erythrocytes in myeloid metaplasia. Blood *14*, 668 (1959).

POLLYCOVE, M., WINCHELL, H. S., LAWRENCE, J. H.: Classification and evolution of patterns of erythropoiesis in polycythemia vera as studied by iron kinetics. Blood *28*, 807 (1966).

PRIBILLA, W., WOLFERS, H.: Das Verhalten des Serumeisens bei Polycythämikern vor und während der Behandlung mit radioaktivem Phosphor. Klin. Wschr. *33*, 960 (1955).

SHARNEY, L., SCHWARTZ, L., WASSERMAN, L. R., PORT, S., LEAVITT, D.: Pool systems in iron metabolism; with special reference to polycythemia vera. Proc. Soc. exp. Biol. Med. *87*, 489 (1954).

SZUR, L., SMITH, M. D.: Red-cell production and destruction in myelosclerosis. Brit. J. Haemat. *7*, 147 (1961).

TELFER, N., SCHIFFMANN, N. L.: The differential diagnosis of the polycythemias by plasma iron turnover determination. Nucl.-Med. (Stuttg.) *3*, 137 (1963).

TUBIANA, M., LOISEAU, J. P., VALLÉE, G.: L'évolution des polyglobulies essentielles. Intérêt des épreuves successives par le chrom et le fer radioactifs. Nouv. Rev. franç. Hémat. *5*, 397 (1965).

VARELA, J. E., ROCHNA, V. E. M., CARMENA, A. O., ETCHEVERRY, M. A., KREMENCHUZKY, S.: Polycythaemia vera. Results of repeated radioisotope studies in 53 patients during a five-year period. Nucl.-Med. (Stuttg.) *3*, 1 (1963).

WASSERMAN, L. R., RASHKOFF, I. A., LEAVITT, D., MAYER, J., PORT, S.: The rate of removal of radioactive iron from the plasma — an index of erythropoiesis. J. clin. Invest. *31*, 32 (1952).

CHAPTER 27

IRON OVERLOAD

The total iron content of the adult human organism is estimated at 4–5 g, of which about 1.0–1.5 g represents the iron reserve. The amount of storage iron normally remains relatively constant throughout life, the stores of males being in general greater than those of females.

Hypersiderosis is a pathological condition in which the total iron content of the organism is increased. It has to be distinguished from conditions in which there is a redistribution of the iron within the organism with consequent increase in storage iron but without a rise in total iron content as, for example, occurs in the anemia of infection.

Extra iron may enter the body by several routes (Charlton et al., 1973).

The intestinal control of iron absorption may be ineffective so that inappropriate amounts of dietary iron are allowed to enter the body (idiopathic hemochromatosis, certain anemias with a considerable degree of ineffective erythropoiesis). In these cases iron overload develops even when a normal diet is consumed.

The controlling mechanisms can be overwhelmed if very large amounts of iron in an absorbable form are ingested, and long-term exposure to a diet containing excessive iron can lead to iron overload in otherwise normal subjects (dietary iron overload).

On the other hand, large amounts of parenteral preparations of iron inappropriately prescribed, or repeated blood transfusions for refractory anemias, may result in the accumulation of excess iron in the body (transfusional siderosis).

Iron overload develops in any circumstance in which there is a prolonged positive iron balance. Most of the surplus iron is deposited in the reticuloendothelial system and/or in the parenchymal cells. It is only when there is significant parenchymal cell involvement that organ damage occurs.

When the surplus iron is derived from excess iron absorption from the gastrointestinal tract, it tends to be deposited in the parenchymal cells, whereas parenterally introduced iron, as from transfusion or parenteral iron treatment, is deposited, at least at first, in the reticuloendothelial system. In the former case the pathological condition is described as hemochromatosis and in the latter, hemosiderosis, although the distinction is somewhat arbitrary because the

Table 27/1

Classification of hypersideroses (from Dagg et al., 1971 — modified)

Generalized		Local
Mainly parenchymal	Mainly reticuloendothelial	
Primary idiopathic hemochromatosis	Chronic refractory anemias	Idiopathic pulmonary hemosiderosis
Liver cirrhosis	Hemolytic anemias	Goodpasture syndrome
Siderosis secondary to portocaval anastomosis	Transfusion siderosis	Paroxysmal nocturnal hemoglobinuria (PNH) (renal hemosiderosis)
Congenital atransferrinemia	Excessive parenteral iron therapy	
Congenital hypersiderosis (Vitale et al., 1969)		
Bantu siderosis		

distribution of iron is subject to changes in the course of the pathological processes (Oliver, 1959).

In some conditions the iron accumulates only in certain tissues of the organism; for example, prolonged intravascular hemolysis, as occurs in paroxysmal nocturnal hemoglobinuria, results in iron deposition in the kidneys, and repeated intra-alveolar hemorrhages produce hemosiderin accumulation in the lungs. The classification of hypersideroses is given in Table 27/1.

IDIOPATHIC HEMOCHROMATOSIS (IRON STORAGE DISEASE)

Primary hemochromatosis is a genetically determined, relatively rare disease, characterized by hepatic fibrosis leading to cirrhosis, pancreatic damage leading to diabetes, pigmentation of the skin, and heart failure.

The condition was first described by Trousseau (1865). Hanot and Chauffard (1882) described the three characteristic manifestations of the disease: pigmentation of the skin, diabetes mellitus, and cirrhosis of the liver. Sheldon (1927) was the first to call attention to hereditary predisposition to the disease, and later (1935) he gave a detailed description of the clinical and pathological picture, stressing the significance of iron accumulation. MacDonald and Mallory (1960) emphasized the importance of environmental factors. Finch and Finch (1955), Dillingham (1960),

Brick (1961), Johnson and Frey (1962), Dreyfus and Schapira (1964) and Lloyd et al. (1964) studied the hereditary features. Until 1955, about 1200 cases had been reported in the literature (Sheldon, 1935; Finch and Finch, 1955) but the condition has received much attention since then (Bothwell and Finch, 1962). Although the condition occurs in both sexes, the onset is delayed in women, probably because of the protective effect of their menstrual loss, so that there is an apparent preponderance of males to females of 10:1 of clinically detectable cases (Sheldon, 1935; Finch and Finch, 1955; Caroli and André, 1964).

Although the defect in iron absorption is hereditary, the first symptoms of idiopathic hemochromatosis are usually not apparent until 40–60 years of age (Finch and Finch, 1955). If it does become apparent at a younger age, the metabolic disturbance is more severe and the course of the disease more rapid (Bothwell and Alper, 1951).

SYMPTOMATOLOGY

The first clinical signs are usually connected with the development of diabetes (Fig. 27/1). Weakness, lassitude, fatigue, and loss of weight occur in about half of the patients by the time the diagnosis is established (Finch and Finch, 1955). Loss of weight may occur even without diabetes. Abdominal pain, particularly in the epigastrium and right hypochondrium, has been a feature in 25–30% of cases (Boulin, 1945; Desforges, 1949; McClatchie et al., 1950; Finch and Finch, 1955). Ascites, pancreatitis, and perisplenitis may contribute to the abdominal pain but often the cause cannot be established. The pain may be sufficiently severe to simulate an acute abdomen (McClatchie et al., 1950; Taylor, 1951; Jones, 1962).

The clinical features are pigmentation of the skin, enlargement of the liver and spleen, lack of body hair, and atrophy of the testes. Features of pathological liver function, such as palmar erythema, ascites, or spider naevi, may also be present. About one-third of the patients develop signs of heart failure (see Fig. 27/1).

The liver is enlarged and firm in almost all cases (Sheldon, 1935; Boulin, 1945; Althausen et al., 1951; Finch and Finch, 1955), and splenomegaly occurs in about 50% (Finch and Finch, 1955). There is a gradual deterioration in hepatic function resulting in weakness, weight loss, and finally in cachexia. Portal hypertension is not as common as in Laënnec's cirrhosis, and demonstrable ascites or esophageal varices are seldom seen (Althausen et al., 1951; Finch and Finch, 1955; Caroli and André, 1964).

Although the liver is obviously involved, liver function tests may not be grossly abnormal. Finch and Finch (1955) found bromsulphalein excretion to be diminished in only 42%, the serum bilirubin elevated in only 24%, the colloidal gold tests positive in only 11–30%, prothrombin time prolonged in 25%, and the alkaline phosphatase activity of the serum raised in only 20%. There is presumably some correlation between the degree of severity of the hepatic lesion and the body hair loss, atrophy of the testes, gynecomastia, and impotence.

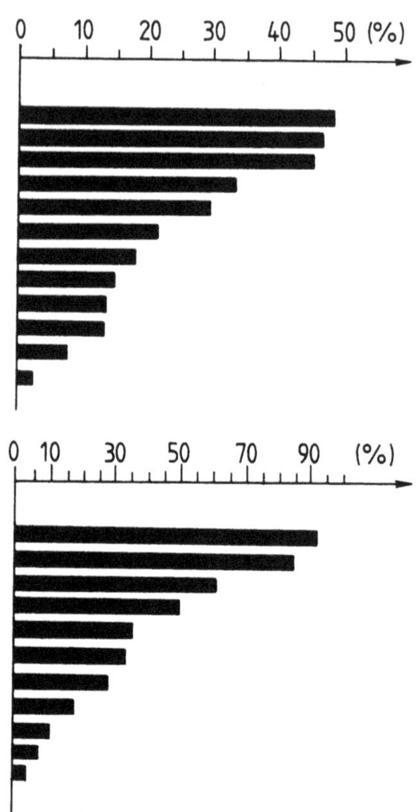

Fig. 27/1. The incidence of the clinical symptoms and signs of idiopathic hemochromatosis (after Finch, S. C. and Finch, C. A.: Medicine *34,* 381, 1955)

In about 14% of cases primary carcinoma of the liver develops, i.e., about three times the incidence found in Laënnec's cirrhosis (Berk and Lieber, 1941; Willis, 1941; MacSween and Jackson, 1966).

Diabetes mellitus develops in about 80% of patients and is usually insulin-dependent (Sheldon, 1935; Boulin, 1945; Finch and Finch, 1955), but in some cases the disturbance of sugar metabolism is mild and can only be detected with glucose loading tests (Althausen et al., 1951). The diabetes may be difficult to control, and hypoglycemic reactions and insulin resistance may be troublesome (Werner, 1942; Eskind et al., 1953; McAlpine, 1959). The vascular complications of diabetes such as retinopathy are rare (Sheldon, 1935; Bell, 1955; Hudson, 1953; Becker and Miller, 1960), probably because most of the patients die before these complications can develop (Bothwell and Finch, 1962).

Endocrine disturbances such as decreased libido, impotence, amenorrhea, and loss of body hair are found in 20% of the patients and are particularly conspicuous in the younger individuals (Boulin, 1945; Bothwell and Alper, 1951; Finch and Finch, 1955). The beard and axillary and pubic hair may be entirely absent, and the loss of body hair may precede the development of the other symptoms by some years (Althausen et al., 1951). Defects in pituitary, thyroid, and adrenal cortical function have been described (Boulin, 1945; Rogers, 1950; Glaser and Smith, 1950; de Gennes, 1952), but the disturbed androgen-estrogen balance, which is related to the cirrhosis, is characteristic (Morrione, 1944). The cardiac complications are the most severe feature, and once heart failure has become evident most patients do not survive more than a year (Bothwell and Alper, 1951; Bothwell et al., 1952).

The pigmentation of the skin is present in 90% of patients at the time of diagnosis. It is more pronounced in the areas exposed to light. When the pigment is due mainly to melanin, it is a bronze color, but if large amounts of iron-containing pigment are also deposited in the skin, the color may be a metallic gray (Sheldon, 1935). In about 10% of the patients the oral mucosa may be discolored (Finch and Finch, 1955).

There is no characteristic blood picture. A mild macrocytosis, leucopenia, and thrombocytopenia may be encountered. If leucocytosis develops without any other demonstrable cause such as infection, a complicating hepatoma should be suspected (Bothwell and Finch, 1962). The hepatic lesion may occasionally give rise to hypoprothrombinemia, but usually other coagulation factors are not affected.

The relatively normal blood picture may be helpful in differentiating true idiopathic hemochromatosis from secondary hypersiderosis.

THE DISTURBANCE OF IRON METABOLISM IN HEMOCHROMATOSIS

The plasma iron concentration in most cases ranges between 200 and 300 μg/dl, but in about 35% it may drop below 200 μg/dl (Fig. 27/2). Occasionally even healthy adults may have values between 160 and 190 μg/dl, so it is not always easy to decide whether or not the iron level is pathological. Dreyfus and Schapira (1964) drew attention to the fact that the serum iron levels of both healthy individuals and patients with hemochromatosis range widely (Table 27/2), and stressed that values falling into the normal range do not necessarily exclude the possibility of hemochromatosis. However, a very high serum iron level in the absence of anemia is suggestive of the diagnosis.

The total iron-binding capacity of the plasma may be diminished. Vannotti and Blanc (1963) found the transferrin level to be 76–130 mg% as compared with the normal of 200–300 mg%. The transferrin is usually fully or almost completely saturated, the saturation coefficient being between 0.8 and 1.0 (see Fig. 7/11).

Following an iron loading test there is little change in the level of plasma iron (Fig. 27/3), the iron that is absorbed being rapidly removed from the circulation.

Table 27/2
Serum iron in normal subjects and in 86 patients with hemochromatosis
(from Dreyfus and Schapira, 1964)

Source of data	Males		Females	
	Mean	+2s	Mean	+2s
Normal values (from literature)	129	193	110	168
Normal values (Dreyfus and Schapira)	134	185	123	179
Hemochromatosis (Dreyfus and Schapira)	225	385	213	331

Ferrokinetic investigations require caution in evaluation. Because of the high saturation of the transferrin, radioiron must be of high specific activity and must be injected very slowly. In spite of these precautions, the iron disappears rapidly from the circulation in the few minutes following injection, the rate of decline in activity not showing the usual exponential curve at the beginning. From the later part of the

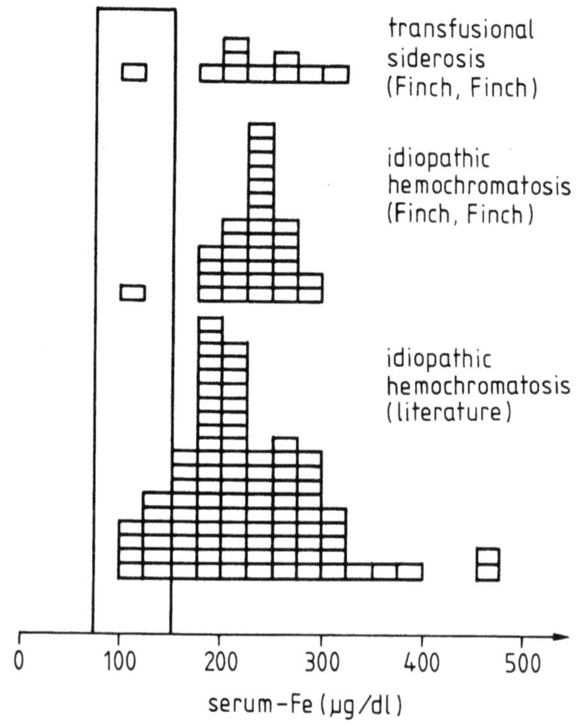

Fig. 27/2. Serum iron level in idiopathic hemochromatosis and transfusional siderosis (after Finch, S. C. and Finch, C. A.: Medicine *34*, 381, 1955)

curve, which is exponential, the T/2 of outflow is longer than normal, 118–244 minutes as opposed to 70–130 minutes, and the plasma iron transport rate as a result of the high plasma iron pool is increased to 46–56 mg/24 hr as opposed to 30–37 mg/24 hr in the normal. As a result of therapeutic venesection the plasma iron transport rate rises still further.

The curve of radioiron incorporation is flattened (see Fig. 14/8), but since the ratio of incorporation is inversely proportional to the size of the iron pool the absolute

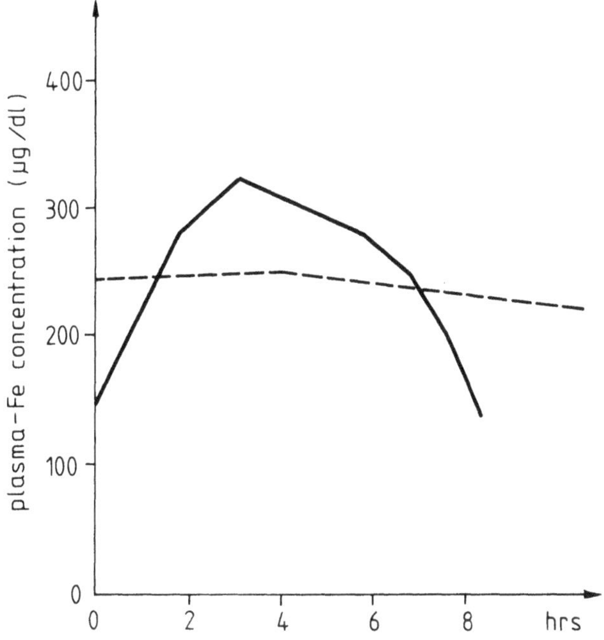

Fig. 27/3. Oral iron loading curve in healthy man (———) and in patients with hemochromatosis (– – –) (after Dreyfus, J. C. and Schapira, G.: in Gross, F.: Iron Metabolism. Springer, Berlin–Göttingen–Heidelberg 1964)

amount of iron utilized for hemoglobin synthesis is not reduced. The average life-span of the red cells is also normal.

The excretion of iron is slightly increased, but this does not compensate for the enhanced iron absorption. The injection of desferrioxamine produces a marked increase in iron excreted in the urine.

The combination of the clinical picture, the higher plasma iron, and saturation of transferrin is usually diagnostic of the condition. The demonstration of iron overload in the reticuloendothelial cells of the bone marrow may be useful. The Prussian blue reaction reveals a large amount of hemosiderin in the reticuloendothelial cells (Fig. 27/4). The iron is accumulated in the macrophages as coarse clusters in random distribution. Although some increase in the RE iron may be seen in other conditions such as infections, thermal injury, and sideroblastic anemia, the deposits in hemochromatosis are usually greatly in excess of

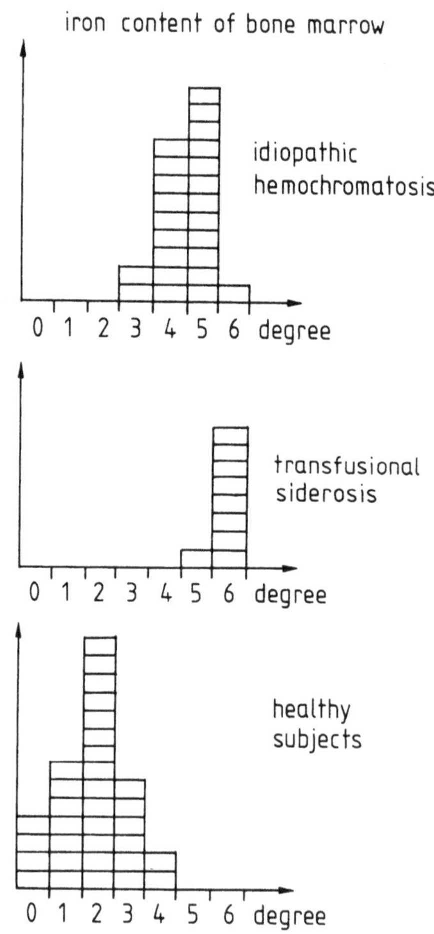

Fig. 27/4. Iron content of the RE cells in idiopathic hemochromatosis and transfusional siderosis (after Finch, S. C. and Finch, C. A.: Medicine *34*, 381, 1955)

those seen in other diseases (Rath and Finch, 1948; Mouriquand, 1961; Bothwell and Finch, 1962).

The cytological examination of the urinary sediment is a useful and simple procedure for the diagnosis of hemochromatosis, the increased iron being demonstrable in desquamating cells from the kidney and ureter (Rous, 1918; Finch and Finch, 1955).

Skin biopsy has sometimes been recommended, but this is not a reliable procedure, and about half the patients are found to have no excess iron in this area (Sheldon, 1935). On the other hand, iron originating from subcutaneous hemorrhages may give rise to a false positive result (Finch and Finch, 1955).

Gastric biopsy has also been recommended for the diagnosis of hemochromatosis (Althausen et al., 1951), but it is no more informative than the bone marrow biopsy or examination of the urinary sediment and does not therefore appear to be justified.

The iron excretion in a 24-hour urine sample following an injection of 500 mg of desferrioxamine intramuscularly gives an indication of pathological iron storage (Moeschlin, 1959, 1963; Wöhler, 1964; Schmid et al., 1964). In the case of normal iron stores the urinary iron excretion does not exceed 0.6–1.2 mg, while in hypersiderosis the excretion of 12 mg or more is usual.

Liver biopsy provides the most conclusive evidence for hemochromatosis, as this shows the extent of iron deposition in the parenchymal cells and the liver damage (Fig. 27/5). Liver biopsy is, however, usually not essential to the diagnosis in most

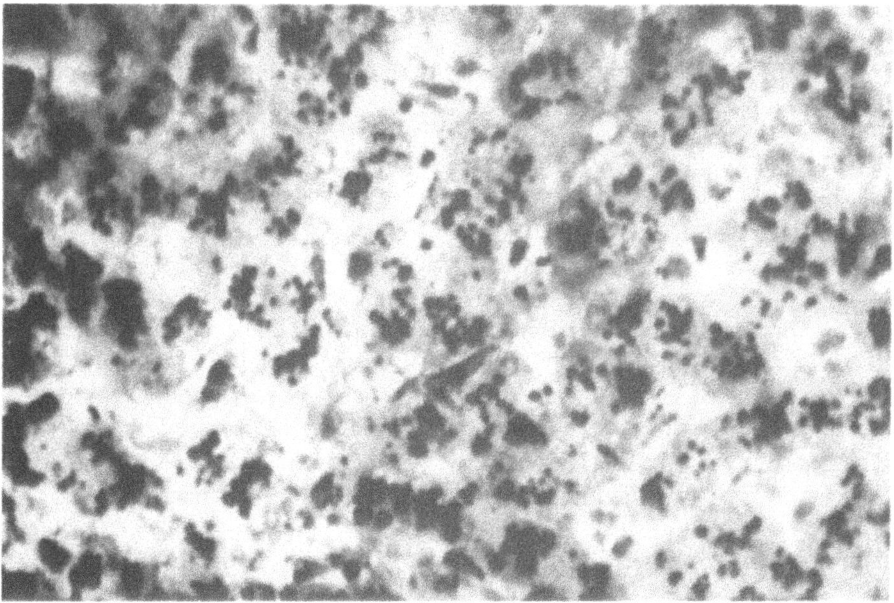

Fig. 27/5. Iron accumulation in the liver in hemochromatosis (biopsy study, Prussian blue reaction)

cases. It is particularly useful where the diagnosis is suspected but not confirmed by biochemical investigations (Frey et al., 1961) or if the transferrin saturation is decreased on account of infection or malignant tumor (Bothwell and Finch, 1962). It can also be useful to identify relatives in whom signs of overt disease have not yet appeared. Ferrokinetic studies can also aid the establishment of the diagnosis in asymptomatic patients; for example, Brunner et al. (1962) describe a case of a 13-year-old boy whose father died of hemochromatosis at the age of 35. The T/2 of the plasma iron disappearance in the boy was 140 min and the plasma iron transport rate 33 mg in 24 hours. The total iron-binding capacity of the plasma was low (195 μg/dl), and the transferrin was completely saturated. The liver uptake of radioactivity was considerably in excess of that over the spleen and the bone marrow. Following venesection, by which means 800 mg of iron had been removed, the boy showed no sign of anemia, indicating pathological iron storage.

In summary, the diagnosis of hemochromatosis can be established usually on the basis of physical examination and the determination of the plasma iron and transferrin saturation. A bone marrow examination for iron and a desferrioxamine test may be confirmatory, but occasionally a liver biopsy may be required to substantiate the diagnosis.

PROGNOSIS

The mean survival following diagnosis is about 4.4 years (Bothwell and Finch, 1962), although 10- to 20-year survivals have also been reported (Sheldon, 1935; Finch and Finch, 1955). The prognosis may have improved since regular venesection and desferrioxamine treatment but there are few recent reliable data concerning the average life-span. Finch and Finch (1955) found the most frequent causes of death to be heart failure in 30%, hepatic coma in 15%, hemorrhage from esophageal varices in 15%, hepatoma in 14–15%, pneumonia in 13%, and diabetic coma in 3%. Hepatocellular carcinoma now accounts for approximately 30% of all deaths in idiopathic hemochromatosis, as compared with about 14–15% in earlier studies (Bomford and Williams, 1976). This increasing frequency is probably related to the longer survival of patients after phlebotomy treatment. Idiopathic hemochromatosis is one of the most important risk factors for the development of hepatocellular carcinoma (Johnson et al., 1978; Powell and Halliday, 1980). The heart failure affects mainly the younger patients, while carcinoma of the liver kills primarily the elderly. There is no predilection for any age group for the other causes of death (Sheldon, 1935, Boulin, 1945, Finch and Finch, 1955).

By the time of presentation the iron content of the subject is 5–10 times the normal value. The increase is attributable to increased iron absorption (Greenberg et al., 1964; Powell, 1965; Losowsky and Wilson, 1967; Smith et al., 1969). The increase is most marked in younger patients, while in older individuals when the stores are already overloaded absorption may appear normal (MacDonald, 1964). It should, however, be remembered that apparently normal absorption may be quite inappropriately high for a grossly iron-overloaded subject.

Sheldon (1935) was the first to suspect a genetically determined metabolic disturbance. The familial occurrence is well recognized. There is a demonstrably higher incidence of liver cirrhosis and diabetes among relatives of patients with hemochromatosis, and abnormal skin pigmentation is also frequent. Of 26 parents of patients examined, 1 had proven hemochromatosis, 5 developed the clinical syndrome, in 2 the subclinical form was present, and 3 had diabetes and 1 hepatomegaly. Of 107 siblings examined, there were 25 cases where the clinical diagnosis was evident, in 9 cases it was confirmed by histology, and there were a further 6 probable cases. Ten other individuals had a high plasma iron level, 6 suffered from diabetes and 7 from hepatomegaly, and 1 had carcinoma of the liver, indicating that approximately half of the siblings were affected. Of the symptom-free relatives, 20–30% showed an elevated plasma iron and increased transferrin saturation (Finch and Finch, 1955). A number of other observers have studied the genetic features of hemochromatosis (Debré et al., 1958; Dreyfus et al., 1960; Schapira and Dreyfus, 1959; Schapira et al., 1962). The results from Dreyfus and Schapira's (1964) study of 40 patients and 161 direct descendants are shown in Fig. 27/6. The mean plasma iron concentration of the patients was 250 μg/dl compared with a mean of 140 μg/dl for healthy controls. In boys under 15 years of age the mean level was 136 μg/dl, whereas at over 15 it was 195 μg/dl. The rise in the mean value in the over-15-year-old boys is due to the fact that about half the boys had a strikingly elevated plasma iron level and increased transferrin saturation. In the two groups of girls below and above 15 years of age the difference between the two mean values was not statistically significant. This is consistent with the delayed onset of iron overload in girls because of the menstrual loss.

Investigations based on measurement of the serum iron and calculation of the coefficient of saturation have revealed anomalies in half of the sons over 15 years of age. They point to dominant transmission, if not of the disease, then at least of the disorder affecting iron metabolism.

Biopsy from eight children with high plasma iron all showed excessive iron in the liver (see Fig. 27/5).

Williams et al. (1962) detected disturbance of iron metabolism in 75% of the family members of their patients and concluded that the inheritance was of an intermediary type, heterozygotes being the carriers and homozygotes the patients.

This would fit the experience that the disease is more frequent among siblings than descendants (Davison, 1961; Morgan, 1961; Bothwell et al., 1959; Brick, 1961; Johnson and Frey, 1962; Pirart and Gatez, 1958; Conte et al., 1958b). On the other hand, if Williams's hypothesis were correct, the children of the homozygotes, at least the boys, would develop some kind of metabolic disturbance, which was not found by Dreyfus and his collaborators.

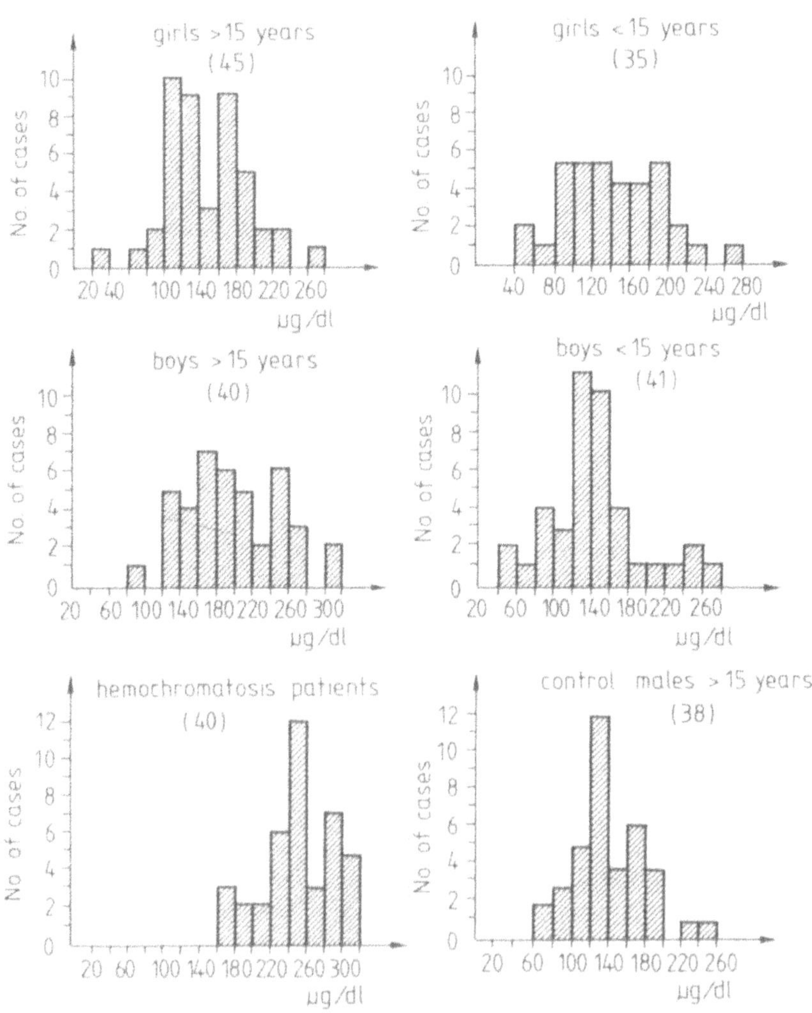

Fig. 27/6. Iron level of the serum in patients with hemochromatosis and in their children (after Dreyfus, J. C. and Schapira, G., in: Gross, F.: Iron Metabolism. Springer, Berlin–Göttingen–Heidelberg 1964)

The screening of relatives has clearly indicated that the manifest disease of hemochromatosis is only the final result of an inherited metabolic disturbance that has been present since birth (Balcerzak et al., 1966). As a result of the genetic abnormality an excess of iron is absorbed from the gastrointestinal tract, but the full clinical picture may only develop in a proportion of the affected individuals (Fig. 27/7). Several factors influence the development of clinical disease, among which the following are the most important:

1. the extent of iron absorption,
2. the iron content of the diet,
3. duration of increased iron absorption, i.e., the age of the patient,
4. the extent of iron loss.

These factors explain the higher incidence in males and the relatively late onset of overt disease.

MacDonald and Mallory (1960) suggested that alcohol consumption played a primary role in the development of hemochromatosis, but the majority of authors accept the genetic basis of the disease, with alcoholism being only a contributory factor that may make the clinical syndrome obvious at an earlier age.

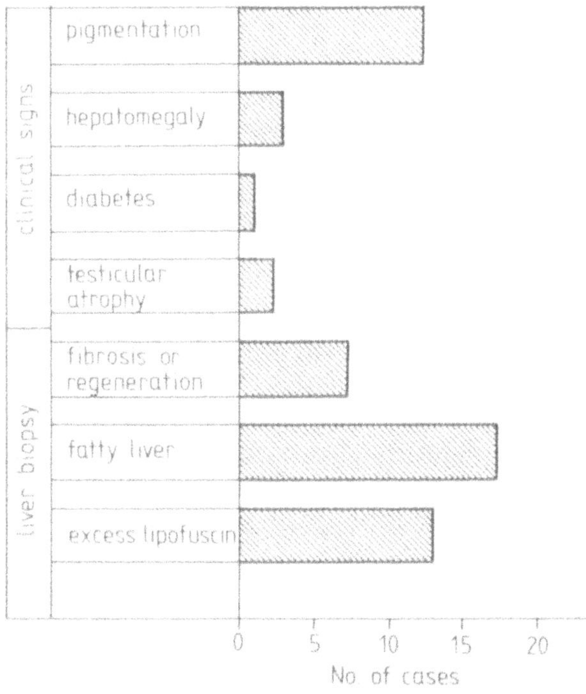

Fig. 27/7. Clinical signs and liver histology in 28 hypersiderotic relatives of patients with hemochromatosis (after Scheuer, P. J. et al.: J. Path. Bact. *84*, 53, 1962)

In the earlier stages of the disease hemosiderin is deposited mainly in the liver cells. It is found in the form of fine granules, first at the biliary pole of the cells but later filling the entire cell. Hemosiderin accumulates first in the peripheral cells of the liver lobules but later involves all hepatic cells. The fine granules ultimately aggregate to form large iron clusters. Liver siderosis leads to cell destruction and is associated with fibrosis. Fibrosis is at first perilobular but later invades the parenchyma and goes on to true cirrhotic transformation by the time the condition becomes manifest. There is occasionally some cellular infiltration with lymphocytes and plasma cells. With progression of the disease, iron accumulates also in the Kupffer cells. This is in contrast to the picture with transfusion hemosiderosis, when iron accumulates first in the Kupffer cells and later in parenchymal cells (Bothwell and Finch, 1962).

Some 10–20 grams of iron can be found in the liver, which is 50–100 times higher than normal. Similar heavy iron deposition is seen in the pancreas, while in the thyroid there is about 25 times the normal amount of iron, and in the heart and adrenal 10–15 times. It is mainly the glomerular zone of the adrenal that is affected (Sheldon, 1935). Iron deposition is less marked in skin, spleen, kidney, and stomach. The parathyroid glands and the anterior pituitary may be affected as well as the salivary and lacrimal glands and the secretory glands of the respiratory tract (Dubin, 1955). The sweat glands, the endothelial cells of the blood vessels, and the connective tissue of the dermis may also show excess iron (Althausen et al., 1951), but the pigmentation of the skin is largely due to melanin accumulation in the deeper layers of the epidermis (Althausen et al., 1951; Dubin, 1955).

The iron-free pigment lipofuscin is also increased in hemochromatosis (Pearse, 1953). It is found in the wall of the blood vessels and in the muscle tissue (Sheldon, 1935). This pigment stains well with Mallory's basic fuchsin and yields a positive PAS reaction (Dubin, 1955). A similar pigment can be demonstrated in other hypersideroses (Higginson et al., 1953) and in old age and cachexia (Dubin, 1955). It is not likely to have any specific significance in hemochromatosis (Bothwell and Finch, 1962).

The liver is usually enlarged, with a mean weight of about 2400 grams, and is a rusty red color (Sheldon, 1935). Cirrhosis is a constant finding, but in about one-quarter of the cases it is still in an early phase. Fatty degeneration is seen only in those cases in which there is a previous alcoholic history (Dubin, 1955). Where a malignant tumor occurs it is usually a hepatoma, rarely a cholangioma (Dubin, 1955).

The pancreas is hard and its color rusty red. Fibrosis is usually present and the number of Langerhans islands reduced. The destroyed epithelial cells are replaced by connective tissue (Sheldon, 1935), and there is heavy deposition of hemosiderin, most of the pigment being found in the acini and connective tissue, although in about 80% of the cases it can also be seen in the cells of the Langerhans islands (Dubin, 1955).

The weight of the heart is some 300–500 grams (Sheldon, 1935; Dubin, 1955); however, in patients who died of heart failure it always exceeded 400 grams (Bothwell and Alper, 1951). There is increased hemosiderin deposition in 90% of cases.

The spleen is usually slightly enlarged, with not very pronounced hemosiderin deposition. There is also only a moderate increase in iron in the kidneys, the testes, and the striated muscles. Only those lymph nodes that are connected to the important hemosiderin-storing organs contain excess iron (Dubin, 1955). Testicular atrophy is frequent, and there may be some atrophic changes in the epidermis, follicles, and sebaceous glands (Althausen et al., 1951).

TREATMENT

The prognosis for patients with hemochromatosis was particularly poor in the preinsulin era, and patients died within weeks or months of their diabetes being recognized (Sheldon, 1935). Diabetic control has been one landmark in the treatment of hemochromatosis, but the principal goal is to remove the enormous iron surplus that has been the cause of the clinical manifestations. Finch (1948, 1949) introduced regular therapeutic venesection. He demonstrated that iron could be mobilized from the depots for hemoglobin formation and that patients tolerated regular venesection well (Davis and Arrowsmith, 1950, 1952, 1953; Finch et al., 1950; Blackburn et al., 1953; Houston, 1963; Warthin et al., 1953; Davey et al., 1954; Bothwell et al., 1955; Finch and Finch, 1955; Myerson and Carroll, 1955; Pirart and Carpent, 1955; Lind, 1957; MacGregor and Ramsay, 1957; Conte et al., 1958a; Crosby, 1958; Deckert, 1958; McAlpine, 1959; Frey et al., 1961). Up to 500 ml of blood can be taken once or twice weekly. During the first 1–2 weeks the hematocrit value and hemoglobin concentration drop by approximately 10–15%, but this period is followed by enhanced erythropoiesis, and the hematocrit and hemoglobin level become constant or may even rise (Fig. 27/8). With each venesection, 200–250 mg of iron are mobilized from the storage depots; since in hemochromatosis an average of 20–40 grams of iron has accumulated, the weekly venesections must be continued for at least 2–3 years, or until the plasma iron concentration falls (Bothwell and Finch, 1962; Finch and Finch, 1955) (Fig. 27/9). A fall in plasma iron indicates the reduction of the stores, and the venesections can then be spaced to 2- to 3-week intervals. When all the mobilizable iron has been removed, the patient may only require 1–2 liters of blood to be removed each year. The hematocrit value should be determined before every venesection, and the plasma iron should be checked once a month. The iron depletion can be monitored also by serial plasma ferritin determinations. Patients undoubtedly usually feel better; they gain weight, show less pigmentation, and have some improvement in liver function (Williams et al., 1969). They may also require less insulin. In some cases the cardiac symptoms have improved (Evans, 1959; Grosberg, 1961). In most instances the biopsy studies do not show any influence on liver cirrhosis (Finch and Finch, 1955), although

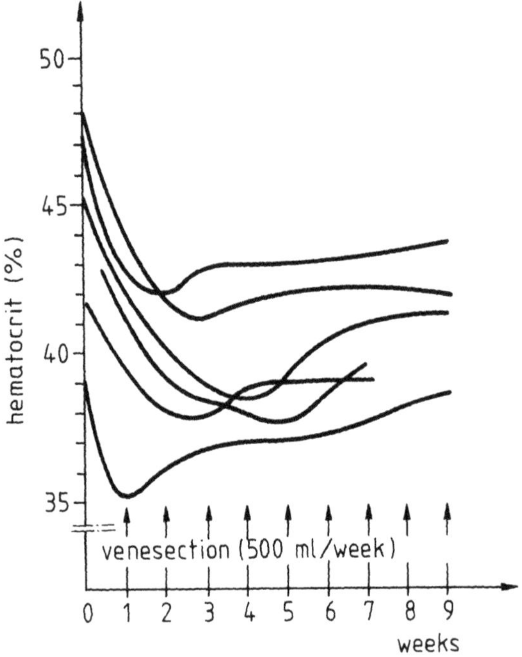

Fig. 27/8. Changes in the hematocrit with weekly venesection in idiopathic hemochromatosis (after Finch, S. C. and Finch, C. A.: Medicine *34,* 381, 1955)

others have found a reduction in liver fibrosis and occasional cases have been reported in which the liver structure has become normal (Pirart, 1964; Knauer et al., 1965; Weintraub et al., 1966; Powell and Kerr, 1970).

According to MacDonald (1965) the treatment with venesection does not essentially alter the prognosis in hemochromatosis. However, Williams et al. (1969) compared the survival of 40 patients treated by venesection with 18 untreated patients. The mean survival in the venesection-treated group was 8.2 years as opposed to 4.9 years in the control group. Less than half of the former died as a direct consequence of the hemochromatosis, while in only 3 of the 18 control patients was death not directly connected with the hemochromatosis.

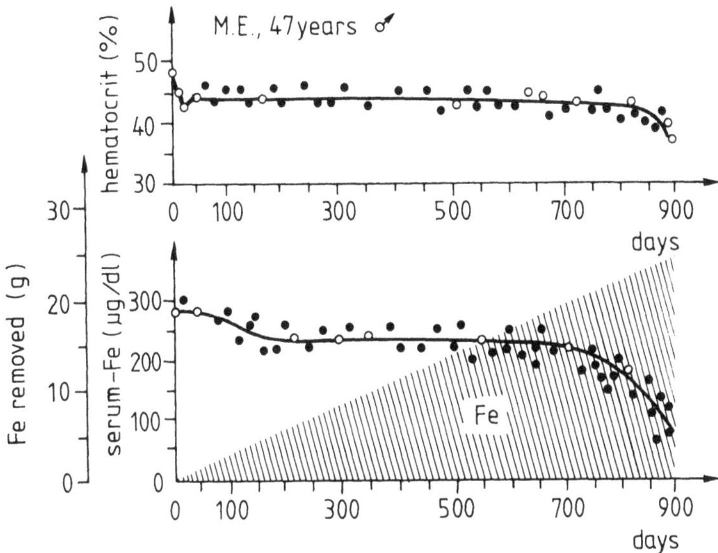

Fig. 27/9. Changes in hematocrit and serum iron in relation to cumulative iron removal through phlebotomy. The serum iron level starts to decline only when storage iron is exhausted (after Finch, S. C. and Finch, C. A.: Medicine *34,* 381, 1955)

The use of iron chelating agents in hemochromatosis is debatable. Desferrioxamine is the most suitable chelating agent since it is highly selective for iron (Table 27/3), is devoid of toxic effects, and enhances iron excretion from the organism. It can be given either intramuscularly, intravenously, or by subcutaneous infusion. Although its daily use could get rid of 10–20 grams of iron in a year and

Table 27/3

Stability constants of the various DFO-B metal complexes and the corresponding values for EDTA and DTPA complexes (from Schwarzenbach, 1964; Prelog, 1964)

Metal ions	DFO-B	EDTA	DTPA
Mg^{2+}	10^4	10^9	10^9
Ca^{2+}	10^2	10^{11}	10^{10}
Sr^{2+}	10	10^9	10^{10}
Fe^{3+}	10^{31}	10^{25}	10^{29}
Co^{2+}	10^{11}	10^{16}	10^{19}
Ni^{2+}	10^{10}	10^{19}	10^{20}
Cd^{2+}	10^8	—	—
Zn^{2+}	10^{11}	10^{16}	—
Cu^{2+}	10^{14}	10^{18}	—

regular DFO injections have been shown by Wöhler (1964) to reduce the liver iron, it is an extremely expensive form of treatment requiring regular injections, and on the whole it is probably considerably less effective than treatment with regular venesection.

SECONDARY HEMOCHROMATOSIS ASSOCIATED WITH CIRRHOSIS OF THE LIVER

Some patients with primary liver cirrhosis develop siderosis, and the differentiation between idiopathic and cirrhotic hemochromatosis is not always easy. Caroli and André (1964) have drawn attention to the differences in clinical signs and laboratory findings and the histological changes found in the two pathological processes.

In cirrhosis the male to female ratio is 4:1, as compared to the roughly 12:1 incidence in the genetically determined disease. Diabetes, neuroendocrine disturbances, and cardiac complications are all less frequent in association with cirrhosis, occurring in 30%, 50%, and 0–10%, respectively, as compared with 54%, 69%, and 31–46% in idiopathic hemochromatosis. Portal hypertension occurs in 40% of the cirrhotic patients with iron overload, while it is rare in idiopathic hemochromatosis (Bothwell and Finch, 1962; Caroli and André, 1964), although Brick (1964) reported on several patients with idiopathic hemochromatosis in whom portocaval shunt operation had to be performed. In idiopathic hemochromatosis the liver is always large, in contrast to the cirrhotics where the liver is either atrophic or normal, or, if hepatomegaly develops, it is of moderate degree (Caroli and André, 1964). Pigmentation of the skin is usual in both diseases. Liver function tests are grossly abnormal in the cirrhotic patients with iron overload but are less frequently abnormal in idiopathic hemochromatosis. Although the plasma iron may be high in cirrhosis, it is usually not as high as in the idiopathic disease, and the transferrin is less often saturated.

Crosby (1963) called attention to the rapid onset of anemia following venesection treatment in cirrhotic patients with secondary hemosiderosis. This is one indication that the iron is more difficult to mobilize from the liver in this condition, and it is confirmed by the fact that injections of desferrioxamine produce little or no increase in iron excretion (Moeschlin, 1964; Vannotti, 1964; Verloop, 1964; Smith et al., 1967) (Figs. 27/10 and 27/11).

The functional iron pool in the iron overload of liver cirrhosis may not exceed the normal level. Rosselin (1960) contrasted the ferrokinetic data obtained in idiopathic hemochromatosis with that in cirrhotics with iron overload. In the latter he found a normal plasma iron disappearance rate, increased iron incorporation, and a normal rate of iron uptake by the liver, indicating that in alcoholic cirrhosis the bulk of the

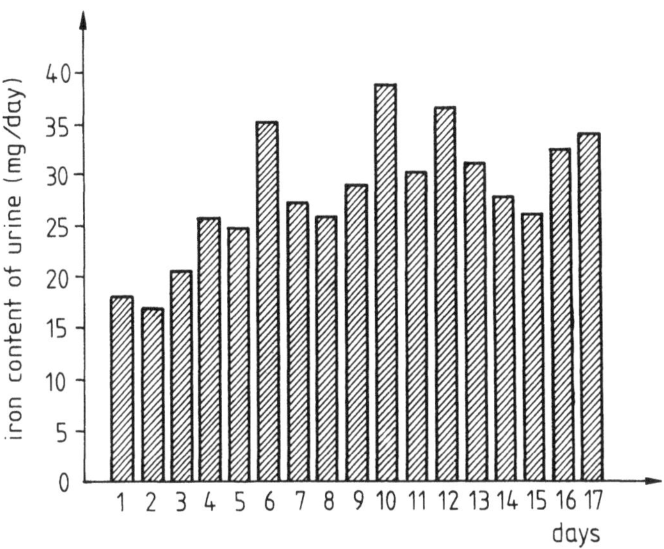

Fig. 27/10. Changes in the amount of urinary iron excreted following intramuscular injection of desferrioxamine-B (1 g/day) in hemochromatosis (after Wöhler, F., in: Gross, F.: Iron Metabolism. Springer, Berlin–Göttingen–Heidelberg 1964)

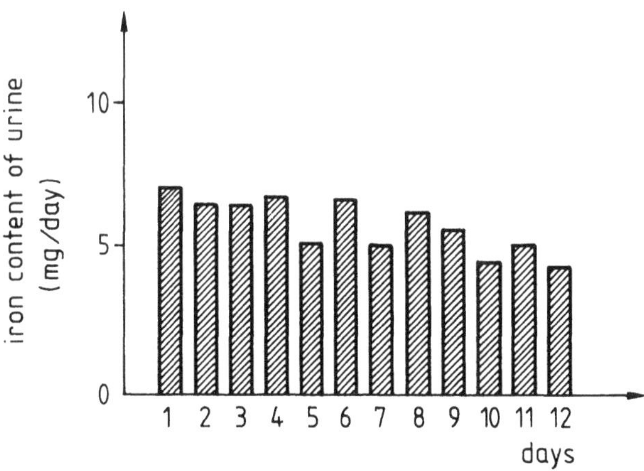

Fig. 27/11. Daily urinary iron excreted following the i.m. injection of desferrioxamine-B (1 g/day) in liver cirrhosis (after Wöhler, F., in: Gross, F.: Iron Metabolism. Springer, Berlin–Göttingen–Heidelberg 1964)

accumulated iron did not participate in the metabolic process and constituted a stable iron pool.

The basic difference between the two diseases is that in idiopathic hemochromatosis the siderosis precedes the development of the cirrhosis, whereas in secondary iron storage disease the enhanced iron deposition develops on the background of the cirrhosis. The development of iron overload in alcoholic cirrhosis is ascribed to the effect of several factors:

1. some alcoholic beverages are very rich in iron (Bothwell and Finch, 1962; MacDonald, 1964),
2. the protein-deficient diet of many alcoholics may promote the development of chronic pancreatitis (Sherlock, 1964), which in turn may lead to high iron absorption (Davis and Badenoch, 1962),
3. chronic liver disease itself is associated with enhanced iron absorption (Malpas and Callender, 1964; Williams et al., 1964),
4. the life-span of the red cells may become shorter or hemorrhages may occur requiring transfusion,
5. the development of collateral circulation and the formation of surgical shunts may play a role in the development of siderosis (Sherlock, 1964; Williams et al., 1964).

CONGENITAL ATRANSFERRINEMIA

This syndrome was described by Heilmeyer et al. (1961, 1966) in a 7-year-old girl with retarded somatic development. She was 106 cm tall (the average height for her age: 126 cm) and weighed 15.6 kg (normal: 22.6 kg). Her anemia was recognized at 3 months of age and proved totally refractory to treatment, and she required blood transfusions to maintain life. At the time of admission the hemoglobin was 9.1 g/dl and RBC 4.1 million. The mean hemoglobin content of the red cells was 22.5 pg, and the red cells were conspicuously hypochromic with many target cells in the blood smear. The diagnosis of thalassemia was considered but excluded.

In the bone marrow there was a preponderance of immature nucleated red cells but no sideroblasts or available iron in the reticulum cells. In contrast, the liver biopsy showed cirrhosis and marked siderosis. The serum iron level was 14–20 μg/dl and the total iron-binding capacity 20–33 μg/dl.

Following the intravenous injection of radioiron incubated with the patient's own plasma, the plasma activity dropped to half within 5 minutes (Fig. 27/12a). The plasma iron transport rate was increased 3- to 4-fold of normal, and the highest concentration of radioactivity appeared over the liver. There was some activity over the spleen but minimal activity over the bone marrow. Only 12 percent of the ^{59}Fe administered was incorporated into erythrocytes after 8 days (see Fig. 27/12b). Immunoelectrophoresis of the serum showed the absence of transferrin.

The patient was treated with plasma infusions, which resulted in an increase in reticulocyte count attributable to the infusion of transferrin in the plasma.

The child died suddenly of acute circulatory collapse, and autopsy revealed severe siderosis in various organs, particularly liver, pancreas, kidney, cardiac muscle, and

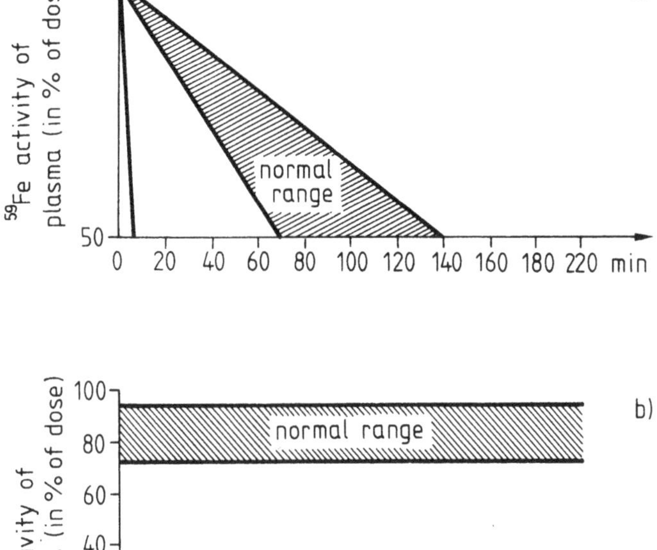

Fig. 27/12. Plasma iron clearance (a) and iron incorporation (b) in congenital atransferrinemia (after Heilmeyer, L.: Human hyposideraemia. In: Gross, F.: Iron Metabolism. Springer, Berlin–Göttingen–Heidelberg 1964b)

thyroid gland. The spleen was little affected, and the bone marrow was completely devoid of iron. It could be concluded that because of the lack of the specific transport protein, a paradoxical situation developed in which iron-deficiency anemia arose in an organism that was saturated with iron.

The iron absorption was apparently unimpaired, which suggests that plasma transferrin does not play a role in the mechanism of absorption. The half-life of transferrin appeared to be 3–5 days.

The transferrin concentration in the parents was only about half the normal amount, although they were healthy, and presumably the child was homozygous for an autosomal recessive gene.

Congenital atransferrinemia is extraordinarily rare, having been reported only 4 times (Heilmeyer et al., 1961; Čap et al., 1968; Sakata, 1969; Goya et al., 1972).

CONGENITAL (FAMILIAL) HYPERSIDEROSIS

Vitale et al. (1969) described two siblings with pathological iron storage associated with damage to the liver parenchymal cells, hypotonia of the musculature, and minor developmental abnormalities. They excluded hemolysis, exogenous iron overdosage, or transferrin anomaly in the etiology. The children both died in infancy of respiratory disease, and the etiology remains unknown. The iron metabolism of both parents appeared normal.

NUTRITIONAL SIDEROSIS — BANTU SIDEROSIS

Strachan (1929) noted that the tissues in the majority of the South African Bantu were frequently affected by siderosis, and this finding was later confirmed by numerous other observers (Higginson et al., 1953; Wainwright, 1957; Bothwell and Bradlow, 1960; Bothwell and Finch, 1962). The high incidence of siderosis is not restricted to South Africa but is also frequently found among the Bantus in Rhodesia (Gelfand, 1955), Bechuanaland, Nyasaland, and Mozambique (Higginson et al., 1953), as well as in Ghana (Edington, 1954, 1959). The same condition does not affect the white population or the Hindus in the same geographical regions (Fig. 27/13).

The incidence of Bantu siderosis is 40–88%. In most cases it becomes evident at the end of adolescence, reaching its severest form in general between 40 and 60 years of age (Wainwright, 1957; Bothwell and Bradlow, 1960; Bothwell and Isaacson, 1962). Both males and females are affected, although in women it is less common and usually less severe than in males (Wainwright, 1957; Bothwell and Isaacson, 1962). The hypersiderosis is attributed to the high iron content of the Bantu diet (Walker and Arvidsson, 1953). A considerable proportion of the iron comes from contamination from the iron cooking pots (Walker and Arvidsson, 1953), and particularly high values of iron are found in the various alcoholic beverages that are brewed in these vessels. Bothwell et al. (1964) have shown that the iron content varies between 0.5 and 15 mg/dl (mean 4 mg/dl), and Bantu males may consume

50–100 mg of iron daily in the beer alone. Most of the iron in the beer is present in a ferric form, from which about 2–3% is absorbed (Bothwell et al., 1964, 1979).

Isotope studies have shown that the iron overload is attributable to the high iron intake and that there is no intrinsic abnormality in the absorption mechanism.

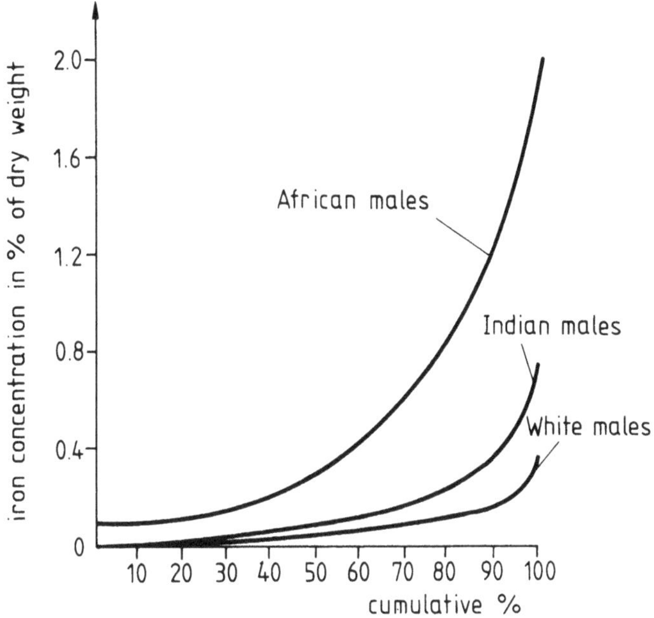

Fig. 27/13. Iron in specimens of liver obtained at necropsy in 240 Bantu males (average age 34 years), 40 Indian males (average age 48 years) and 60 white men (average age 40 years) (after Bothwell, T. H., in: Gross, F.: Iron Metabolism. Springer, Berlin–Göttingen–Heidelberg 1964)

Dietary iron overload occurs in other parts of the world; for example, in Brescia in Italy the red wines from the vineyards in this area have a high iron content, and Perman (1967) found that, of 22 patients with hemochromatosis, 8 consumed more than 2 liters of wine per day. Familial occurrence of iron overload here is not due to hereditary factors but to similar dietary habits and regular wine drinking.

Very rarely a clinical picture of hemochromatosis may arise after prolonged oral iron therapy (Turnberg, 1965; Johnson, 1968), but such cases are exceptional. In the early phases of dietary iron overload, the iron appears first in the form of hemosiderin granules in the liver parenchymal cells, mainly at the periphery of the liver lobules (Bothwell and Bradlow, 1960). Later the iron becomes demonstrable in all the liver cells, and there is a simultaneous accumulation in the Kupffer cells. If the iron content of the liver is 5–10 times normal, both the parenchymal cells and the Kupffer cells abound in iron. With greater accumulation of iron it is found in the tissues around the portal vein. There is also some accumulation of iron in the spleen, and according to Bothwell and collaborators (1965) the splenic and Kupffer cell iron exceeds that found in idiopathic hemochromatosis. Buchanan (1969a, b) attributes the increased iron deposition in the RE system to frequent chronic infections and other causes.

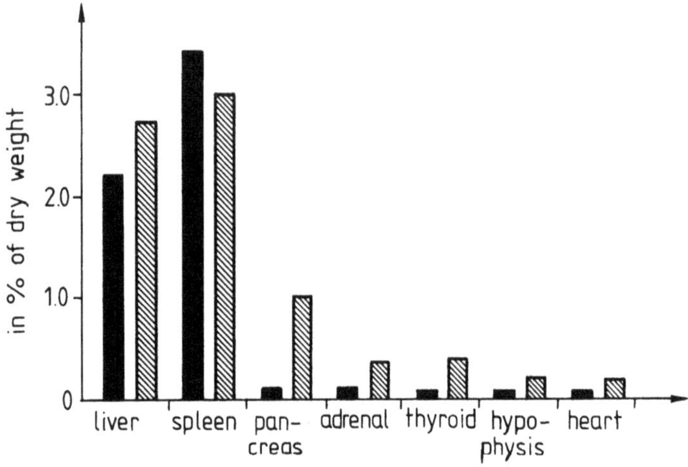

Fig. 27/14. Iron concentration of different organs in Bantu siderosis with or without cirrhosis of the liver (after Isaacson, C. et al.: J. Lab. clin. Med. *58*, 845, 1961). (■) severe siderosis; (▨) severe siderosis + cirrhosis

Wapnick et al. (1969) showed that Bantus with hypersiderosis had a high incidence of ascorbic acid deficiency, and Lipschitz et al. (1970) have demonstrated that ascorbic acid plays an important role in the mobilization of iron from the RES.

Glover et al. (1972) showed that in guinea pigs the percentage of iron retained in the spleen is higher in scorbutic than in normal animals; thus ascorbic acid deficiency may be a factor in the massive iron storage in the reticuloendothelial system in the Bantu siderosis. On the other hand, it has also been suggested that siderosis, scurvy, and osteoporosis are causally linked. There is some evidence that siderosis is the primary condition, that this produces a state of chronic ascorbic acid deficiency, and that the ascorbic acid deficiency is responsible for the osteoporosis commonly found in Black males with severe siderosis (Charlton et al., 1973).

In those cases of Bantu siderosis where liver cirrhosis has developed, other organs may be involved (Fig. 27/14), and the iron distribution may be similar to that found

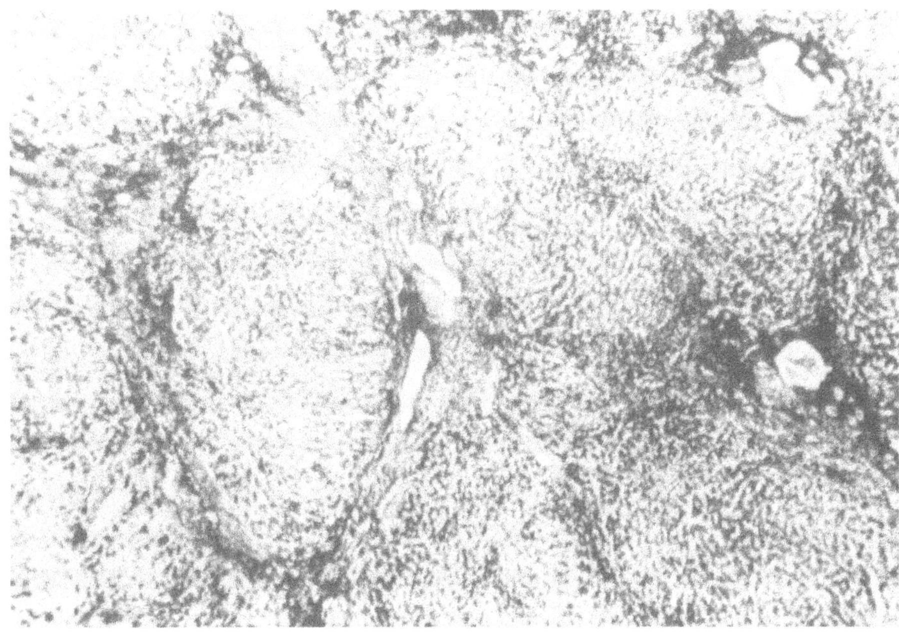

Fig. 27/15. Fully developed hemochromatosis in Bantu siderosis. The morphological picture of the liver corresponds to severe cirrhosis with massive iron accumulation (magnification: × 30) (after Isaacson, C. et al.: J. Lab. clin. Med. *58*, 845, 1961)

in idiopathic hemochromatosis. The reason for the difference in iron distribution in such cases is not known. It has been suggested that the high saturation of transferrin (Seftel et al., 1961) may play a role, as the tissues take up iron more readily when transferrin is saturated with iron (Jandl et al., 1959; Bothwell, 1964). Transferrin saturation is not usually so high when the siderosis is not associated with liver cirrhosis (Hathorn et al., 1960).

In a proportion of the Bantus who die of siderosis the post mortem findings cannot be distinguished from those of idiopathic hemochromatosis (Fig. 27/15). More than 20% of such patients have diabetes (Isaacson et al., 1961), while about 70% of the diabetic Bantus sooner or later develop the full clinical syndrome of hemochromatosis (Seftel et al., 1961). The incidence in males is twice as high as in females, and the majority of patients are between 40 and 60 years of age. Nearly all are emaciated and have massive hepatomegaly, and their diabetes requires insulin medication. Without exception they are regular consumers of alcohol. The prognosis of Bantu hemochromatosis is poor, and the cause of death is mostly hepatic insufficiency and/or portal hypertension. In the Bantu cases other factors, such as alcohol and malnutrition, contribute to the hepatic lesion (Bothwell and Isaacson, 1962), which is in agreement with the experimental observation that pathological iron storage renders the liver sensitive to various noxious nutritional and metabolic effects as well as to toxic agents (Golberg and Smith, 1960).

SIDEROSIS DEVELOPING IN REFRACTORY ANEMIAS ASSOCIATED WITH INEFFECTIVE ERYTHROPOIESIS

Massive tissue siderosis may develop in refractory anemias associated with ineffective erythropoiesis. The absorption of iron from the gastrointestinal tract is increased in these conditions and contributes to the iron overload. The plasma iron is high while the free iron-binding capacity is reduced (Garby et al., 1957; Crosby and Sheely, 1960), which promotes iron deposition in the parenchymatous organs.

Most of these patients also receive regular blood transfusions, which contribute to the iron overload, although the latter may occur with little or no transfusion (Houston, 1951; Currin, 1954; Heilmeyer et al., 1960; Kent and Popper, 1960; Bowdler and Huehns, 1963). The high erythropoietic activity stimulates iron absorption in spite of the substantial iron storage, and in some cases the increase in iron absorption may be considerable (Heilmeyer, 1964a).

The anemias associated with siderosis are usually those caused by disturbance in globin production or heme synthesis (Caroli et al., 1957; Heilmeyer et al., 1960). In severe congenital anemias, such as thalassemia, all the features of idiopathic hemochromatosis may be found — liver cirrhosis, diabetes, pigmentation, and heart failure — whereas in other less severe cases, such as some of the sideroblastic anemias, the siderosis may be accompanied by only moderate functional disturbance.

TRANSFUSIONAL SIDEROSIS

Patients with refractory anemia who require regular blood transfusions will sooner or later develop siderosis (Zeltmacher and Bevans, 1945; Schwartz and Blumenthal, 1948; Cottier, 1952; Paterson, 1952; Aufderheide et al., 1953; Bothwell, 1953; Platzer et al., 1955; Davies, 1955; Pengelly and Jones, 1956; Nay, 1957; Cappell et al., 1957; Hughes and Truelove, 1958; Oliver, 1959; Kent and Popper, 1960; Modell, 1975; and others).

The iron liberated from the transfused cells cannot be excreted, and it accumulates in the cells of the reticuloendothelial system, in the liver, spleen, lymph nodes, and bone marrow. Later, as larger amounts are deposited, there is iron accumulation in the parenchymal cells, producing organ damage and the picture of secondary posttransfusional hemochromatosis (Kent and Popper, 1960). In some refractory anemias associated with a hyperplastic bone marrow and ineffective erythropoiesis, increased iron absorption plays an additional role in the development of siderosis, but in those cases in which the bone marrow is hypoplastic, iron absorption is minimal and the increased iron storage is almost entirely due to the iron received by transfusion (Bothwell et al., 1958). The ultimate effect of the iron overload, whether it be due to enhanced absorption or to regular transfusions, is very similar, although disturbance of organ function may not be significant as long as the iron is primarily stored in the RE system. Later when the parenchymal cells become involved, pancreatic damage may be substantial and may even lead to the development of diabetes. Fibrosis of the liver is also encountered. Long-term treatment with desferrioxamine to remove the excess iron may result in reduction of hepatic fibrosis. Hunt et al. (1979) have suggested that this may be a direct effect on collagen synthesis rather than solely the effect of reduced iron stores.

Siderosis may also affect the kidneys, thyroid gland, adrenal gland, and cardiac muscle (Cleton and Blok, 1964; Jacobs, 1977). Changes in iron distribution from the primary reticuloendothelial iron to parenchymal iron overload are ascribed to the high saturation of transferrin, which provides favorable conditions for uptake of iron by parenchymal cells. According to Laurell (1952), free transferrin protects the tissues from siderosis, and, as already noted, Heilmeyer et al. (1961) found severe siderosis in the rare condition of congenital atransferrinemia.

Siderosis has been produced experimentally either by parenteral iron injections or by blood transfusions. The iron accumulates in the early stages primarily in the reticuloendothelial system (Finch et al., 1950; Brown et al., 1957; Cleton and Blok, 1964; Nath et al., 1972). The parenteral iron complexes used to produce siderosis are relatively large molecules that are taken up selectively by the reticuloendothelial cells, and the functions of the liver, pancreas, and heart are not affected (Brown et al., 1957); glucose tolerance tests, bromsulphalein excretion, and serum proteins all remain normal (Golberg et al., 1957). Increased iron storage outside the reticuloendothelial system was not found by Brown et al. (1957) even in chronic experiments lasting for years, and the iron storage capacity of the RES appears to be very considerable (Bothwell and Finch, 1962). In the animal experiments siderosis by itself did not appear to be cytotoxic, nor did it enhance the injurious effect of other toxic agents on the liver (Nath et al., 1972).

Sturgeon and Shoden (1964) produced hypersiderosis in rabbits by the intravenous administration of iron dextran and saccharated iron oxide. Following the injection of iron dextran, little, if any, histochemically demonstrable iron in the form of hemosiderin could be detected either in the Kupffer cells or in the parenchymal cells until the iron content was more than 100 mg. Up to a level of about 150 mg, 90% of the iron dextran was deposited in the form of ferritin. The

Table 27/4

Biochemical and morphological features of hypersiderosis produced in rabbits by various doses of iron dextran injections (from Sturgeon and Shoden, 1964)

	Amount of iron injected (mg)						
	0	50	100	200	400	800	1600
Total iron content of liver (mg)	7	30	54	98	210	408	1048
Water-soluble iron (mg)	6	28	51	88	191	290	494
Water-insoluble iron (mg)	1	2	3	10	19	118	554
Iron granules in the parenchymal cells	0	0	0	±	+	+ + +	+ + + +
Iron granules in the Kupffer cells	0	0	0	±	±	±	±

Table 27/5

Biochemical and morphological features of hypersiderosis produced in rabbits by various doses of saccharated iron oxide injections (from Sturgeon and Shoden, 1964)

	Amount of iron injected (mg)						
	0	50	100	200	400	800	1600
Total iron content of liver (mg)	7	38	49	94	213	495	1055
Water-soluble iron (mg)	6	20	28	65	152	354	534
Water-insoluble iron (mg)	1	18	21	28	61	141	521
Iron granules in the parenchymal cells	0	0	0	0	±	+ +	+ + +
Iron granules in the Kupffer cells	0	+	+	+	+ +	+ +	+ + +

ratio of hemosiderin began to rise between 150 and 200 mg. Above this level hemosiderin increased mainly in the parenchymal cells, although most of the iron in the liver was still ferritin, which cannot be demonstrated by histochemical methods (Table 27/4). The ferritin and hemosiderin iron deposits reached about equal proportions when there was a total of about 800 mg, but above this only the hemosiderin continued to rise. Following the injection of saccharated iron oxide (Table 27/5) hemosiderin could be demonstrated in the Kupffer cells after only 50 mg of iron. After the injection of 400 mg of iron most of the Kupffer cells contained hemosiderin, and some began to be evident in the parenchymal cells. At the same time the liver contained 213 mg of iron, of which 152 mg was ferritin and only 61 mg hemosiderin.

Sturgeon and Shoden suggest that the parenchymal cells contain the iron in the form of ferritin dispersed in the cytoplasm initially. This becomes concentrated in the form of crystals or clusters, and denaturation of the apoferritin matrix occurs, in the course of which the ferritin loses its water solubility and becomes "pre-

Table 27/6
Comparison in iron-loaded rabbits of the shifts from ferritin to hemosiderin of iron and radioactivity during 67 days (from Sturgeon and Shoden, 1964)

	Iron (mg)			^{59}Fe (cpm × 10^{3})	
	Control	day 16	day 67	day 16	day 67
Ferritin	497	415	373	1090	515
Hemosiderin	760	900	968	690	1200
Total	1257	1315	1341	1780	1715

hemosiderin." The iron then aggregates into micelles, becoming true hemosiderin. Ferrokinetic studies are in keeping with this theory in that initially the radioactivity of ferritin is higher than that of hemosiderin and the proportions shift later toward the hemosiderin (Table 27/6).

RENAL HEMOSIDEROSIS

Isolated renal siderosis is usually a complication of diseases associated with chronic intravascular hemolysis, for example, paroxysmal nocturnal hemoglobinuria. If the amount of blood pigment liberated in the course of intravascular hemolysis exceeds the hemoglobin-binding capacity of the circulating haptoglobin, the free blood pigment is filtered through the glomeruli (Laurell, 1958). Some of the blood pigment is resorbed by the tubules; the iron is split off from the hemoglobin and is stored in the form of ferritin (Hampton and Mayerson, 1950). The hemoglobin that is not resorbed will be excreted with the urine. Substantial iron may accumulate in the cells of the tubules, and hemosiderin originating from the desquamated cells of the tubules may be demonstrated in the sediment of the urine (Crosby, 1953). Hemoglobinuria and hemosiderinuria occur most frequently in paroxysmal nocturnal hemoglobinuria (PNH), but it may also occur in other hemolytic diseases such as thalassemia and sickle cell anemia, as well as genetically determined haptoglobin deficiency.

In PNH, the daily excretion of iron in the form of hemosiderin may amount to as much as 3–5 mg. This iron loss may be considerably augmented by the excretion of hemoglobin, and a state of iron-deficiency anemia may be superimposed on the PNH while there is substantial siderosis in the kidneys (Ellenhorn et al., 1951). The loss of iron must be replaced. One of the complications of iron therapy may be an exacerbation of hemolysis, due to the rapid production and subsequent destruction of a greater number of complement-sensitive cells. If increased erythropoiesis is suppressed by transfusion, iron may be given without any increase in the amount of hemolysis (Rosse, 1972). Renal function is not usually affected (Bradley and Bradley, 1947; Leonardi and Ruol, 1960; Hutt et al., 1961) because of the very rapid turnover of the renal tubular cells (Crosby, 1953).

IDIOPATHIC PULMONARY HEMOSIDEROSIS

The etiology of this rare disease is still unclear. Clinically it is characterized by recurrent pulmonary hemorrhages, with iron accumulation in the lungs. This may be so excessive that occasionally there is an increase in the total iron content of the organism (Bothwell and Finch, 1962). The disease is usually restricted to childhood, but it may manifest itself in young adults as well (Boyd, 1959). Soergel and Sommers (1962) collected 112 cases with ages ranging between 4 months and 47 years. Most of the children were between 1 and 7 years, and most of the adults were younger than 30. In the children the sex distribution appeared to be equal, but in the adults the ratio of males to females was 2:1. There was no familial occurrence. The course of the disease is characterized by attacks of rapid onset accompanied by faintness, dyspnea, cyanosis, cough, and increasing pallor. The sputum is stained with blood, but massive pulmonary hemorrhage may also occur. The episode is usually accompanied by fever, and there may be enlargement of the liver and spleen and some jaundice. The episode usually subsides leaving the patient anemic. Occasionally death occurs in the first attack, but episodes may occur over a number of years, although the average survival is of the order of 3 years (Soergel, 1957). The cause of death is internal hemorrhage, cor pulmonale as a result of the massive pulmonary fibrosis, or renal insufficiency (Reye, 1945; MacGregor et al., 1960). Diagnosis is confirmed by the radiological appearances (Wyllie et al., 1948; Tait and Corridan, 1952; Boyd, 1959), and some authors suggest a lung biopsy (King, 1949; Gellis et al., 1953; Schuler and Flesch, 1955). The histopathology has been described by Wyllie et al. (1948), Chatgidakis (1955), Schuler and Flesch (1955), and others. The alveoli of the lungs are filled with red blood cells, hemosiderin-containing macrophages, and randomly distributed hemosiderin granules. Similar cells and free hemosiderin can be found also in the interstices of the lung. In early attacks there is little evidence of fibrosis, but later considerable changes occur. The dilated alveolar capillaries become more tortuous and have thicker walls, and the elastic fibers become swollen, fragmented, and surrounded by iron-containing pigment. The fragmented fibers form aggregates, which are then phagocytosed by foreign-body giant cells. The collagen and smooth muscle tissues in the walls of the small and medium-sized arteries and of the veins undergo hyaline degeneration, while the larger blood vessels remain relatively intact. Hemosiderin may also be found in the epithelial cells of the bronchioli and small bronchi, and hemosiderin-containing macrophages are found in the adjacent lymph nodes. Iron accumulation is restricted to the lungs, and there is no other organ involvement.

HEMATOLOGICAL CHANGES

The anemia may be pronounced during the attacks (Bothwell and Finch, 1962). During remissions the blood may return to normal, but it is more usual for an iron-deficiency type of anemia to be found. Following the acute hemorrhage, reticulocytosis may develop, as well as an increased white cell count.

Apt et al. (1957) made ferrokinetic studies in two patients with idiopathic pulmonary hemosiderosis and in their cases found the iron kinetics to be completely normal during remission of the disease. In addition, the life-span of the patients' red cells in a healthy recipient and the survival of the red cells of the healthy donor in the patient's blood were normal. During an exacerbation of the disease the plasma iron transport rate trebled. The bone marrow activity reached a maximum in 1 hour, with a subsequent rapid decline in activity and a simultaneous rapid increase in

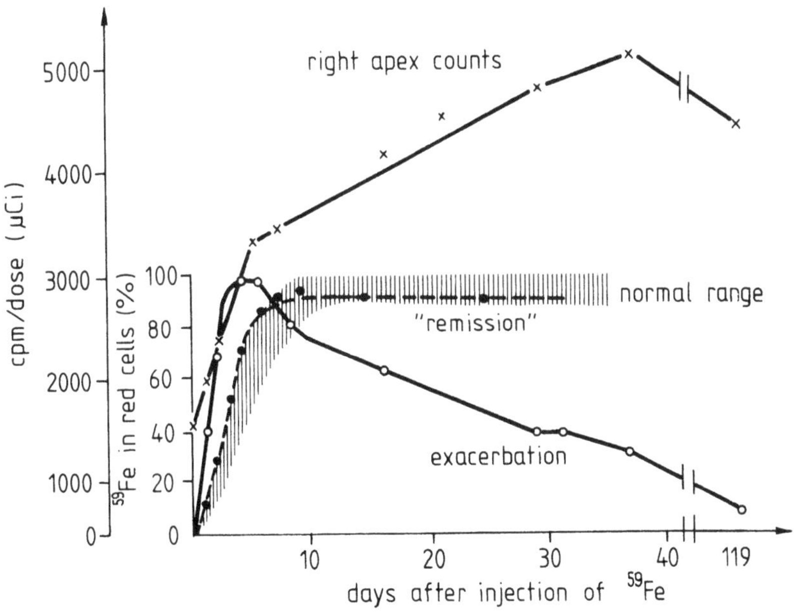

Fig. 27/16. Gradual decline (o – o – o) in the radioactivity of the red cells and a related gradual rise (× – × – ×) in the activity of the lungs after the i.v. injection of ^{59}Fe in idiopathic pulmonary hemosiderosis. In remission, the iron incorporation was normal (after Apt, L. et al.: J. clin. Invest. *36*, 1150, 1957).

activity in the circulating blood cells. Four days after injection of the radioiron, 99% of the dose was recovered in the peripheral blood. In the subsequent 4 months there was a gradual decline in activity of the red blood cells but an accumulation over the lungs (Fig. 27/16). The series of pulmonary hemorrhages during this period explains this finding. Others have made similar observations in such cases (Hamilton et al., 1960; de Gowin et al., 1968; Samuels and Bass, 1969). The ferrokinetic data are characteristic of iron deficiency secondary to pulmonary hemorrhages, the deficiency state being due to the fact that the hemosiderin in the lungs is not mobilizable for hemoglobin formation.

Some authors (Hanssen, 1947; Wiesmann et al., 1953; Steiner, 1954) had suggested that there might be a hemolytic factor contributing to the development of anemia in idiopathic pulmonary hemosiderosis. There is an apparent decline in survival of ^{51}Cr-labeled red blood cells, but this is only because blood is accumulating in the lung and there is no evidence for a reduced red cell survival.

Very rarely the picture presents as a simple iron-deficiency type of anemia, and the clinical and radiological signs may be absent (Gillman and Zinkham, 1969).

TREATMENT OF IDIOPATHIC PULMONARY HEMOSIDEROSIS

The hypochromic anemia may be relieved by iron medication, and the success of iron therapy indicates that absorption is unaffected. In contrast to other states of iron overload, mobilization of the iron in the lungs with the use of desferrioxamine is disappointing. Dagg et al. (1971), with treatment over a year, failed to remove even as much as 1 gram of iron.

Severe hemorrhages have to be treated by blood transfusion or parenteral iron (Dagg et al., 1971). Various other types of therapy have been tried without avail, e.g., splenectomy (Paterson, 1946; de Castro Freire and Cordeiro, 1948; Wyllie et al., 1948; Cordeiro, 1952; Steiner, 1954, 1959), ACTH and cortisone therapy (Sandøe, 1954; Burckhardt and Vogel, 1955; Halvorsen, 1956; Browning and Houghton, 1956; Irvin and Snowden, 1957; Soergel, 1957; Soergel and Sommers, 1962), and immunosuppressive therapy (Mutz and Schrofl, 1971).

GOODPASTURE SYNDROME

Pulmonary hemorrhage associated with glomerulonephritis, known as Goodpasture syndrome, differs in several aspects from idiopathic pulmonary hemosiderosis. It occurs in a different age group and does not affect children. Severe hemoptysis is rare, the histological picture of the lungs is different, and the most frequent cause of death is renal insufficiency. Respiratory insufficiency is due to the combined disturbance of ventilation and diffusion.

The antigenicity of the alveolar capillary and the glomerular epithelium is identical (Goodman et al., 1955; Mellors et al., 1955; de Gowin et al., 1963). The simultaneous occurrence of acute glomerulonephritis and pulmonary hemorrhage suggests the possibility of a common pathogenesis, for example, an autoimmune mechanism elicited by a viral infection (Benoit et al., 1964). Immunofluorescent studies of the renal tissues have shown that the immunoglobulin forms bandlike deposits along the glomerular basement membrane (Duncan et al., 1965), which is not so indented and uneven as in streptococcal glomerulonephritis (Hammer and Dixon, 1963). Sturgill and Westerveldt (1965) showed the presence of immunoglobulin in the alveolar septa, but this did not prove to be a constant finding (Scheer and Grossman, 1964; Duncan et al., 1965). Treatment by immunosuppression or plasmaphoresis may prove beneficial in some cases.

BIBLIOGRAPHY

ABBOTT, D. F., GRESHAM, G. A.: Arthropathy in transfusional siderosis. Brit. med. J. *1*, 418 (1972).

ALTHAUSEN, T. L., DOIG, R. K., WEIDEN, S., MOTTERAM, R., TURNER, C. N., MOORE, A.: Hemochromatosis: An investigation of twenty-three cases with special reference to nutrition, to iron metabolism and to studies of hepatic and pancreatic function. Arch. intern. Med. *88*, 553 (1951).

APT, L., POLLYCOVE, M., ROSS, J. F.: Idiopathic pulmonary hemosiderosis. A study of the anemia and iron distribution using radioiron and radiochromium. J. clin. Invest. *36*, 1150 (1957).

AUFDERHEIDE, A. C., HORNS, H. L., GOLDISH, R. J.: Secondary hemochromatosis; transfusion (exogenous) hemochromatosis. Blood *8*, 824 (1953).

BALCERZAK, S. P., WESTERMAN, M. P., LEE, R. E., DOYLE, A. P.: Idiopathic hemochromatosis. Amer. J. Med. *40*, 857 (1966).

BARRY, M.: Iron overload: clinical aspects, evaluation, and treatment. In: CALLENDER, S. T. (ed.): Clinics in Haematology, Vol. 2/2, p. 405. Saunders, London–Philadelphia–Toronto 1973.

BARRY, M., SHERLOCK, S.: Measurement of liver iron concentration in needle biopsy specimens. Lancet *1*, 100 (1971).

BARRY, M., CARTEI, G. C., SHERLOCK, S.: Measurement of iron stores in cirrhosis using diethylenetriamine penta-acetic acid. Gut *11*, 899 (1970).

BECKER, D., MILLER, M.: Presence of diabetic glomerulosclerosis in patients with hemochromatosis. New Engl. J. Med. *263*, 367 (1960).

BELL, E. T.: Relation of portal cirrhosis to hemochromatosis and to diabetes mellitus. Diabetes *4*, 435 (1955).

BENOIT, F. L., RULON, D. B., THEIL, G. B., DOOLAN, P. D., WHATTEN, R. H.: Goodpasture's Syndrome. Amer. J. Med. *37*, 424 (1964).

BERGDAHL, U., BERGE, T., JOHANSSON, S.: Pulmonary haemosiderosis and glomerulonephritis. Acta med. scand. *186*, 199 (1969).

BERK, J. E., LIEBER, M. M.: Primary carcinoma of liver in hemochromatosis. Amer. J. med. Sci. *202*, 708 (1941).

BICKEL, H., GÄUMANN, E., KELLER-SCHIERLEIN, W., PRELOG, V., VISCHER, E., WETTSTEIN, A., ZÄHNER, H.: On iron containing growth factors, sideramines, and their antagonists, the iron containing antibiotics, sideromycins. Experientia *16*, 129 (1960).

BINFORD, C. H., LAWRENCE, R. L., WOLLENWEBER, H. L.: Hemochromatosis with primary carcinoma of the liver. Arch. Path. *25*, 527 (1938).

BLACKBURN, C. R. B., MCGUINNESS, A. E., KALDOR, I.: Removal of excess body iron in hemochromatosis by repeated venesection. Aust. Ann. Med. *2,* 202 (1953).
BOMFORD, A. B., WILLIAMS, R.: Quart. J. Med. *45,* 611 (1976).
BOTHWELL, T. H.: The relationship of transfusional haemosiderosis to idiopathic haemochromatosis. S. Afr. J. clin. Sci. *4,* 53 (1953).
BOTHWELL, T. H.: Iron overload in the Bantu. In: GROSS, F. (ed.): Iron Metabolism. Springer, Berlin–Göttingen–Heidelberg 1964.
BOTHWELL, T. H., ALPER, T.: The cardiac complications of haemochromatosis. S. Afr. J. clin. Sci. *2,* 226 (1951).
BOTHWELL, T. H., BRADLOW, B. A.: Siderosis in the Bantu. A combined histopathological and chemical study. Arch. Path. *70,* 279 (1960).
BOTHWELL, T. H. et al.: Radioiron studies in hemochromatosis. The effects of repeated phlebotomies. J. Lab. clin. Med. *45,* 167 (1955).
BOTHWELL, T. H. et al.: Iron Metabolism in Man. Blackwell, Oxford–London–Edinburgh–Melbourne 1979.
BOTHWELL, T. H., FINCH, C. A.: Iron Metabolism. Little, Brown and Co., Boston 1962.
BOTHWELL, T. H., ISAACSON, C.: Siderosis in the Bantu. Brit. med. J. *1,* 522 (1962).
BOTHWELL, T. H., ABRAHAMS, C., BRADLOW, B. A., CHARLTON, R. W.: Idiopathic and Bantu hemochromatosis. Arch. Path. *79,* 163 (1965).
BOTHWELL, T. H., CHARLTON, R. W.: Dietary iron overload. In: KIEF, H. (ed.): Iron Metabolism and its Disorders. Excerpta Medica, Amsterdam–Oxford, American Elsevier, New York 1975.
BOTHWELL, T. H., COHEN, I., ABRAHAMS, O. L., PEROLD, S. M.: A familial study in idiopathic haemochromatosis. Amer. J. Med. *27,* 730 (1959).
BOTHWELL, T. H., PIRZIO-BIROLI, G., FINCH, C. A.: Iron absorption. I. Factors influencing absorption. J. Lab. clin. Med. *51,* 24 (1958).
BOTHWELL, T. H., SEFTEL, H., JACOBS, P., TORRANCE, J. D., BAUMSLAG, N.: Iron overload in Bantu subjects; studies on the availability of iron in Bantu beer. Amer. J. clin. Nutr. *14,* 47 (1964).
BOTHWELL, T. H., VAN LINGEN, B., ALPER, T., DU PREEZ, M. L.: Cardiac complications of hemochromatosis. Amer. Heart J. *43,* 333 (1952).
BOULIN, R.: Étude statistique de 70 cas de diabète bronzé. Presse méd. *53,* 326 (1945).
BOULIN, R., BAMBERGER, J.: L'hémochromatose familiale. Sem. Hôp. Paris, *29,* 3153 (1953).
BOWDLER, A. J., HUEHNS, E. R.: Thalassaemia minor complicated by excessive iron storage. Brit. J. Haemat. *9,* 13 (1963).
BOYD, D. H.: Idiopathic pulmonary haemosiderosis in adults and adolescents. Brit. J. Dis. Chest. *53,* 41 (1959).
BRADLEY, S. E., BRADLEY, G. P.: Renal function during chronic anaemia in man. Blood *2,* 192 (1947).
BRICK, I. B.: Liver histology in six asymptomatic siblings in a family with hemochromatosis. Gastroenterology *40,* 210 (1961).
BRICK, I. B.: In: GROSS, F. (ed.): Iron Metabolism, pp. 340–341. Springer, Berlin–Göttingen–Heidelberg 1964.
BROWN, E. B., DUBACH, R., SMITH, D. E., REYNAFARJE, C., MOORE, C. V.: Studies in iron transportation and metabolism. X. Long-term iron overload in dogs. J. Lab. clin. Med. *50,* 862 (1957).
BROWNING, J. R., HOUGHTON, J. D.: Idiopathic pulmonary haemosiderosis. Amer. J. Med. *20,* 374 (1956).
BRUNNER, H. E., FRICK, P. G., HITZIG, W. H.: Familiäre Hämochromatose — Nachweis der Stoffwechselstörung in der Frühphase mit Fe^{59}. Schweiz. med. Wschr. *92,* 343 (1962).
BUCHANAN, W. M.: Aetiology of extrahepatic epithelial iron deposits in Bantu siderosis. J. clin. Path. *22,* 296 (1969a).
BUCHANAN, W. M.: Bantu siderosis — a review. Cent. Afr. J. Med. *15,* 105 (1969b).
BURCKHARDT, D., VOGEL, C.: Ein Beitrag zur idiopathischen Lungenhämosiderose. Wschr. Kinderheilk. *103,* 455 (1955).
BYRD, R. B., COOPER, T.: Hereditary iron-loading anemia with secondary hemochromatosis. Ann. intern. Med. *55,* 103 (1961).

ČAP, J. et al.: Kongenitálna atransferinémia u 11-mesačného dietata. Čs. Pediat. *23,* 1020 (1968).

CAPPELL, D. F., HUTCHISON, H. E., JOWETT, M.: Transfusional siderosis: the effects of excessive iron deposits on the tissues. J. Path. Bact. *74,* 245 (1957).

CAROLI, J., ANDRÉ, J.: Surcharge ferrique dans les cirrhoses (à l'exclusion de l'hémochromatose idiopathique). In: GROSS, F. (ed.): Iron Metabolism, pp. 326–339. Springer, Berlin–Göttingen–Heidelberg 1964.

CAROLI, J., BERNARD, J., BESSIS, M., COMBRISSON, A., MALASSENET, R., BRETON-GORIUS, J.: Hémochromatose avec anémie hypochrome et absence d'hémoglobine anormale: Étude au microscope électronique. Presse méd. *2,* 1991 (1957).

CHARLTON, R. W., BOTHWELL, T. H., SEFTEL, H. C.: Dietary iron overload. In: CALLENDER, S. T. (ed.): Iron Deficiency and Iron Overload (Clinics in Haematology), Vol. 2/2, p. 383. Saunders, London–Philadelphia–Toronto 1973.

CHATGIDAKIS, C. B.: Primary pulmonary hemosiderosis (Ceelen's disease). S. Afr. J. Lab. clin. Med. *1,* 166 (1955).

CHOU, A. C. et al.: Abnormalities of iron metabolism and erythropoiesis in vitamin E-deficient rabbits. Blood *52,* 187 (1978).

CLETON, F. J., BLOK, A. P. R.: Post-transfusional haemosiderosis. In: GROSS, F. (ed.): Iron Metabolism. Springer, Berlin–Göttingen–Heidelberg 1964.

CLETON, F. J., FINCH, C. A. (1961), cit. BOTHWELL, FINCH, 1962.

COHEN, I., BOTHWELL, R. H.: Haemochromatosis in a young female. S. Afr. med. J. *32,* 629 (1958).

CONN, H. O.: Portocaval anastomosis and hepatic hemosiderin deposition: a prospective, controlled investigation. Gastroenterology *62,* 61 (1972).

CONTE, M., RISTELHUEBER, J., JULIEN, C.: Hémochromatose grave traitée par les saignées répetées et améliorée. Bull. Soc. méd. hôp. Paris *74,* 270 (1958a).

CONTE, M., RISTELHUEBER, J., MALVEZIN, J. C., JULIEN, C.: Les hémochromatoses familiales et héréditaires. Bull. Soc. méd. hôp. Paris *74,* 267 (1958b).

CORDEIRO, M.: Un cas d'hémosidérose pulmonaire idiopathique guéri par splénectomie. Helv. paediat. Acta *7,* 501 (1952).

COTTIER, H.: Transfusionssiderose und allgemeine Hämochromatose. Schweiz. med. Wschr. *82,* 873 (1952).

COTTIER, P.: Über ein der Hämochromatose vergleichbares Krankheitsbild bei Neugeborenen. Schweiz. med. Wschr. *87,* 39 (1957).

CROSBY, W. H.: Paroxysmal nocturnal hemoglobinuria. Relation of the clinical manifestations to underlying pathogenic mechanisms. Blood *8,* 769 (1953).

CROSBY, W. H.: Treatment of haemochromatosis by energic phlebotomy. One patient's response to the letting of 55 litres of blood in 11 months. Brit. J. Haemat. *4,* 82 (1958).

CROSBY, W. H.: The control of iron metabolism: Lecture given at the Boston City Hospital, Boston, 16th April, 1963, cit. KATZ, J. H.: In: GROSS, F. (ed.): Iron Metabolism. Springer, Berlin–Göttingen–Heidelberg 1964.

CROSBY, W. H., SHEEHY, T. W.: Hypochromic iron loading anemia. Brit. J. Haemat. *6,* 56 (1960).

CURRIN, J. F.: Occurrence of secondary hemochromatosis in patient with thalassemia major. Arch. Intern. Med. *93,* 781 (1954).

DACIE, J. V.: Transfusion of saline-washed red cells in nocturnal haemoglobinuria. Clin. Sci. *7,* 65 (1948).

DACIE, J. V.: The haemolytic anaemias, 2nd ed., Parts 1 and 2. Grune and Stratton, New York–London 1960 and 1962.

DAGG, J. H., CUMMING, R. L. C., GOLDBERG, A.: Disorders of iron metabolism. In: GOLDBERG, A., BRAIN, M. C. (eds.): Recent Advances in Haematology. Churchill–Livingstone, Edinburgh–London 1971.

DAGG, J. H., SMITH, J. A., GOLDBERG, A.: Urinary excretion of iron. Clin. Sci. *30,* 495 (1966).

DAVEY, D. A., FOXELL, A. W. H., KEMP, T. A.: Treatment of haemochromatosis by repeated venesection. Brit. med. J. *2,* 1511 (1954).

DAVIES, D. M.: Secondary haemochromatosis. Lancet *2,* 1064 (1955).

DAVIS, A. E., BADENOCH, J.: Iron absorption in pancreatic disease. Lancet *2*, 6 (1962).
DAVIS, W. D., ARROWSMITH, W. R.: The effect of repeated bleedings in hemochromatosis. J. Lab. clin. Med. *36*, 814 (1950).
DAVIS, W. D., ARROWSMITH, W. R.: The effect of repeated phlebotomies in hemochromatosis; report of 3 cases. J. Lab. clin. Med. *39*, 526 (1952).
DAVIS, W. D., ARROWSMITH, W. R.: The treatment of hemochromatosis by massive venesection. Ann. Intern. Med. *39*, 723 (1953).
DAVISON, R. H.: Inheritance of haemochromatosis. Brit. med. J. *2*, 1262 (1961).
DEBRÉ, R., DREYFUS, J. C., FRÉZAL, J., LABIE, D., LAMY, M., MAROTEAUX, P., SCHAPIRA, F., SCHAPIRA, G.: Genetics of hemochromatosis. Ann. hum. Genet. *23*, 16 (1958).
DE CASTRO FREIRE, L., CORDEIRO, M.: Hémoptyse sub-aiguë récidivante par diathèse hémorragique thrombopénique. Splénectomie. Guérison. Helv. paediat. Acta *3*, 255 (1948).
DECKERT, T.: Behandling af haemochromatose med venaesektion. Nord. med. *60*, 1751 (1958).
DE GENNES, L.: Le syndrome endocrinien des cirrhoses bronzées. Acta gastro-ent. belg. *15*, 208 (1952).
DE GOWIN, R. L., ODA, Y., EVANS, R. H.: Nephritis and lung hemorrhage. Arch. Intern. Med. *111*, 16 (1963).
DE GOWIN, R. L., SORENSEN, L. B., CHARLESTON, D. B., GOTTSCHALK, A., GREENWOLD, J. H.: Retention of radioiron in the lungs of a woman with idiopathic pulmonary hemosiderosis. Ann. intern. Med. *69*, 1213 (1968).
DESFORGES, G.: Abdominal pain in haemochromatosis. New Engl. J. Med. *241*, 485 (1949).
DILLINGHAM, C. H.: Familial occurrence of hemochromatosis. New Engl. J. Med. *262*, 1128 (1960).
DREYFUS, J. C., SCHAPIRA, G.: The metabolism of iron in haemochromatosis. In: GROSS, F. (ed.): Iron Metabolism, p. 296. Springer, Berlin–Göttingen–Heidelberg 1964.
DREYFUS, J. C., SCHAPIRA, G., SCHWARZMANN, V., ARBOU, R.: Surcharge ferrique hépatique. Forme inapparente chez un fils d'hémochromatosique. Presse méd. *68*, 577 (1960).
DUBIN, I. N.: Idiopathic hemochromatosis and transfusion siderosis; review. Amer. J. clin. Path. *25*, 514 (1955).
DUNCAN, D. A., DRUMMOND, K. N., MICHAEL, A. F., VERNIER, R. L.: Pulmonary haemorrhage and glomerulonephritis. Ann. intern. Med. *62*, 920 (1965).
DYMOCK, I. W., CASSAR, J., PYKE, D. A., OAKLEY, W. G., WILLIAMS, R.: Observations on the pathogenesis, complications and treatment of diabetes in 115 cases of haemochromatosis. Amer. J. Med. *52*, 203 (1972).
EDINGTON, G. M.: Haemosiderosis and anaemia in the Gold Coast African. W. Afr. Med. J. *3*, 66 (1954).
EDINGTON, G. M.: Nutritional siderosis in Ghana. Cent. Afr. J. Med. *5*, 186 (1959).
ELLENHORN, M. J., FEIGENBAUM, L. Z., PLUMHOF, C., METTIER, S. R.: Paroxysmal nocturnal hemoglobinuria with chronic hemolytic anemia. Report of a case with postmortem observations. Arch. Intern. Med. *87*, 868 (1951).
ESKIND, I. B., FRANKLIN, W., LOWELL, F. C.: Insulin-resistant diabetes mellitus associated with hemochromatosis. Ann. Intern. Med. *38*, 1295 (1963).
EVANS, J.: Treatment of heart failure in hemochromatosis. Brit. med. J. *1*, 1075 (1959).
FAHEY, J. L., RATH, C. E., PRINCIOTTO, J. V., BRICK, I. B., RUBIN, M.: Evaluation of trisodium calcium diethylene-triamine-pentaacetic acid in iron storage disease. J. Lab. clin. Med. *57*, 436 (1961).
FIGUEROA, W. G., TUTTLE, S. G.: A study of iron excretion following CaDTPA administration. Clin. Res. *9*, 92 (1961).
FINCH, C. A.: Iron metabolism in hemochromatosis. Abstr. Int. Soc. Hemat. 1948.
FINCH, C. A.: Iron metabolism in hemochromatosis. J. clin. Invest. *28*, 780 (1949).
FINCH, C. A., HEGSTED, M., KINNEY, T. D., THOMAS, E. D., RATH, C. E., HASKINS, D., FINCH, S. C., FLUHARTY, R. G.: Iron metabolism. The pathophysiology of iron storage. Blood *5*, 983 (1950).
FINCH, S. C., FINCH, C. A.: Idiopathic haemochromatosis, an iron storage disease. Medicine *34*, 381 (1955).
FINCH, S. C., HASKINS, D., FINCH, C. A.: Iron metabolism. Hematopoiesis following phlebotomy. Iron as a limiting factor. J. clin. Invest. *29*, 1078 (1950).

FREY, W. G., MILNE, J., JOHNSON, G. B., EBAUGH, F. G.: Management of familial hemochromatosis. New Engl. J. Med. *265*, 7 (1961).

GALE, E., TORRANCE, J., BOTHWELL, T. H.: The quantitative estimation of total iron stores in human bone marrow. J. clin. Invest. *42*, 1076 (1963).

GARBY, L., SJÖLIN, S., VAHLQUIST, B.: Chronic refractory hypochromic anaemia with disturbed haem metabolism. Brit. J. Haemat. *3*, 55 (1957).

GELFAND, M.: Bantu siderosis. Trans. roy. Soc. trop. Med. Hyg. *49*, 370 (1955).

GELLIS, S. S., REINHOLD, J. D. L., GREEN, S.: Use of aspiration lung puncture in diagnosis in idiopathic pulmonary hemosiderosis. Amer. J. Dis. Child. *85*, 303 (1953).

GELPI, A. P., ENDE, N.: An hereditary anemia with hemochromatosis. Studies of an unusual hemopathic syndrome resembling thalassemia. Amer. J. Med. *25*, 303 (1958).

GILLMAN, P. A., ZINKHAM, W. H.: Severe idiopathic pulmonary haemosiderosis in the absence of clinical or radiological evidence of pulmonary disease. J. Pediat. *75*, 118 (1969).

GILLMAN, T., HATHORN, M., LAMONT, N. M.: Liver disease in Durban Africans. Histopathological findings in biopsies. S. Afr. J. med. Sci. *23*, 187 (1958).

GLASER, R. J., SMITH, D. E.: Hemochromatosis versus Addison's disease. Amer. J. Med. *9*, 383 (1950).

GLOVER, J. M., JONES, P. R., GREENMAN, D. A. et al.: Iron absorption and distribution in normal and scorbutic guinea pigs. Brit. J. exp. Path. *53*, 295 (1972).

GOLBERG, L., SMITH, J. P.: Iron overloading and hepatic vulnerability. Amer. J. Path. *36*, 125 (1960).

GOLBERG, L., SMITH, J. P., MARTIN, L. E.: The effects of intensive and prolonged administration of iron parenterally in animals. Brit. J. exp. Path. *38*, 297 (1957).

GOODMAN, M., GREENSPON, S. A., KRAKOWER, C. A.: The antigenic composition of the various anatomic structures of the canine kidney. J. Immunol. *75*, 96 (1955).

GOYA, N. et al.: A family of congenital atransferrinemia. Blood *40*, 239 (1972).

GREENBERG, M. S., STROHMEYER, G., HINE, G. J., KEENE, W. R., CURTIS, G., CHALMERS, T. C.: Studies in iron absorption. III. Body radioactivity measurements of patients with liver disease. Gastroenterology *46*, 651 (1964).

GREENWALT, T. J., AYERS, V. E.: Calcium disodium EDTA in transfusion hemosiderosis. Amer. J. clin. Path. *25*, 266 (1955).

GROSBERG, S. J.: Hemochromatosis and heart failure: Presentation of a case with survival after three years treatment by repeated venesection. Ann. Intern. Med. *54*, 550 (1961).

HALVORSEN, S.: Cortisone treatment of idiopathic pulmonary hemosiderosis. Acta paediat. (Uppsala) *45*, 139 (1956).

HAM, T. H.: Chronic hemolytic anemia with paroxysmal nocturnal hemoglobinuria. Arch. Intern. Med. *64*, 1271 (1939).

HAMILTON, H. E., SHEETS, R. F., EVANS, T. C.: Erythrocyte destruction in the lungs as the major cause of anemia in primary pulmonary hemosiderosis. Proc. Cen. Soc. Clin. Res. 33, C. V. Mosby Co., St. Louis 1960.

HAMMER, D. R., DIXON, F. J.: Experimental glomerulonephritis. II. Immunologic events in the pathogenesis of nephrotoxic serum nephritis in the rat. J. exp. Med. *117*, 1019 (1963).

HAMPTON, J. K., MAYERSON, H. S.: Hemoglobin iron as a stimulus for the production of ferritin by the kidney. Amer. J. Physiol. *160*, 1 (1950).

HANOT, V., CHAUFFARD, A.: Cirrhose hypertrophique pigmentaire dans le diabète sucré. Rev. Méd. *2*, 385 (1882).

HANSSEN, P.: Haemosiderosis pulmonum. Acta paediat. *34*, 103 (1947).

HARTMANN, R. C., AUDITORE, J. V.: Paroxysmal nocturnal hemoglobinuria. Amer. J. Med. *27*, 389 (1959); J. clin. Invest. *38*, 702 (1959); J. appl. Physiol. *14*, 589 (1959).

HARTMANN, R. C., JENKINS, D. E.: Paroxysmal nocturnal hemoglobinuria: current concepts of certain pathophysiologic features. Blood *25*, 850 (1965).

HARTMANN, R. C. et al.: Paroxysmal nocturnal hemoglobinuria: clinical and laboratory studies relating to iron metabolism and therapy with androgen and iron. Medicine *45*, 331 (1966).

HATHORN, H., GILLMAN, T., CANHAM, P. A. S., LAMONT, N. M.: Plasma iron and iron-binding capacity in African males with siderosis. Clin. Sci. *19*, 35 (1960).

HEILMEYER, L.: Die Störungen der Bluthämsynthese. Thieme, Stuttgart 1964a.

HEILMEYER, L.: Human hyposideraemia. In: GROSS, F. (ed.): Iron Metabolism, pp. 201–213. Springer, Berlin–Göttingen–Heidelberg 1964b.

HEILMEYER, L., KEIDERLING, W., MERKER, H., CLOTTEN, R., SCHUBOTHE, H.: Die Anaemia refractoria sideroblastica und ihre Beziehungen zur Lebersiderose und Haemochromatose. Acta haemat. (Basel) *23*, 1 (1960).

HEILMEYER, L., KELLER, W., VIVELL, O., KEIDERLING, W., BETKE, K., WÖHLER, F., SCHULTZE, H. E.: Kongenitale Atransferrinämie bei einem sieben Jahre alten Kind. Dtsch. med. Wschr. *86*, 1745 (1961); Acta haemat. (Basel) *36*, 40 (1966).

HIGGINSON, J., GERRITSEN, T., WALKER, A. R. P.: Siderosis in the Bantu of Southern Africa. Amer. J. Path. *29*, 779 (1953).

HOUSTON, J.: Haemochromatosis and refractory anaemia. Guy's Hosp. Rep. *100*, 355 (1951).

HOUSTON, J.: Phlebotomy for haemochromatosis. Effect of removing 52 pints of blood in 16 months. Lancet *1*, 766 (1963).

HOWARD, R. B., BALFOUR, W. M., CULLEN, R.: Extreme hyperferremia in two instances of hemochromatosis with notes on treatment of one patient by means of repeated venesection. J. Lab. clin. Med. *43*, 848 (1954).

HUDSON, J. R.: Ocular findings in haemochromatosis. Brit. J. Ophthal. *37*, 242 (1953).

HUGHES, J. T., TRUELOVE, L. H.: Transfusional haemosiderosis simulating haemochromatosis. J. clin. Path. *11*, 128 (1958).

HUNT, J., RICHARDS, R. J., HARWOOD, R., JACOBS, A.: The effect of desferrioxamine on fibroblast and collagen formation in cell cultures. Brit. J. Haemat. *41*, 69 (1979).

HUTT, M. P., REGER, J. F., NEUSTEIN, H. B.: Renal pathology in paroxysmal nocturnal hemoglobinuria. Amer. J. Med. *31*, 736 (1961).

HWANG, Y. F., BROWN, E. B.: Evaluation of desferrioxamine in iron overload. Arch. Intern. Med. *114*, 741 (1964).

IRVIN, J. M., SNOWDEN, P. W.: Idiopathic pulmonary hemosiderosis. Report of a case with apparent remission from cortisone. Amer. J. Dis. Child. *93*, 182 (1957).

ISAACSON, C., SEFTEL, H., KEELEY, K. J., BOTHWELL, T. H.: Siderosis in the Bantu. The relationship between iron overload and cirrhosis. J. Lab. clin. Med. *58*, 845 (1961).

JACOBS, A.: Iron overload — clinical and pathological aspects. Seminars in Haematology *14*, 89 (1977).

JANDL, J. H., INMAN, J. K., SIMMONS, R. L., ALLEN, D. W.: Transfer of iron from serum iron-binding protein to human reticulocytes. J. Clin. Invest. *38*, 161 (1959).

JOHNSON, B. F.: Hemochromatosis following prolonged iron therapy. New Engl. J. Med. *278*, 1100 (1968).

JOHNSON, G. B., FREY, W. G.: Familial aspects of idiopathic hemochromatosis. J. Amer. med. Ass. *179*, 747 (1962).

JOHNSON, P. J. et al.: Gut *19*, 1022 (1978).

JONES, N. L.: Irreversible shock in hemochromatosis. Lancet *1*, 569 (1962).

KABOTH, W.: Idiopathische Lungenhämosiderose. In: BEGEMANN, H. (ed.): Klinische Hämatologie, p. 298. Thieme, Stuttgart 1970.

KENT, G., POPPER, H.: Secondary hemochromatosis. Its association with anemia. Arch. Path. *70*, 623 (1960).

KERR, L. M. H., RAMSAY, W. N. M.: Plasma iron in a terminal case of haemochromatosis. Biochem. J. *57*, 22 (1954).

KING, A. B.: Pulmonary haemosiderosis. Proc. roy. Soc. Med. *42*, 87 (1949).

KINNEY, T. D., HEGSTED, D. M., FINCH, C. A.: The influence of diet on iron absorption. I. Pathology of iron excess. J. exp. Med. *90*, 137 (1949).

KLOTZ, H. P., AVRIL, J., PARIENTE, R.: Aggravation rapide de deux cirrhoses bronzées après l'essai thérapeutique d'un chélateur (EDTA calcique). Bull. Soc. méd. Paris *73*, 1001 (1957).

KNAUER, C. M., GAMBLE, C. N., MONROE, T. S.: Reversal of hemochromatic cirrhosis by multiple phlebotomies. Gastroenterology *49*, 667 (1965).

LAURELL, C. B.: Plasma iron and the transport of iron in the organism. Pharmacol. Rev. *4*, 371 (1952).
LAURELL, C. B.: Iron transportation. In: WALLERSTEIN, R. O., METTIER, S. R. (eds.): Iron in clinical Medicine, pp. 8–23. University of California Press, Berkeley 1958.
LAURENDEAU, T., HILL, J. E., MANNING, G. B.: Idiopathic neonatal hemochromatosis in siblings. Arch. Path. *72*, 410 (1961).
LEONARDI, P., RUOL, A.: Renal hemosiderosis in the hemolytic anemias; diagnosis by means of needle biopsy. Blood *16*, 1029 (1960).
LEVIN, E. B., GOLUM, A.: The heart in hemochromatosis. Amer. Heart J. *45*, 277 (1953).
LIND, A.: The treatment of hemochromatosis with phlebotomy of 35,000 cc. of blood. N. Y. med. J. *57*, 4027 (1957).
LIPSCHITZ, D. A., SEFTEL, H. C., CHARLTON, R. W., WAPNICK, A. A., LYNCH, S. R., BOTHWELL, T. H.: The role of ascorbic acid in iron transport. Abstr. XIIIth Internat. Congr. Haemat., München 1970.
LLOYD, H. M., POWELL, L. W., THOMAS, M. J.: Idiopathic haemochromatosis in menstruating women. Lancet *2*, 555 (1964).
LOPERENA, L., DORANTES, S., MEDRANO, E., BERRON, R., VEGA, L., CUARON, A., RODRIGUEZ, C., MARQUEZ, J. L.: Atransferrinemia hereditaria. Bol. Med. Hosp. Infant. Mex. *31*, 519 (1974).
LOSOWSKY, M. S., WILSON, A. R.: Whole body counting of the absorption and distribution of iron in haemochromatosis. Clin. Sci. *32*, 151 (1967).
LUNDIN, P., LUNDWALL, O., WEINFELD, A.: Iron storage in alcoholic fatty liver. Acta med. scand. *189*, 541 (1971).
LYNCH, S. R., BERELOWITZ, I., SEFTEL, H. C., MILLER, G. B., KRAWITZ, P., CHARLTON, R. W., BOTHWELL, T. H.: Osteoporosis in Johannesburg Bantu males. Its relationship to siderosis and ascorbic acid deficiency. Amer. J. clin. Nutr. *20*, 799 (1967).
MACDONALD, R. A.: Idiopathic hemochromatosis. Arch. Intern Med. *112*, 184 (1963); *116*, 381 (1965).
MACDONALD, R. A.: Hemochromatosis and hemosiderosis. Thomas, Springfield 1964.
MACDONALD, R. A., MALLORY, G. K.: Hemochromatosis and hemosiderosis. Autopsy study of 211 cases. Arch. Intern. Med. *105*, 686 (1960).
MACGREGOR, A. G., RAMSAY, W. N. M.: Iron metabolism during treatment of idiopathic haemochromatosis. Lancet *273*, 1314 (1957).
MACGREGOR, C. S., JOHNSON, R. S., TURK, K. A. D.: Fatal nephritis complicating idiopathic pulmonary haemosiderosis in young adults. Thorax *15*, 198 (1960).
MACSWEEN, R. N. M.: Acute abdominal crisis, circulatory collapse, and sudden death in haemochromatosis. Quart. J. Med. N. S. *35*, 589 (1966).
MACSWEEN, R. N. M., JACKSON, J. M.: Haemochromatosis; a clinico-pathological review of 37 cases. Scot. med. J. *11*, 395 (1966).
MALLORY, F. B., PARKER, F., NYE, R. N.: Experimental pigment cirrhosis due to copper and its relation to hemochromatosis. J. med. Res. *42*, 461 (1921).
MALPAS, S. T., CALLENDER, S. T. (1964), cit. CALLENDER, S. T.: Digestive absorption of iron. In: GROSS, F. (ed.): Iron Metabolism, p. 94. Springer, Berlin–Göttingen–Heidelberg 1964.
MANCHESTER, R. C.: Chronic hemolytic anemia with paroxysmal nocturnal hemoglobinuria. Ann. intern. Med. *23*, 935 (1945).
MCALPINE, S. G.: Haemochromatosis in a family, and its occurrence in two women. Brit. med. J. *2*, 618 (1959).
MCCLATCHIE, S., TAYLOR, H. E., HENRY, A. T.: Acute abdominal pain and shock associated with haemochromatosis. Canad. med. Ass. J. *63*, 485 (1950).
MCMAHON, F. G.: Comparison of the effect of Fe-3-specific versenole and calcium disodium versenate on urinary iron excretion in patient with hemochromatosis. J. Lab. clin. Med. *48*, 589 (1956).
MELLORS, R. C., SIEGEL, M., PRESSMAN, D.: Analytical pathology. Histochemical demonstration of antibody localization in tissues, with special reference to the antigenic components of kidney and lung. Lab. Invest. *4*, 69 (1955).
MENGEL, C. E., KANN, H. E., O'MALLEY, B. W.: Increased hemolysis after intramuscular iron administration in patients with paroxysmal nocturnal hemoglobinuria. Blood *26*, 74 (1965).

MODELL, C. B.: Transfusional haemochromatosis. In: KIEF, H. (ed.): Iron Metabolism and its Disorders. Excerpta Medica, Amsterdam–Oxford; American Elsevier, New York 1975.

MOESCHLIN, S.: Klinik und Therapie der Vergiftungen, pp. 378–401. 3rd ed., Thieme, Stuttgart 1959.

MOESCHLIN, S.: In: GROSS, F. (ed.): Iron Metabolism, p. 344. Springer, Berlin–Göttingen–Heidelberg 1964.

MOESCHLIN, S., SCHNIDER, U.: Treatment of primary and secondary haemochromatosis. In: GROSS, F. (ed.): Iron Metabolism. Springer, Berlin–Göttingen–Heidelberg 1964.

MORGAN, E. H.: Idiopathic haemochromatosis. A familial study. Aust. Ann. Med. *10*, 114 (1961).

MORRIONE, T. G.: Effect of estrogens on testis in hepatic insufficiency. Arch. Path. *37*, 39 (1944).

MOURIQUAND, C.: L'exploration des réserves de fer par les techniques histologiques. Path. Biol. (Sem. hôp. Paris) *9*, 1557 (1961).

MUTZ, J., SCHROFL, P.: Zur Therapie der idiopathischen Lungenhämosiderose. Wien. klin. Wschr. *83*, 885 (1971).

MYERSON, R. M., CARROLL, I. N.: Treatment of hemochromatosis by massive venesection; report of case treated by removal of 40 litres of blood in 28 months. Arch. Intern. Med. *95*, 349 (1955).

NAJEAN, Y. et al.: Erythrokinetic studies in myelofibrosis: their significance for prognosis. Brit. J. Haemat. *40*, 205 (1978).

NATH, I., SOOD, S. K., NAYAK, N. C.: Experimental siderosis and liver injury in the rhesus monkey. J. Path. Bact. *106*, 103 (1972).

NAY, C. P.: Exogenous hemochromatosis; a case of typical hemochromatosis caused by repeated transfusions of blood. N. Y. med. J. *57*, 3003 (1957).

OHLSSON, W. T. L.: Therapy of hemochromatosis. Nord. med. *51*, 706 (1954).

OLIVER, R. A.: Siderosis following transfusions of blood. J. Path. Bact. *77*, 171 (1959).

PATERSON, D.: Pulmonary haemosiderosis. Proc. roy. Soc. Med. *39*, 131 (1946).

PATERSON, J. C. S.: Haemosiderosis in aplastic anaemia. Postgrad. Med. *28*, 313 (1952).

PEARSE, A. G. E.: Histochemistry, Theoretical and Applied, p. 360. Little, Brown and Co., Boston 1953.

PENGELLY, C. D. R., JONES, P.: Acquired haemochromatosis following multiple blood transfusions for hypoplastic anaemia. Lancet *2*, 445 (1956).

PERKINS, K. W., MCINNES, I. W. S., BLACKBURN, C. R. B., BEAL, R. W.: Idiopathic hemochromatosis in children. Amer. J. Med. *39*, 118 (1965).

PERMAN, G.: Hemochromatosis and red wine. Acta med. scand. *182*, 281 (1967).

PETIT, D. W.: Hemochromatosis with complete heart block, with discussion of cardiac complications. Amer. Heart J. *29*, 253 (1945).

PIRART, J.: Douze ans de récul dans le traitement de l'hémochromatose par saignées répétées. Méd. dans le Monde *40*, 170 (1964).

PIRART, J., CARPENT, G.: Aspects dynamics de l'hémochromatose. Acta gastro-enterol. belg. *18*, 7 (1955).

PIRART, J., GATEZ, P.: L'étiologie de l'hémochromatose nontransfusionnelle, étude de l'hérédité dans 21 familles. Sem. hôp. Paris *34*, 1044 (1958).

PIRZIO-BIROLI, G., FINCH, C. A.: Iron absorption. III. The influence of the iron stores on iron absorption in the normal subjects. J. Lab. clin. Med. *55*, 216 (1960).

PLATZER, R. F., YOUNG, L. E., YUILE, C. L.: Hemosiderosis resembling hemochromatosis following multiple transfusions. Acta haemat. (Basel) *14*, 185 (1955).

PLOEM, J. E. et al.: Idiopathic haemosiderosis. Scand. J. Haemat. *2*, 3 (1965).

POWELL, L. W.: Iron storage in relatives of patients with haemochromatosis and in relatives of patients with alcoholic cirrhosis and haemosiderosis. Quart. J. Med. N. S. *34*, 427 (1965).

POWELL, L. W., HALLIDAY, J. W.: Idiopathic haemochromatosis. In: JACOBS, A., WORWOOD, M. (eds.): Iron in Biochemistry and Medicine, Vol. 2. Academic Press, London–New York 1980.

POWELL, L. W., KERR, J. F. R.: Reversal of "cirrhosis" in idiopathic haemochromatosis following long-term intensive venesection therapy. Austral. Ann. Med. *19*, 54 (1970).

PRELOG, V.: Iron-containing compounds in microorganisms. In: GROSS, F. (ed.): Iron Metabolism. Springer, Berlin–Göttingen–Heidelberg 1964.

RATH, C. E., FINCH, C. A.: Sternal marrow hemosiderin. A method for the determination of available iron stores in man. J. Lab. clin. Med. *33*, 81 (1948).
REYE, D.: Pulmonary haemosiderosis. Med. J. Aust. *1*, 35 (1945).
RISDON, R. A., BARRY, M., FLINN, D. M.: Transfusional iron overload: the relationship between tissue iron concentration and hepatic fibrosis in thalassaemia. J. Path. Bact. *116*, 83 (1975).
ROGERS, W. F.: Familial hemochromatosis: with comments on adrenal function in hemochromatosis. Amer. J. med. Sci. *220*, 530 (1950).
ROSSE, W. F.: Paroxysmal nocturnal hemoglobinuria. In: WILLIAMS, W. J., ERSLEV, A. J., RUNDLES, R. W. (eds.): Hematology, p. 460. McGraw-Hill, New York–St. Louis–San Francisco 1972.
ROSSELIN, G.: Comparative study of the behaviour of radioactive iron in idiopathic hemochromatosis and alcoholic cirrhosis. Path. Biol. (Sem. hôp. Paris) *8*, 1833 (1960).
ROUS, P.: Urinary siderosis. Hemosiderin granules in the urine as an aid in the diagnosis of pernicious anaemia, hemochromatosis and other diseases causing siderosis of the kidney. J. exp. Med. *28*, 645 (1918).
SAKATA, T.: A case of congenital atransferrinaemia. Shonika Shiuryo *32*, 1523 (1969).
SAMUELS, L. D., BASS, J. C.: ^{51}Cr lung scan in idiopathic pulmonary haemosiderosis. J. nucl. Med. *10*, 106 (1969).
SANDØE, E.: Essential pulmonary haemosiderosis: Account of 2 cases, 1 treated experimentally with ACTH. Dan. med. Bull. *1*, 175 (1954).
SCHAPIRA, G., DREYFUS, J. C.: In: KEIDERLING, W. (ed.): Eisenstoffwechsel, p. 238. Thieme, Stuttgart 1959.
SCHAPIRA, G., DREYFUS, J. C., SCHWARZMANN, V., ETÉVE, J.: Hypersidérémie et surcharge ferrique hépatique chez les descendants d'hémochromatosiques. Rev. franç. Étud. clin. biol. *5*, 485 (1962).
SCHEER, R. L., GROSSMAN, M. A.: Immune aspects of the glomerulonephritis associated with pulmonary hemorrhage. Ann. intern. Med. *60*, 1009 (1964).
SCHEUER, P. J., WILLIAMS, R., MUIR, A. R.: Hepatic pathology in relatives of patients with haemochromatosis. J. Path. Bact. *84*, 53 (1962).
SCHMID, J. R. et al.: Ein einfacher 6stündiger i.m. Test mit Desferal zur Diagnose der Hämochromatose. Schweiz. med. Wschr. *94*, 1652 (1964).
SCHULER, D., FLESCH, I.: Über die Aetiologie und Pathogenese der essentiellen Lungenhämosiderose. Ann. Paediat. *185*, 96 (1955).
SCHWARTZ, S. O., BLUMENTHAL, S. A.: Exogenous hemochromatosis resulting from blood transfusions. Blood *3*, 617 (1948).
SCHWARZENBACH (1964), cit. MOESCHLIN, SCHNIDER, 1964.
SCOTT, R. B., ROBB-SMITH, A. H. T., SCOWEN, E. F.: The Marchiafava-Micheli syndrome of nocturnal haemoglobinuria with haemolytic anaemia. Quart. J. Med. *7*, 95 (1938).
SEFTEL, H. C., ISAACSON, C., KEELEY, K. J., BOTHWELL, T. H.: Siderosis in the Bantu. The clinical incidence of hemochromatosis in diabetic subjects. J. Lab. clin. Med. *58*, 837 (1961).
SEFTEL, H. C., MALKIN, C., SCHMAMAN, A., ABRAHAMS, C., LYNCH, S. R., CHARLTON, R. W., BOTHWELL, T. H.: Osteoporosis, scurvy and siderosis in Johannesburg Bantu. Brit. med. J. *1*, 642 (1966).
SEVEN, M. J.: Observations on the toxicity of intravenous chelating agents. In: Metal-binding in Medicine, pp. 95–103. Lippincott, Philadelphia 1960.
SHELDON, J. H.: Iron content of tissues in haemochromatosis, with special reference to the brain. Quart. J. Med. *21*, 123 (1927).
SHELDON, J. H.: Haemochromatosis. Oxford University Press, London 1935.
SHERLOCK, S.: Introduction to the general discussion on iron overload. In: GROSS, F. (ed.): Iron Metabolism, pp. 392–399. Springer, Berlin–Göttingen–Heidelberg 1964.
SIMON, M. et al.: Idiopathic hemochromatosis. Demonstration of recessive transmission and early detection by family HLA-typing. New Engl. J. Med. *297*, 1017 (1977).
SKIKNE, B. S., LYNCH, S. R., BEZWODA, W. R. et al.: Franconi's anaemia, with special reference to erythrokinetic features. S. Afr. med. J. *53*, 43 (1978).

SMITH, P. M., GODFREY, B. E., WILLIAMS, R.: Iron absorption in idiopathic haemochromatosis and its measurement using a whole body counter. Clin. Sci. *37*, 519 (1969).

SMITH, P. M., STUDLEY, F., WILLIAMS, R.: Assessment of body-iron stores in cirrhosis and haemochromatosis with the differential ferrioxamine test. Lancet *1*, 133 (1967).

SOERGEL, K. H.: Idiopathic pulmonary hemosiderosis. Review and report of two cases. Pediatrics *19*, 1101 (1957).

SOERGEL, K. H., SOMMERS, S. C.: Idiopathic pulmonary hemosiderosis and related syndromes. Amer. J. Med. *32*, 499 (1962).

SOLOMON, L. (1972), cit. CHARLTON, BOTHWELL, SEFTEL, 1973.

STEINER, B.: Essential pulmonary haemosiderosis as an immunohaematological problem. Arch. Dis. Child. *29*, 391 (1954).

STEINER, B.: The value of splenectomy in the treatment of essential pulmonary haemosiderosis. Acta med. Acad. Sci. hung. *14*, 211 (1959).

STEINER, B.: Ethylenediaminetetraacetic acid (EDTA) in treatment of essential pulmonary haemosiderosis. Helv. paediat. Acta *16*, 97 (1961).

STRACHAN, A. S.: Haemosiderosis and haemochromatosis in South African natives, with a comment on the etiology of haemochromatosis. Thesis, Glasgow 1929.

STRÖDER, U.: Infantilismus und Myokardfibrose bei der Hämochromatose. Dtsch. Arch. klin. Med. *189*, 141 (1942).

STURGEON, P., SHODEN, A.: Mechanism of iron storage. In: GROSS, F. (ed.): Iron Metabolism, pp. 121–147. Springer, Berlin–Göttingen–Heidelberg 1964.

STURGILL, B. C., WESTERVELDT, F. B.: Immunofluorescent studies in a case of Goodpasture's syndrome. J. Amer. med. Ass. *194*, 914 (1965).

TAIT, G. B., CORRIDAN, M.: Idiopathic pulmonary haemosiderosis. Thorax *7*, 302 (1952).

TAYLOR, H. E.: The possible role of ferritin in the production of shock in hemochromatosis. Amer. J. clin. Path. *21*, 530 (1951).

TROUSSEAU, A.: Clinique Médicale de l'Hôtel Dieu de Paris, 2nd ed. Ballière, London 1865.

TURNBERG, L. A.: Excessive oral iron therapy causing haemochromatosis. Brit. med. J. *1*, 1360 (1965).

VANNOTTI, A.: In: GROSS, F. (ed.): Iron Metabolism, p. 343. Springer, Berlin–Göttingen–Heidelberg 1964.

VANNOTTI, A., BLANC, B.: Compartement de la transferrine dans l'hémochromatose et dans la lésion hépatique. Schweiz. med. Wschr. *93*, 1189 (1963).

VERLOOP, M. C.: In: GROSS, F. (ed.): Iron Metabolism, p. 342. Springer, Berlin–Göttingen–Heidelberg 1964.

VITALE, L., OPITZ, J. M., SHAHIDI, N. T.: Congenital and familial iron overload. New Engl. J. Med. *280*, 642 (1969).

WAINWRIGHT, J.: Siderosis in the African. S. Afr. J. Lab. clin. Med. *3*, 1 (1957).

WALKER, A. R. P., ARVIDSSON, U. B.: Iron "overload" in the South Africa Bantu. Trans. roy. Soc. trop. Med. Hyg. *47*, 536 (1953).

WALKER, R. J. et al.: Relationship of hepatic iron concentration associated with iron overload. Gut *12*, 1011 (1971).

WAPNICK, A. A., LYNCH, S. R., CHARLTON, R. W., SEFTEL, H. C., BOTHWELL, T. H.: The effect of ascorbic acid deficiency on desferrioxamine-induced urinary iron excretion. Brit. J. Haemat. *17*, 563 (1969).

WAPNICK, A. A., LYNCH, S. R., SEFTEL, H. C., CHARLTON, R. W., BOTHWELL, T. H., JOWSEY, J.: The effect of siderosis and ascorbic acid depletion on bone metabolism, with special reference to osteoporosis in the Bantu. Brit. J. Nutr. *25*, 367 (1971).

WARTHIN, T. A., PETERSON, E. W., BARR, J. H.: The treatment of idiopathic hemochromatosis by repeated phlebotomy. Ann. intern. Med. *38*, 1066 (1953).

WEINTRAUB, L. R., CONRAD, M. E., CROSBY, W. H.: The treatment of hemochromatosis by phlebotomy. Med. clin. N. Amer. *50*, 1579 (1966).

WEISS, E. B., EARNEST, D. L., GREALLY, J. F.: Goodpasture's syndrome. Case report with emphasis on pulmonary physiology. Amer. Rev. resp. Dis. *97*, 444 (1968).

WERNER, H.: Insulinresistenter Bronzediabetes. Zbl. inn. Med. *63*, 753 (1942).
WIESMANN, W., WOLVIUS, D., VERLOOP, M. C.: Idiopathic pulmonary hemosiderosis. Acta med. scand. *146*, 341 (1953).
WILLIAMS, R., SCHEUER, P. J., SHERLOCK, S.: The inheritance of idiopathic haemochromatosis. A clinical and liver biopsy study of 16 families. Quart. J. Med. *31*, 249 (1962).
WILLIAMS, R., SMITH, P. M., SPICER, E. J. F., BARRY, M., SHERLOCK, S.: Venesection therapy in idiopathic haemochromatosis. Quart. J. Med. N. S. *38*, 1 (1969).
WILLIAMS, R., WILLIAMS, H. S., PITCHER, C. S., SHERLOCK, S. (1964), cit. SHERLOCK, 1964.
WILLIAMS, R., WILLIAMS, H. S., SCHEUER, P. J., PITCHER, C. S., LOISEAU, E., SHERLOCK, S.: Iron absorption and siderosis in chronic liver disease. Quart. J. Med. N. S. *36*, 151 (1967).
WILLIS, R. A.: Haemochromatosis, with special reference to supervening carcinoma of liver. Med. J. Aust. *2*, 666 (1941).
WINTROBE, M. M.: Clinical Hematology. Lea and Febiger, Philadelphia 1967, 1975.
WISHINSKY, H., WEINBERG, T., PRÉVOST, E. M., BURGIN, B., MILLER, M. J.: Ethylenediaminetetraacetic acid in the mobilization and removal of iron in a case of haemochromatosis. J. Lab. clin. Med. *42*, 550 (1953).
WÖHLER, F.: The treatment of haemochromatosis with desferrioxamine. In: GROSS, F. (ed.): Iron Metabolism, pp. 551–567. Springer, Berlin–Göttingen–Heidelberg 1964.
WYATT, J. P.: Patterns of pathological iron storage; exogenic siderosis in chronic anemia due to peroral medication. Arch. Path. *61*, 56 (1956).
WYLLIE, W. G., SHELDON, W., BODIAN, N., BARLOW, A.: Idiopathic pulmonary haemosiderosis (essential brown induration of the lungs). Quart. J. Med. *17*, 25 (1948).
ZELTMACHER, K., BEVANS, M.: Aplastic anemia and its association with hemochromatosis. Arch. Intern. Med. *75*, 395 (1945).

AUTHOR INDEX

A

Aalam, F. 128
Aasa, R. 80
Adams, E. B. 299
Addison, G. M. 93, 99, 237
Agner, K. 93, 251
Alder, A. 246
Aldrich, R. A. 257
Alekseev, G. A. 231
Al-Kassab, S. 221
Allen, D. W. 159
Allerton, S. E. 79
Alpen, E. L. 145, 287
Alper, T. 341, 343, 353
Altemeier, W. A. 291
Althausen, T. L. 341 - 343, 347, 352, 353
Alvarez, A. S. 243
Andersson, N. S. E. 251-253
André, J. 341, 356
Apt, L. 369
Apte, S. V. 50
Archdeacon, J. W. 151
Arkun, N. S. 233
Árky, I. 331
Arrowsmith, W. R. 353
Arvidsson, U. B. 27, 360
Aschkenasy, A. 299
Asenjo, C. F. 27
Ashby, W. 186
Aufderheide, A. C. 364
Azari, P. R. 79

B

Badenoch, J. 39, 40, 229, 244, 358
Baehner, R. L. 74
Bainton, D. F. 275
Baird, I. M. 38, 222, 245, 253
Balcerzak, S. P. 39, 105, 351
Baldini, M. 76
Bang, O. 302
Bannerman, R. M. 44, 243, 315, 316, 320
Banwell, J. G. 39
Barbezat, G. O. 258
Barkan, G. 78
Barr, G. D. 258
Barrett, P. V. D. 164, 194, 195
Bass, J. C. 370
Basta, S. 232
Bearn, A. G. 78, 79
Beck, K. 313
Becker, C. E. 255
Becker, D. 342
Beebe, R. T. 229
Belcher, E. H. 175
Bell, E. T. 342
Ben Ishay, Z. 209
Benoit, F. L. 371
Berde, B. 49
Bergsagel, D. E. 171
Berk, J. E. 342
Berk, P. D. 194, 195
Berlin, N. I. 186, 190, 194, 335

Bernát, I. 41, 43, 44, 51–53, 61, 74, 77, 82, 104, 105, 117, 167–169, 171, 172, 175, 177, 178, 184, 189–192, 201, 206–208, 210–212, 219, 220, 223–228, 232, 247, 285–295
Bertenchamps, A. 75, 173
Bessis, M. C. 91, 94, 96, 151, 159, 160, 164, 201, 205, 206, 208, 210, 211, 234, 275, 312, 316, 317, 318, 320
Bethell, F. H. 314
Betke, K. 124, 125, 256
Beutler, E. 19, 22, 24, 25, 29, 49, 77, 82, 84, 96, 103, 105, 159, 173, 216, 217, 232, 248, 249
Bevans, M. 364
Beveridge, B. R. 219, 230, 243
Bezwoda, W. R. 106
Bickel, H. 10
Bickers, J. N. 314
Biggs, J. C. 39, 46, 240
Bilger, R. 234
Birgegård, G. 108
Bishop, R. C. 314
Björkman, S. E. 309, 312, 313
Bjorn Rasmussen, E. 47
Blackburn, C. R. B. 353
Blackwell, L. H. 331
Blaisdell, R. K. 232
Blake, J. 245
Blanc, B. 343
Blaud, P. 246
Blazar, A. S. 255
Blix, G. 27
Blok, A. P. R. 365
Bloom, W. 329
Blumenthal, S. A. 364
Boddington M. M. 223
Boender, C. A. 54
Boenecke, I. 72
Boggs, D. R. 144
Bogorad, L. 151
Bokhari, S. M. 216
Bolin, T. 39
Bomford, A. B. 93, 348
Bond, V. P. 145
Bonnet, J. D. 41
Bonser, M. 255
Borsook, H. 155
Bothwell, T. H. 25, 41, 47, 49–51, 72, 77, 82, 95–98, 100, 103, 104, 114, 117, 118, 122, 123, 134, 167, 170–174, 176, 178, 183, 184, 188, 193, 196, 197, 240, 249, 250, 254, 305, 335, 341, 343, 347, 348, 350, 352, 353, 356, 358, 360–365, 368, 369
Böttner, H. 302
Boulin, R. 341–343, 348
Bourne, M. S. 314
Bousser, I. 313
Bowdler, A. J. 364
Bowman, W. D. 202, 211, 317
Box, H. C. 188
Boyd, D. H. 368
Boyett, J. D. 318
Bradley, G. P. 368
Bradley, S. E. 368
Bradlow, B. A. 360, 361
Brady, G. W, 91
Braithwaite, F. 291
Brazhnikova, M. G. 10
Brendstrup, P. 77, 82, 167
Breton- Gorius, J. 94, 96, 151, 160, 201, 205, 208, 210, 211, 234, 275, 312, 316–318, 320
Brick, I. B. 341, 350, 356
Bridgeforth, F. 128
Brise, H. 41, 42, 247, 248
Brittin, G. M. 50
Brøchner-Mortensen, K. 167, 275
Bronson, W. R. 258
Bronte-Stewart, B. 77
Brooks, F. 285
Brown, A. 285
Brown, A. K. 185
Brown, E. 124, 126, 128
Brown, E. B. 38, 50, 253, 365
Brown, E. B. Jr. 39, 52
Brown, E. G. 149
Brown, G. B. 44

Brown, W. D. 219
Browning, J. R. 370
Brunner, H. E. 335, 348
Brüschke, G. 51, 52, 201, 202, 234
Bryce, C. F. A. 92
Buchanan, W. M. 362
Büchmann, P. 301, 303
Buchthal, F. 24
Bunge, G. 246
Burckhardt, D. 370
Burger, T. 335
Burman, D. 232
Burnham, B. F. 151, 154
Burté, B. 159, 160
Buttenwieser, E. 49
Butterworth, C. E. 318

C

Callender, S. T. 38–40, 44, 230, 244, 248, 358
Calvin, M. 15, 16
Campbell, B. C. 277
Čap, J. 360
Cappell, D. F. 101, 364
Caroli, J. 341, 356, 364
Carpent, G. 353
Carroll, I. N. 353
Carter, B. N. 291
Carter, R. L. 255
Cartwright, G. E. 74, 77, 81–84, 150, 167, 275–277, 313, 314
Cattau, D. 41
Chanarin, I. 123
Chandra, R. K. 233
Chanutin, A. 74
Chappelle, E. 49, 116
Charache, S. 317
Charley, P. J. 41, 79
Charlton, R. W. 27, 40, 339, 362
Chase, M. S. 51
Chatgidakis, C. B. 368
Chauffard, A. 340
Cheney, B. 51
Chisholm, M. 221, 223, 224, 231
Chisolm, J. J. 318
Chodos, R. B. 44
Chou, A. C. 300
Chown, B. 131
Cleton, F. J. 365
Cline, M. J. 194
Coburn, R. F. 161
Coghill, N. F. 230
Coleman, D. H. 77, 103
Conrad, M. E. 23, 38, 44–47, 255
Conrad, M. E. Jr. 64, 117
Conte, M. 350, 353
Cook, J. D. 38, 106, 239, 240
Cook, J. S. 331
Cooke, W. T. 244
Cookson, G. H. 151
Cooper, R. A. 188
Cope, O. 253, 291
Cordeiro, M. 370
Cornides, I. 189, 190
Corridan, M. 368
Cottier, H. 364
Cowan, B. 230
Coward, W. 299
Crafts, R. C. 142
Craig, J. O. 257
Cranmore, D. 145
Crichton, R. R. 92, 93, 94
Croft, D. N. 57, 58, 59
Crosby, W. H. 53, 54, 55, 56, 58, 175, 201, 202, 311, 353, 356, 364, 367, 368
Crowley, J. 251
Cumming, R. L. C. 54
Currin, J. F. 364

D

Dabski, H. 223
Dacie, J. V. 201, 202, 312
Dagg, J. H. 38, 46, 105, 117, 230, 232, 236, 237, 240, 249, 258, 340, 370

Dallman, P. R. 22, 24, 232
Damasio, E. 208
Dameshek, W. 215
Danilenko, S. S. 223
Darby, W. J. 123
Davey, D. A. 353
Davidson, L. S. P. 216
Davidson, S. 250
Davidson, W. M. B. 229
Davies, D. M. 364
Davies, J. W. L. 173, 287
Davis, A. E. 39, 40, 358
Davis, L. J. 201
Davis, L. R. 240
Davis, P. S. 38
Davis, W. D. 353
Davis, W. M. 287
Davison, R. H. 350
Dawkins, S. 237
Dawson, R. B. 46
Dearing, W. H. 186
Debré, R. 349
de Castro Freire, L. 370
Deckert, T. 353
de Gennes, L. 343
de Gowin, R. L. 370, 371
de Gruchy, G. L. 256
Delachaux, A. 301
de Leeuw, N. H. W. 122, 123
Deller, D. J. 38, 40
del Riego, M. G. 255
de Maat, C. E. M. 277
Denborough, M. A. 230
den Hartog, C. 32
de Potter, E. 243
de Raadt, M. F. 75
Dervichian, D. 125
Desforges, G. 341
de Vries, S. I. 243
Diamond, L. K. 233
Dietrich, M. R. 76
Dillingham, C. H. 340
Dinant, H. J. 277
Dintzis, H. M. 155
Dixon, F. J. 371
Doering, P. 336
Donaldson, G. U. 194
Donati, R. M. 299
Doniach, J. 201, 202
Donohue, D. M. 193
Douglas, A. S. 201, 202
Dowdle, E. B. 51
Drabkin, D. L. 25
Dresel, E. I. B. 149, 317
Dreyfus, J. C. 24, 25, 74, 81, 82, 133, 175, 189, 336, 341, 343–345, 349, 350
Drysdale, J. U. 96
Dubach, R. 49, 62, 113, 114, 116, 118, 161, 168
Dubin, I. N. 352, 353
Ducci, H. 76
Dugdale, A. E. 258
Duncan, D. A. 371
Dyment, P. G. 219
Dymock, I. W. 115

E

Edington, G. M. 360
Ehrenstein, G. 203, 207
Eigner, E. A. 96
Elin, R. J. 275
Ellenhorn, M. J. 367
Elmlinger, P. J. 169, 173
Elwood, P. C. 27, 29, 248
Entwhistle, C. C. 223
Eriksen, L. 317
Erlandson, M. E. 76, 99, 100, 130, 256
Erslev, A. J. 142, 143
Eskind, I. B. 342
Estren, S. 244
Evans, J. 354
Evans, L. A. J. 252

F

Fairbanks, V. F. 24, 25, 29, 173, 216
Falk, J. E. 149, 151, 317
Farrant, J. L. 205
Feeney, R. E. 79
Fehérvári, T. 287
Felder, S. L. 242
Feldman, F. 317
Felicetti, L. 155
Fenton, V. 124, 130
Fielding, J. 96, 251, 255, 257
Figueroa, W. G. 255
Finch, C. A. 25, 41, 47–49, 51, 63, 74, 75, 77, 82, 94–97, 100, 101, 103, 104, 113, 123, 134, 167, 168, 174, 176, 178, 183–185, 188, 193, 196, 197, 239, 240, 249, 250, 254, 275, 301–303, 305, 335, 341–344, 346–349, 352–356, 358, 360, 365, 368, 369
Finch, S. C. 60, 74, 77, 94, 95, 103, 340–344, 346–349, 353–355, 365
Fineberg, R. A. 53
Fischer, D. S. 50
Fisher, M. 240
Fleischmann, O. 227
Flesch, I. 368
Fletcher, J. 81, 101
Fliedner, T. M. 329
Flores, M. 32
Fontès, G. 246
Forristal, T. 255
Foss, O. 243
Foy, H. 118
Fraser, D. K. 258
Freireich, E. J. 77, 104, 171, 172, 175, 178, 276, 277
Frenchman, R. 118
Frerichs, H. 313
Frey, W. G. 341, 348, 350, 353
Fritsch, F. 246
Fuhrmann, G. 305

G

Gabrio, B. W. 96
Gabuzda, T. G. 96
Gajdos, I. A. 317
Gale, G. E. G. 97, 103
Garby, L. 125, 194, 216, 239, 252, 309, 310, 364
Gardner, F. H. 153, 301
Gasser, C. 231
Gatez, P. 350
Gause, G. F, 10
Gehrmann, G. 313
Gelfand, M. 360
Gellis, S. S. 368
Gerritsen, T. 124
Geszti, O. 331–333
Gewitz, R. 301
Ghitis, J. 299
Ghosh, S. 229
Giannopoulos, P. P. 171
Giblett, E. R. 78, 79, 174, 184, 186, 195
Gibson, K. D. 149, 154, 317
Gillman, P. A. 370
Girdwood, R. H. 250
Gitlin, D. 80
Glaser, R. J. 343
Glevitsch, E. 218
Glover, J. M. 362
Goetsch, A. T. 254, 255
Golberg, L. 364, 365
Goldberg, A. 152, 236, 240, 249, 252, 317, 318
Goldeck, H. 167, 301
Goldstein, G. W. 162
Göltner, E. 72, 122, 243
Goodman, M. 81, 371
Gosden, M. 215
Gouttas, A. 245
Govan, A. D. T. 251
Goya, N. 360
Granick, S. 15, 25, 93, 94, 96, 143, 151, 154, 162

Gray, C. H. 163, 164
Green, D. E. 24
Green, R. 58, 113–115, 117, 118
Greenberg, D. M. 53
Greenberg, L. D. 77
Greenberg, M. S. 39, 49, 349
Gribble, T. J. 155
Grimes, A. J. 252
Grinstein, M. 164, 318
Groden, B. M. 252
Grosberg, S. J. 354
Grossman, M. A. 371
Grotepass, W. 318
Grüneberg, H. 201
Gubler, C. J. 81, 128
Guest, G. 124, 126, 128
Gumińska, S. 306
Günther, B. 301
Gupta, S. P. 230
Gutnisky, A. 277

H

Habte, D. 124
Haddow, A. 255
Haeger-Aronsen, B. 317, 318
Hagberg, B. 71, 72, 76, 81, 82, 133, 275
Hagedorn, A. B. 251
Hahn, P F. 41, 49, 51, 96
Hall, R. 309
Hallberg, L. 34, 41–43, 47–52, 54, 55, 105, 119, 185, 216, 244, 247, 248
Hallgren, B. 77
Halliday, J. W. 348
Halsted, J. A. 230
Halvorsen, S. 370
Ham, T. H. 287
Hamilton, H. E. 302, 370
Hamilton, L. D. 72–74
Hammer, D. R. 371
Hammersten, J. F. 336
Hammond, D. 240
Hampton, J. C. 96
Hampton, J. K. Jr. 95, 367
Hampton, M. C. 32
Hanot, V. 340
Hanssen, P. 370
Hara, M. 24
Hardisty, R. M. 216
Hargreaves, R. M. 238
Harris, H. 78
Harris, J. W. 76, 159
Harrison, P. M. 91, 92
Harrison, P. R. 96
Hartwig, Q. L. 75
Harvalik, Z. 78
Harvey, J. E. 171, 178
Haskins, D. 249
Hathorn, H. 363
Hauck, H. M. 32
Haurani, F. J. 276, 277
Haurowitz, F. 20
Hausmann, K. 201
Hawkins, C. F. 303
Hayhoe, F. G. 202, 313
Heath, C. W. 254, 255, 299
Hebbert, F. J. 246
Hedenberg, L. 105
Hegsted, D. M. 44
Heilmeyer, L. 33, 34, 52, 75, 83, 91, 93, 96, 151, 152, 154, 155, 186, 203, 215, 238, 246, 275–279, 301–303, 305, 309–312, 314, 316, 317–319, 358–360, 364, 365
Heimpel, J. 238
Heinrich, G. 245
Heinrich, H. C. 47, 64, 106. 122, 240
Heistø, H. 243
Hemmeler, G. 72, 76
Henderson, F. 258
Henderson, P. A. 252
Henry, J. A. 335
Herbut, P. A. 39
Herschko, C. 277
Hesseltine, C. W. 10

Heubner, W. 246
Higginson, J. 77, 352, 360
Hiller, O. 78
Hillman, R. S. 252
Hirvonen, M. 275
Hoag, M. 131
Hoffbrand, A. V. 313
Hofwander, Y. 97, 98
Hollán, S. R. 128
Holländer, L. 243
Hollingsworth, D. R. 277
Hollingsworth, J. W. 129, 277
Holmberg, C. G. 82, 167
Holt, J. M. 242
Hoppe, I. 167
Horsfall, W. R. 78
Horst, W. 82, 335
Houghton, J. D. 370
Houston, J. 353, 364
Howard, J. 55, 301
Howe, R. B. 194
Høyer, K. 72
Hudson, J. R. 342
Huehns, E. R. 81, 155, 364
Huff, R. L. 78, 167 - 169, 171, 173, 178, 301, 305, 335
Hughes, J. T. 364
Hunt, J. 365
Hurley, T. H. 186
Huser, H. J. 238
Hussain, R. 118
Hutt, M. P. 368
Hutt, M. S. R. 252
Hwang, Y. F. 44
Hyman, G. A. 171, 178

I

Ikkala, E. 229
Irvin, J. M. 370
Isaacson, C. 360, 362, 363, 364
Israels, L. G. 164
Ito, K. 299
Izak, G. 103

J

Jackson, J. M. 342
Jackson, R. L. 131, 132
Jacobi, H. 41, 125, 127, 128, 238
Jacobs, A. 38, 55, 57, 133, 223, 230, 232, 237, 238, 240, 365
Jacobs, J. 258
Jacobs, P. 38, 54
Jacobson, L. O. 329
Jager, B. V. 81
Jalili, M. A. 221
Jandl, J. H. 79, 81, 83, 85, 151, 159, 188, 210, 363
Janoff, A. 277
Jasinski, B. 22, 49, 77, 218, 219, 223, 231, 232, 249
Jeffrey, M. R. 275
Jennison, R. F. 240
Joffey, J. M. 209
Johnson, B. F. 361
Johnson, G. B. 341, 350
Johnson, P. J. 348
Johnston, F. A. 72, 118
Jones, N. L. 341
Jones, P. 364
Jones, R. F. 223
Josephs, H. W. 41, 61, 76, 128
Joske, R. A. 229
Joynson, D. H. M. 233
Jung, F. 201
Justus, B. W. 38

K

Kähler, H. J. 287
Kaldor, J. 94, 95
Kalinin, V. I. 232
Kaltwasser, J. P. 106

Kampschmidt, R. F. 232, 277
Kaplan, E. 202, 209, 234
Kaplan, O. 246
Kappest, P. 302
Karibian, D. 155
Karlefors, T. 252
Kassenaar, A. 317
Kassirski, I. A. 231
Katz, J. H. 79–81, 210
Kaufman, N. 38
Kaufmann, O. 230
Kavin, H. 39
Kaznelson, P. 215
Kehl, R. 301
Keiderling, W. 77, 173, 178, 275–277, 335,
Keller-Schierlein, W. 10
Kelly, A. M. 130
Kench, J. E. 162
Kent, G. 364, 365
Kerr, D. N. S. 91, 205, 250
Kerr, J. F. R. 354
Kerr, L. M. H. 103
Kiely, J. M. 335
Kikuchi, G. 149
Kilpatrick, G. S. 216
Kind, A. 82
King, A. B. 368
Kingston, P. J. 320
Kinney, T. D. 38
Kirkman, H. 131
Kjellberg, S. R. 223
Kleihauer, E. 128
Klopper, A. 82
Knauer, C. M. 354
Kniseley, H. Jr. 232
Knüttgen, H. 305
Koechlin, B. 79
Kohn, R. 317, 319
Kondi, A. 118
Konijn, A. M. 277
Konopka, L. 313
Kovács, E. 41, 43, 44, 51–53, 61, 74
Kovács, L. 231
Krantz, S. 49
Krauss, H. J. 313
Krebs, H. A. 78
Kreiner-Birnbaum, M. 318
Krimsky, I. 258
Kruh, J. 149, 155
Kuhn, I. N. 44
Kümmerle, F. 242
Künzer, W. 124, 125, 127, 128
Kurnick, J. E. 277

L

Laache, S. 22
Labardini, J. 239
Lajtha, L. G. 144, 168, 172, 193, 331, 335
Lange, J. 82
Lascelles, J. 151, 154
Laub, R. 231
Laurell, C. B. 74, 75, 82, 84, 167, 275, 305, 365, 367
Laver, W. G. 149
Lawrence, J. H. 173, 335
Lawrence, J. S. 185
Layrisse, M. 47, 64, 242
Lederer, J. 287
Lee, G. R. 275, 276
Lees, F. 229, 230
Leibel, R. 10
Leiken, S. 258
Lelkes, G. 206, 207, 210, 211
Leonardi, P. 368
Lester, R. 162
Levere, R. D. 143
Levine, P. H. 56
Le Xuan Chat 299
Lichtman, H. C. 317
Lieber, M. M. 342
Lind, A. 353
Lindner, E. 201
Lindvall, S. 252
Lintzel, W. 131

Lipschitz, D. A. 107, 108, 362
Lloyd, H. M. 341
Lochhead, A. G. 10
Lockner, D. 203, 207
Loeb, V. 168
Loftfield, R. B. 96
Löhr, G. W. 159
London, I. M. 153, 155, 159, 162–164, 188, 192, 302, 335
Lorenz, B. 336
Loria, A. 238
Losowski, M. S. 309, 349
Louwagie, A. C. 327–331
Lübbers, D. 21
Lucas, J. E. 251
Ludewig, S. 74
Ludwig, K. 285, 295
Luke, C. G. 38
Lund, C. 128
Lundin, P. M. 255

M

MacDonald, R. A. 27, 340, 349, 351, 354, 358
MacDougall, L. G. 233
MacGregor, A. G. 353
MacGregor, C. S. 368
Mack, R. B. 258
MacKay, H. M. M. 216
MacSween, R. N. M. 342
Mahler, H. R. 24
Maier, C. 314
Mallett, B. 72, 171
Mallory, G. K. 340, 351
Malpas, J. S. 39, 40, 44
Malpas, S. T. 358
Mándi, E. 329, 331
Manis, J. 54
Manolidis, L. 223
Mantz, J. J. C. 32
Markson, J. L. 229
Marmond, A. 208
Marshall, S. R. 219
Marti, H. R. 277
Martin, L. E. 252
Martinez-Torres, C. 64
Mason, D. Y. 94
Masuya, T. 219
Mathorn, M. 77
Mauzerall, D. 162
Mayerson, H. S. 367
Mayet, F. G. H. 97, 98
Mazur, A. 51, 76, 92, 258
McAlpine, S. G. 342, 353
McCance, R. A. 246
McClatchie, S. 341
McCrea, P. C. 103
McCurdy, P. R. 252, 253, 255
McDonald, R. 219
McEmery, J. T. 258
McFadzean, A. J. S. 201
McGibbon, B. H. 309, 313
McGuigan, J. E. 244
Meerkreebs, G. 248
Meier, W. 97, 98
Meincke, H. A. 142
Mellors, R. C. 371
Melville, G. S. 75
Menon, K. 215
Merker, H. 95, 105, 210, 280, 313
Meulengracht, E. 246
Miescher, P. 207
Miles, L. E. M. 237
Millar, J. A. 56, 57
Miller, A. 77
Miller, M. 342
Miller, V. 252
Minnich, V. 219
Mirand, E. A. 142
Mitchell, J. 83
Modell, C. B. 364
Moeschlin, S. 347
Mollin, D. L. 309, 313

Moncrieff, A. 238
Monsaingeon, A. 285
Moore, C. V. 41, 49, 130, 161, 246, 301, 303
Moore, F. D. 285, 287, 291, 295
Morawitz, P. 22
Morczek, A. 305
Morell, H. 155
Morgan, E. H. 79, 81, 98, 350
Morrione, T. G. 343
Morrow, J. J. 258
Mortimer, R. 145, 170
Mouriquand, C. 209, 347
Moutier, F. 219, 231
Muir, A. R. 205
Muir, I. F. K. 287
Muirhead, H. 148
Munro, H. N. 93
Murphy, A. 240
Murphy, W. P. 301
Murray, M. J. 39
Mutius, I. 33, 34
Mutz, J. 370
Myerson, R. M. 353
Myhre, E. 301
Myhrman, G. 77

N

Naegeli, O. 22, 215, 246
Naiman, J. L. 229, 230, 232
Najean, Y. 170, 336
Nakajima, H. 162
Nath, I. 365
Nathan, D. G. 153, 301, 335
Naughton, M. A. 155
Nay, C. P. 364
Neale, F. C. 78, 82
Neilands, J. B. 10
Német, K. 124
Neve, R. A. 151, 152, 155
Newcombe, R. 255
Newton, M. 123
Nilsson, L. 119
Nissim, J. A. 247
Nizet, A. 184
Norden, A. 252
Norrby, A. 44, 249
Noyes, W. D. 96, 232
Nylander, G. 77

O

Oberhoffer, G. 82
Oertel, J. 106
Oetzel, W. 186
Ohira, Y. 22, 232
Oliver, R. A. 101, 340, 364
Olsson, K. S. 99, 105
Opitz, E. 21
Orskov, S. I. 302
Orten, A. U. 299
Orten, J. M. 299
Osgood, E. E. 144
Ostrow, J. D. 163
Owen, G. M. 217

P

Palmer, H. 74
Pannacciulli, I. M. 41, 42
Pappenheimer, A. M. 201
Parker, W. C. 78, 79
Pass, I.' J. 162
Paterson, D. 370
Paterson, J. C. S. 74, 171, 364
Patwardhan, V. N. 215
Pearse, A. G. E. 201, 352
Pearson, P. B. 299
Pekkarinen, M. 32
Pengelly, C. D. R. 364
Penner, J. A. 186
Perls, M. 201
Perman, G. 361

Perutz, M. F. 147, 148
Peter, H. 124, 127
Pinkerton, P. H. 56, 320
Pirart, J. 350, 353, 354
Pirzio-Biroli, G. 44, 48, 49, 82
Platzer, R. F. 364
Plaut, G. W. E. 150
Ploem, J. E. 105
Plötner, K. 215, 246, 275, 276, 301
Podmore, D. A. 253
Policard, A. 160, 206
Pollitt, E. 10
Pollycove, M. 105, 145, 169, 173, 239, 335
Ponder, E. 302
Poppen, K. J. 313
Popper, H. 364, 365
Powell, L. W. 258, 348, 349, 354
Prasad, A. S. 221
Prato, V. 153
Prelog, V. 10–12, 355
Pribilla, W. 335
Price, D. S. 50
Pringle, A. 252
Pritchard, J. A. 122, 123, 254

Q

Quaglino, D. 202, 313
Quincke, H. J. 201

R

Rabinowitz, M. 155, 317
Racker, F. 258
Rademaker, W. 313
Ramalingaswami, V. 215
Ramsay, N. W. 252
Ramsay, W. N. M. 77, 353
Rapaport, S. 159
Rasch, C. A. 238
Rath, C. E. 75, 77, 94, 103, 347
Rath, C. R. 275
Rawson, A. 229, 230
Rechenberger, J. 76, 82, 98
Rechnitzer, P. A. 245
Reid, J. D. 215
Reiff, R. H. 184, 185
Reimann, F. 216, 227, 231–233, 246
Reissman, K. R. 76, 299
Remy, D. 167, 301
Reye, D. 368
Reynafarje, C. 49, 75, 173
Reynolds, D. M. 10
Reynolds, J. W. 32
Reynolds, R. D. 219
Rhoads, C. P. 302
Rich, A. 141
Richert, D. A. 24
Richter, G. W. 93, 94, 205
Rifkind, D. 82
Rigas, J. 193
Riley, H. 131
Rimington, C. 151
Rinehart, J. F. 77
Rios, E. 130
Rittenberg, D. 149, 186, 188, 191
Roberts, F. D. 276
Robinson, S. H. 164, 239
Robsheit-Robbins, F. S. 291, 299
Rogers, W. F. 343
Roine, P. 32
Rosenthal, F. D. 229, 230
Rosenthal, R. L. 329
Ross, J. F. 60
Rosse, W. F. 368
Rosselin, G. 356
Roth, O. 22, 49, 77, 99, 218, 219, 223, 231, 232, 249
Rothen, A. 92
Rous, P. 347
Roy, L. M. H. 277
Rubin, D. 75
Rubino, J. F. 318
Rudzki, Z. 38

Rumball, J. M. 76
Ruol, A. 368
Rybo, G. 119–123, 239, 243
Ryss, E. S 229

S

Saarinen, U. M. 106, 133
Sachtleben, P. 125
Saddi, R. 128, 130
Sahli, H. 22
Saita, G. 76
Saito, H. 113, 301
Sakata, T. 360
Salmon, H. A. 232
Salomon, K. 96
Saltman, P. 79
Samuels, L. D. 370
Sandøe, E. 370
Sano, S. 151, 154, 317
Saunders, J. B. 246
Saylor, L. 63
Schabert, J. 159
Schachter, D. 54
Schade, A. L. 77
Schäfer, K. H. 72, 74, 82, 130, 275
Schairer, E. 98
Schapira, G. 15, 24, 25, 74, 81, 82, 128, 130, 133, 175, 189, 336, 341, 343–345, 349, 350
Scharpf, H. 77, 275
Schatz, A. 10
Scheer, R. L. 371
Scheuch, D. 159
Scheuer, P. J. 351
Schiffmann, N. L. 335
Schlegel, B. 302
Schloesser, L. L. 233
Schmeltzer, W. 201
Schmelzer, M. 335
Schmid, R. 161, 162, 164, 301, 302, 318, 347
Schmidt, H. A. E. 276
Schmidt, M. B. 215, 228, 231
Schneider, W. 243
Schrofl, P. 370
Schubothe, H. 128, 287
Schuler, D. 368
Schulman, J. 128
Schulten, H. 215, 246
Schultz, G. A. 277
Schulz, J. 44, 130
Schwartz, E. 74
Schwartz, H. 159
Schwartz, H. C. 155
Schwartz, S. 195, 318, 364
Schwarzenbach 355, 364
Schweet, R. 155
Scott, D. E. 122
Scott, J. M. 251
Seeleman, K. 125
Seftel, H. C. 255, 363
Seibold, M. 216, 218
Seip, M. 184
Sevitt, S. 287, 291
Shahidi, N. T. 233, 319
Shapiro, N. 258
Sharney, L. 170, 335
Sharp, A. A. 238
Sheely, T. W. 311, 364
Sheldon, J. H. 340–343, 347–349, 352, 353
Shemin, D. 149, 186, 188, 191
Shen, S. C. 287
Sherlock, S. 358
Shoden, A. 93, 94, 365–367
Shooter, E. M. 155
Shorr, E. 258
Siegrist, J. 295
Siimes, M. A. 106, 107, 133
Simonovits, I. 216
Simpson, W. L. 82
Sinniah, R. 72
Sisson, T. 128, 258
Siurala, M. 229
Sjölin, S. 128, 252
Sjöstrand, T. 161, 185, 287

Skikne, B. S. 320
Skouge, E. 276
Slack, H. G. B. 251
Sladić-Simić, D. 321
Slater, L. 21
Smith, D. E. 343
Smith, J. A. 53
Smith, J. P. 364
Smith, M. D. 41, 42, 336
Smith, N. J. 44, 99, 130
Smith, P. M. 38, 349, 356
Smithies, O. 78
Snowden, P. W. 370
Snyder, A. L. 162
Snyderman, E. 313
Sobel, H. D. 40
Soergel, K. H. 368, 370
Sölvell, L. 44, 50–52, 54, 55, 249, 250
Somers, K. 232
Sommers, S. C. 368, 370
Spencer, S. 313, 314
Speyer, B. E. 96
Srikantia, S. G. 233
Starkenstein, E. 78, 246
Stein, K. S. 275
Stein, N. 39
Steiner, B. 370
Stevens, A. R. 38, 252
Stewart, W. B. 49, 51
Stoeckenius, W. 160
Stone, W. D. 230
Stott, G. 216, 242
Strachan, A. S. 360
Strangeway, A. K. 32
Strohmeyer, G. 62, 64
Strumia, M. M. 188
Sturgeon, P. 72, 93, 94, 122, 240, 365–367
Sturgill, B. C. 371
Suit, H. D. 331
Summerskill, W. H. J. 243
Surgenor, D. M. 78, 79
Suter, P. E. N. 81
Suzman, M. M. 223
Szelényi, J. G. 128
Sztanyik, L. 329, 331, 332
Szur, L. 336

T

Tafari, N. 124
Taft, L. I. 124, 222
Tait, G. B. 368
Takeda, Y. 24
Tamaki, H. T. 39
Tanaka, Y. 210
Taylor, A. 28
Taylor, C. R. 94
Taylor, F. H. L. 299
Taylor, H. E. 341
Taylor, J. 38
Telfer, N. 335
Temperley, I. J. 238
Theorell, H. 21
Thiessen, R. 230
Thivolle, L. 246
Thoenes, F. 275
Thorell, B. 142
Topley, E. 287
Torrance, J. D. 275
Torun, B. 218
Tötterman, L. E. 74, 167
Troell, L. 287
Trousseau, A. 246, 340
Truelove, L. H. 364
Tubiana, M. 335
Tunessen, W. W. 230
Turnberg, L. A. 39, 361
Turnbull, A. 38, 41, 44, 46, 78, 79

U

Umeda, T. 277

V

Vahlquist, B. C. 72, 149, 275, 301, 318
Valassi, K. V. 32
Valberg, L. S. 237, 240
van der Heul, C. 96
van Dyke, D. 49, 277
van Eijk, H. G. 124
van Hoek, R. 64
van Kreel, B. K. 93
Vannotti, A. 218, 232, 301, 303, 343
Varela, J. E. 335
Veall, N. 169
Vecchi, G. P. 32
Vellar, O. D. 216–218, 240
Ventura, S. 82, 238
Verhoef, N. J. 81
Verloop, M. C. 54, 124, 236, 239, 240, 313
Verwilghen, R. 309
Vest, M. 125
Vetter, H. 169
Vitale, L. 360
Viteri, F. E. 218
Vogel, C. 370
Volwiler, W. 244
Vries, A. 103
Vuylsteke, J. 314

W

Wack, J. P. 38
Wagenknecht, C. 159
Wainwright, J. 360
Waksman, S. A. 10
Waldenström, J. 22, 72, 77, 149, 167, 218, 223, 301, 303, 318
Walker, A. R. P. 27, 124, 360
Wallenius, G. 78
Waller, H. D. 159
Wallerstein, R. O. 105
Wallinsten, S. 245
Walser, A. 10
Walters, G. O. 106
Walters, M. N. I. 98
Wapnick, A. A. 362
Warburg, O. 78
Ward, H. P. 277
Warthin, T. A. 353
Wasserman, L. R. 301, 335
Watson, C. J. 153, 301, 318
Watson, W. C. 230, 231
Waxman, H. S. 155, 317
Waye, J. D. 40
Wazewska-Czyzewska, M. 306
Weatherall, D. J. 317
Weicker, H. 125
Weinfeld, A. 49, 91, 94, 95, 97–99, 101-103, 203, 243
Weinstein, I. M. 277
Weintraub, L. R. 46, 354
Weisman, R. 186
Weissbecker, L. 280
Werner, E. 106
Werner, H. 342
West, R. 164
Westall, R. G. 149
Westerfeld, W. W. 24
Westerman, M. P. 317
Westerveldt, F. B. 371
Wetherley-Mein, G. 252
Wharton, M. A. 32
Whitehead, R. G. 299
Whitten, C. F. 257, 258
Widdowson, E. M. 122, 246
Wiese, W. C. 151
Wiesmann, W. 370
Wilander, O. 77
Wilkinson, J. F. 251
Will, G. 252
Williams, G. 248
Williams, R. 38, 348, 349, 354, 358
Willis, R. A. 342
Wilson, A. R. 349
Wilson, E. 131

Wilson, G. H. 38
Wilting, W. F. 117
Winterhalter, K. H. 155
Wintrobe, M. M. 77, 81–84, 123, 125, 193, 229, 233, 234, 249, 275–277, 287, 313, 314
Witt, M. 255
Wittenberg, J. 149
Witts, L. J. 229
Witzleben, C. L. 315
Wöhler, F. 38, 41, 52, 91, 93, 95, 105, 106, 240, 277, 315, 347, 356, 357
Wolfers, H. 335
Wood, J. K. 252
Woodruff, C. 128
Worwood, M. 106, 130, 132, 133, 237, 238
Wretlind, A. 27, 28, 32–34
Wyatt, J. P. 38, 315
Wyllie, W. G. 368, 370
Wynter, C. V. A. 38

Y

Yamada, H. 96
Yao, A. C. 131
Yetgin, S. 233
Young, L. E. 185

Z

Zähner, H. 10
Zalusky, R. 143
Zeltmacher, K. 364
Zinkham, W. H. 370
Zipursky, A. 128, 164

SUBJECT INDEX

A

Absorption,
  iron *see* Iron absorption
Achlorhydria and iron absorption, 38, 45
Acquired sideroblastic anemias, 312–319
Acute iron intoxication, 257–258
  treatment, 258
Acute iron poisoning *see* Iron poisoning
Acute radiation injury, 327–333
  blood values, 327–328
  bone marrow, 327–329
  iron metabolism, 329–332
Alcohol,
  iron overload and, 360, 363
Alcoholic beverages, 360
Anaphylactic reaction due to parenteral iron, 255
Anemia,
  aplastic, plasma iron concentration, 329
  coeliac syndrome, 244
  erythrokinetic classification of anemias, 196
  Fanconi's, 320
  ferrokinetics, classification of anemias, 178–179
  ferrokinetics in, *see* Ferrokinetics
  folate deficiency, 299
  hemolytic, 305
  in atransferrinemia, 358
  in protein deficiency, 299
  in radiation injury, 327
  iron deficiency, 215
  iron overload in anemias,
    congenital atransferrinemia, 358
    hemolytic, 305
    idiopathic pulmonary hemosiderosis, 368
    pernicious, 301
    sideroblastic, 313
    thalassemia, 315
    with ineffective erythropoiesis, 339, 364
  megaloblastic *see* Anemia, pernicious
  microcytic hypochromic in mice, 320–321
  myelophthisic *see* Radiation injury, Acute radiation injury
  nutritional *see* Protein deficiency
  of Belgrade laboratory rats, 321
  of infancy, 130–131
  of infection, 257
  of thermal injury, 258
  pernicious, 301–303
  pyridoxine-responsive, 313
  refractory, 364
  Shahidi-Nathan-Diamond, 319–320
  sideroblastic, 309–319
Angular stomatitis, 221
Antibacterial and antitoxic effect of transferrin, ferritin, and hemosiderin, 277
Apoferritin synthesis, 56, 94; *see also* Ferritin
Ascorbic acid,
  deficiency in iron overload, 362
  iron absorption and, 41, 42

Ascorbic acid,
storage iron release, 362
Aspirin, gastrointestinal blood loss due to, 243
Atransferrinemia, congenital, 358–360

B

Bacteriostatic effect of transferrin, ferritin, and hemosiderin, 277
Bantu siderosis, 360–363
ascorbic acid deficiency, 362
incidence, 360
iron content of the diet, 360
iron distribution, 362–363
osteoporosis, 362
Binding capacity of iron *see* Iron-binding capacity
Biochemical evolution, heme-type enzymes, 15–17
Blood donors,
iron status of, 49, 100, 101, 243, 244
non-heme iron in bone marrow, 101
Blood loss,
gastrointestinal, 117, 242
iron deficiency and, 242–243
menstrual, 119–122, 243; *see also* Iron loss
Blood smear,
changes with iron deficiency, 233
Blood transfusion,
siderosis due to, 364; *see also* Transfusional siderosis
Blood volume in pregnancy, 122
Body iron distribution,
effect of transferrin saturation, 363, 365
Body iron loss, 113–114; *see also* Iron loss
Body iron stores,
at birth, 128
decrease, causes of, 97
estimation of, 103–108
factors influencing, 97
increase, causes of, 97
in infancy, 99, 100
in normal adult females, 97
in normal adult males, 97, 99
in regular blood donors, 99, 100
location of, 97
see also Iron stores; Storage iron
Body surface radioiron measurements, 176–178
extramedullary hemopoiesis, 177
hemolytic anemias, 177
hypoplastic bone marrow, 177
iron deficiency, 238
Bone marrow,
cytochemical evaluation of, 201–203
cytochemical examination of, 103–105
E/M ratio, 184
hemosiderin, 94, 95, 103, 104
iron,
in Björkman anemia, 104
in chronic disease, 104
in iron deficiency, 104
in iron overload, 105
parenteral iron therapy, effect on bone marrow iron, 103
radioiron transit time, 168

C

Carbon monoxide,
endogenous, 185
production, measurement of red cell destruction, 185
Carcinoma, liver, in idiopathic hemochromatosis, 348, 352
Cardiac failure in idiopathic hemochromatosis, 343, 348, 353
Causes of death in idiopathic hemochromatosis, 348
Ceferro, 249
Cell protein synthesis, 141
Cellular immunity,
impairment of, in iron deficiency, 233

Cheilosis, 221
Chelate therapy *see* Desferrioxamine
Children,
  body iron stores, 99
Chloramphenicol,
  effect on serum iron concentration, 75
Chlorophyll, 15
Chronic pancreatitis,
  iron absorption in, 39
Circadian rhythm,
  plasma iron concentration, 72, 73, 74
  total iron-binding capacity, 82
Cirrhosis of the liver,
  Bantu siderosis, 362, 363
  idiopathic hemochromatosis, 341, 352
  iron absorption in, 39
  iron overload and, 340, 341, 352
  secondary hemochromatosis, 356
Cobalt,
  effect on erythropoietin production, 280
  excretion test, 240, 241
Concentration of heme precursors in the bone marrow, red blood cells, and urine, 154
Conferon, 249
Congenital atransferrinemia, 358–360
Contraceptives, oral, 122
Coproporphyrin, 151, 153
  concentration in red cell, 237, 277, 310, 312, 318
Coproporphyrinogen, 150, 151
Counting, *in vivo, see In vivo* surface measurement
  whole body, *see* Whole body counting
Cytochemical investigations, 201–203
  sideroblasts, 202–203
    normal, 202
    pathological, 202
    ringed, 202
  siderocytes, 201–202
  sideromacrophages, 203
Cytochrome oxidase, 21
Cytochromes, 21

D

Delta-aminolevulinic acid, 149, 151
  dehydrase, 151
  synthetase, 149, 150
Desferrioxamine,
  stability constant with iron, 12, 355
    with various metal ions, 355
  treatment of acute iron intoxication, 258
    iron overload, 355, 356, 357, 365, 370
Desferrioxamine therapy,
  in idiopathic hemochromatosis, 355–356
  in idiopathic pulmonary hemosiderosis, 370
  in secondary hemochromatosis associated with cirrhosis of the liver, 356–357
Desferrioxamine urinary iron excretion test, 105
  in hemochromatosis, 345, 347, 357
  in iron deficiency, 240, 241
  in iron overload secondary to cirrhosis of the liver, 356–357
Developmental abnormalities in iron-deficient children, 233
Dextran, iron *see* Iron dextran
Diabetes,
  idiopathic hemochromatosis, 342, 348, 351
  nutritional siderosis, 363
  secondary hemochromatosis associated with liver cirrhosis, 356
Dietary iron, 27–34
  caloric intake and, 30–31, 34
  contamination, 27, 360–361
  heme, 44, 46–47
  intake, 27, 32
  iron content of various foods, 30–31
  iron split from various nutrients, 33
  requirement in, 28, 29
Dietary iron absorption, 34, 44, 47
  factors affecting, 34, 38–40, 44, 45, 46, 47
  gastric juice, 37, 38
  heme, 44, 45, 46, 47

Disappearance half-time of plasma iron *see* Plasma iron, clearance half-time
Distribution, body iron *see* Body iron distribution
Distribution of iron in nature, 9–12
Diurnal variation of plasma iron concentration, 72–74
Double-isotope method for iron absorption measurement, 63
Drugs, iron deficiency due to, 243
Duodenum, site of iron absorption, 38, 39

E

Early labeled bile pigment, 163–164, 195
Effective erythropoiesis, 174
Electron microscopic investigations, 205–212
  erythroblast island, 208
  erythrophagocytosis, 206–207
  ferritin, 205
  hemosiderin, 205–206
  iron transport, 207–209
  rhopheocytosis, 209–210
  ringed sideroblasts, 211, 212
  sideroblasts and siderocytes, 211–212
  siderosomes, 211
Electron transfer, 20–21
E/M ratio in bone marrow, 184
  factors affecting, 184
  normal values, 184
  significance of, 184
Endocrine abnormalities,
  in idiopathic hemochromatosis, 343
  in iron deficiency, 231
Enteropathy in iron deficiency, 230–231
Epithelial changes in iron deficiency, 219
Erythrocyte coproporphyrin, normal values, 154
Erythrocyte iron turnover,
  infection, 276
  iron deficiency, 238–239
  normal values, 173, 178
  thermal injury, 291, 294
Erythrocyte protoporphyrin, normal values, 154
Erythrokinetics, 183–197
  classification of anemias, 196–197
  E/M quotient, 184
  endogenous carbon monoxide, 185
  iron used for erythropoiesis, 184
  life-span of red cells, 186–195
  reticulocyte count,
    absolute or corrected, 184–185
Erythroleukemia, transformation of sideroblastic anemia into, 313
Erythropoiesis, 141–145
  control of, 142–144
  effective, 173–174, 184
  extramedullary ,177
  generation time, 144
  ineffective, 163–164, 175, 178, 301
  total, 184
Erythropoietin, 142
Exchange of transferrin between plasma and extravascular space, 83
Exocytosis, 210
Extramedullary erythropoiesis, 177, 336, 337

F

Fanconi's anemia, 320
Fecal urobilinogen, 183, 186, 195
Fe-isotopes, 9
Ferricholine isocitrate, 248
Ferric iron, absorption, 247
Ferricitrate, 248
Ferrioxamines, 10
Ferrisulfate, 248
Ferritin, 24, 91–95, 205
  content of epithelial cells, 53
  electron microscopy, 96, 205

Ferritin,
ferritin–hemosiderin functional interrelationship, 95
ferritin/hemosiderin ratio, 94–95
heterogeneity, 96
in intestinal mucosa, 53–54
isoferritins, 93, 96
molecular weight, apoferritin, 91–92
rhopheocytosis, 209–210
shift from, to hemosiderin, 367
synthesis of apoferritin, 56, 94
see also Plasma ferritin
Ferritin concentration in plasma (serum),
children, 133
chronic disorders, 107, 108
infancy, 133
iron deficiency, 107, 241
iron overload, 107
myocardial infarction, 108
normal values, 107, 241
pregnancy, 124
sex difference, 241
Ferriversenate, 248
Ferrlecit 100, 249
Ferro 66, 249
Ferrokinetics, 167–179
aplastic anemia *see* Ferrokinetics, radiation injury
classification of anemias, 178–179
hemolytic anemia, 170, 177, 305, 306
hypoplastic anemia, 177
idiopathic hemochromatosis, 344–345
ineffective erythropoiesis, 175
infection, 175, 276, 277
iron clearance rate, 170–171
iron deficiency, 170, 174, 238–239
iron incorporation, 168, 169, 174–175
iron overload, 174
mean marrow duration time, mean marrow transit time, 168
megaloblastic anemia, 178, 303
myelofibrosis, 336
normal values, 178
outflow constant, 168
plasma iron clearance rate, 170–171
plasma iron transport rate, 168, 171–173
polycythemia vera, 335, 336
protein deficiency anemia, 299
radiation injury, 329, 330, 331, 332
radioiron reflux, 172
red cell iron turnover rate, 168
red cell life-span, 186–195
secondary polycythemia, 336, 337
sideroblastic anemias, 310, 311
surface counting, 169, 176–177
thermal injury, 291, 294
Ferronicum, 249
Ferro-Redoxon, 249
Ferrostabil, 249
Ferrous citrate, 247
Ferrous fumarate, 247
Ferrous iron, absorption, 247, 248
Ferrous gluconate, 247
Ferrous glutamate, 247
Ferrous glycine sulfate, 247
Ferrous lactate, 247
Ferrous pyrophosphate, 247
Ferrous succinate, 247
Ferrous sulfate, absorption, 247
Ferrous tartarate, 247
Focal iron overload *see* Local hypersiderosis
Food intake, effect on plasma iron concentration, 74
Food iron absorption, 38, 44
heme, 38, 44, 46–47
Free erythrocyte protoporphyrin *see* Erythrocyte protoporphyrin
Free iron-binding capacity of plasma, 81

G

Gastrectomy, partial *see* Partial gastrectomy
Gastric biopsy, 229

Gastric juice, effect on dietary iron absorption, 38
Gastric mucopolysaccharides, 38
Gastritis, atrophic, iron loss due to, 58
Gastritis, in iron deficiency, 58, 229
Gastroferrin, 38
Gastrointestinal blood loss, 117, 242
  aspirin, 243
  butazolidine, 243
  hiatus hernia, 242
  hookworm infestation, 242
  iron deficiency due to, 242
  steroids, 243
Gastrointestinal iron loss, 114, 117
Globin, biosynthesis of, 155
Glossitis, and atrophy of the mucosa of the tongue, 222, 223
Glutathione, and reductase, 159
Glycine, 149, 150
Golberg's enzyme, 152
Goodpasture's syndrome, 370–371
Granulocyte function, impairment of, in iron deficiency, 233
Growth retardation in iron deficiency *see* Iron deficiency, retarded somatic development

H

Heart failure in idiopathic hemochromatosis, 348
Heme,
  absorption of, 44, 46–47
  biosynthesis of, 149–155
  enzymes, 21, 24, 25
Heme iron compounds, 23–24
Heme synthesis, pathway, 149–155
Heme-type enzymes, 21, 24, 25
  biochemical evolution of, 15
Hemochromatosis, idiopathic, 340–356
  chelating agents, 355–356
  clinical signs in relatives of patients, 351
  clinical symptoms and signs, 342
  diagnosis, 346–348
  etiology and pathogenesis, 349–351
  genetic features, 349–350
  iron metabolism, 343–346
  iron overload in, 349, 352–353
  liver biopsy in relatives of patients, 351
  pathology and histopathology, 352–353
  prognosis, 348
  symptomatology, 341–343
  treatment, 353–356
Hemochromatosis, secondary, associated with liver cirrhosis, 356–358
Hemochromatosis, transfusional *see* Transfusional siderosis
Hemoglobin
  A, A_2, F, 149
  carbon monoxide, 161
  catabolism, 161–164
  concentration in blood, 240
    adults, 240
    after birth, 124
    children, 124, 217, 240
    females, 240
    infants, 124, 240
    males, 240
    normal values, 240
    pregnant females, 240
    premature infants, 124
    sex difference, 124, 240
    umbilical cord, 124
  synthesis, 147–155
Hemoglobinuria, 287, 367
Hemolytic anemias, 305
  effect on iron absorption, 49
  ferrokinetics, 306
  incorporation of radioiron, 305, 306
  plasma iron clearance, 305
  plasma iron transport rate, 305
  serum iron-binding capacity, 305
  serum iron concentration, 75–76, 305
  storage iron, 305
Hemorrhage in idiopathic hemochromatosis, 348

Hemosiderin,
bone marrow, in the differential diagnosis of anemias, 104
chemical composition, 94
iron content of, 94
urinary, *see* Hemosiderinuria
see also Bone marrow hemosiderin
Hemosiderinuria, 367
Hemosiderosis *see* Siderosis; Hemochromatosis; Idiopathic hemochromatosis; Idiopathic pulmonary hemosiderosis; Renal hemosiderosis; Transfusional siderosis; Nutritional siderosis
Hepatic coma in idiopathic hemochromatosis, 348
Hepatic fibrosis, 354
Hepatic insufficiency in Bantu siderosis, 363
Hepatoma (primary carcinoma of the liver) in idiopathic hemochromatosis, 342, 348
Hepatomegaly
in cirrhosis of the liver, 356
in idiopathic hemochromatosis, 341, 352
Hereditary sideroblastic anemia, 309–312
Hexokinase, 159
Hiatus hernia, bleeding from, as cause of iron deficiency, 242
High altitude, 75, 78
Hookworm infestation, as cause of iron deficiency, 242
Humoral factors in iron absorption, 49
Hydroxamic acid, 10
Hypersideremia, 74–77
Hypersiderosis, 339–371
associated with cirrhosis of the liver, 356
classification, 340
congenital (familial), 360
definition, 339
Goodpasture's syndrome, 370–371
idiopathic hemochromatosis, 340–356
in hemolytic anemias, 340
pulmonary, 368–370
hematological changes, 369–370
treatment, 370
renal, 367–368
transfusional, 364–367
Hyperthyroidism, in iron deficiency, 231
Hypertransferrinemia, 82
Hypo- and atransferrinemia, 82
Hypogonadism
in idiopathic hemochromatosis, 341
in iron deficiency, 231
Hypopituitarism
in iron deficiency, 231
Hyposideremia, 77–78
Hypoxia, effect on plasma iron, 77

I

Idiopathic hemochromatosis *see* Hemochromatosis, idiopathic
Idiopathic pulmonary hemosiderosis, 368–370
Ineffective erythropoiesis,
iron overload, 101, 364
pernicious anemia, 301, 303
In vivo surface measurement, 176–178, 238
Iron absorption, 37–64
ascorbic acid and, 41, 42
control of, 47
chronic pancreatitis and, 39
corporeal factors, 44, 45
Crosby's concept, 53, 55
desferrioxamine and, 44
dioctylsulfosuccinate sodium and, 41
dose of iron, effect on, 41, 42
effect of food intake on, 44
endotoxin and, 45
erythropoiesis and, 47, 49, 50
factors influencing, 38, 44, 45
ferric salts, 41, 247
ferritin content of epithelial cells of the intestinal mucosa and, 53–54
ferrous salts, 41, 247
from different foods, 44
from various gastrointestinal segments, 39

Iron absorption,
gastric secretion and, 38
Granick's "mucosal block" theory, 51
heme, 44, 46–47
idiopathic hemochromatosis, 339, 341
in atransferrinemia, 359
in infants, 130
intraluminal factors, 44, 45
iron-binding capacity of the plasma and, 50
iron deficiency, 49, 241, 235
iron stores, effect on, 47, 48, 49
iron therapy and, 249
liver disease and, 39, 40
measurement of, 60–64
double-isotope method, 63
from a non-heme and a heme iron pool, 47
from the diet, 64
isotope balance technique, 62
whole-body counting, 63–64
mucosal factors, 44, 45
oral iron loading, effect on plasma iron concentration, 43
pancreas and, 39, 40
partial gastrectomy and, 38
phosphates and phytates, 41
physiological limits of, 47
plasma ferritin and, 106
plasma iron tolerance test *see* Oral iron loading
rate of erythropoiesis, 49
role of the liver, 39
role of the stomach, 33, 37
site of, 38, 39, 54–55
succinic acid and, 41, 43
transfer of iron from the intestinal lumen to the plasma, 40, 55
xanthinoxidase, 46, 51
Iron, amount of, split by gastric juice, 33
Iron-binding
gastric factor, 38
mucopolysaccharides, 38
transferrin, 79, 81
Iron-binding capacity,
after Billroth II operation, 101
healthy adults, 82
hemolytic anemia, 82
idiopathic hemochromatosis, 343
infancy, 81–82
infectious and tumorous anemia, 241
iron deficiency, 235, 241
nephrotic syndrome, 82
newborn, 81
normal values, 82, 241
pregnancy, 82
rheumatoid arthritis, 82
thermal injury, 82, 288, 291
total, 241
see also Transferrin
Iron chelating agents *see* Desferrioxamine
Iron clearance, plasma, *see* Plasma iron clearance rate; Plasma iron disappearance rate
Iron-containing compounds,
biological significance of, 19–22
iron content of, 25
tissue respiration and, 19–22
total amount of, 25
Iron-containing enzymes, 25
Iron-containing macrophages *see* Sideromacrophages
Iron-containing metabolites, 10
Iron content,
beer, 27, 360
bone marrow, 104
dietary *see* Dietary iron
ferritin, 91, 92
fetus, 128, 129
hemosiderin, 94
liver, 133
milk, 30
newborn, 128
reticuloendothelial cells, 104

Iron content,
  sideromacrophages, 104
  various foods, 27, 30–31
  wine, 27
Iron cost of pregnancy, net, 123
Iron deficiency, 215–257
  achlorhydria, 229
  "anémie ferriprive pseudo-aplastique" 234
  antibodies, 230, 231
  atrophy of the nasal mucosa, 224
  bilirubin concentration in plasma, 234
  biochemical changes, 234–235
  blood donors, 243; *see also* Blood donors, iron status of; Blood donors, non-heme iron in bone marrow
  blood smears, 233
  bone marrow, 234
    sideroblasts in, 235
  clinical picture, 218–232
  copper level of plasma, 235
  coproporphyrin and uroporphyrin levels of red blood cells, 237
  crocus Martis, 246
  developmental abnormalities, 233
  diagnosis, 239–241
  differential diagnosis, 242
  dysphagia, 223–224
  effects of, epithelial changes, 219–231
  etiology, 242–245
    blood loss, 242–243
    disturbance in iron absorption, 243, 244
    esophageal varices, bleeding, 242
    gastric and duodenal ulcers, 242
    gastrointestinal blood loss, 242
      aspirin, 243
      hemorrhoids, 242
      hiatus hernia, 242
      hookworm, 242
      tumors, 242
      ulcerating processes, 242
    hemorrhagic teleangiectasia, 242
    idiopathic pulmonary hemosiderosis, 370
    increased iron requirements, 119–123, 130, 131–132, 245
    malabsorption, 244
    menorrhagia, 122, 243
    menstrual blood loss, 119–122
    nutritional deficiency, 299
    parasitic infestation, 242
    partial gastrectomy, 245
    pathological iron loss, 245
    tumors of stomach and intestines, 242
    ulcerating polyps, 242
    ulcerative colitis, 242
  excretion of betanin, 230
  ferrokinetics, 238–239
  histological changes, 223, 224–225
  hypochromia of red cells, 233–234
  impaired bactericidal functions of granulocytes, 233
  impaired cellular immunity, 233
  incidence, 215–218
  ineffective erythropoiesis, 238
  infancy and childhood, 232–233
  intrauterine life, 227–230
  iron absorption in, 49, 241, 235
  iron concentration of plasma, 235, 236
  iron-binding capacity of plasma, 235, 236
  iron in bone marrow, 234
  iron metabolism, indices, 241
  laboratory findings, 233–238
    in infants and children, 238
  mean cell volume, 238
  mean corpuscular hemoglobin, and concentration, 233
  menstrual blood loss, 119–122
  mucosal epithelial changes, 222–231
  neuroendocrine disorders, 231
  oral iron loading, 235
  osmotic resistance of red cells, 233
  ozaena, 224–228
  Paterson-Kelly syndrome, 223
  pathogenesis, 242

Iron deficiency,
pica sideropenica, 218
plasma ferritin, 107, 237
plasma iron-binding capacity, 235, 236
plasma iron concentration, 235, 236
plasma iron disappearance rate, 170, 238
platelet count, 233
Plummer-Vinson syndrome, 223
prevalence,
adolescence, 218
blood donors, 243–244
children, 217
females, 217
infants, 217
males, 217
pregnants, 217
protoporphyrin concentration in red cells, 236, 237
reduced activity of tissue enzymes, 232
retarded somatic development, 231
reticulocytes, 233
rhinitis atrophicans, 224–228
serum ferritin, 107, 237
symptoms, 218
therapy, 245–257 *see also* Iron therapy
dosage of iron, 248–249
duration of treatment, 249
effect of, 256–257
ferrous salts, 247–248
historical review, 245–247
in childhood, 256
oral, 247–250
parenteral, 251–255
principles of, 247–248
side effects, 250
slow-release preparations, 248
toxicity of parenteral preparations, 255
transferrin saturation, 235
transient leucocytosis, 233
trophic changes, 219–227, 220–229
Iron deficiency anemia,
frequency
in Hungary, 216
in London, 216
in Scotland, 216
in Western Europe, 216
Iron dextran, 252
Iron dextrin, 251
Iron disappearance, plasma, *see* Plasma iron disappearance rate; Plasma iron clearance rate
Iron excretion,
in idiopathic hemochromatosis, desferrioxamine-induced, 347, 355–357
in iron overload associated with liver cirrhosis, desferrioxamine-induced, 356, 357
Iron incorporation *see* Ferrokinetics, iron incorporation
Iron intake, 27–34
recommended, 28–29
Iron intoxication, 257, *see also* Iron poisoning
Iron isotopes, 9
Iron kinetics *see* Ferrokinetics
Iron loading test,
double oral, 52–53
in idiopathic hemochromatosis, 343, 345–346
intravenous, 241
oral, 43, 48, 52, 60
Iron loss, 113–122
associated with desquamation of epithelium, 57–58, 114, 118
at parturition, 123
average obligatory, 113–114
from the skin, 118
gastrointestinal tract, 114
in iron deficiency, 114, 116
in iron overload, 114, 115
in the bile, 114
in women, 119–124
measurement of, 113
menstrual, 119–122
lactation, 123
obligatory, 113, 114

Iron loss,
of the organism, 113–114
pregnancy, 122, 123
through exudation, and exfoliation of epithelial cells of the intestinal tract, 58–59
urinary, 117
Iron overload, 339–371
ascorbic acid deficiency and osteoporosis in, 362
atransferrinemia, congenital, 359
cardiac failure in, 341
classification, 340
definition, 339
detection of, 346–348
distribution of iron in, 361–362, 365–367
ferrokinetics, 174, 344–345, 348
Goodpasture's syndrome, 370–371
hemolytic anemias, 305, 315
idiopathic hemochromatosis, 340–356
diagnosis, 346–348
etiology and pathogenesis, 349–352
iron metabolism, disturbance of, 343–348
pathology and histopathology, 353–354
prognosis, 348
symptomatology, 341–343
treatment, 353–356
in refractory anemias, associated with ineffective erythropoiesis, 364, 365
in sideroblastic anemias, 309, 313
in South African Blacks, 361, 362
nutritional, 360–364
osteoporosis, and, 362
pancreas and, 352–353, 362, 363
plasma ferritin, 99, 107
plasma iron concentration, 343, 344, 350, 353
changes in relation to iron removal through phlebotomy, 355
pulmonary, 368–370
renal, 367–368
secondary, associated with liver cirrhosis, 356–358
Shahidi-Nathan-Diamond anemia, 319–320
transferrin saturation, 343
transfusional, 364–367
see also Hypersiderosis
Iron poisoning, 257–258
clinical features, 257
enzyme systems, inhibition, 258
hepatocellular damage, 257
mortality, 257
treatment, 258
Iron porphyrin proteids, 9, 20
Iron release from transferrin, 79
Iron requirements,
changes from birth until completion of growth, 134
for growth, 130, 131, 132
for replacing menstrual iron loss, 119
for replacing obligatory iron losses, 113, 114
infancy, 130, 132
pregnancy, 122–124
Iron sorbitol–citric acid, 252
Iron springs, 9
Iron stores *see also* Storage iron
during infancy, 99
evaluation of, 103–108
factors influencing, 97
in Bantu siderosis, 100, 361
in idiopathic hemochromatosis, 100, 349, 352–353
iron absorption and, 47, 48, 49, 101
quantitative aspects of, 97–101
Iron therapy, 245–257
complication of, in paroxysmal nocturnal hemoglobinuria, 369
oral, 247–250
dose, 248–249
dose for children, 256
duration of treatment, 249
historical review, 245–247
Iron therapy, oral,

iron absorption, 249
placebo, 250
principles of, 247
side effects, 250
slow release preparations, 248
parenteral, 251–256
anaphylactic reactions, 255
dose, total, 253
iron dextran, 252
iron dextrin, 251
iron sorbitol citric acid, 252, 253
lost in the urine, 253
pain, 255
saccharated iron oxide, 251
side effects, 250
total dose infusion, 253
toxicity of parenteral iron preparations, 255
Iron tolerance test *see* Oral iron loading
Iron transferase, 56
Iron transfer by mucosal cells, 54–56
Iron transport, 71–84
hypersideremia, 74–77
hypertransferrinemia, 82
hypo- and atransferrinemia, 82–83
hyposideremia, 77–78
transferrin saturation, 84–85
Iron turnover *see* Plasma iron transport rate
Iron uptake by red cell precursors, 79, 81
Isoferritins, 93, 96

K

Kendural C, 249
Koilonychia, 219–221
Kwashiorkor, 299

L

Labile iron pool, 25
Lactation, iron loss, 123
Lead poisoning, 317
Life-span of the red cells, 145
determination of, 186–195
in the neonate, 129
see also Red cell, determination of life-span
Liver biopsy,
idiopathic hemochromatosis, 347
Liver disease,
hemochromatosis and, 341, 356
iron absorption and, 358
iron overload and, 348, 358, 362–364
plasma iron concentration, 356
Liver iron concentration, 98, 99, 101
Local hypersiderosis, 340, 367, 368, 370
Loss of hair, in iron deficiency, 221, 257
Lung biopsy, in idiopathic pulmonary hemosiderosis, 368

M

Malabsorption, iron deficiency due to, 244
Marrow, bone, *see* Bone marrow
Marrow radioiron transit time, 168
Marrow sideroblasts, 202, 211–212, 234, 241
Maturation of reticulocytes, 185
Meal, effect of, on iron absorption, 44
Mean corpuscular hemoglobin, and concentration, in iron deficiency, 233
Measurements, ferrokinetic, *see* Ferrokinetics
Meat, iron content of, 31
Megaloblastic anemia, 301–303
ferrokinetics in, 301, 303
Melanin in idiopathic hemochromatosis, 352
Menorrhagia, 122
Menstrual blood loss, 119–122
Microbacterium lacticum ATCC 8181, 11
Microcytic hypochromic anemia,
of Belgrade laboratory rats, 321
of mice, 320
Microscopy, electron, *see* Electron microscopic investigations

Milk,
effect of, on iron absorption, 24
iron concentration of mother's milk, 130
iron content of, 30
Mitochondrial damage in sideroblastic anemias, 212, 310, 318
Mucopolysaccharides, iron-binding, in gastric juice, 38
Mucosal factors affecting iron absorption, 44, 45
Mucosal ferritin, 52, 53, 54
Mucosal iron transfer, 54
Muscle iron, 24
Myeloblastic leukemia, transformation of acquired sideroblastic anemia into, 313
Myelofibrosis, ferrokinetics in, 336
Myelophthisic anemia *see* Radiation injury; Acute radiation injury
Myoglobin, affinity for oxygen, 19

N

Necrotizing effect of iron on the gastrointestinal mucosa, 257
Nephrotic syndrome, iron-binding capacity, 82
Newborns, serum ferritin, 130
^{15}N-glycine, 186, 188–192
Non-heme iron,
in bone marrow, 98, 99
after Billroth II operation, 102
in liver, 98, 99
Non-heme iron compounds, 24
Nutritional anemia,
folate deficiency in, 299
iron deficiency in, 299
vitamin B_{12} deficiency in, 299
Nutritional siderosis, 360–364, *see also* Bantu siderosis

O

Occult blood in feces, 117
Oral contraceptives, 122
Oral iron loading, 54, 60
Oral iron therapy *see* Iron therapy, oral
Osteoporosis,
ascorbic acid deficiency and, in Bantu siderosis, 362
Overload, iron, *see* Iron overload
Oxidoreductive processes, 20
Oxyhemoglobin, dissociation curve of, 20
Oxymyoglobin, dissociation curve of, 20

P

Pancreas,
idiopathic hemochromatosis, 352
iron absorption and, 38–40
iron overload and, 38, 352, 362
Parenchymal iron overload, 339, 347, 362, 365–366, 367
Parenteral iron therapy *sse* Iron therapy, parenteral
Parietal cell antibodies, 230
Paroxysmal nocturnal hemoglobinuria, 367
Partial gastrectomy,
iron absorption, 38
iron content of bone marrow, 99, 101
iron deficiency following, 245
total iron-binding capacity of plasma, 99, 102
Phlebotomy,
in idiopathic hemochromatosis, 353–354
iron mobilization, 353, *see also* Venesection, therapeutic
Pica sideropenica, 218
Pigmentation of skin in idiopathic hemochromatosis, 341, 343, 352

Plasma ferritin *see also* Serum ferritin concentration
age and, 133
childhood, 133
chronic disorders and, 107, 108
infancy, 133
in females, 241
inflammation and, 107–108
in males, 241
iron absorption and, 106
iron deficiency, 107, 237
iron overload, 107
iron stores and, 106–108
normal values, 107, 241
pregnancy, 124
sex difference, 241
Plasma hemoglobin, 287
Plasma iron, 71–78
clearance half-time, 170, 172
clearance rate, 170, 172
tolerance test *see* Oral iron loading
Plasma iron-binding capacity *see* Iron-binding capacity
Plasma iron concentration, 71–78, *see also* Hypersideremia; Hyposideremia
atransferrinemia, 358
changes throughout life, 71–72
diurnal variation, 72, 73
drug-induced erythropoietic disturbances, 75
food intake, effect on, 43, 44
following acute hemorrhage, 76, 77
hemolytic diseases, 75, 76
hypoxia, 77
idiopathic hemochromatosis, 343, 344, 350
infancy and childhood, 71–72
inflammatory diseases, 77, 241, 275, 276, 279
ionizing irradiation, 74, 329
iron deficiency, 234–235, 241
iron overload, 343, 344, 350
newborn, 71
normal adult values, 71–72
pregnant women, 124
pyridoxine deficiency, 76
sex difference, 72
stress situations, 74
thermal injury, 288, 291
Plasma iron disappearance rate, 170, 172
Plasma iron transport rate, 168, 171–172
clinical usefulness, 173
hypoplastic and aplastic anemias, 173
marrow damage, 171–172
normal values, 171–172
Plasma iron turnover, *see also* Plasma iron transport rate 171–173
Plasma radioiron disappearance rate *see* Plasma iron disappearance rate
Poisoning, iron, *see* Iron poisoning
Polycythemias, secondary, 336–337
Polycythemia vera,
iron metabolism, 335–337
transition of, to myelofibrosis, 336
Porphobilinogen, 149, 150
Porphyrin synthesis *see* Heme synthesis
Portocaval shunt, 356, 358
Postcricoid esophageal webs, 223–224
Pregnancy,
anemia in, 123, 124
iron requirement of, 122–124
serum ferritin, 124
serum iron concentration, 124
serum total iron-binding capacity, 124
Protein deficiency, 299
Protoporphyrin, 151, 152, 153
concentration in red cell, free, 236, 237, 277, 285, 286, 292, 310, 312, 318
in anemia of infection, 236, 279
in hereditary hypochromic sideroblastic anemia, 310, 312
in iron deficiency, 236, 237, 279
in lead poisoning, 318
in response to iron therapy, 237

Protoporphyrin concentration
in sideroblastic anemias, 236, 312
in thermal injury, 285, 286, 292
normal values, 154, 236
Protoporphyrinogen, 151
Pulmonary hemosiderosis *see* Idiopathic pulmonary hemosiderosis
Pyridoxine-responsive anemia *see* Sideroblastic anemias

R

Radiation injury, 327–333, *see also* Acute radiation injury
Radioiron
disappearance from the plasma, 170–171
in hemolytic anemia, 170
in iron deficiency, 170, 238
in radiation injury, 329
incorporation, 173–175
reflux, 172
transit time, marrow, 168
Red cell
average life-span, 145
count,
infancy, 124
neonates, 125
sex difference, 124
umbilical cord, 124
destruction, 159, 160, 185–195
determination of life-span, 186–195
based on bilirubin production, 194
cohort labeling, 186, 187–192
random labeling, 186, 192–194
using ^{14}C-glycine, 188
using ^{15}N-glycine, 186, 188–192
iron turnover, normal values, 178
production of, 141
protoporphyrin of, *see* Protoporphyrin, concentration in red cell
utilization of radioiron, 173, *see* Radioiron incorporation
infection, 276
iron deficiency, 238
normal values, 178
thermal injury, 291, 294
Red cell mass in pregnancy, 122
Reflux of iron, 172
Renal hemosiderosis, 367–368
Requirements, iron, *see* Iron requirements
Resoferrix, 249
RE-system *see* Reticuloendothelial system
Reticulocyte count, in iron deficiency, 233
Reticulocytes, maturation time in bone marrow, 145
Reticuloendothelial system,
activation of, 291
iron block, 277
iron release from, 277
Rheumatoid arthritis, 277
Rhopheocytosis, 209–210
Ringed sideroblasts, 201, 211

S

Saccharated iron oxide, 251
Sarcoma due to parenteral iron, 255
Saturation, transferrin, *see* Transferrin, saturation of
Serum ferritin concentration,
evaluation of iron stores, 106–108
in children, 133
in inflammations, 107–108
in iron deficiency, 107, 237, 241
in iron overload, 107
normal values in adults, 107, 241
see also Plasma ferritin
Serum iron-binding capacity *see* Iron-binding capacity
Serum iron concentration,
in hemochromatosis, 343, 344, 345
in hemolytic anemias, 74, 75
in iron deficiency, 234–235, 241
in radiation injury, 74, 329
normal values, 72, 133, 241
see also Plasma iron concentration

Shahidi-Nathan-Diamond anemia, 319–320
Sideramines, 10
Sideroblastic anemias, 309–319
  acquired, 312
  due to antituberculous drugs, 319
  hereditary hypochromic, 309
  lead poisoning, 317
  pyridoxine-responsive, 313
  thalassemia, 316
Sideroblasts, marrow, *see* Marrow sideroblasts
Siderochromes, 10
Siderocytes, 201, 211
Sideromacrophages, 203
Sideromycins, 10
Siderophilin *see* Transferrin
Siderosis,
  alcohol and, 358, 360, 363
  ascorbic acid deficiency and, 362
  atransferrinemia, 359
  congenital (familial) hyposiderosis, 360
  Goodpasture's syndrome, 370
  idiopathic pulmonary, 368
  in South African Blacks, 360–364
  nutritional, 360–364
  osteoporosis and, 362
  renal, 367
  transfusional, 364–365
    see also Iron overload; Hypersiderosis
Siderosomes, 201, 202, 211
Skin
  in idiopathic hemochromatosis, 341, 352
  in iron deficiency, 219
  iron loss, 118
Skin biopsy, 347
Spinach, iron content of, 30
Stability constant of
  desferrioxamine + Fe^{3+}, 12
  EDTA + Fe^{3+}, 12
  transferrin + Fe^{3+}, 79
  various DFO-B metal complexes, 355
Steatorrhoea, 244
Storage iron, 24, 91–108
  after Billroth II operation, 101
  estimation of 103–108
  in bone marrow, 97, 98
  in liver, 97, 98
  quantitative aspects, 97–101
    see also Iron stores
Stores, iron, *see* Iron stores; Storage iron
Succinic acid, 149, 150
Surface counting *see In vivo* surface measurement
Survival, red cell, *see* Red cell destruction
Sweat, iron loss, 118

T

Thalassemia, 315–316, 364
Therapy, iron, *see* Iron therapy
Thermal injury, anemia of, 285–295
Thrombocytosis in iron deficiency, 233
Tissue hemins, 9
Total erythropoiesis *see* Erythropoiesis, total
Total iron-binding capacity *see* Iron-binding capacity, total
Toxicity of parenteral iron preparations, 255
Toxins, vaccines, and protein breakdown products, effect on serum iron concentration, 275
Transferrin, 78–85
  biological half-life, 79–80
  determination of plasma level, 81, *see* Iron-binding capacity
  distribution, 83
  factors influencing the transferrin concentration of plasma, 83
  genetic variants, 78–79
  half-life, 79–80
  in idiopathic hemochromatosis, 343
  in iron deficiency, 82, 235, 240, 241
    infection and, 82
  iron absorption and, 50
  iron uptake by erythroid cells, 79, 81

Transferrin,
malignant tumors and, 82
molecular weight, 79
plasma concentration of, *see* Iron-binding capacity
saturation of, 84–85
stability constant, 79
Transfusional siderosis, 364–367
Transport iron, 24
Two-pool model, dietary iron absorption, 47

U

Umbilical cord, ligation of, at birth, 124
Unsaturated iron-binding capacity *see* Free iron-binding capacity of plasma
Uroporphyrin, 237
Uroporphyrinogen, 150, 151
Utilization of radioiron by normoblasts and reticulocytes *see* Red cell, utilization of radioiron

V

Venesection, therapeutic, 353, 354
Vitamin B_1, 313
Vitamin B_2, 313
Vitamin B_6, 313–314
Vitamin B_{12}, 299, 302
effect of, upon erythrokinetics in pernicious anemia, 302
Vitamin C, 313, *see also* Ascorbic acid
Vitamin E deficiency, 300

W

Water, iron content of, 27
Wheat, iron content of, 31
Whole-body counting, 63–64
Wine, iron content, 27
Work capacity, reduction of, in iron deficiency, 218

X

Xanthinoxidase, 46, 92

GPSR Compliance
The European Union's (EU) General Product Safety Regulation (GPSR) is a set of rules that requires consumer products to be safe and our obligations to ensure this.

If you have any concerns about our products, you can contact us on

ProductSafety@springernature.com

In case Publisher is established outside the EU, the EU authorized representative is:

Springer Nature Customer Service Center GmbH
Europaplatz 3
69115 Heidelberg, Germany

www.ingramcontent.com/pod-product-compliance
Ingram Content Group UK Ltd.
Pitfield, Milton Keynes, MK11 3LW, UK
UKHW051127260726
13967UKWH00010B/2910
* 9 7 8 1 4 6 1 5 7 3 0 9 8 *